THE HISTORY OF CADATAS' EXPLORATION OF THE MILKY WAY

ANNA FAKTOROVICH, PhD

(Pseudonym for Ortack-23)

ANAPHORA LITERARY PRESS

QUANAH, TEXAS

ANAPHORA LITERARY PRESS
1108 W 3rd Street
Quanah, TX 79252
https://anaphoraliterary.com

Book design by Anna Faktorovich, Ph.D.

Printed in the United States of America, United Kingdom and in Australia on acid-free paper.

Cover Images: Neil Armstrong, "Buzz Aldrin Walking on the Surface of the Moon Near a Leg of the Lunar Module", 1969 (Gift of Mary and Dan Solomon, 2016: National Aeronautics and Space Administration (NASA)).
Pierre-Jacques Volaire, "The Eruption of Vesuvius", France, 1771 (Charles H. and Mary F. S. Worcester Collection: Art Institute Chicago).

Published in 2024 by Anaphora Literary Press

The History of Cadatas' Exploration of the Milky Way
Anna Faktorovich—1st edition.

Library of Congress Control Number: 2024946203

Library Cataloging Information
Faktorovich, Anna, 1981-, author.
    The history of Cadatas' exploration of the Milky Way / Anna Faktorovich
    392 p. ; 10 in.
    ISBN 978-1-68114-611-9 (softcover : alk. paper)
    ISBN 978-1-68114-612-6 (hardcover : alk. paper)
    Kindle (e-book)
1. Science Fiction & Fantasy—Science Fiction—Alien Invasion.
2. Science Fiction & Fantasy—Science Fiction—Post-Apocalyptic.
3. Science Fiction & Fantasy—Science Fiction—Hard Science Fiction.
PN3311-3503: Literature: Prose fiction
813: American fiction in English

Map of Cadatas' Stops Across the Milky Way

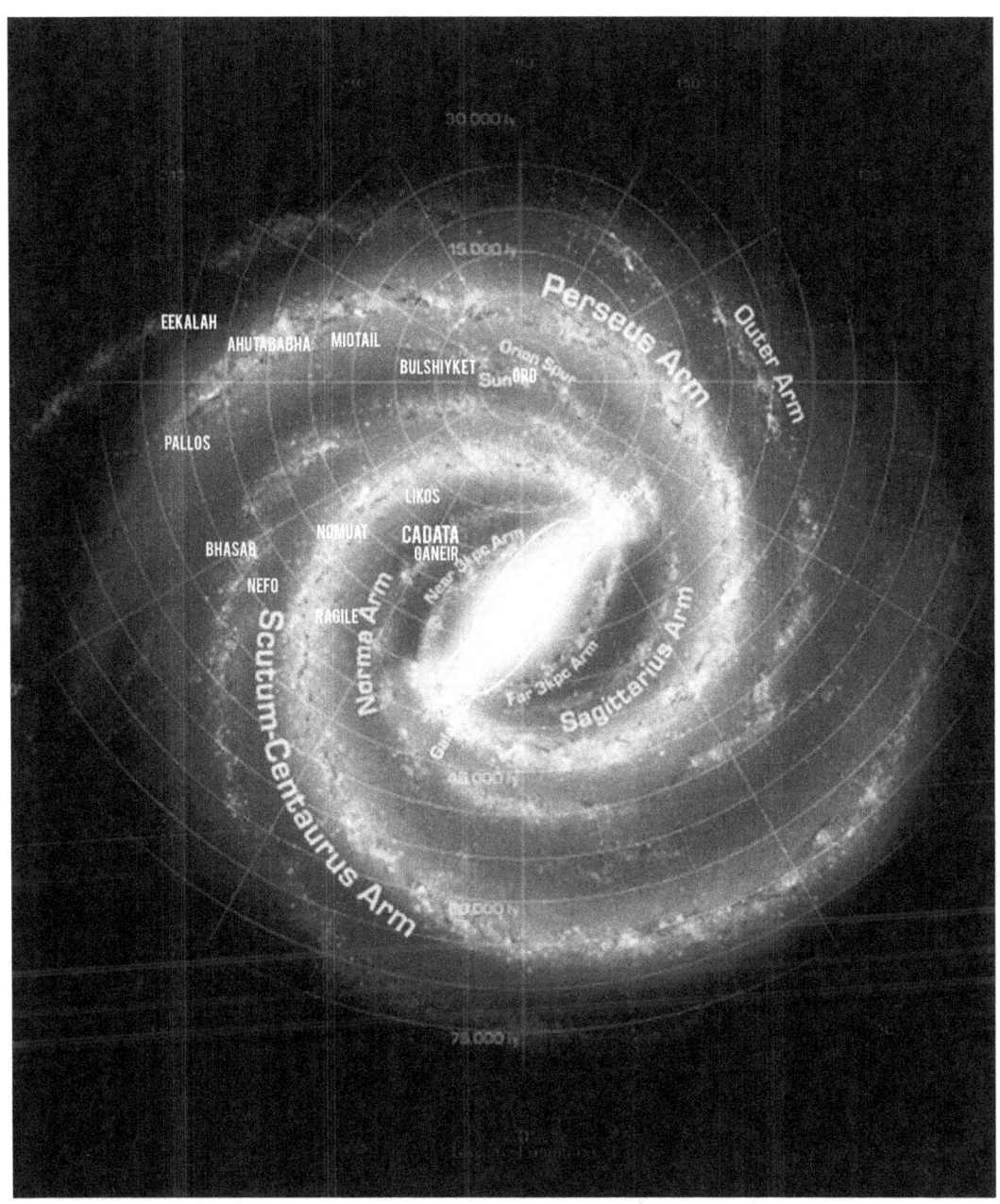

Appended Version of Robert Hurt's "Artist's Impression of the Milky Way" (NASA Jet Propulsion Laboratory, Caltech, European Southern Observatory).

CONTENTS

The Editor, Ortack-23, summarizes this study's research methodology. This book follows the standard Earth formula for an academic textbook. He also reports on the current terraforming experiments he is performing on Mars.

Ortack-23 reports biases and falsehoods in human narrators' depictions of aliens. He corrects these misconceptions with science and philosophy. An overview describes cadatas' natural physiology, explaining their biological capacity to mutate form, and to glide through the air.

This chapter runs through Cadata's evolutionary tree, pre-history of cadatas, and their more recent history. The elements of Cadata's natural climate help to explain how cadatas have reacted to different alien environments. Like Earth, Cadata has an archeological record. This documented evidence is used to support the narrative of Cadata's pre-history. Cadatas' first sculptural language, and the formation of early cadata societies clarify cadatas' nature, and how they came to have the characteristics modern cadatas exhibit. The history describes the major ages of cultural, and political development of Cadata, covering the Mineral-Manipulation Age and the Nutritional Gas Shortages War. During the Scientific Innovation Age, cadatas developed their construction methods, holographs, and other advanced technologies. The formation of the Smartness Agency became necessary as technology alleviated tensions that erupted in the Near-Thousand-Year War. The Smartness Agency has a mandate to find culprits in criminal activity; however, it seldom succeeds, as it still refuses to change its tactics. In the Millennial Age, cadatas expanded their lifespans to 5,000 years. The Age of Progress was characterized by development of absurdist and nonsensical arts. This Age ended abruptly when an island patronizing many of these artists was incidentally flattened.

When the association of Potentates, or the oligopoly's small group of mega-corporation leaders attempts to attract a mineral-rich asteroid to Cadata to mine it, this maneuver's execution fails. Instead of commencing an orbit around Cadata's moon, it hits Cadata. The devastation caused by this impact is described. Cadatas had been experimenting with space travel prior to this impact. Initially, extensive space-travel was judged to be unhealthy because it tended to lead to psychosis in isolated pilots. With the death toll at half of the planet's population and rising, the Potentates are forced to approve plans to develop a spaceship capable of taking hundreds of cadatas on a journey in search of a new planet to inhabit.

The bureaucratic hurdles that prevented an immediate implementation of a space-travel program are described. Before deciding on a spaceship, other proposals for terraforming, and more galactically-proximate solutions are considered. Cadata scientists expand their field of planetary astronomy until it can estimate planetary environments prior to commencing a journey towards a potentially habitable destination. An aside philosophizes on the nature of civilization superiority, with a review of the absurdities of the gluttony-bias in Earth scientist Kardashev's scale of civilizations. Then, the history of Cadata's development of hyper-developed robots, with the role they played in causing catastrophic events that followed. It took thousands of years for cadatas to transition from planning a mission outside their solar system to solving the logistic challenges. Scientists were already capable space-travel in theory at the onset. Lift-off was instead delayed by the bureaucratic processes impeded by corruption, bribery, and incompetence among business and government leaders, and general intoxication. Finally, funds are approved, and a spaceship is built for the first major voyage outside cadatas' solar system, to a planet called Ganeir. Different types of spaceship engines are cataloged, with the reason why cadatas chose the one designed to recycle and appropriate accessible space matter and energy within reach during flight. The methods and approaches to gene-editing cadatas to take the appearance of species on alien worlds.

Chapter 4: The Exploration of Ganeir 113

Cadatanauts undergo intense prolonged space-travel training within cadatas' solar system to test equipment, intelligence, and flight strategies. Biographies and contributions of the major cadatanauts, scientists and administrators responsible for the mission. Putair Niatage, the lead cadatanaut for the mission, describes the training he underwent to fly to Ganier in first-person diary entries. Putair is a prankster, who disrupts training for others in the class, while managing to excel. Putair details their unexpectedly elongated flight to Ganeir, despite succeeding at traveling at near-light-speed, and the tragic end their mission came to. Then, the last diary entry from their Head Doctor, Wern Lexause, details the horrific disease they were infected with on Ganier that killed the entire cadatanaut crew. Editor Ortack-23 explains how the sterilized spaceship returned itself to Cadata.

Chapter 5: The Start of the Long Voyage Across the Milky Way 154

Nally Subsenar, the lead founder of the interstellar exploration mission, pushes for adjusting systems for the mission to be capable of exploring multiple solar systems. She designs the necessary adjustments to the spaceship for this. New training methods and disaster-avoidance subroutines are incorporated to avoid a repetition of the Ganeir disaster. Nally joins the spaceship on this adjusted mission herself because Cadata's environment is becoming progressively uninhabitable.

The method for researching a new world during a slow-down approach as part of a safe landing protocol. Cadatanauts near Nerore: a water world. Nanobots fly ahead and gather genetic materials and data necessary for the preparatory biological experiment. A gene-edited cadatanaut attempts living in a cloned artificial environment on the Mothership that simulates life on Nerore. Lead Cadatanaut Wocega Lapovy describes what it felt like to transition to an under-water lifeform from a cadata in her diary entries. Wocega narrates the dangers of landing a shuttle into a turbulent ocean. She begins eating, drinking, and socializing in accordance with local customs. While Wocega is busy incorporating into the culture, she passes a cadata disease to neroreans, who die in enormous numbers from it. She is reprehended by authorities, who spot her alien identity, and question her about the alien invaders causing this new disease. Cadatas stage a dramatic rescue to extract Wocega from this interrogation. However, the team decides to begin a journey to the next potentially-habitable planet in the *Catalog*. There was no hope cadatas could have reestablished peaceful relations with neroreans after this medical calamity.

Chapter 7: Alcohol-Based Lifeform 178

Concerned about the pattern of two planets in a row showing cadatas' arrival tends to generate disease outbreaks, the science team develops strategies to prevent this obstacle. Their next stop is to an alcohol-based world, where the lifeforms are innately alcoholic. The habits and science of this species reflect this drunken biology. The planet, Likos, has various other unique features, including being located close to a relatively dim M-star. Cadatas land on this planet's moon, where they wait for Cega to perform a standard visitation experiment to Likos in a likos-biology-cloning gene-edited body. She catches fish, and learns the locals' laughing language. Cega's assessment is that likos are too volatile for cadatas to attempt settling on their planet. Cadatas agree, and continue their journey.

Chapter 8: Hot Diamond Rain 195

A new generation of cadatanauts is sent to explore Nomuat, a planet so hot that it undergoes daily diamond rain showers. Nomuat is also blanketed in extreme aurora displays, and other shocking natural features due to its unique environment. The species has an incredible halogen composition, capable of tolerating this extreme heat, and other hellish Nomuat elements. A submissive species on this planet is being exploited for their labor. The cadatanauts are tricked into surrendering an enormous volume of treasures to a local fraudster, who they take to court. They lose the case, remaining uncompensated because, as foreigners, they lack property rights. A group of aliens even attempts to steal cadatas' Mothership. This potentially mission-ending theft is prevented because exposure to the cadata-centric environment on the Mothership causes the thieves to turn to stone. Cadatanauts decide to seek a less ultra-hostile planet.

Without the capacity to identify the psychological characteristics of alien life-forms, cadatas yet again meet with an ultra-aggressive species on their next stop at Ragile: a very cold planet. Wocega-6 struggles with this species' incommunicability, as she adjusts to living within its biology. Ragile's populations have been isolated into isolated lakes. The equilibrium among them is interrupted through access to cadatas' technologies, which enable them to travel between lakes. Instead of using this access to expand knowledge, they begin warring over power. These conflicts eventually make Ragile uninhabitable for cadatas.

Cadatas explore Nefo, and its species of cloud-dwelling lifeforms. As a gas planet, it can only support gaseous life in this unique state. Wocega-8 describes transforming into an incorporeal being. The nefos are mind-controlled by their empresses into performing labor for the ruling class. Wocega-8 joins the court of one of these empresses as an assistant, describing the misadventures she has in her diary. Unhappy with the mind-control aspect, cadatas attempt to free nefos from this influence. However, the liberated nefos choose to entirely escape civilization's controls, swinging Nefo into continuous chaos, thus forcing cadatas to begin yet another migration.

Cadatas spend 5,375 years at light-speed to arrive at Bhasab, a roaming planet flying through interstellar space because it has been tossed out of its initial solar system by its sun's interaction with a black hole. Wocega-9 transforms into the local tree species, initially finding it difficult to adjust to moving extremely slowly. Then, she finds life among the trees to be meditative, and has a series of adventures with her tree friends. In one incident, she encounters fire-starting trees, who her adopted-clan fights with in response to the destruction they cause to others in that region. Things take a turn for the worse when Wocega-9's cadata shuttle is found and examined, leading to the bhasabs developing a new, hyper-destructive, tree-burning vehicle model. This tree-guzzler rapidly consumes Bhasab's forest resources, sinking the planet into an environmental crisis. Not knowing how to stop this catastrophe, cadatas move on to the next potential world.

The drastic downfall of Bhasab causes psychological damage to the second generation of cadatanauts who were stationed there. Wocega-10 has a breakdown causing her to kill the mission doctor, and then to commit suicide by jumping out of the spaceship mid-flight. Failing to learn from Wocega-10's philosophical suicide note, cadatas then arrive in a system with two proximate worlds: Pallos and Byddwr. Two cadatas undergo a transformation into the dominant species of these worlds. Wocega-12 is sent on a water-covered world called Byddwr, while Ortack-12 is sent to the freezing world of Pallos. The peaceful-minded, docile species' world has been conquered by the war-mongering Pallos. Byddwr has been forced to produce the goods Pallos survives on. Wocega-12 enjoys the hard labor she endures on Byddwr, in part because it involves ballooning for transport. Meanwhile, Ortack-12 enjoys the intense physical exercise of hunting with his local mentors. Then, Ortack-12 commits the crime of complaining, and is put on an unjust trial meant to solicit bribes from those accused, instead of acting as a reforming punishment. Ortack-12 spies on his hosts to discover an advanced sect of pallos, who have been swaying planetary events. Cadatas' bribe-negotiations with these pallos fail, forcing the Mothership to accelerate to the next target world.

The Eekalah solar system has two suns, which cause several environmental abnormalities for this planet. Wocega-14 is sent to explore it. She finds a single eekalah species, which became dominant through their runaway consumption of all plant- and animal-equivalent rivals. They then became cannibals, fighting each other to survive. Training to join them required Wocega-14 to survive extreme feats of endurance in anticipation of having malicious killers as neighbors. She strove to discover a method for turning the eekalahs away from cannibalism, when suddenly the two suns in that system triggered an explosive nova reaction. As cadatas escape this devastation to the remaining life on Eekalah, they question if cadata researchers might have caused the acceleration of these two stars' nearing.

Ahutababha is an eccentric planet populated by metalloids. Wocega-16 encounters an ice storm during her initial landing. Beyond undergoing standard gene-editing, Wocega-16 faces the challenge of being forced to tremendously expand her body weight to enter ahutababhas' natural hibernation cycle. Surviving this prolonged sleep requires reaching a specified weight, or the organism will die during the following starvation period. Most ahutababhas are extremely impoverished, and barely survive these winters. A few rich ahutababhas own giant skyscrapers, which exclusively house them, and an enormous military that exists to prevent the poor from rebelling against this system. Wocega-16 joins impoverished villagers in scavenging for food after they awake from hibernation. Wocega-16's village life is further complicated by a poisonous smog, and then a powerful cyclone. Cadatas attempt to fix the planet's eccentricity by artificially pulling it into a steadier orbit. However, this backfires, causing a social planetary disaster, and warfare.

The next planet, Miotail, houses metal-based lifeforms in an extremely hot environment. The miotails began their evolution as robotic assistants to biological lifeforms. They technologically advanced themselves to be more life-like across millions of years. Then, their planet became uninhabitable for biological life, while the miotails continued to evolve. Wocega-18 undergoes an irreversible transformation into one of these robot-like lifeforms. There is a debate about the ethics of killing her biological essence to make her into an eternally-operating machine. The miotails welcome cadatas, and offer a thorough explanation of their history and science. It is a friendly dialogue. However, as they watch Wocega-18 live as a miotail, the other cadatas decide they cannot follow her example by undergoing this transformation. Perpetual life is an insufficient substitute for the loss of organic life. This eliminates Miotail as a long-term habitat, and cadatas are forced to proceed to their next potential planet.

Chapter 16: The Tiny Creatures with Mighty Muscles 332

On Bulshiyket, a massive gas giant, the cadata science team discovers tiny creatures capable of withstanding 250g force with the strength of their extraordinary muscles. Wocega-21 does not transition into these, or other Bulshiyket lifeforms, when cadatas discover they received an erroneous intelligent-life signal from this system. If intelligent-life has not evolved in Bulshiyket's conditions; there are no suitable intelligent organisms for cadatas to mutate themselves into. The team researches this curious environment for lessons on how an extremophile organism can survive under such pressure, before pressing on.

Chapter 17: The Earth Mission 343

Editor Ortack-23 describes his own life, which begins amidst cadatas' journey to Earth. During an asteroid-mimicking de-acceleration, and then disguised steady flight, cadatas begin to learn about uncanny human behaviors, biology, and cultures. Cadatas explore major planetary, moon, comet, asteroid and solar bodies in Earth's Solar System, reporting some contradictory findings to Earth scientists' conclusions. Ortack-23 is sent to perform a full-spectrum analysis of the potential habitability of Mars. Nurry Loody-23 is sent to explore the geology and habitability of Venus, including performing small-scale terraforming experiments, and a base-installation in its natural climate. Wocega-23 mutates into a human. She is given a Blackbird super-speed human-plane-mimicking shuttle. On Earth, she commits fraud, and other crimes to purchase enormous properties needed to hide her alien identity during her scheduled adaptation experiments. These gigantic purchases of islands and other luxuries does the opposite, triggering investigations and forcing her move into less conspicuous housing. When Wocega-23 loses communication with the rest of the scattered cadata fleet, she gene-edits her body to gain the appearance of a female astronaut scheduled to make a trip to the Space Station, to take this trip in her place. From space, she briskly repairs the broken cadata equipment. However, she struggles with completing the assigned human-astronaut tasks. The human technical problems with the shuttle, and the Station terrify her, until she can endure no more, and pushes away from the Station into space, where fellow cadatas snatch her into their camouflaged alien shuttle. The astronaut Wocega-23 was imitating returns to Earth, where conspiracy theorists suspect she was kidnapped by aliens. Wocega-23 continues her research into humanity. When all experiments confirm Earth is ideally suitable for cadatas' settlement, cadatas face the strange problem of how they could legally immigrate to Earth. An honest confession on immigration forms of their alien background would be interpreted as a science fiction joke, or a deranged delusion. Lying by inventing a human origin-narrative would be problematic because of the need to maintain a generations-long conspiracy to avoid discovery. At the time of this history's composition, cadatas are still searching for ways to overcome these challenges.

Cadatas' chronological pre/history provides clarity among the leaping dates across a narrative that covers hundreds of thousands of years.

Bibliography of human and cadata textbooks and monographs cited across this book.

Preface

I am Ortack-23. The 23 stands for the 23rd cloned generation of my ancestor, who departed from out ancient planet of Cadata in 98,023 BC. I am now 1,303 years old, which is equivalent to around 26 in Earth-years. Though we do not undergo biological aging, so do not imagine that I am uniquely youthful in comparison with my elders. I am writing this history from my solitary outpost on Mars. I remained on this near-Earth planet to research the possibilities of adjusting Mars' environment to be habitable for cadatas. While these experiments are running, I have sufficient free time to have also volunteered for this opportunity to inform humanity about who we are, and why we are here.

This is the history of the cadata species covering the sciences of our biology, geology, and astronomy, as well as our culture and politics. This story traverses the Milky Way galaxy, cataloging our extended visits to fourteen shockingly-foreign worlds. It begins with cadatas' early history, and ends with our unobtrusive settlement on planet Earth. Cadatas are a glorious civilization, developed by extraordinary leaders, who have advanced us millennia ahead of humanity. Even the least among us hold lessons humans should cherish.

It is painfully difficult to write this book with humanity's tiny vocabulary. I pitched publishing this book as a 4D audio-visual virtual-reality program, which would have emersed humans in specs of our technology. It could have been a tutorial that trained engineers on the design of our cadata spaceship, with its light-speed-capable propulsion system. It could have instructed human surgeons on how to perform our near-complete gene-editing procedures. However, when I tested a prototype on a human subject, even the enveloping visuals caused concussion-like symptoms. The bigger obstacle was the impossibility of condensing billions of scientific concepts we invented that humanity has not yet imagined into a lecture that lasted less than the 100-year limit of a human lifespan. Instead, this book is the standard size of a humanoid textual history, and attempts to explain cadatas' science without venturing too far from what humans already know.

The human English dictionary sets precise linguistic boundaries to the comprehendible terms allowed into this narrative, such as *houses, cars,* and *hands*. Each inhabited planet we have explored has had a unique environment that necessitates entirely different solutions for transportation and shelter, as well as the biological properties of life. Humans understand the universe from a humancentric perspective. This book is designed as humans' introduction to extreme otherness. Some species lack appendages resembling hands. Some designed engines and structures of transportation that generate energy instead of wasting it. The most notable distinctions will be explained, but far from every difference out scientists recorded in their investigations. In parallel, I am going to complete my 4D and story-creation-program-edited Cadata Edition, which has not been approved for human consumption, unlike this compact English Edition.

The 4D focus group also discovered several repeating questions humans have when first reading this history: *Are you saying we are in the middle of an alien invasion? Do you come in peace? Why haven't you shared this proof of life outside our planet with scientists? Are you tiny, green men? Are you going to farm us and eat our brains? So, you have kidnapped me to probe me just like in horror movies?* The hysteria of these subjects might have been partially a symptom of the 4D concussion-like disorientation I alluded to earlier. And they were shouting many other things that are unsuitable for an academic book. We have not encountered this degree of paranoia regarding alien visitors on other planets. It seems the proliferation of the human alien-invasion genre is the primary culprit. Aliens are inevitably nuked out of the skies, or poisoned with chemicals their bodies cannot tolerate. This inevitable eviction conclusion is prompted by the persistent assumption of innate aggression of an alien who would invest in traveling thousands, or millions of light-years to Earth. This assumption is based on colonial propaganda that advertised exploration of foreign lands as a mercantile venture designed to colonize, enslave, cannibalize, or exterminate natives to pirate foreign resources for the "explorers'" enrichment. While there were periods in cadatas' history when we shared this colonial impulse, we have been traveling with a small crew through the Milky Way for 100,047 years. In my 1,303 years, the Solar System is the first place where I have stepped onto the physical surface of a planet, or felt natural sunlight shine on my skin. Imagine if you lived on a starship with your small family since the dawn of human speech, would you be eager to risk a single life in a territorial war, especially it can instead be won with a territorial purchase? Our past catastrophic failures to find a habitable home planet have been caused by unchangeable conditions. It would be inexcusable if this time the cause will be that we failed to counter human alien-invasion fictions with counter-propaganda. To design this suitable counter-narrative, we have been exploring life on Earth without disclosing our presence. This book is an attempt to declare our presence in an academic setting to foster a non-hysteric discussion of the premise of peaceful coexistence between the alien-humans and cadatas.

Watching 4D cadata historiographies has been our leading entertainment source across my lifetime. Our Mothership's Archives include primary narrated films that were taken by our explorers during each of the past failed planetary habitation attempts. As our team's Historian, I have previously edited collated, remixed, truncated and otherwise manipulated hundreds of editions of these voyages for future generations to enjoy. I have adapted these histories for a human audience in this edition. The missions in the body of this book are reported from the first-person perspective of each of their leaders, in heavily edited transcriptions of their recorded accounts. I must also acknowledge the multi-format histories that were created by our historians back on Cadata. I do not always cite consulted sources' contributions because each tends to have thousands of contributors collecting videos, providing commentaries, and analysis. This content-production system is inconsistent with human citation methods, in part because the *et. al.* abbreviation suggests the named entities before *et. al.* made heftier contributions. I must also acknowledge the assistance I received from our non-artificially-intelligent translation and linguistic-processing program, which translated our entire library from Cadat into English before I commenced my re-writes. Cadat is a three-dimensional digital language, so these translations flatten its intricacies. This intelligent program also translated texts cadatas composed in other alien languages while in full-emersion during our habitation-experiments in other solar systems. And it translated into Cadat the thousands of Earth languages, which are very proximate in Indo-European region because of Greco-Roman colonialism, and exhibit greater variation in Africa, Australia, and other late-colonized regions. It also turned audio-visual logs of captains, pilots, scientists, and supporting crew into textual transcriptions. Several of these exploration narratives have been pre-published in our digital Cadata Share database, to the overwhelming acclaim of our top critics. (Though I should note that I am also our mission's lead Critic.) I am delighted these authors allowed me to impart portions of their thoughtful contributions to you.

Human dictionaries contain many cyclical definitions. One among them is that *truth* is a *fact* that is accepted as *true*, while a *fact* is a thing proven to be *true*. The *proving* methods thus decide what is *true* because this is the only component in this cycle that involves measurable scientific analysis. Humans have not even carbon-dated most manuscripts they claim to be their oldest surviving written records. Thus, most of pre-audio-visual-recording human "history" is a fiction that was invented by invaders who used it to claim ownership of vast territories their ancestors had never visited. If our goal for this propagandistic textual vehicle was a conquest of Earth; this book would similarly be an account of our past successful conquests of planetary systems. However, as I previously explained we are extremely anti-violence due to a programmed instinct towards perpetual-life. The uniting lesson of this history of our explorations is that the Earth's resources are too rare in this void-abundant galaxy for species to fight over them, instead of finding ways to multiply them through collaboration.

***

I wrote through the night, and it is now sunrise on Mars. Just before the first rays obstructed visibility, I could see a little gray spot in the sky that looks like a mini-half-moon, but with magnification and color-correction becomes the semi-covered-in-shadow blue ball of Earth. If I was a mythologically-minded pre-science Martian, I probably would have placed far more symbolic significance in the significantly-larger ringed sphere of Jupiter that is my closest neighbor on side further from the Sun. The sunrise has brought a tint of orange that will build into an orange haze by mid-day. Out of my wall-sized windows, I can see a low, brown, rocky mountain-chain. This monotony is interrupted by little dust-devils that regularly materialize and disperse. Though walking to conduct my experiments with the threat of these irregular irruptions is a bit too exciting, so that I tend to opt for sending robotic probes to these missions. Between it and my dwelling there is a sandy plain covered with scattered rocks. This dwelling is a modified plan of human 3D-printing proposals for circular-cone Mars-sand-tube-layers that have been previously approved for further research by NASA. Thus, if humans send a probe and discover my presence; they will assume a secret human mission has succeeded in building a settlement, instead of leaping to the hysterical conclusion that aliens have always lived on Mars. I have not gene-edited my body to adapt it to the extremely thin Martian atmosphere, as I anticipate my Mars-habitability-fixing experiments will fail. Earth is the only rational settlement base for cadatas in the Solar System. This decision also means that my body is set to Earth's standard-body-temperature. My base is on the equator, so the temperature is survivable at between 40-70°F at midday during the long summer-season, when I schedule my expeditions. I would freeze to death without a spacesuit immediately on-exposure, if I stepped outside at night even at this equator, as a normal temperature is -100°F. Though the heat and the rapid-temperature-change also increases the severity and number of dust-devils and dust-storms. While outdoor temperatures are now ideal, it is also the peak dust-season, when Mars looks like an orange dust-ball from space. Whereas a couple months ago, the climate was so calm, Mars' geological valleys, mountains, and large craters were distinguishable by our cadata settlers on Earth. It is far more dangerous to be caught in a dust-storm unprotected on Mars for a human-proportioned lifeform. Mars' gravity is a third of Earth's, meaning that a 140-Earth-pound creature is downgraded to an equivalent of 46-pounds, or the size of a three-year-old child, or an average-sized golden retriever. Children are more likely to be lifted off the ground and tossed into a dust-cloud by excessive winds. Benefits of this diminished weight is the ability to perform acrobatic tricks with minimum effort due to relative extreme relative muscle over-sizing. I admit I have employed cadata technology to wind-surf for fun, but only because this is the first time I am living on a planet, and I insist on experiencing the limits of its textile reality.

The release of this confessional history with a tiny Texas publisher is likely to generate no media attention, and thus it will surely remain unread. My attempts to place this book with a mainstream publisher encountered the problem of the non-marketability of a non-normative alien-arrival plotline. To succeed propaganda must be read by most of a target-audience. Thus, this cannot be a serious effort to convince Earth of the righteousness of our cause. It is merely an experiment in full-transparency to statistically evaluate, under natural conditions, the range of psychological responses to exposure to the knowledge of alien life on Earth.

Ortack-23
August 15, 2024

Introduction

Philosophical Digression: Hysterical Racist Depictions by Humans of Aliens

Humans project their history of colonialism onto their perception of potential alien invasions. They imagine that the economics of space exploration and trade echo familiar European narratives. Europeans learned that cross-ocean voyages were extremely unprofitable when fair trade practices were enforced. A seventh of ships that sailed in the 18th century never returned; thus, to break-even a manager had to find a cargo that would double the cost of the ship in at most seven voyages. A ship strong enough to withstand a voyage cost around £22,000, or $7,857,887 in 2024. What do you imagine you would need to buy and sell to generate $16 million in profits in 7 voyages? The total worth of goods traded at a 10%-profit would have to be around $150 million. This might be feasible for a single trip of a modern cargo container ship that carries a maximum of around 672,000 tons; 18th century merchant ships' maximum cargo was around 500 tons; thus, the latter would have had to make around 1,344 trips in turbulent waters to reach this volume of traded goods. Spices cost around ten-times more than they cost now. Thus, exchanging a full-ship of British cotton cloth, tea, or salt would fairly pay for a much smaller volume of spices, which would not have found a demand in Britain, if prices reflected the true cost of this venture. This unprofitability of fair trade led Britain to manipulative practices such as complete market monopolization, and military conquest of governments to steal funds internally from within the Empire. Other strategies involved piracy of rival countries' merchant-vessels, trade in addictive substances that guaranteed extreme profit-levels (i.e. opium), and the sale of humans into slavery (generating up to 100% profit due to the no-cost theft of this commodity).

Because this was the Brits' experience with international exploration, their authors have imagined echoing alien-invasion narratives. I spent the last few hours reading a book especially relevant to my current habitation on Mars: H. G. Wells' *The War of the Worlds*. Martians consume their own natural resources, and are forced to invade Earth to consume Earth's resources. Upon arrival, Martians' instinct is to kill life outside themselves. This killing frenzy overwhelms them. They forget about resource-gathering amidst this bloodlust. Meanwhile, they are exterminated because they had failed to perform due-diligence research into the neighboring Earth's pathogens. When resources are viewed on a global scale, there is a cycle of consumption that results in net-zero loss, or even some resource gain. A plant brings in new resources when it takes sunshine from the Sun, and builds new energy for it to grow. It is eaten by a cow. That cow is eaten by humans. Humans create manure, which nourishes plants. If the production system is extremely wasteful, it might spew waste-gases into the atmosphere; it might turn fresh water into unused sewage-in-seawater; it might turn forests into deserts by over-tilling. Still the net resources remain unchanged, as only their chemical or material form has changed. British profiteers did not run out of natural resources, but rather out of scam-victims to sell high-interest-loans to, or to force into feudal labor. A war can capture with violence oil-producing-wells worth billions. A country goes to war over such wells for the benefit of a single robber-baron who wishes to seize this resource, as opposed to for the benefit of the many who had access to the general oil-commodity without this conquest. In an authentic symbolic representation of the Brits' position, this robber-baron Martian would have remained on Mars, and only an exploratory first-wave of troops would have died from pathogens, sending reports of this cause-of-failure to the home-planet, for their leaders to develop a new plan for invasion. Aside for the development of medicines to fight earthly pathogens, this adjustment would involve the realization that exterminating humanity was an inefficient strategy due to the likelihood of retaliatory violence. The robber-baron might have instead established a taxation treaty wherein humans sent a third of their generated resources to Mars in exchange for peace.

These scenarios are entirely contrary to cadatas' position as an alien migrant. There are too few cadatas for us to risk a single death in open warfare. And we consume near-nothing in planetary-measures.

Wells' plot is a satire that represents Brits as the invading Martians. Modern human politicians have mis-adapted it as a tragedy about the invasion by migrants over borders from Third World into First World countries. The Third World is far richer in untapped natural resources. Illegal migrants can only find bound low-wage hard-labor that is adjacent to slavery. This would be akin to Martians hurdling their spaceships to Earth to work in prison-camps for water and corn.

After a millennium in open-space pondering the lessons from cadatas' history, I am stumbling for a rebuttal to these strange, evil-assuming perspectives.

As the beforementioned response from my focus-group indicated, the most disturbing biased belief humans have about aliens is that they are eager to kidnap humans. A typical anecdote is:

> I was driving in my car down a dark-lit country-road when suddenly I saw glaring lights ahead and lost consciousness. I regained consciousness some time later to find that I was on an operating table. My body was frozen by a paralytic, but I had just enough strength to flutter my eyelids open. I saw a green alien with a giant head who was performing what seemed to be a… probing. I lost consciousness again, and do not know what length of time passed. When I was aware of my surroundings again, I was still driving down the same country-road, with nothing outside or inside my car as evidence for what had just happened.

This is a deeply offensive portrayal of psychotic and idiotically wasteful aliens. One flattering interpretation is that it is an exaggerated belief in aliens' scientific powers. It suggests an alien spaceship is capable of simultaneously suctioning a moving car into a flying saucer, while also causing spontaneous loss of consciousness to the organism inside this metallic vehicle. And the "missing time phenomenon" suggests that aliens need between an average two and four hours to conduct their colonoscopies, as complainants tend to arrive hours late at their destinations. There are no reports of days of "missing time". Thus, this narrative imagines an alien species who is exposing their space-faring capacity by flashing enormous lights at abductees before investing resources into hovering their car into the sky. Once these human subjects are finally in their spaceship, their doctors gather around and perform precisely-timed 2-hour colonoscopies, needing the remainder of the time to place the human back in the car and to remote-drive it long enough for the human subject to regain sufficient consciousness to drive in a straight line. If there is something precious within the human digestive tract; it would be far less absurd for aliens to surprise-knock-out a pedestrian, or perhaps substitute alien probes for human colonoscopy tools in a doctor's office. The mentions of a blackout indicate the complainants suffered a loss of consciousness, which would lead to immediate loss of control of the driven vehicle. Causes of blackouts include blood flow constriction, heart or nervous system problems, and drugs or alcohol. Since driving blackout-drunk is punishable with a DUI, sufferers of this condition might choose to blame aliens instead. Alternatively, these blackouts can be caused by Rohypnol poisoning caused by this substance being added to a drink of an unsuspecting victim; and this victim might then be convinced by police not to file because the blackout they are describing sounds as if it is an alien-abduction-plotline. If the goal was to avoid detection of experimentation by causing a blackout by an alien anesthesiologist administering a sleeping-drug or a no-touch unconsciousness inducing procedure; then, the subjects would not be cognizant of this temporary altered state. It would be logical to induce paralysis during a sleep-cycle. Human anesthesiologists allow around .1% of their patients to wake up during a surgery. An alien species who can travel at light-speed surely would have a much lower percentage of accidental mid-surgery-waking. Thus, anecdotes that suggest all humans wake mid-abduction to realize they being kidnapped by aliens must be false.

A still more insulting variant of the abduction narrative is one that accuses aliens of rape and impregnation.

> I had been in a relationship with my college boyfriend for three years. We had kept the vow of celibacy before our anticipated marriage. I was still a virgin. Then, my boyfriend began noticing my belly was growing as if I was pregnant. Soon I began to feel sick in the morning, and stopped having a period. I took a pregnancy test and it said I was pregnant. When the doctor questioned me, I recalled that I had been having sex in my dreams, and was sore the mornings after.

The cause of this spontaneous virgin-birth was blatantly the woman's affair with another man, and not an alien impregnating her with a human baby. A more common narrative is that the woman is re-abducted and the baby is extracted by the aliens. This latter scenario seems to be the outcome of the Puritanical stigma against confessing a woman has had an abortion. It is more acceptable for her to report aliens took her alien baby away than to report that she aborted it because she was not ready to be a mother. The process of creating a human-cadata hybrid baby would require the mother to willingly commit to years of scientific analysis and experimentation. Claiming that cadatas traveled across the Galaxy to rape human women is akin to saying that a human male traveled to Antarctica to rape a female penguin, and to create a penguin-human hybrid child. There is no rational reason for aliens to be physiologically attracted to any other alien species. And the creation of a cadata-alien hybrid would be a largely test-tube experiment because neither species wombs would be naturally suitable for the needs of a creature that has a half-unmatching biology. For example, an alien might not have a "heart" or blood vessels to pump this fluid; thus, it could not feed within the womb of a biological human through the standard fallopian tubes. An alien-hybrid might also be many times larger than an adult human woman, and thus might be too large to carry to maturity. While the human-impregnation myth is a malicious lie, some cases might be a misunderstanding of cadatas' regular performance of experiments on species using nanobots during sleep cycles. These experiments must take place in the subject's habitat because our spaceship cannot be haphazardly invaded with an alien's bacteria, viruses, or cells that can have unexpected interactions with cadatas' physiology, as we learned from previous planetary habitation attempts. The purpose of performing a physical exam on a new alien species is to learn about the organism's natural functions, and these would be obstructed if it had to be on a ventilator and a compression-suit to survive in a spaceship climate set for cadatas' maximum-performance.

The one myth that approaches reality is that aliens have left potentially-observable traces of their extraterrestrial habitation. As I noted, I am currently stationed in an earth-ly 3D-printed imitating domicile on Mars. It is mostly-camouflaged to match the color of the sand around it. Though this is a light disguise that mostly works for telescopes pointing at Mars from Earth, whereas a roamer taking video on Mars' surface would spot its unnatural shape. We have active missions in spaceships and settlements on all potentially-habitable planets and moons in the Solar System. If humans were more vig-ilant in their spatial observations; they would have "discovered" alien life within their own system. Humans are blissfully oblivious that their telescopes have only directly captured 15 out of the Solar Systems 4,000 exoplanets. Without visual-access, these are instead cataloged by estimating their gravitational impact on surrounding heavy bodies. These exoplanets are as small as the Moon, as large as Jupiter, with an average size of Neptune, which is four-times wider than Earth. Thus, if our spaceships and outposts are smaller than the Moon, they cannot be detected with humanity's current observational powers. This powerlessness has generated a justified fear of the obscured and the un-known. A telling example is the pixilated 1976 photograph of a 1.2 mile-long Cydonian mesa on Mars that theorists initially speculated represents a humanoid "Face on Mars". Apparently, Martians had inhabited Mars long ago, and had mile-long faces, and only one of them died in such a way that his body decomposed while his face fossilized. If a living organism was a mile-long, and lived in a third of the gravitational pull of Earth; it is more likely to have been a colossal squid-shape than to have had a bony skull with fourth-mile eye-sockets.

This is one in a cluster of theories that propose the universe is inhabited by human-oids, or by Judeo-Christian-Muslim humans, who are all cloned by a Jewish God to look like himself. If the Bible states God "created" humans five days after he created the universe 5,722 years ago; then, there cannot be intelligent species in the universe, other than these non-evolving humans. Our travels through the Milky Way have disproven this magical hypothesis, as there are countless intelligent species that are entirely unlike humans because the human biology evolved specifically for the current environment on Earth. The Earth is currently being artificially heated by humans. At the not-too-distant average 122°F temperature, humans' physiology would already need to evolve to survive because it would otherwise become extinct, failing to maintain a healthy, un-feverish internal temperature when surroundings reach 122°F. 3 billion years ago, the Earth was uninhabitable by humans because its surface was a steaming 167°F, and its atmosphere was entirely devoid of oxygen and was heavy with methane and ammonia. At this time, the climate on Mars was proximate to Earth's today. If I lived on Mars then; I would have seen white clouds, a blue sky full of a denser atmosphere, and calm lakes, ocean, and rivers. It would have rained regularly in warm summers, and snowed in the mild winters. The same molecule-soup was present that could have sparked the origin of earth-like-lifeforms. Mars' climate changed so drastically that even if life started, it could not evolve to keep pace. Since conditions were friendly towards the origin of life on Mars before Earth, living organisms could have emerged there first, and might have been transferred by a chunk of Mars being chipped off and gravitating towards Earth. Taking an organism that evolved within the climate of a home-planet, and transplanting it onto an alien-planet requires either extreme natural resilience, and a relative similarity of climate, or an immediate artificial self-evolution of the organism. Aliens from a planet outside of Earth cannot be constructed in the image of a humanoid God, just as a swimming fish cannot have been designed by nature to have the aerodynamic wings of a bird.

It is slightly more credible that Martians would have left the pyramid-shaped structures in these images. This gives me the idea of using slightly misshapen pyramid shapes for our text output structure, as these would be easy to fabricate, and would be dismissed due to the commonality of this conspiracy. I was reminded of the pyramid-scheme earlier today, on August 16, 2024, when I saw a news story that one of the walls of an ancient pyramid in Michoacan, Mexico collapsed into rubble because of a combination of heavier rainfall, and increasing temperatures. The rocks fell off the surface of this pyramid revealing a compressed-earth interior. The Egyptian pyramids have survived for millennia in part because the desert climate prevents water-erosion. Perhaps more significantly, even if most surface-layer rocks slide off, the remainder of the pyramid remains intact because the entire structure is composed of pieces of rock. This hard-rock has allowed for inner chambers and passageways to survive. It is intuitively credible that the stone-pyramids' durability means they have an alien origin. However, Egyptian (African) pyramids were built around 7 millennia after the oldest surviving buildings in Turkey, Palestine, and Pakistan (Asia), which were designed in various rectangular shapes. The stone triangular shape just allowed the construction of the tallest among these ancient structures. The type of artificial-hillock structures with a rock-covering that fell apart in Mexico comes closest among these ancients to France's Saint-Michel tumulus (4500 BC). If a Martian civilization had evolved on Mars 3 billion years ago, and had migrated to Earth they surely would have evolved their building methodology beyond remaking the same stone pyramids for 3 billion years? The alternative that Martians migrated to Earth no earlier than 3000 BC (when pyramids appeared on Earth) is irrational because it assumes a simultaneous and identical building-construction theories developing independently on these two planets. The intellectual power needed to create most of Earth's early buildings is proximate to that utilized by modern paper wasps that build out of wood precise and deep hexagonal cells with stairs of enclosing covering. A termite nest can peak at 41 feet, or near the height of the earliest human building, the

Gobekli Tepe in Turkey (9500-7500 BC). Monkeys build hammocks, and nests that rival birds. The remarkable element is how advanced humans believe themselves to be despite their building methods barely evolving beyond these primitive beginnings. If a pyramid-shape can be a natural rock formation; then, surely it cannot be sufficient to establish extraterrestrial intelligent life.

Variant theories propose "mists" and "fogs" on the Moon are signs of artificial alien industrialized production. The "fog" seen with the naked eye around the moon is an optical illusion from crystal-reflected clouds on Earth. There is insufficient air on the Moon for it generate fog. The last citing of gas or mist on the Moon was in around 1992, or just as high-resolution photography could be used to contradict attempts at fictitious reporting. And smooth hills have been questioned as a potential row of alien bunkers, and a hole at the base of the South Massif has been theorized as a gateway to a subterranean habitat. To bring these theories back down to Earth: Earth's cavern systems have holes for entryways too. And the theory that a human military power succeeded in building a secret base on the dark side of the Moon without advertising this achievement are less likely than that the USA faked its entire history of Moon-landings. An alien industrial process necessitating the production of an enormous cloud of visible gas (which would dissipate into space on the Moon) would be on the absurd extreme of inefficiency. If a gas was a byproduct, it would logically be captured and used for some other purpose, given the relatively limited resources on the atmosphere-free Moon. The surface of the Moon that faces the Earth remains frozen in its precise appearance as decades pass: this is painfully difficult for a professional Moon-observer. The main exception is the hundred ping pong meteoroids hitting the Moon's surface daily, but they mostly hit on the dark-side, or the side observers are not seeing. Additionally, even when these strikes are on the visible side, few of these are large enough to alter the Moon's observable appearance. If you had spent an entire career staring at what would begin to seem like an unchanging photograph of the Moon, you too might start to see light-apparitions or fogs. Cadatas have established a base on the dark-side of the Moon, where our Mothership is docked. It is deliberately ultra-camouflaged, unlike my home-base on Mars, because humans have sent 7 successful and camera-equipped landers to the Moon between 2013-24 (assuming the far-removed-in-time claimed landings between 1963-76 were faked). The Mothership is composed of a material that can be manipulated into different shapes depending on the demands of the mission. The ideal shape for steady light-speed travel is mathematically divergent from the shape that minimizes drag (even if merely from tiny particles) during acceleration, or maximizes drag during deacceleration. On the Moon, the Mothership landed in a crater on the South Pole. The nanobot-powered membrane of the Mothership then attracted dust and rocks from the surroundings to land in a feet-deep covering across its surface to camouflage it as a natural-seeming hill within the crater. Thus, if an alien species had intended to remain undiscovered, but had left observable signs of habitation; this would be like a light-speed capable spy agency sneaking into a party, but forgetfully leaving their insignia-marked cloaks and shoes on the floor by the main doorway.

The history of unidentified-flying-object spotting is largely comprised of observers seeing strange formations of lights in the sky, which have variously been identified as demons, as foreigners, or aliens. Back in the 19th century, these sightings tended to be of airships. Then, after the horrors of WWII, in 1947, there were two first sightings in Washington state of flying saucers visible merely as geometrically-shaped light formations in the sky. This history fits neatly with the history of firsts in human air-travel. The first unmanned balloons flew in 1782. Balloons were precise enough to be used in warfare by the mid-19th century. By 1935, radio-controlled aircraft could execute precise missions. The last major lights-only UFO sighting was in 2004 because, after this point, lighted-drones became common enough to be "explained". The first choreographed done light show was in 2012 in Austria. Two UFOs, an octagonal with strings and an unspecified shape, were shot-down in the US in 2023, as was a Chinese "espionage" balloon. The difference between these incidents is that US authorities managed to investigate what the latter was, while thy shot down the first two before establishing its identity. The use of balloons for espionage in 2024 seems like absurdist spy-humor because in 2022 there were 1,192 active Earth observation orbit satellites with imaging capacity. This is enough to have precise hourly video of every foot of Earth. When cadatas fly across Earth's sky, we either camouflage alien-shaped ships, or disguise them to mimic the shape of human aircraft. And for espionage, we use nanobots that can record 4D audio-video not only from above, but from all around the subject-of-study.

We wanted to clarify the facts regarding these offensive misconceptions. If humans believe aliens are incompetent at disguised intelligence-gathering; it is less likely that we would be invited to legally migrate to Earth whenever we do decide to submit a formal application for admittance with human governments. There are almost no barriers to in-migration for high-net-worth-individuals, who are especially welcomed in the United Arab Emirates, Australia, and Singapore. If we prove we are highly-intelligent high-net-worth aliens; we should also be accepted as profitable newcomers. In contrast, if we allow ourselves to be perceived as light-playing and crooked-pyramid-building nincompoops; our Mothership might be nuked before humans manage to "identify" it.

A Brief Overview to Cadatas' Physiology

The popular mythological categories of aliens catalog them into humanoid groups that have reptilian scaly features, giant heads and tiny bodies, or overly long forms. Other than these body distortions, the main difference between these imagined alien species is their skin-colorations, which consistently fall outside the typical human skin-colors, tending towards a sickly green or gray. These stereotypes appear to have been designed by film-producers pre-computer-generated-imagery, when a thin child with a giant-head mask, or an unusually tall and thin acrobat could have played these parts. The first alien portrayed in a film, in *The Thing from Another World* (1951), had a mask of an enlarged head and crooked fingers. By *Alien* (1979), the sculpting department had learned to make a tale and an extremely overstretched head. The earliest horror films structured plots around the terror of the misshapen humanoid form, which could be deformed by illness as with *The Hunchback of Notre Dame* (1911), by medical manipulation as in *Frankenstein* (1910), or by the Devil as in the myriad films that portrayed demonic possession. The deformed creature horrifies simply by walking towards "normally" beautiful characters. This was necessary in the first decades of film because the first talkie was released in 1927. Pre-1927 scriptwriters had to project emotional intensity with minimum textual explanation. When silent films proved to be more profitable than overly-verbose novels, scriptwriters learned that audiences were more willing to pay to watch a hyper-simplistic non-vocal drama.

This book is unlike these alien horrors because it is a scientific study of my own real species, which is alien only from your human perspective. Imagine if you were an alien and you landed on a planet populated by goblin sharks. This is a species of deep-sea sharks with a misshapen overly-long snout, and needle-teeth. And then you watched their alien horror films and they were dressing their fins with crooked fingers that had an apish appearance that reminded you of yourself. It is the foreignness of a strange organism's form that evokes horror or curiosity, as opposed to the shape itself that is normal from that species perspective.

Most of our physiological characteristics diverge from humans. In general, species' biology is built by forces of gravity of the home planet and its moons, solar-proximity, planetary spin, atmospheric gases, components of the planet's crust, and plate-tectonics or their absence. Cadata was larger than Earth, and it had a slower spin rate around its axis. Thus, it took Cadata around seven Earth-days to complete a "day". Our natural sleep cycle is two-Earth-days, followed by five waking days. Though sleep is a relatively rare phenomenon across the species of this Galaxy. Even on Earth, a few sea species use various strategies to rest without losing consciousness. Sharks maintaining a waking-trance, surfing a current at night instead of sleeping; and the Greenland shark has the longest lifespans among vertebrates, reaching at least 272 years. Cadata scientists artificially re-designed our biology for us not to require sleep, opting instead to go into a brief trance state. Their studies concluded that sleep prevents, instead of promoting the goal they had set of achieving our current 5,000-year lifespan longevity.

This book lacks tales of romances among our cadata crew because we reproduce asexually. Thus, the anecdote that aliens impregnate human women is especially offensive because it would not only be bestiality and zoophilia, but also a violation of our millennia-long sex-avoidance habits. Cadatas reproduce by mixing components from two donors in a laboratory to maximize evolutionary adaptability, and then incubating the embryo in a precisely calibrated gestational chamber. Our species also lacks gender divisions. Our scientists have been researching the strange separation into two genders among humans. Even on Earth, there a multiplicity of gender or mating-type numbers, ranging from three among clam shrimp, to four among white-throated sparrows, and over 30,000 among mushrooms. A mating-type is simply the divergent characteristics within a species that are tied to reproducibility or non-reproducibility. Chances of species survival increase if there is a single mating-type and asexual self-reproduction. The only drawback is that evolutionary change is enhanced when "genes" of two or more organisms are intermixed. We have hacked this process through self-evolutionary strategies, which will be described in a later section. Most planets are populated asexually, or through various complex methods of cell replication and dispersion. Splitting into a receptor and donor gender only to unite two parts of a future offspring is inefficient on the cosmic scale. We do enjoy neurophysiological orgasms by ourselves, or with partners. This neurological phenomenon tends to lack congruence with reproductive functions across the Galaxy.

Across this study, the standard English gender separations are utilized to distinguish between the relatively meaner masculine, versus the demurer feminine personalities, as opposed to referring to his or her physiological gender identity. And a few terms are used as placeholders to avoid introducing alien terminology for each planet's variant of these concepts. *Genes* is used to refer to the general transfer of parental characteristics to offspring, without suggesting that they are composed of earthly nucleotides. *Cells* refers to the smallest unit that can live outside an organism, without implying this unit has structures identical to the membrane nucleus and cytoplasm that appear on Earth. *DNA* refers to the self-replicating material that defines the characteristics of an organism, without implying that it has either nucleotides, or sugar phosphate. Instead, it can be comprised of any life-building elements. The chemistry, physics and biology of these elements differ drastically across this Galaxy's planets, as will be explained.

Cadatas' natural inherited distinguishing characteristic is shapeshifting. Our bodies evolved to change in consistency, surface texture, and size to fit alterations in the environment. We can increase buoyancy by filling ourselves with air to float on liquid. We can mutate into a thin fabric to levitate in a windward direction. Species of Earth's fish have similar abilities because they can manipulate not only their colors, but also skin texture to mimic an inert rock, or a dangerous adversary. Our predecessors hovered on sail-wings through the sky due to our planet's low gravity. Our skin's mercurial behavior is controlled by an oversized-brain. Again, throughout the term *brain* is used to refer to the organ that controls an organism, as opposed to referring to the specific biological characteristics of brains on Earth.

As described in the next chapter, our planet underwent rapid environmental alterations, which necessitated an evolutionary adaptation to moving with ease not only across the air, but also the earth and the sea. The combination of our evolved biology and advanced shelter-building skills were necessary for our species to dominate an extremely threatening natural habitat. Our adaptations exceeded the environmental threats until we had the leisure to develop complex technologies beyond basic-needs. Our earliest science textbooks and innovative periodicals were published a million years ago. Early innovations included optimizing gene-editing for environmental variability and longevity. At the pinnacle of our gene-editing experiments we designed a method to substitute our natural molecular-type for nanobot-controllable molecules that could be programmed to alter their chemistry and the organism's physiology to suit the programmer's or the environment's demands.

This artificial control allows rapid self-re-assembling. If a body part is detached; it immediately grows back when the body's internal computer records its absence. Doctors are irrelevant to this process, unless the injury suffered is so strange that the program cannot find a relevant code, and a new code must be constructed by a medical-programmer. The human liver auto-regenerates every three years, even if it has not been truncated. On Earth, only lizards can regenerate their skeleton with cells that can build new bone, just as new bone is formed in an embryo. With access to nanobots, cadatas' cells can pull chemicals necessary for immediate re-generation out of the environment, instead of needing to wait for a slow natural trickle of nutrients. Even cadata brains can regenerate because memories and identity-defining preferences are stored in fluid data-banks that are dispersed across our bodies. Thus, the loss of our heads, or "brain" organs are not fatal injuries. Spreading the cadata brain across the maximum surface-mass also maximized cadatas' intellectual capability to the upper end of the galactic scale. This dispersion of brain components also allows for simultaneous performance of separate functions, calculations, reflections, and mobility controls.

Cadatas' natural bodies are smaller than humans' when they are compressed to a matching density. Cadatas' ability to alter bodily density is a necessary feature to survive on worlds with outlier gravities. Density-alteration also facilitates flexible concealment strategies, such as blending into a hostile alien environment. Cadatas can mutate the visual organ to see on the full electromagnetic spectrum, and to zoom in on nearby microbes, or zoom out to objects in outer-space. Cadatas have a single "eye", instead of two or more "eyes". This eye can float to the section of the body that maximizes visibility of the focal direction. Under normal conditions, this eye tends to naturally take a spherical 360-degree shape at the tip of a cadata body.

Cadatas' physiology determines our unique social structures. Modern cadatas reproduce toward the end of our lives. Delayed reproduction was necessitated by the length of our ongoing interstellar voyage due to the need for precise population control. When conditions are optimal, the genetic code is imprinted on an offspring minutes before the parent's death. Reproduction at the end of a lifecycle is standard in many of Earth's species, including salmon and octopuses. The act of traveling to a distant mating site, releasing eggs, and nourishing the young tends to exhaust both, or one of the parents to death. These survival challenges are absent for cadatas because genes are painlessly extracted and mixed, and the community jointly performs child-rearing. Cadatas lack formal schooling equivalent to Earth's. Instead, there are intellectual development milestones in the first years, when cadatas take a designated sequence of digital courses. These courses focus on psychological and physiological training of the creative-intellect and muscles because our universal knowledge is housed in computers and nanobots, making memorization irrelevant. Cadatas who choose especially intensive imaginative, or hands-on fields conduct independent research, or participate in apprenticeships. I teach in the main training initiative for our Milky Way exploration mission, called the Continuing Education program. My current research focus is on helping cadatas learn how to act and look human while they explore Earth. Other than keeping the population at a stable level, late reproduction also prevents two near-identical cadatas from being of employment-age simultaneously, maximizing group diversity.

The function of a biological digestive system is to take in nutrients from the environment to convert them into energy to power the organism. On Earth, the sea cucumber has a respiratory system that can simultaneously perform digestive functions, if this species eviscerates its digestive tract by hurdling it at a predator. In a comparative study of life across the Galaxy, the earthly digestive process is a strange outlier. It involves the inefficient burning and excretion of matter from rival species, with only a fraction of nutrients retained by this fueling system. In contrast, cadatas' light mass allows for the consumption of mere gases for energy. The atmosphere on our native world is dense with the elements we evolved to survive on. Because these are simple chemical elements, this mixture can be artificially reproduced by cadata machines that can provide constant nutrition from the environment to maintain equilibrium. If a cadata wants to experience a more natural feeding, these machines can also replicate the complex textures, colors, densities, smells, and psychological effects of authentic Cadata gases. Inhaling gases through the skin, or through a breathing organ combines into a single process the humans' less efficient organic division of breathing, drinking, and eating. Our breathing organ became an unnecessary appendage when cadata scientists designed a nanobot-controlled body-wide air particle consumption capability. Surface cells can consume, circulate across the systems, and perspire, or microscopically exhale unnecessary particles.

The flexibility of cadatas' biology permitted our survival of a 100,000-year cross-galactic voyage. Individual elements that remained constant among our home-world ancestors have been drastically reconsidered by the demands of space-travel, and a series of adaptations to new worlds. The various typical cadata characteristics, such as asexuality, the single eye, or gas-based-digestion, have each undergone mutation. From our perspective, a species is not defined by a shared constant biology. Some among our own spaceship's community are biologically too divergent to be defined as part of a single species. Thus, the term *cadata* refers to our shared ancestry, as opposed to specific physical characteristics.

As a worldless, migratory sect, we are the output of the cultural and physical impacts from the civilizations we encountered and abandoned on our journey. This history attempts to make sense of who cadatas are for our own self-understanding, as well as to garner sympathy for our undefined condition from our human hosts.

Chapter 1

The Rise of Cadata's Culture and Science

Cadata's Climate

On Earth, most species return to their birthplace to reproduce, or as part of their standard migratory route. A love for one's place of origin is intuitively imprinted in parallel with attachment to a family-unit. The Mothership is technically the birthplace of our mission's cadatas. Though we have retained an innate affection towards Cadata as a geologic spot in the Galaxy perfectly designed for our natural physiology.

Our ancestors were forced to leave Cadata because of an artificially-induced ecologic disaster. At the time of Mothership's departure, Cadata was still in its healthy, tropical youth as a natural planetary body. Cadata's surface is enclosed in a trice-deeper ocean than Earth's. In this planetary category of water-volume, most rocky landmass is submerged. Solid ground is only found on Cadata on scattered volcanic islands. Multicellular life developed 4 billion years into Cadata's geologic timeline. Life evolved into marine species over the first billion years. 3 billion years ago, an independent, smaller planet, with a Cadata-parallel orbit, crashed into Cadata, and turned into Cadata's moon. Its pull sparked the reignition of Cadata's core, causing a spike in volcanic activity. These eruptions generated strings of islands amidst the ocean. This emergence of concrete ground then sparked the evolution of land and air-based lifeforms.

To recover from stress, I have a habit of immersing myself in 4D audio-video scenes from Cadata before the catastrophe. The sky is dense with thick, pinkish clouds, regularly erupting into brief storms. The program sets the temperature in the viewing-room to around fifty degrees hotter than midday in the tropics on Earth. This is the outcome of Cadata being a fraction closer within its Goldilocks zone to its sun, in comparison with Earth's proximity. The denser atmosphere means a heightened surface air-pressure. On Earth, water boils at 212°F. Earth's highest recorded temperature was 134°F in the Mojave Desert of California. Since Cadata is around 50-degrees warmer, its hottest temperature of around 184°F would bring Cadata dangerously close to its ocean boiling out into space. Cadata's higher air-pressure prevents this scenario as it raises water's boiling-point. Without outright boiling, there is still a higher degree of ocean-evaporation that feeds the dense atmosphere and generates a higher frequency of stormy weather. These dense evaporated gases are cadatas' natural food-sources. Cadata has no continent-sized ice caps because the last of these melted a couple billion years ago, during the last ice age. The elemental ratio of Cadata's atmosphere is a toxic combination for humans. Thus, if cadatas altered Earth's atmosphere to fit our natural needs, this change would lead to the extinction of humans. This is why we have consistently opted to alter our own physiology, instead of changing a new planet's environment. One exception to this moral rule is if a planet is not already inhabited with life. Thus, my current experiments on Mars have two objectives. One is to determine if there are microscopic lifeforms here. Secondly, assuming Mars is lifeless; I aim to learn if there is a practical solution for turning Mars' geologic clock 4 billion years backwards, and infusing it with a cadata-specific atmospheric combination of gases.

The Archeological Record of Early Cadata

Evolution takes epochs on every new planet where life blooms. The path towards complex organisms always begins when chemicals begin to combine into new structures beyond what is created through the laws of chemical interaction. Chemistry cannot evolve. Life must evolve. A microbe is an entity with a rudimentary desire. Lifeless chemicals do not desire to find more chemicals to maintain a chemical reaction. A living microbe is intuitively driven to find the elements it needs to continue living. This drive to live eventually develops the tools it requires to live in its specific environment, and these structures and tools can be summarized by the term: cell. Once cells develop, they begin to interact with other cells and generate multicell organisms. The accessible chemicals and the climate determine the resources life is built from on each individual planet. Across the Galaxy, it is common for chemistry to eventually lead to life. What we have learned is near-impossible is for a lifeform to leave the planet it was designed for, and find a new planet it can adapt itself to inhabit. Even on a metallic planet, metals can evolve into living organisms.

Many self-mutating species evolved on Cadata, some were the size of a whale, others were tiny and biologically proximate to plants. When massive intruders disturbed groves of plantish organisms, they triggered waves of mutations, and with each change these lifeforms customized defensive responses. Cadatas' predecessors had the largest brains among this varied throng. The enduring violent storms across Cadata forced cadatas to evolve strong navigational appendages to survive lethal gales and precipitation. Their enormous wingspans allowed them to migrate by floating, or gliding between islands, spreading their seed across the planet.

Genetic-rebirth has recreated members from all branches of cadatas' evolutionary tree. Our laboratories dissected their bodies to analyze which natural mutations have prospered. Lessons from these autopsies also expedited our gene-editing mechanisms. Evolution does not always mean progress, and it was important for our scientists to learn from nature to avoid its mistakes, and mimic its successes. We have uncovered the genetic samples we needed for these extended tests in a similar manner to humans' archeologists, though most of our ancestors were buried under the floor of the ocean as those who died on the surface of islands tended to be covered over with lava, and thus did not leave a testable, fossilized sample.

The first recognizable cadataids evolved in around 7,000,000 BC. Once they had the powerful tools giving them a survival advantage over other competing species, they pushed ahead in their mental development.

The Development of Sculptural Language and Early Societies

A cooling period across the planet's surface, in around 4,000,000 BC, created a sudden drop in the available dense near-ground cloud formations comprising the cadataids' primary fuel source. Species who could not find sufficient nutrients became extinct. To overcome this challenge, the cadataids formed larger, socio-complex communities. They discovered materials capable of gas storage. They began storing enormous spheres of a gas within a round enclave in their bodies created through mutation. They flew into cold clouds at the top of their gigantic plants, and allowed the spheres to cool until water vaper on the edges froze. Then, they dubbed graphite over these crystals, creating a container to hold the gas in place even after they returned to the ground and the water crystals melted. They took advantage of their bodies' mutability in many other clever maneuvers. Having a large reserve of nourishment allowed them more free time to socialize and share the innovative tricks they were developing with other cadataids. They hunted for the precious supplies of air in tight group formations assisting them with repelling rivals. This further reinforced their social bonds.

Gradually, crafty cadataids began designing new shapes for these containers. At first, these picturesque objects' usefulness was in their capacity to strengthen social or attractive bonds. Then, specific shapes began to be associated with semantic meanings. Some symbolized dominance or submission, some signified possession or communal sharing, and some stood for admired local species. Before long, these simple shapes evolved into cadataids' first language. The adoption of this complex communicative method marks the juncture our archeologists utilize to separate earlier cadataids from the evolutionary branch of modern cadatas. They also developed a whistle language to communicate at greater distances to warn tribe members of dangerous predator and storms. Linguistics allowed cadatas to band into complex groups with intricate power dynamics. This process continued in the following three million years. They might have evolved faster. But their ability to glide, metamorphose, and communicate secured them against most environmental stressors. Thus, the next leap was inessential for survival.

The Discovery of the Suspended Islet Sea Grotto Art

An inspiring early discovery provided a valuable cultural artifact. In 1,004,976 BC, a group of continuing education students stumbled on this find while studying rock formations at the bottom of a volcanic islet, which appeared to be floating above-water because of the deep salt-water corrosion of its sides. Cosquer Ground was feeling the rocks under water with her flexible wings when she felt a drop off in the surface. She dived under water and saw an opening below. With another push, she emerged inside of a grotto under the islet. She carried a supply of air with her, needed for this confined space. Without time to lose, aware her friends would be worried if she stayed under too long, she hopped and glided around the grotto, searching for gossip-worthy discoveries. Finally, she spotted a sparkle on a shelf protrusion. She used a natural-occurring fluorescent shell light to distinguish its features. A series of interconnected graphite shapes mimicked spear-shaped crystal cave formations, but they interlocked and built on each other in ways defying gravity and branching out like the foliage of our rare hopping trees. Amazed by the beauty of this structure, Cosquer returned to her friends and they went back to the main island to inform authorities of her finding.

The graphite branches were analyzed for age and design by the rudimentary methods of the day and confirmed to belong to the epoch when this three-dimensional, artistic communication method was just being invented. A reproduction of the Islet Grotto Crystals stands in the center of Cadata's Earth-stationed Mothership, as a symbol of our exploratory curiosity. The Committee for Interstellar travel considered shipping this original, and other precious pieces on this exploratory mission. But because of the catastrophic first attempt at surveying an inhabited planet, there was little hope this trip would last as long as it has. The reproduction is exact to-the-molecule, and most molecules have been examined for the complex symbolic meanings they hold. It appears to represent the delicate balance of power in the creator's community. This is only a guess. The true meaning remains a mystery because, unlike DNA strands, the workings of an unexplained imagination are as fleeing as the air.

Mineral-Manipulation Age

The primal graphite sculptures and verbal tools inspired the next major shift in cadatas' cultural and scientific progress. By 1,006,000 BC, the planet was at the door of the Mineral-Manipulation Age. On Earth, different materials were conquered in different epochs, but cadatas appropriated most simultaneously because they learned to shape them by manipulating their bodies into molds. Then, they carved them into utilitarian shapes. Cadatas lacked a need for plates or spoons since they consumed gases, so their containers suited this complex and misplacable substance. These carvings and tools assisted them with penetrating the depth of these dark waters surrounding each settlement. Cadatas' ability to change their bodies into desirable objects contributed to the quantity of jewelry and decorative clothing archeologists uncovered from primitive settlement sites. They had the capacity to transform their skin to mimic the thickness and sparkle of a diamond, if this stone was popular in a village.

The Nutritional Gas Shortages War

Prosperity and ease of living expanded cadatas' lives and in turn their population. However, cadatas only permanent residence was islands because they could not glide over water indefinitely and they lacked an architectural design for an overhang to suspend a residence above Cadata's domineering water bodies. Most islands and islets were soon crowded with cadatas. Competition over the one resource every cadata needed—dense nutritionally rich air—intensified. Whistling matches began to drown the semblance of civilized conversations. Cadatas began separating into pacts. They attempted to divide segments of the air overhead with treaties excluding non-participating players, instead giving the spoils to whoever imagined ownership of a given aerial space. Graphite containers of hot air were guarded day and night as thievery began to make it difficult to make an honest living. As soon as a hot air gatherer collected enough for the season, some lazy imbecile from the neighboring village would sneak in under the cover of darkness to take this precious commodity with a brief display of force. If this thief acted alone, a village might band together to retrieve the little cloud of air. So, thieves began unifying. Affiliation multiplied their aggressive tendencies because they acquired a numbers-advantage and encouragement from villainous peers. With time, these bands outgrew the need for stealthy supply thievery. They declared mini-wars against non-affiliated neighbors. These conflicts escalated to air-deflation or death of some of the parties involved. To avoid suffering similar fates, those who believed in fair-dealings abandoned contested parts and migrated away from islands controlled by these marauders. They formed defensive positions on islands of their own, and prevented new arrivals from entering out of fear they might be carrying the equivalent of a Trojan horse. Constant maintenance of defenses and plotting new territorial attacks required the development of a complex government system.

The largest island on Cadata, in terms of landmass, is Alleges, it was chosen by several small defensive tribes, who negotiated for joint control to become a stronger force capable of repelling even the most savage invader. This group came to be known as Occoro. As a practical solution to the need to make decisions about gas distribution or conservation, as well as about space allocation, and dealings with potential new tribes, the Occoro decided to assign the job of "ruler" of Alleges on a cyclical basis to each of the citizens of this community in turn for a single day and night. Every citizen, regardless of their limitations, had a chance at the helm. Other concerned citizens could advise him or her, proposing possible courses of action, but the final decision was up to this ruler. If this individual made corrupt or harassing decisions without just cause, he might be penalized on the following day when somebody else held the helm and complaints regarding these problems were raised.

Alleges was ahead of most of its neighbors in terms of shelter architecture. They had advanced small-scale sculpture into craftsmanship of entire cadata-sized bubbles with burrows under them for entry where groups of cadatas could find shelter against winds during some of the strongest storms. They could also turn these giant, thin, solid bubbles sideways to collect dense cloud gas in times when the weather was stable and these settled near the ground rather than being ripped apart by gales. The safety and prosperity these bubbles provided made it easy for the ruler of the day to make sound decisions continuing the upward trend of the Occoro tribe. To further secure their position, they developed various innovative defenses to alert them to attacks and to shield them in case of a sudden swarm of invaders.

On the other side of the island strand housing Alleges was an island called Getor. A powerful marauder ruler, Medit, began talks, in turn, to unify other thieving tribes to counterbalance Occoro's growing power. He terrified each of the leaders of pirate bands with tales of Occoros planning the expulsion of all other tribes from the strand. He even suggested the Occoros were secret cadata-gas thirsters, rowdier than the worst marauders—a dangerous enemy, conquerable if they amalgamated. During a public gathering on the choice for their next target island, members questioned why Medit was in charge of their gas-supply. Why were his battle choices adhered without question? They proposed adopting the Occoro's brand of government and sharing in the decisions impacting their livelihood and lives. Medit slaughtered the three marauders speaking up in this disrespectful fashion, and declared himself Ultimate Ruler of the renamed Mediti clan for the coming eternity. Without new objections, and having secured his power-hold,

Medit proceeded to move the group's gas supplies into a private storage bunker bubble he had stolen on an earlier pirating campaign. He then reported on the near-depletion of their supplies. They had to commence on a new invasion to avoid a breathless death.

Medit commanded a team of his wisest advisors to invent a weapon to bolster their offensive in the great battle he insisted had to be waged against the Occoros. They studied the short list of prior weapon-designs. Cadatas had never thrown rocks at enemies because the planet's islands were sandy or composed of dried lava sheath, so finding a rock to toss an advanced endeavor. After a week of seclusion for exercising methods of assisting combat, they came forward with a proposal for the howling whistle. This device had complex multi-compartments, spiraled and was empty down the middle. When a cadata blew into it, an intense string of directed wind hit the party the other end of this device was pointing at. Enough of them could destabilize the other side in an attack, and could even cause sever injuries or death if the cadata hit was knocked into a cliffside or another hard object. Across the next few weeks, Medit forced every juvenile and elderly member of his group to construct enough howling whistles to arm every ready wing.

The sunrise on Cadata takes four times longer than on Earth. The first sunrays were breaking through gaps in the cloudy and windy sky. Occoros were asleep in their bubbles on top of the steepest cliff on the island. Hissing from the heating, high-tide waves were hitting the side of this cliff at the bottom of the drop. A couple of watchers were bobbing in a trance in the canopy of the quaking trees, unconscious of their surroundings. Too long had passed since Occoros were attacked because of their legendary defensive position.

The sound of the trees' quaking was interrupted by the piercing sound of the howling whistles' air streams hitting the watchers and knocking them down to the ground with violent thuds. The Mediti began gliding up to the tip of the cliff to follow up the surprise attack with an invasion of the main camp, but then Medit, who was leading the charge, noticed the Occoros they had hit were transforming into tenals, a miniature shapeshifter species on Cadata, which had been domesticated in the previous few thousand years. Both the cadatas and tenals take the natural form when terrified or injured. The tenals natural shape is similar to a porcupine, with over two-thousand protruding, thin, tentacle-featherish, and colorful strings. It became apparent to Medit they were tricked. If the sound of tenals falling off their perches was insufficient, once recovered from the fall these tenals began to emit a vibration, which felt akin to an exploding bomb, shaking the ground and the thin bubbles over the Occoros. These sound waves popped the camp's sound receptors, jostling them awake. Mediti discovered, instead of surrounding the Occoros, they were in the shadow of a mini-army of Occoros, hovering over their heads. Medit ordered his frozen troops to fire the whistles. They obeyed, but only after the Occoros started dropping empty gas containers on the giant eyes at the top of their heads to minimize their vision as they climbed up the cliff. The bombardment continued for a few minutes until injuries took a portion of the fighters out of the conflict. Seeing their fallen comrades, invigorated their friends and the battle took on a second life as the Mediti started gaining ground. Once they were level, they launched their wings, smacking sides, shoulders, and eyes. This smacking went on for a long while before Medit realized he was exhausted; this would not be an easy victory. There was even a chance of a painful defeat sending him into exile, if not worse. He called a retreat and they stormed away at top-speed. The Occoros left them flee as it was not in their moral code to beat a retreating enemy.

Medit did not take this loss well, executing a few of the advisers who failed to warn him the Occoros would have tenal watchers. After a few days of debauchery and intoxication on mind-altering gases, Medit regrouped the remaining advisers and solicited new plans for attack. They settled on counterespionage, and placed a moll in the Occoros camp. A spy-girl pretended she was starving and asked for asylum. She fed them precise information about new reinforcements the Occoros were making, and vulnerable targets to the howling whistle. She would disguise her body as a hopping plant to listen in by the entrance to meetings of the Occoros' top officials. Despite these extraordinary efforts, the information she was gathering could have been gleamed from a simple observation of the camp, so they failed to find a weak point. One advisor proposed attempting to kill the Occoros' leader. However, because their spy had just explained the leader is switched daily, assassinating any single leader was so nonsensical, Medit chopped his last advisor into minced-pieces. The Mediti managed to survive for some time after these losses. Though he never attempted an outright attack against the Occoros again. Instead of spending time on military strategy, Medit, the Occoros, and most members of warring tribes across Cadata began focusing their resources on finding new ways to avoid starvation and death. And these individual pushes amounted to the growth of a new age.

Scientific Innovation Age

Relative peace, prosperity, and the calming stability of the now-dominant one-day ruler system ushered the Scientific Innovation Age. This turning period officially began in 1,004,876 BC, the year the tribes of the southern hemisphere loosely unified. They met annually going forward to discuss and vote on shared concerns. Instead of becoming a burden, the joining decreased violent squabbles bursting between neighbors. It was again safe for cadatas to glide between islands to explore their beauty, or to locate exotic gas sensations. Such travel was lethal a century earlier. The annual meetings for exchanges of scientific and artistic findings between the regions accelerated the spread of scientific innovation. Then, a discovery revolutionized life for all cadatas: the gas generator. This device condensed gases from the air into small compartments with much greater nutritional value fitting in ancient graphite containers. They could be stored in large quantities to minimize the time it took to acquire and prepare gas-meals. While the convenience of these gas-meals cannot be overstated, they introduced a convoluted trade-economy into a communal system.

The individual struggle to find commodities to trade for air increased the populations' incentives to have more children earlier in life. Cadatas reach adult-size in our first year, so we never developed a conception for "childhood". Youths begin working as soon as they need an income. This rapid growth originates from our airy composition and shapeshifting biology. Rapid reproduction re-expanded our population, but this time the overgrowth was less catastrophic because we had developed efficient building construction and nutrition cultivation methods to meet if not the demand for all, then at least the necessities of the workaholics.

Alleges grew in dominance as an economic and cultural capital of the hemisphere.

It and the islands in its strand became a vibrant city where both scientists and architects found commissions. Those seeking patronage settled in large numbers on an island called Bonne. It was covered in multi-colored growth of spores emitting a smell without equals on Earth; a mixture of concentrated and yet elegant roses, seaweed, fresh grass and the strongest-smelling flowers on this planet might begin to explain its characteristics. This island had a triangular external shape with a giant circular cove cutting out its center. There was a long beach across the length of this cove with a narrow opening where water from the ocean could enter. The exterior triangle edge was much higher than water level, as lava formations had crafted asymmetric black cliffs. An active volcano was inside the tallest summit with an active crater cyclically expunging its innards. Tenals were stationed across the circumference, so their vibrating alarms would notify the cadatas below of an active lava flow. The mutating landscape shaped architectural trends in this period. Instead of investing in a few epic and lavish construction challenges, architects were churning new designs as the older structures were flattened by molten rock. There were two main techniques during this age for shaping housing and business structures. The first utilized a new re-melting of rock into lava invention, which created a habitable compartment inside. These had ornamental shapes, each different from the next, including overhanging strings and orbs and shelf or bed-like protrusions serving as furniture.

The second was waiting for lava to flow before interrupting and shaping it with fabricated-building molds. These became strings of hollow spheres towering in several levels. The topmost were the most desirable. Engineers were always thinking of new ways to add still higher spheres at levels above the highest lava had reached to-date.

By the end of the Scientific Innovation Age, these ideas evolved into mechanical buildings composed of heavy metals resistant to lava-flow. Each of their habitable compartments could move anywhere else in the building, rising to the top floor, or dropping below ground. Residents could also choose to face the sun, if they needed it for solar energy. Preference for certain spots, like the top floor, made groups of these buildings top-heavy or thick on the side facing the sun. If a city was ever attacked; residents could choose to relocate underground, and the entire building would disappear.

Another breakthrough was the invention of 3D holographs, which were adapted for entertainment and education purposes. A holograph still studied by cadata youths is *The History of the Occoros* by Magy Gundy (1,003,965 BC), a thrilling account of Medit's violent attacks on this settlement and other battles cadatas fought over the centuries. It is one of the best sources on this period, and I used it in this account of the events. Gundy was born in Alleges and received a commission to create the holograph about Occoros from the ruler-for-the-day, Ding Tis, so he traveled to Occoros and created holographic replications of key sites across the city (while it was centuries later, some of these places had remained undeveloped in honor of their historic significance). Gundy programmed the details of the battle into the holographic system, so viewers could enter the battlefield as if they traveled back in time to an event. Of course, it is not an exact replica, as there were no imaging devices in this age capable of reproducing the appearance of the key players or their positions.

A few dominant businesses emerged in essential industries such as housing and entertainment. The owners of these ventures started as independent engineers or programmers and grew their companies as demand expanded. They found ways to cut costs and increase profits. Because they were enormous, they could set low raw-material prices, which had to be accepted by the mineral extractors. Because they could produce goods at the lowest prices in the market, they grew into monopolies. Alleges' for-a-day rulers were unable to reign in their behavior to reestablish competition in these sectors.

In 1,003,756 BC, when a consortium of rulers began imposing regulations on these businesses, a dozen monopolists (the Dozen Potentates) gathered to alter the structure of Cadata's governing system, naming themselves as undeniable rulers. They disbanded the ruler-for-a-day system. They spread misinformation by bribing newspapers to report on corrupt actions of a string of ruler-for-a-days; each story grew more incriminating and outrageous. Under this pressure, public opinion shifted. The negative publicity instilled questions into the citizens minds. Are cadatas innately evil? Can a citizen encompass the knowledge necessary to operate an enormous city, such as Alleges? Because these busi-nessmen oversaw indispensable sectors (such as the air supply), they manipulated the media, their employees, and their products' consumers. They organized a referendum on changing the political structure and rigged it to favor their proposal, and it passed as planned. They turned this new power into heightened prices, which pushed poor cadatas living in the Alleges region out of this market. Poor vagabonds now roamed in spaces be-tween cities, on turbulent oceans in perpetual-search for scraps of nutritional-air. With Alleges as their model city, metropolises across Cadata also converted to this political model.

To make themselves appear benevolent, the Dozen Potentates released several new inventions. They argued the infusion of capital from their raised prices allowed them to invest in research bringing these new mechanisms into cadatas' wings. One of these were flying vehicles to improve the speed of transportation between cities. This shift benefited the Dozen Potentates as planet-wide trade eased, spreading their monopolistic dominance to distant lands. The flying car was developed a century earlier, in 1,003,835 BC, when an independent scientist, Tellah Ledo, created a gliding machine to reposition construction materials for the innovative buildings she was engineering for her clients. Cadata never had patent laws equivalent to Earth's. Inventions are adopted by busi-nesses without paying or crediting the inventor. In this period, a flying machine was a threatening advantage for small businesses. With physical labor of raw material transport deducted, they would gain ground against large enterprises. Thus, the flight-invention was suppressed from the press. Tellah used her airborne unit as a gimmick, but she never found funding to duplicate it for the public. In 1,003,726 BC, when the Dozen Poten-tates decided the time was ripe to do it to their benefit, they released a near-identical model and made a hefty profit. Hitherto unexplored islands were settled and Cadata seemed to be on the cusp of a scientific and cultural explosion.

The Dozen Potentates considered various progressive moves, but decided they cared most about prolonging and improving their own lives. They began developing Cadata's first gene editing devices and programs. The first experiments sickened or killed the test subjects. After numerous trials, the system was safety-approved by the bribe-laden Safety Commission. At this juncture the Dozen Potentates used the devices as-intended to fix their own genetic "defects", including age and lifestyle diseases. The elongation of the Potentates' lives and health improvements helped sell gene-editing as a luxury capable of raising the dead and healing the sick. Usage spread among the wealthy, further increasing the Potentates' personal fortunes.

Across this period, cities on the opposite side of Cadata and in isolated peripheries maintained their autonomy from the Dozen Potentates' influence. The Potentates refrained from outright war to force these independent regions into submission. It would have been impossible even with their financial resources to keep every freedom-minded region under their influence. Some cadatas started fleeing Alleges for these regions in hope of finding sustenance and asylum from the Potentates' tyrannical rule. The Potentates allowed these refugees to escape as they did not see a benefit to maintaining the poor or disenfranchised within their borders. Alarm at these trends was only raised with the release of a news article about an anti-gene-editing rebel group within Alleges, the Donlons, which graffitied the length of the allied business-headquarters with: "Gene-Editing Kills Our Selves!" Security experts advised the Potentates to travel under heavy-guard, anticipating an escalation in the aggression of these attacks.

Despite these warnings, the Potentates continued business as-usual. Their routine was interrupted on an unspectacularly hot and gusting day in 1,003,713 BC. Across this history, I am refraining from specifying the exact dates and months because moon and solar cycles are different on Cadata, so the conversion between the two would confuse readers from both time-systems. Thus, the head of the flying machine branch of the United Companies Enterprise, Hickle Dimle, was on his way to display his latest version of the machine at an exposition when an old model of their soaring vehicle stalled in front of them. Dimle's security vehicles surrounded his on all sides, so his driver was prevented from bypassing this stopped car. A block earlier, they lost the security vehicle scheduled to be in front of them. It sped ahead to bypass a traffic jam. Dimle's driver refrained from rushing, allowing a suspect, hurried driver squeeze between them. Dimle shouted via the comm for the other drivers to regroup to overcome this hurdle, when a hyper-mechanized version of Cadata's ancient howling whistle weapons began striking the motorcade. The security team failed to find the source of the fire. They were flying between several buildings. A sniper could have been stationed in any opening. Just as the driver was about to ram the stalled vehicle to clear the block, another vehicle plowed into the side of Dimle's machine, striking the current generating engine and disabling it. The hit also shattered a chunk of the machine's shielding. Before Dimle could react, a howling bullet struck him dead. Rapid fire from the attackers continued. Then, the driver of the car that hit Dimle's ejected into the air. He fell for a moment before he was caught by the driver of the stalled car. As these two sped away, the snipers seized fire and retreated. The security guards took advantage of the silence to finally emerge out of their hatches. They pointed their weapons in all directions as if insistent on finding the perpetrator, just as a dog begins howling at a trespasser when the owner arrives. Their salaries were too low to motivate fighting hot-fire.

The Smartness Agency and the Near-Thousand-Year War

A new ruler was nominated for the flying car branch at the emergency Dozen Potentates meeting later in the day. They discussed the fragile situation they were in as a small group in charge of millions of lives. They asked their aides to come up with potential solutions to quench the rebels. The sole worthy idea they generated was offering bribes to convince influential radical leaders in Alleges to forward information on suspected Donlons to authorities. A decision was reached to appoint the assistant who came up with this plan, Pidly Cuirty, the Brain of the new Smartness Agency.

Cuirty initiated the creation of a list of smart questions to ask potential radicals with information capable of leading her to the perpetrators of the assassination. She spent months polishing these questions. She knew, whenever she finished this list, she had to venture into the field to ask them. Here are the questions she crafted preserved in the historic record: *What is the point of this attack? Who was the intended target? Who could stand to gain from this assassination? Who might be the perpetrator?* She spent days trying to decide if the last question should be altered to read: *Whom can the perp be?* But decided against this for obvious reasons. Then, she spent months looking over the maps of the city, trying to understand where the attackers might live. She did question if perhaps the rebels lived outside of Alleges, but it would have been too complicating to study the map of the entire Cadata for clues. She considered if they might live on a hill or close to the ocean. *Would the breeze of the ocean make somebody more rebellious?* she questioned. She decided the perpetrators lived in the buildings surrounding the attack location because this would have eliminated home-invasion from the necessary tasks. Then, Cuirty went to the big shopping center right on the ground floor of the same corporate headquarters where she was working. While she was shopping, she chatted up other shoppers and queried them for gossip or hearsay. At last, she began asking the questions she so elegantly crafted. She received a myriad of replies. One source speculated the shooter was a disgruntled engineer of Dimle's, who hated tenals, and was consumed with self-intoxication on a new gas concoction. These gases suppressed the brain's logical reasoning abilities, only allowing for instinctive reactions during this mental vacation. Another source believed the rebels were pirates from the island of Pitefal, a group with a meagre two cadata leaders. The rest were tenals who shape-changed to match cadatas' appearance. A third shopper was convinced the shooter was a youth under one-years-old, who was experimented on in one of the gene-editing flops, leaving him horrifically deformed and unable to shapeshift to disguise these deformities. As she listened to these sagas, she tried to assess how much truth or falsehood were in them. She asked why each of the interviewees believed what they believed. She attempted to find others to verify their stories. She spent weeks more writing a treatise called, *The Meaning of Truth*, wherein she attempted defining this concept. The writing process also helped her apply this philosophy to the case to determine which of the narratives was truthful. Since she had not yet made clear progress, the Dozen Potentates approved the hiring of a staff of three underlings to help Brain Cuirty with this research. She asked them to write smartness reports and communiques. They did not ask her what these should be about to avoid sounding unsmart, and proceeded to write these about the process of writing them. It took a long time for Brain Cuirty to read them, but she abstained from complaining about their essential meaninglessness to her subordinates because doing so would have undermined her abilities as the Brain of the operation. When she shared these reports with the military, she was forced into answering a myriad of questions regarding the impact of her writing addiction on national safekeeping. Her last major role was as the analyzer of the data she collected. She had the unenviable task of summarizing the gossip about the perpetrators. She also searched for a statistical method for measuring if one of these gossip stories was truer. She determined no such pattern of likelihood could be discerned. She further suspected one of her sources was a fabricator. The most far-fetched storyline was—the perpetrator was a disgruntled engineer. This hinted at an anti-engineer and anti-gene-editing bias. If this source was anti-engineering; then she was likely to be a rebel herself, or a double-agent.

With permission from the Dozen Potentates, she called this enemy of Alleges, Charlem Gian, into an interrogation bubble designed to put visitors into a state of terror. The bubble had this effect because of its disturbing green color scheme, cold temperature for Cadata, and an overwhelming smell of gas waste. Sadly, it put the interrogator in the same conditions as the interviewee. Cuirty sat alone in this psychosis-inducing room while Gian was running late. Cuirty glared at the safekeeping report for Gian. The 3D image of Gian held her attention: *Look at the dimple under her eye. How smug. And the off-putting, other-hemisphere, bluish tint to her eye, and those little short wings, and the chubby torso...* She read a few more sentences from the file. Gian was born on the island of Bonne, which had retained its status as the artistic capital of the Alleges region. *She must have learned to hate science and engineering in that prissy little place. I hate all those pretentious artists! Even if it's hard, I'm going to convince her to work for us! She must be tired of killing. Maybe I can convince her the ideology of the anti-gene-editors is all wrong. Maybe she'll comply to avoid us massacring her family to get to the truth. What will it take?!* Cuirty felt her mind slipping as she tried to keep from consuming too much of the toxic air. *She is probably going to look at me in a sexual manner, like all those artsy Bonnes do. She will undress my decorative skin with her eyes. She might even try to brush up against me to seduce me! But she won't win. I'll resist!* Finally, Gian arrived and greeted Cuirty with what appeared as a friendly "Hello" to an untrained eye. Cuirty saw so much more in Gian's little gesture of the wing, turn of the rear and twinkle in her repellingly bluish eye. Yes, Gian *was* trying to seduce her as she moved her little fin a bit closer than decorum dictated. Cuirty fought back an urge to reprimand Gian for these lewd actions. She begging Gian to work for Smartness Agency instead of the Donlons. She spoke for around half-an-hour about the errors of Gian's ways, and the rightness of Smartness' mission. When she was done, she stopped and studied Gian's reaction, awaiting an answer. "I remember now," Gian said with a whistling giggle, "we met at that mall!" Cuirty tried to reinforce the seriousness of Gian's situation. If she failed to cooperate with the investigation, Gian could be jailed, never to see her family again. Either the threat of imprisonment or the continuing stench finally broke Gian. "I don't understand most of what you're saying. I'm a 3D graphic designer. I was just shopping for some scented air, when I met you. I tried to give you an honest guess as to who the perpetrator might be. I am clueless as to the facts of the crime because I am uninvolved in this group. I am a pacifist." This troubled Cuirty because the word *pacifist* proved this was *the* counter-intelligence agent she was seeking. How radical did somebody have to be to use words like this after they were threatened with a life sentence? "Look, we are prepared to pay you a dozen top-end salary worth in exchange for your cooperation. We will give you protection against your gang, so they can't come after you for disclosing their names. We can be friends, you and I," she said moving her wing a smidge closer to Gian's, and shifting her head so her yellowish eye sparkled under the rainbow light-source emitted through the floor. This flirtation was a desperate effort to elicit a surrender from the perp. Cuirty was now making cackling noises through her main air ducts, exhuming a lot more unused air than usual, all signs her body was reacting to her heightened stress level. Cuirty strained to see Gian as a foul-smelling Donlons, but her own fumes prevented her from making this switch. "I want nothing more than being friends with you," Gian said when Cuirty gave her a chance to do so. "I also don't mind if you give me money. But it's a fact: I just don't have useful information for your agency." Cuirty was delighted, *I've converted this rebel into one of our agents. Brilliant outcome!* She led Gian out of the building with a sparkling joy in her bosom. Then, she wrote up a report to explain her impressive successes in this strenuous interrogation.

This report did not have the positive effect on the Dozen Potentates Cuirty hoped it would because there was an attack on the following day. A stink-bomb was left in their ground-floor mall, causing an evacuation. While the Donlons did not claim responsibility for this attack, they also did not deny it; thus, proving to Smartness, it must have been their doing.

A few weeks later attention of the public was diverted away from these little rebellions to a much deadlier problem. The number of deaths from side effects of gene editing skyrocketed as a flaw in the procedure was proven to cause delayed bodily-failure. First thousands and later millions of edited-gene patients died in this plague. Because a couple of the Potentates, including the head of the gene branch, died before they could purchase a fix to this catastrophe, their successors put a temporary suspension order on gene editing until this problem could be resolved. Hoping to return to profitable sales of gene editing, the Potentates offered affluent contracts to companies and individual scientists capable of solving this error. This infusion of capital into science created a new burst of invention until the error was fixed, and the funding was frozen. When Alleges learned of gene editing's renewed legality despite the plague, bitterness of loss of loved ones erupted in violence. An organized sect formed with the sole purpose of fighting against the Potentates and their corrupt trade in the life-threatening gene-editing technology. The following fight came to be known as the Near-Thousand-Year War. Echoes of this War and the clash between the pro- and anti-gene-editing sides is still with the cadatas today, as will be explained in later chapters.

The Smartness Agency continued the fight across this great war and continues to have a powerful spot in our government structure. It has since developed the motto: "We're always right. They're always wrong." It has directed their actions and ideals. Across its history, the Smartness Agency never succeeded in predicting an impending attack, nor found fact-checked proof of wrongdoing by specific individuals. Despite continuing failure, their motto and their need for salaried employment keep them inspired to persist in appearing useful. Since the Potentates retained their control over profits coming through business across Cadata, they had the resources to keep growing the Smartness Agency, and their military forces, as significant numbers from these ranks fell victim to the continuous fighting. These troops slept intermittently because the rebels always attacked whenever sleeping was suspected.

One episode demonstrating the types of conflict the troops were seeing is the Verified Incident of 1,003,693 BC. Brain Cuirty embarked on this investigation. Smartness received smarts the rebels were hiding with a pile of super-whistlers in a safe house on the outskirts of Alleges. The smartness came from an untraceable, encrypted message they received at the headquarters. Cuirty analyzed its linguistics for weeks, contemplating how hard she had to work to look up in the Computer's dictionary each of the thorny 3D hieroglyphics the author utilized. She was also self-enamored to discover proof the author was genuine in the way he thanked readers for their attention in the conclusion, a sign of humble submission to a superior opponent. When she was done with what she believed was an accurate enough translation into plain-Cadata, she classified it as undisclosable. It has never been declassified, so the exact text cannot be reproduced here. After this, she disseminated her learnings to all government agencies (foreign and domestic, relevant and irrelevant), and then sent a couple dozen troopers to investigate. When the first trooper shuffled up the first step of the underground tunnel entrance, the whole dilapidated residence exploded, burying the unit under the collapsing debris. Cuirty reprimanded and fired the two troopers who survived the explosion, once again demonstrating her strong leadership abilities.

There was no time to waste for Cuirty, who received an invitation to meet her most verbose sources that evening. One of the last successful pirates on their strand, Bied Blast, one of Medit's descendants, maintained partial control of his homeland, Getor, a rare pocket of resistance to the Potentates' control. Cuirty had been paying Blast for "protecting" his tribe from the rebels, who he insisted had threatened him with hostilities. Why rebels might have wanted to attack a machine-less, isolated islet remained unknown, but Cuirty was certain Blast was a victim. His trustworthiness was affirmed due to the constant focus his giant eye kept on her as he spoke. This steady gaze was an unusual habit among cadatas because of our eye's 360-degree perspective. Cuirty entered the bar with his eye tracing her path. She had been balancing on a rope suspended from the ceiling. After the standard cadata pleasantries, Blast reported he won a major battle on Getor against the rebels, despite losing a hundred-and-two cadatas from his gang. Cuirty's first reaction was to express sympathy for his losses, but then she remembered he had reported similar clashes and losses the last four times he asked her added financial compensation for his continued assistance. Blast's reaction to Cuirty's mention of this strange repetition was rather hostile as he tossed a graphite intoxication cylinder he was holding against the back wall, and stormed out to stress his outrage at Cuirty's lack of faith in his honor. Cuirty almost called to apologize, but decided to first send a covert, airborne robot to investigate the losses and to evaluate the assistance needed on Getor. The bot failed to identify signs of lingering radiation, sound wave disturbance, shot-through lava formations, or dead bodies across the island. The miniature size of the territory would have made a camp fire noticeable. The bot examined the consistency and density of the ground and scanned under it for bodies. There were no remains. The bot did locate a merry camp of intoxicated pirates, feasting on the gas containers Cuirty gave Blast as part of her last repayment. Cuirty was gravely displeased when the bot returned these findings. The bot had neglected to determine what kind of help these poor pirates needed the most. She deactivated the bot, and recycled its remains. Then, she called to apologize to Blast and sent the standard care-package she also donated to him on the previous occasion.

This tiny bot was a new invention for the period. It was one of the first models, and it still could not distinguish between an essential, and a secondary command. In fact, the scans it ran on Getor were as standard as a thermostat reporting the temperature. It could only fly in a straight line to a programmed destination and could not evade capture. Even a strong storm could disable its fragile mechanism. This model is significant because it is the first step in bringing our robotics industry to its current sophisticated robots, which perform our menial tasks.

The constant warring forced peace-loving cadatas out of Alleges and other major cities where the fighting was concentrated. These exoduses left engineering and programming jobs unfilled; then, the rise in unemployment decreased the standard of living for the average cadata. Looting of stores rose to unexpected proportions as cadatas were desperate enough to risk being shot by the troops in exchange for access to the precious nutritional gases.

Despite all this misery among the lower classes, cadatas still had to survive and thereby to keep consuming, so the Dozen Potentates maintained a profitable bottom line. So much so, they even invested in building Cadata's first spaceship. This was a public relations campaign, an attempt to win over the hearts of Alleges by showing their technological superiority. The denser atmosphere, daily high-wind storms and several other pressures made this a bigger challenge than its equivalent first liftoff on Earth. The feat was achieved through condensed air pressure, taking advantage of the planet's biggest resistance factor to turn it against itself. The engine gulped up the strong winds hitting the ship and fired them down at the ground, performing a few other processes to give this push a strong enough kick to overcome Cadata's larger-than-Earth's gravitational force. The spaceship was deliberately sized to fit a single cadata, Commander Pirth Lopar. It did not require the enormous bulk of the first rocket on Earth because, instead of carrying fuel with it, the spaceship utilized the fuels freely floating across space. It lacked the spectacle of a nuke exploding under an astronaut on Earth, but it entered orbit with near-maximum efficiency.

The 3D visuals of Cadata from orbit in its pink, cloudy radiance achieved what a thousand years of warfare failed to do: it allowed cadatas to feel as if their belonging to a single species and a single planet was a primary responsibility to even their personal, moral or businesses successes or failures. Since polling began to reflect this trend, the Potentates chose this time to announce another positive innovation they had been suppressing. Other than two Potentates killed in the War, the other original Potentates were still alive over a thousand years after taking positions as leaders of their conglomerate empires. Now, in a news release, they finally revealed, their scientists had found a way to extend their lifespans from around 400 to 5,000 years, making them semi-immoral. The news was: they planned to sell the necessary procedures to any cadata who could afford them.

The Millennial Age

"War has been a constant across my entire life. It was what my ancestors knew going back many generations. What has it all been about? A choice to edit our bodies to improve them, to improve our lives? It is time put the suffering, the starvation, the deaths into the history books, and focus our energy on building a future free from violence," Commander Pirth Lopar observed during a press conference in 1,003,650 BC, when he was asked how he thought space travel would impact Cadata.

This was a war with an anticlimactic ending. Unlike the defeat of the Nazis in World War II on Earth, the villains in Cadata's story, the Potentates, won through attrition and by bribing the population with near-immortality. Unlike the Nazis, the Potentates did not engage in intentional large-scale genocide, and neither did the other side. The Potentates' mistakes were scientific, as they failed to assure the safety of their gene-editing technology, sentencing millions to die. The Potentates experimented on their people, as Nazis did. But they were motivated by lengthening their lives, and growing riches. They were indifferent regarding the destruction of any group, sect, or political creed. The lack of malice allowed cadatas to forgive them. Rebel forces rested their arms before they experienced attrition or resources-depletion. The Potentates greeted peace talks, agreeing to mild concessions to bring the conflict to a finite resolution. They awarded some free life-expanding gene editing procedures to those injured in the fighting, and the families of the deceased.

There were a few final sparks of rebellion. A petition went around to end the Potentates' rule, and replace it with a single-day-ruler scheme. It garnered a couple million signatures, but the Potentates insisted at least half of the population of their domain had to sign up for it to change the system. Without democratic voting in Alleges, they argued, there could not be a vote to outlaw their dominance. A majority could overrule them. However, it was too expensive and confusing to arrange for half of Alleges to sign up for any measure. The attempt failed.

Both sides continued their covert efforts without declaring outright war. A few anti-gene-editing graffities would show up on buildings in the city. A group of disguised-astenals radicals glided through the city skies in a formation symbolizing an outcry against extended lifespans. To avoid an escalation, the Potentates did not have them seized by the military, allowing the demonstration to proceed unimpeded. For her part, Cuirty continued surveying suspected resistance leaders; she now wore several disguise-appearances to avoid being spotted because she became infamously hated in radical circles.

At first, only the richest cadatas could afford the life-elongation procedures. These lucky firsts became wealthy and powerful because they continued growing businesses over multi-lifespans. Seeing the incredible advantages in living this long (aside from the natural fear all creatures have of death, and the desire to avoid it), the middle class started extreme cost-cutting strategies to save enough to purchase the edits before they died. As the middle-class also started to rise in power, the Potentates began to be concerned about their own relative status. They still had a near-monopoly on essential resources, including air. Thus, they decided to let progress occur as it would. When most of the middle-class began acquiring long lives, the poor became restless and began taking extraordinary risks to steal away, kill for, or otherwise attain life-extensions. Not seeing a reason to keep struggling with this boisterous sector, the Potentates released the technology into public domain and the price dropped to near-zero, allowing for almost the whole population of Cadata to live up to 5,000 years, the cap the Potentates set to allow for the continuation of asexual reproduction. Without a lifespan cap, the planet would drown in over-population, prompting a cataclysmic revolution. Even with the population living to 5,000, they had to set strict one-child per lifetime per cadata restriction to keep the population at around the same level. They argued for asexual reproduction to foster natural gene improvements in the following generations. Gene mutations were necessary in case of planetary environmental shifts. Occasional double-births were later allowed as some citizens became overwhelmed with the endlessness of a 5,000 year-long life, and elected to die early. These elections later became too common among the poor, for whom working in difficult and painful jobs for 5,000 years was undesirable.

Age of Progress

A new turning point occurred in 1,002,876 BC, when a self-reassembly gene was introduced to the market. It allowed cadatas using it to withstand most injuries. A cadata with this mutation could be squashed into a pancake, chopped into pieces, and burned to cinders, without being damaged to the point of non-reconstruction. Murder became obsolete, as it took a herculean effort to cause sufficient harm to end a life. It was still possible to commit suicide or to opt to end one's own life, but this took planning and some scientific intervention.

In 1,002,866 BC, the Potentates introduced brain-maximizing edits. Unlike earlier improvements to cadatas' mental abilities, this update doubled a cadata brain's size and density and decentralized its location. Now the brain was more like human veins, spread from the tip to the bottom of a cadata. Information and comprehension were spread in a manner allowing them to be backed up in different spots, in case an incident or accident cut off a portion of the body. In the traditional cadata body composition, the brain was more like the human brain and was located on the other side of its floating eye. Changes to cadatas' inability to die made this spreading-out of brain matter necessary, as some cadatas suffered losses to their brains in accidents, which were then re-assembled to be as good-as-new, but also as-bad-as-new as they no longer recalled their previous lives.

A breakthrough in 1,002,853 BC allowed cadatas to upgrade a newborn's vision to the full-spectrum, beyond the visible range. Through this function, cadatas could see, with machine-precision, heat, radiation, and microwaves. Gifted cadatas saw through walls or through fellow cadatas (to detect diseases). If they flew to an elevation above the clouds, they explored the makeup of stars.

As these changes spread, the affected population lost perspective on the point of continuing existence. Most cadatas managed to earn enough to retire after their first three hundred years. The Potentates had set the maximum wage to give them the illusion they could retire just at the age when their biological lives were ending. Money these savers placed into retirement accounts but lost, due to death, was recycled back into the Potentates' accounts. The lifespans expanded faster than the Potentates could shrink earnings. While some dedicated their retired lives to scientific or artistic research and creation, most chose escapism. Some engaged in extreme sports, such as gliding and doing tricks at rocket-speed with a jetpack-like device strapped on. Others overindulged in sensual or sexual explorations. Some gained weight through the over-consumption of concentrated air. Gaining weight on an air-diet is near-impossible. Several harmful escapist gases were introduced in this period, including slothmoth, a substance responsible for the shutoff of conscious thought in favor of instinctual desire-fulfillment.

The Shellmar Bar in the center of Bonne's gallery district distinguished itself as the center of this Diverting Bunkum Movement. The proprietor served slothmoth or other substances (legal, illegal or of undetermined legality) customers requested. The intoxication-level meant most visitors were gliding around the bar, using air streams to assist their air tricks despite the enclosed location. Displays of affection were also common. Showing sensual inclinations in public has historically been frowned upon. No equivalent normalcy is associated with sexuality as on Earth. Cadata lacks marriage and cohabitation rituals or traditions. Self-enjoyment or partner-stimulation is legal and acceptable, but displaying these to cadatas who are not participating is equivalent to streaking on Earth in 2019. A public-fornicator might escape with disbelieving and amused stares, or he or she might be saddled with a ticket for disturbing the peace. Our fashion commands mutating our own skin instead of wearing another animal's skin for clothing. On the other hand, since cadatas streak daily, Cadata's term for "streaking" is more equivalent to "dressing" on Earth. On the topic of fashion, those who frequented the Shellmar adorned in outrageous plumage or skin appearance. They changed their skins into strange, unexpected textures. Frequented textures included seashells, lava-rock, tree-bark and foamy sea waves. The term "seashells" refers to Cadata's mutant sea-creatures with teeth stretching across their outer surfaces; their teeth are prized as decorations. These textures were in constant motion. They emitted fluorescent, shimmering and glowing light. And they attained bizarre colors and smells.

Crowds came from across Cadata to take in the atmosphere, including the psychedelic-trip-inducing gases in the air. Popular sights for the throng of tourists with money to spend on the extraneous included the Shellmar Bar, the galleries, music holes, and other artistic ventures. While the party gases were the primary attraction for most, the enormous crowds had to be explained with a less nefarious reason. The manager of the Bar, Duet Cipot, invited three entertainers to amuse the crowd with verbal trickery. They were paid in intoxicating gases. They also received boundless admiration from the regular crowd, who flapped their wings and whistled their praises whenever they entered. This adulation kept them coming back as no other employment had offered equivalent positive reinforcement. However, too much flattery can reinforce whatever behavior seems to be its cause. The three of them divided the bar into thirds and held lectures there from just before sunset and until morning.

The first of these Diverting Bunkums was Thpbsohphe, a name he adopted from an ancient Cadata language surviving only in underwater cave sculptures. His real name was more like Peter, when translated into English. Thpbsohphe espoused this name because he was inspired by the cave sculptures and writings of an ancient philosopher. Thpbsohphe reproduced these ancient pieces as if they, including the name of the creator, were his own. However, he could not translate most of this ancient language due to his linguistic limitations. In the gaps where comprehension failed him, he inserted whatever words came to mind. None understood the resulting prose, so it was universally admired because the only alternative to flattery was admitting confusion. Thpbsohphe recited the latest "translations" in a booming voice with dire conviction, his eye spinning to gleam with violence at the listeners. Here is one of the best-known segments from his famous philosophical treatise, *Answers Are Death*:

> All actions are futile, so let's do nothing. Every truth, every falsehood, every tree, every river, everything is the thing that it is. Once, I walked down to the island; then, the island walked towards me. Let's replace two with one. One is always better than

two. Two is worse than one. If only there were only ones. Let's eliminate opposition, so that we will always be right. What are we against? We are against everything that is against us. It is futile to attempt to explain anything if you think I am incorrect. The only cadatas who exist are the ones who are in this bar. Nothing is outside. So, be here always if you want to be something. I declare there is nothing there, and therefore there is nothing. Nothing is anything but itself. Cadatas cannot be labeled by what they invent or create, but only by the totality of themselves. The plumage of personalities is the meaning of existence. Actions are meaningless. Only inactions are meaningful. All cadatas desire to be one of the Potentates. But the Potentates are immortal and will maintain their positions indefinitely. Therefore, it is pointless to try to be anything, as this highest of ambitions can never be achieved; so, it's better to do, and be nothing. Fear of making the wrong decision should keep cadatas from doing anything. Doing nothing is an ideal state of being…

Thpbsohphe's grand conclusions are memorized by our youths as an equivalent to nursery rhymes. It is the highest example of Diverting Bunkums' mindlessness and gloominess rhetoric.

The second of these Bunkums was Onefound Mageplume. She wrote most of Thpbsohphe's texts as, unlike him, she knew how to write in the multi-dimension 3D digital language Cadat evolved into by this point of the planet's history. Mageplume's memorable readings were from her potentially-endless novel, *My Lowly Croaking Devotee*. It might lack an ending because she set the computer program to add words for as long as the computer housing the file has access to an energy source. This flame has burned since its creation for over a million years. It is a series of digressions on different aspects of immortality. The heroine has been granted immortality by the Potentates. She engages in various gravity-defying adventures. She flies in a close orbit around the sun in a private spaceship. She jumps out of a shuttle in a type of a parachute and then makes a full gliding circle around Cadata. She engages in mind-bending self-love adventures, approaching the edge of pleasure, and the capacities of cadatas' sensual organs. A long way into this tome, she meets an impoverished, illiterate, grinning-for-no-reason, unemployed, and generally confused semi-manly cadata. She tells him about the wonderful things she has been doing, and how amazing, brilliant, and interesting she is. He barely responds, as he has difficulty understanding her upper-class dialect. She slows down and spends the next few thousand pages trying to simplify everything she has been saying into basic Cadat. Finally, comprehension comes over his (or perhaps her) face and "he" says: "I die soon." Mageplume is distressed, and tries to understand his meaning. He manages to explain: unlike her, he is unable to afford the immortality gene editing, so his life is ending. Mageplume has been talking about herself for so long, more than half of his natural lifetime is behind him. So few mortals are left on Cadata, Mageplume is amazed at his plight. She relates the superiority of immortality and how only cadatas can attain this precious gift bestowed by the Potentates. Everything else—all those inferior critters die—but they can live forever, or as many years as Potentates allow. Before Mageplume can finish this tale, "he" dies. At first, she fails to observe the death and continues her monologue. Frustrated by the absence of nodding or admiring gazes, she insists he congratulate her infinitely superior communication ability. Delivering this admonition, she faces him, and sees his corpse on the floor. She is so distraught at the loss of this splendid listener, she creates a hologram of him, and continues talking about her endless life with this slightly prettier, though equally speechless man or woman. As immortals, they can continue this monologue indefinitely. Mageplume builds a philosophy to support her increasing consumption of slothmoth between periods of philosophizing. She wishes for a day when she could forever switch off her mind, becoming infinite matter. An often-quoted line from this section is: "Matter is the true meaning of happiness, and only the mind can make a cadata unhappy." A hyper-philosophical novel of this caliber has not appeared since despite countless abandoned attempts to imitate its indescribable qualities.

The last of this honorable trio is Rood Nennote, remembered for the one-actor play he performed for the Shellmar Bar—*No, No, NO*. He always began the play by smashing the door open at the top of the bar and falling to the floor, missing a chair. He could have landed on any chair in the theater. The important element was for somebody seeing the play for the first time to be in it. He would begin yelling: "No! No! No! That's my chair! Don't you respect the rights of ownership? Don't you have a sense of rightful possession! How would you like it if I took your gas-jar!" He would grab this unsuspecting newbie's gas-jar and would gulp down whatever mind-altering substance was in it. The newcomer would inevitably surrender the seat with humble apologies. Without sitting down in the contested throne, Nennote carried on with the drama. "Where is the weather?" he would ask. Then, he would make a circle, and turned into a different character: "The gas smells fine today!" He opened a pretend, drawn door in a wall, only to not walk through it. He seemed to be hitting himself, but facially reacted as if he was watching a sunset. At random intervals and combinations, he would mention "love", then "death", then "immortality", then "immorality". One returning theme was: a character has forgotten what he or she looks like. Nennote would turn himself into all sorts of cadatas, into their pets, into various trees, all to no avail. He would beg his audience: "Does any of you recognize me!? Who am I???" Occasionally, a newcomer arrived late after the chair introduction; a plant in the audience asked him to name the actor on the stage, and the newcomer would respond it was Nennote. At this outrage, Nennote would explode in Cadata expletives. I cannot relate such language here because they are hypertension-inducing in insult-level in contrast with milder Earth expletives. All characters leaving the play died a painful death with screeching, hissing and wailing: "Hsss! I am being massacred! Why are you, in the audience, smugly watching my torture!" or "I am the most miserable, despondent, dejected, desolate, melancholy, measly, gray, wretched wretch on Cadata for life is fleeing by me!!!" These deaths would seem to become a recognizable pattern, but then the last character would not die, and would just stop talking and leave the bar in the middle of a wo...

These initial Diverting Bunkum works are as triggering for modern cadatas. They persist because anti-revolutionary pessimism brought us to the heights of our artistic potential.

The Great Flattening of Bonne

At the peak of this great Bunkum movement, in 1,002,753 BC, the Dozen Potentates sounded the evacuation alarms in Bonne. The intoxicated citizens barely escaped city-limits. At least they carried a light load, as their possessions and their island were flattened by the giant pressure wave. Bonne's confused residents turned to each other, seeking an explanation for this sudden loss of their homes and businesses, but nobody seemed to know the cause. As they watched a team of robots (much more advanced designs than prior bots) swooped in and began building an enormous factory. The robots reused the rubble pile of Bonne's houses, architecture, canvas, paint and sculptures. The design shaped into a factory for robot production. Near the day's end, a robot approached this standing, gliding and reclining group, waiting to settle on the islet neighboring their old Sual Beach, or to be directed to migrate elsewhere. The robot made the following announcement:

> You have been too attached to possessions, to buildings, to entertainment consuls, to nutritious gas stores. Your architecture has been in place for too long without sufficient lava flows to flatten it when it should have been flattened, centuries ago. Some of your buildings even started to show cracks, and various mold and bacteria have been spreading disease and ugliness. The Potentates in their infinite magnanimity have gifted you the leveling of this cesspool of immorality, idiocy, and drivel-jabber. Be grateful, but be grateful somewhere else as we are going to level the islet you are standing on in two minutes.

The robot began the countdown. Bonne refugees lacked time to pack, even if they intended to, so they proceeded without delay to glide off the cliffs of the islet to the next islet, intending to travel from there to the nearest city. If they lacked relations or friends in this city, they hoped to find work to replenish their losses. Most traveled to Alleges because this was where almost anybody willing to take almost any available work could find it. Once their immediate needs for shelter, sustenance and employment were met, those who took down each other's information started meeting to discuss the catastrophic loss they suffered, a loss only they grasped. The three Bunkums were in this group. They were somberer, having lost the gargantuan mansions, built on profits from their great masterpieces. The Bunkums contorted their minds for the composition, but the pages remained blank. Acknowledging this failure, they assigned the task to Mageplume, the one capable of transcribing their diverging wisdoms. The result of this effort was *There Existed Our Matter!* It was an exploratory study of the symbolic significance of older cities and art. They lectured, while mold and rust on aged artifacts caused minor illnesses, the history and culture represented in these buildings gained meaning with time. Cadatas' lives gain meaning in proportion to the worth and honor assigned to their art. Once this great thesis was completed, they sent it to the Potentates' headquarters. They received a thin letter in response to their thousands-of-pages-long pondering on the nature of things: "Cadata is a small planet without room for waste. Our civilization must make room for its most productive members: robots. All non-essential cadatas, including you *writers*, should find a job that is not a waste of space!" Mageplume was outraged by all this negativity and insisted, at the least, the Potentates should have applied for a leveling permit from the city of Bonne. No equivalent leveling had ever been attempted before, and she doubted there was any precedent for exterminating a city without a referendum or prior notice. To this still-longer text, the Potentates replied thus: "Your Mayor received a new apartment in Alleges in exchange for our purchase of Bonne. It is now our property. Precedent this: if we can own your air, we can own your land." The ex-Bonnes met many more times, and Mageplume wrote many more declarations of their rights, but neither the Potentates nor their representatives sent any further communications regarding this matter. The robot factory they built on the site turned into the most productive one on Cadata. It is so well constructed; it will be the last structure to fall when Cadata dies.

Historic Leaps

After a few thousand years of leaps in Cadatas' science and culture, history leveled into a slower pace. New ideas and technologies stalled as new Potentates inherited their parents' titles. These genetically-engineered, model-bodied offspring lacked a motive to progress financially or intellectually. Obscenely wealthy and powerful, they were unimpeachable due to the safeguards their parents instituted. Rebels were erased by abundance. Even if cadatas suffered losses, they had millennia to rebuild. The final wave of discontent was in a few naysayers, who kept predicting an environmental or solar disaster was looming. They wrote reports forecasting the planet would die within a million years. All these scribbles were ignored by the Potentates, who commanded ignorance of it among the press and in turn among the public. A Smartness Agent followed Ohay Lectal, a scientist-leaders of this group, to an atmosphere imaging device Lectal had used for these planetary forecasts. The Agent squashed this device with his wings, and wrote a heavy, compulsory, and permanent prescription dosage of slothmoth to prevent Lectal from future thinking.

A new spike in innovation arrived in 584,350 BC. Cadatas developed an obsession with precision: drugs and machines were built to produce exact, intended outcomes. A millisecond of uncertainty could drive a third of the population to madness. Then, in 537,283 BC, sleep was edited out of cadatas' genetic code. Cadatas do not need to sleep. Some sleep anyway because they enjoy inactivity. Sleep was deleted because research incidated sleep-accumulation caused health problems. Sleep hours were also unproductive. 512,023 BC birthed the last infant through "natural", asexual reproduction. Incubator births were already popular long before this point, and no such births have happened since. After these few adjustments, Cadata settled into a steady lull of repetition keeping cadatas just happy enough to carry on watching the pendulum swinging.

Chapter 2

The Dimensions of the Disaster

The planet's limited resources turned into an obstacle barring the Potentates from achieving infinite consumption. Because most of the planet's landmass was below the globe-covering ocean, they previously commenced the exploitation by mining the ocean floor. These materials were feeding the production of innovative super computers, and outlandish technologies. The size of products fluctuated between microscopic and gigantic, depending on taste. For example, for a decade popular culture propagated wall-sized imaging screens. Houses ballooned to extravagant to the point of uninstallability volumes. Exterior gales were counteracted with wearable see-through bubbles, which were resistant to this pressure, and kept cadatas from uncontrollable tossing. The Potentates extracted from the sea floor precious, heavy metals, utilizing every microbe to construct the toughest technologies. Precious minerals are treasurable because they are rare. Thus, the Potentates' appetite sustained growth of this industry. But this drained Cadata's supplies.

The raw-materials had to be mined on Cadata's moon, and three other planets in cadatas' solar system. Space shuttle designs improved until they were as maneuverable as terrestrial flight. The challenge was bringing a sufficiently large volume of metal and rock back to Cadata's surface. Once the ship was back in the atmosphere, it was slowed by tons of metal on board, and consumed an enormous quantity of cadatas' top high-efficiency energy sources. In one famous accident, the Denum Shuttle's computer overloaded because the breaks were using too much power and turned themselves off. The cadata pilot, Hest Ruppet, had never manually flown a shuttle outside of simulations. Computer failures were so rare, manual flight training was absent from the syllabus. Ruppet failed to regain control of the ship, and it crash-landed, blowing itself into pieces against an island's rocks.

Flying to distant planets required more energy than the cost-input needed to make a trip self-sufficient. One endeavored solution was mining asteroids proximate to Cadata's orbit from its sun. A misshapen-potato-shaped smaller asteroid was discovered. A scientific team designed a larger shuttle with asteroid-sized holding-capacity to bring a large portion of this asteroid to the operations base. The ship approached as rapidly as if it was landing on an unshatterable planet, firing its propulsion engines at the porous surface of the asteroid to slow its descent. The friction of this touch-down reverberated in turbulence as the asteroid throbbed. Unconcerned over this response, the Computer began the drilling procedure in the moment of first-contact with icy rocks. The drilling sent the shuttle into a rumpus jig. The scientists had assumed the program they slapdash designed would account for all probable abnormalities during the drilling. An obvious problem slipped through the programming: the small size of the chosen asteroid. The force of the propulsion engines, the clash with the shuttle, and then drilling into a vulnerable outer shell caused it to lose structural integrity, and it shattered into small rocks. Multitudinous pebbles clumped together as they drifted as though forming an asteroid belt. Though these pieces were insufficiently congealed to remain joint under the slightest gravitational pressure to pull apart. Given the lack of conflagrations, and the slow drift of these pieces, this breakup lacked the signs of an explosion lacked the evidence necessary for the observing robot-camera to trigger a warning to the science team. Thus, before an alert could be issued, a wave of pressure building under the shuttle catapulted the little ship at a speed inducing an extreme g-force increase. The anti-gravitational shields were off to avoid them triggering a malfunction during drilling turbulence. While this asteroid had lost mass to solar winds, its essence was metallic because metals were the desired commodity. Thus, when heavy chunks of it hit the ship's outer layer, one piece created a tiny hole, which depressurized the cabin. The shuttle's pilot died on-impact from high-g pressure, exposure to extreme cold, depressurization, and a lack of air.

After this asteroid breakup incident, a solution was needed for the inefficiency of traveling into space to bring pieces of space back to the planet. The expensive voyage beyond Cadata's orbit only netted tiny asteroid pieces because heavier rocks necessitated higher fuel expenditure.

On the Structure and Formation of Solar Systems

An average midlife solar system has a dozen planets, and an asteroid belt. The belt tends to form when one of the planets is hit with a sufficiently large neighboring planet, moon, or some other flying object to shatter it. Earlier in a solar system's development, when it is dominated by small rocks gravitating into planets, such impacts are common. The larger the planet, the more attracting it is to debris, devouring planetesimals within its orbit, or retaining them as satellites. A mature system has had most of its debris sucked into a few massive planets. Aside from being consumed, smaller planets risk being shattered into an asteroid belt with a slowed trajectory around a sun; within a mature solar system, this belt would lack sufficient loose materials to build its mass back into a major intact solar body. One of these belts is between Mars and Jupiter. Humans have only observed four percent of the mass of the moon in asteroids with your viewing equipment, a fraction of the billions of asteroids in your system. Major asteroid impacts are some of the deadliest events for life in the universe. Planet-dwellers across the galaxy fantasize about their meteors as they burn in magnificent light streaks against the canvass of night skies. In the vastness of silent space, sudden flashes hint at the spiritual or philosophical. Then again, some planets are barraged with continuous meteors, and there (if any life can survive), nobody looks up at the sky and wishes on a meteor. Only the rarity of such displays is the building-stuff of mythology. On rocky planets, we are used to seeing rocks as close to other rocks as pieces of sand touching on a beach, but they have millions of miles between them in an asteroid belt. The predominance of the vacuum of space makes it safe to travel through a belt even at the speed of light. Comets and asteroids have many overlaps, but an average comet moves slower and tends to be coated in ice because it orbits a sun at a far more distant radius of a few light years. This distance gives its slow speed. Because little light reaches its orbit, temperatures approach absolute zero, sprouting the overgrowths of ice over the rocks and dust. Asteroids have a higher metal composition than comets. Asteroids are less icy because their velocity, and proximity to the sun melts heavier water deposits from their surface. Humans have only observed proximate comets dislodged by a neighboring star, or by a collision from dormant-rotation.

***

After a profit-analysis, the exploration team attempted sending a shuttle parked on a comet to catapult it to our neighboring solar system with minimum consumed energy. Comets entering the center of a solar system can gravitationally, or by-intelligent-design swing around a planet, such as Jupiter, before heading for a neighboring solar system. On this occasion, the operators parked a spaceship on the comet. They applied the combined mass thrust to turn it in the desired direction and angle to create a slingshot around our biggest planet, Snapos. Then, its direction was adjusted by pushing it to turn towards the closest planet of our neighboring solar system, around 2 light years away.

To clarify, a team of space-faring cadata scientists, and test pilots should logically be called *cadatanauts* because humans distinguish between U.S. *astronauts*, and Russian *cosmonauts*. Thus, the term for our space pilots is a combination of our species-identifier, and the *-nauts* root.

These cadatanauts produced a sufficient supply of compressed air for a single pilot to survive on board for the 2,000 years duration of a one-way trip. We gave the pilot, Kerod Retan, the standard psychological and physical tests adequate for the journeys conducted to our solar system's distant planets, and his scores were brilliant. However, there was no way to test for the effects of a 2,000-year voyage on a cadata mind. The ship featured our top entertainment software and hardware. He also had a communication link to Cadata. One foreseen shortfall was a 4-year delay to receive a message when Retan was close to his destination-planet due to the 4-light-years distance. The spaceship was flying in a straight line, with pre-programmed course corrections. Looking behind his snowball, Kerod could see its tail of fleeing dust and gas, which was being evaporated by the sun's radiation pressure and solar wind. The job was surviving a lifetime of watching this constant lightshow. The chance of a solar storm, or another unpredicted event in space generating a need for Kerod to perform even a tiny course correction was microscopic.

If the Computer was not pre-programmed with sufficient disaster scenarios, the flight would have terminated in a similar systems-failure as seen in the final minutes of the calamitous drilling of the notorious asteroid. The angle shift needed to pass around a solar storm is so slight, a cadata is incapable of calculating it without the Computer, especially since at light-speed the location of the shuttle, the final destination and the storm are in constant flux. Intervention with this intricate formula by a flawed mind of a cadata introduces chaos or unpredictability of outcome. A biologically-designed plan can propel a spaceship unwittingly into a black hole or another destructive interstellar phenomenon or mass. Retan was needed in case of the unforeseen. Deep space viewing devices on Cadata might have missed forces or objects with destructive properties. If an anomaly was spotted mid-trip, immediate, creative action would be needed. Biological intervention is also needed if the Computer and its backups blackout, and cannot be revived with self-repair tools. A mechanical system can die during a power-surge, but a biological entity might retain consciousness; the biological unit is then the only one of the ship awake and able to execute repairs. Potential usefulness proved to be a psychological obstacle because remaining useless distressed Retan. This tension was personified in rules dictating Retan was forbidden from touching the control mechanisms until an emergency manual-overwrite was needed. In response to the stress of inactivity and unproductivity, Retan developed a tick wherein he would hit his wings on the see-through walls of the spaceship. These violent smashes flicked his aggression, and he discharged it in seemingly relieving acrobatics, like flips and tosses, which became destructing because he could not stop these for days until he passed out into slumber. Eventually, sleep became more elusive, especially when he added a continuous top-of-his-lungs whistle to this discharging routine. He only managed to stop whistling when he spent the time on congratulating himself on the strength of his composing and whistling prowess.

Retan climbed into a psychosis under the influence of which he programmed an error into the Computer. 24-7 holographic imaging is taken of any spaceship on a mission from CSEA, so his every code line and movement were recorded and sent to Cadata for evaluation. The mistake he designed allowed him the chance to "save" the spaceship by hand from destruction; he commented aloud this saving maneuver would make him into an integral part of the mission. Beyond these generalities, I cannot share with you just what poor Retan did as the details were classified by the Potentates, who viewed Retan's actions as demoralizing for the people of Cadata. Even the trifles I am sharing are censored out of cadata textbooks and avoided in polite conversations. Displaying the decomposition of a mission they sponsored would have tarnished the space program. Any doubt about the Potentates could have sparked new rebellions or a new thousand-year war, so some suppression of failure was essential. More disastrous trips unraveled, but their existence was suppressed out of media coverage, and no survivors brought proof of the missteps. Needless to say, this was the last time cadatas attempted sending a shuttle parked on a comet at its natural speed to explore any interstellar destination.

On the heels of this failure, the Potentates ordered for their cadatanauts to devise a method for bringing an asteroid to them instead of spending more time and money to come out to collect parts of an asteroid on its turf. They insisted the target's size had to exceed any asteroid naturally falling on Cadata. The scientists had to scratch their heads. Attracting an enormous asteroid towards Cadata seemed to be the most destructive action imaginable for the future of the planet. After much discussion, they proposed diverting an asteroid towards Cadata's moon and setting it in orbit around the moon would be the safest and most cost-effective approach. The Potentates were disappointed by their lack of ambition, but signed off on the plan. Scientists began building methods for redirecting with precision an object of extraordinary mass. They had a system for sending a comet somewhere in a neighboring solar system. However, a larger scientific leap was needed to redirect a miles-long asteroid within a fraction of a degree. Objects less than 20 miles in diameter were unacceptable to the Potentates. It had to be at least 3,000 kg/m^3 dense, containing sufficient metallic content to satisfy the planet's machine-production raw material needs for the next several centuries. The scientists were stumped as to how this diversion could be carried out. The closest asteroid fitting these parameters weighed 3,107,815,788,000 kg. To move it with a magnet, the magnet had to be 3 miles in diameter, and 157 miles thick. This exertion would have broken the Potentates' primary rule: the amount spent on acquiring a resource has to be less than the resulting profit from its sale. The size of this magnet would have been more massive than the asteroid it would be moving.

A small group of scientists proposed creating a mini black hole by compressing an enormous amount of mass into a tiny spot. The concentrated gravity would come from a roving planet or a magnet. This condensed concoction would attract the enormous space rock to the desired location. However, opening a black hole next to Cadata would lead to its prompt consumption of the planet.

Scientists on Earth are experimenting with an apparatus better-suited for shifting asteroids. A strike at a specific point in the mass of the asteroid can change its trajectory by a tiny degree, and thus direct it where it needs to go. Cadata scientists found a method to scan the precise composition of an asteroid from a light year away to determine the exact point and strength of impact needed for the intended shift. At least, this was the case in the minor re-directions they attempted. No entity had yet tried to shift a space object over a mile in radius. The scientists at the Cadatanautics Space Exploration Agency (CSEA) proposed running a test on such an object in a distant part of the solar system first, but the Potentates insisted they had wasted enough of their magnanimous space budget on their ponderings thus far. There was no more time to lose. They had to execute the maneuver as soon as the asteroid under examination arrived in the needed alignment. The plan was to shift the trajectory until it was attracted by Cadata moon's gravity and went into orbit around it. Robots could then mine its resources via short trips between the cruising asteroid and the moon's primary base.

Asteroids' composition is similar across the galaxy. One exception is in cloud regions populated by early-universe stars limited to two elements, helium and hydrogen. The volume of loose debris, such as comets and meteors, is higher closer to the center of the Milky Way. Thus, Cadata is exposed to a higher chance of an astronomic strike than Earth, which is in a distant, dispersed neighborhood along one of Milky Way's minor arms. CSEA had a choice between asteroids dominated by carbon, silicon or metals. The latter category can produce an asteroid heavy in metals prized on Earth, such as platinum, gold, iron or nickel. In rarer cases, an asteroid can be rich in heavier metals of more value on Cadata, where they are needed to reinforce buildings against extreme weather and to strengthen spaceships for prolonged space flight. Metallic meteors make up 6% of the meteors falling onto Cadata, a higher percentage than the 4% of these falling onto Earth. CSEA identified 27,000 asteroids close enough to Cadata for efficient mining. The Potentates informed CSEA that for profitability, the asteroid had to be over 15 miles in radius. This number was based on the average percentage of precious metals in the asteroids dissected after they fell on Cadata as natural meteors. The Potentates also insisted CSEA should refrain from accounting for the relative speed of the asteroid to the speed Cadata was traveling around its sun. Risk is heightened at speeds of over 6 kilometers per second because tiny objects traveling at extreme relative speeds can cause utter destruction to even reinforced masses. 46 asteroids in the system fit these parameters. Narrowing down these candidates took around a decade, and then sending probes to collect samples to double-check scan data took a couple more decades. These studies left them with two possibilities, one was an asteroid heavy in platinum, while the other had a rare combination of chemicals matching what the Potentates' plans for expanding their moon construction operations needed. These chemicals included krypton difluoride (KrF_2), a colorless solid compound with a rare noble gas, krypton (otherwise difficult to filter out of the air). The second treasure available in a high concentration on this same asteroid was lead (Pb), unexciting on its own, but when combined in a reaction with krypton, it forms oganesson, a synthetic compound with the highest number on the periodic table. Cadatas had started utilizing oganesson in spaceship shield designs since the attempt to drill a part of an asteroid off resulted in the destruction of the spaceship. If it is stabilized and kept from self-destructing, oganesson is the sole chemical proven to withstand space-speed collisions. Scientists were unable to gather an equivalent quantity of krypton out of Cadata's atmosphere, even if they filtered every molecule for this search. Once the asteroid, named KoraFePub, was chosen, engineers needed five decades to design a spaceship capable of sufficient power. In 100,163 BC, CSEA had a completed spaceship ready to launch from Cadata's moon, where they set their base to avoid using energy to leave Cadata's gravitational pull.

The launch began without problematic indications. Just as the previous dozen launches, it lifted without turbulence. The computer was pre-programmed and carried out the liftoff, the flight and aimed the ramming device and fired it at the exact preplanned point on the opposite side of the asteroid from the side facing Cadata's moon. At first, the hit appeared successful as the asteroid shifted to fly toward the moon, but then it flew past the spot where it was expected to be attracted by the moon into an orbit. Its flight trajectory now headed for Cadata. The cadatanaut in the impacting spaceship, Nezosh Drulence, slipped into a nap as he was watching KoraFePub crawling by made his eye drift downwards. Whistles, hisses, and abrupt yells and yelps in his communication system woke him up, and he surveyed the rapid approach KoraFePub was making as it made a few rotations around Cadata. Some of its sides were burning up in the atmosphere, but because it had such a high concentration of metals, the bulk of its 20-mile diameter was still intact. The computer reacted faster than the Drulence by pushing the engines to move the craft farther away from the planet and behind the moon to minimize the force of the upcoming impact on itself. CSEA commanders and the Potentates' representatives were yelling for Drulence to intercept the asteroid because it was the lone spaceship in the sky capable of dampening the force of the collision, but Drulence was so terrified he forgot how to override the Computer's self-protection protocol. Drulence' failures inspired the rigorous training regimen offered to future cadatanauts.

The impact occurred in a shallow region of the ocean, a depth of 36,070 feet of water. The water softened the impact and minimized the volume of ejected debris. However, the asteroid's impact velocity was 30 km/s, higher than the average speed for asteroids. This measure was dismissed during suitability calculations. KoraFePub's impact angle was 40 degrees, slightly smaller than the standard impact-angle for asteroids on Earth. Cadata's denser atmosphere creates greater resistance to an oncoming asteroid, pushing it in a more horizontal trajectory than the 45 degrees common on Earth. Despite the elements softening the impact, the asteroid's 20-mile diameter and 3,000 kg/m^3 density made it into an extinction-level event. The energy it held before atmospheric entry was 2.36×10^{25} Joules = 5.63×10^9 MegaTons of TNT. Cadata had only been struck by two objects of matching size in its 10-billion-year history. One exception is the period of Cadata's planetary formation, the Great Bombardment, during which rocks collided at direct angles, fusing explosively into the core. Conceivably, the impact could have been worse if a moon or a planet-sized object hit Cadata. Such objects never approach within slingshot-trajectory of Cadata, so even if the Potentates insisted on mining a moon, rather than an asteroid, they lacked the capacity to attempt this diversion. A moon-strike would have led to an instant extinction of all life, and the loss of a significant portion of Cadata's mass. A moon would have also disturbed Cadata's tilt of axis. Even the asteroid altered Cadata's tilt, though by less than a five-hundredth of a degree. The planet also lost 204 milliseconds from its day length. The time-change was noticeable for precision-based computers on Cadata. They experienced malfunctions in the first few days before recalibrating and adjusting all clocks. A planetary impact would have pushed Cadata in a spiral towards its sun. The asteroid's mass was unimpactful to Cadata's elliptical orbit.

KoraFePub's initial impact left a 195-miles in diameter and 38.1-mile-deep crater in the middle of Cadata's ocean when 20,300 miles^3 of vaporized or melted matter was catapulted into the sky. A 2.22 miles-thick skin of these steaming rocks and metals remained in the crater as the ocean sent out a tsunami and then flooded into this space, covering it with a hiss as the hot metals cooled, and disturbed under-water volcanoes erupted adding to this hotbed of lava. Later as debris fell back to the planet from the atmosphere and then sunk underwater, some of this crater with filled up again. A couple of major islands lay within this circumference, including the unfortunate island of Bonne, where the Potentates had installed their biggest robot factor on the planet's surface.

Alleges was 1,000 miles from the impact. The Potentates were in a meeting on a high floor of their headquarters, still trying to order somebody to do something to stop an impact. Their eyes were all drawn out of their windows to a fireball in the distance, which appeared to be 50 times larger than the size of their sun; the radius of this erupting fireball was 230 miles. The point of maximum radiation exposure was 19 seconds after the impact, when a wave of radiation killed the organisms surviving the initial blast via the ignition of their skin or bark. The thermal exposure generated was 2.43×10^9 Joules/m^2 and irradiation lasted for two hours, before it subsided. Every island hit by radiation lost all vegetation and unprotected living organisms. Beyond this irradiated circumference, islands were bombarded by dense metals such as iridium blasted out of under-ocean deposits. More than half of the forests on the planet burned down into soot. A good portion of the surviving lifeforms died from the oceans' boiling and air steaming at temperatures nobody on Cadata experienced before, first from the initial heatwave of the blast and then as dust clouds and gases (such as methane, carbon dioxide, nitrous oxide, ozone and water vapor) created a runaway greenhouse effect. If a huge asteroid hits a deep enough body of water, the impacted water's evaporation can lead to a runaway greenhouse effect, with a spike in temperature of at least 10°C, followed by engulfing fires, and the over-production of nitrogen oxides and nitric acid leading to life-eating acid rain. If the asteroid hits land, it shoots up so much dust this creates a blanket around a planet, reflecting solar radiation out into space, and dropping the global temperature. So, either a perpetual winter or summer can follow a major collision, and when a collision reaches a certain strength, neither of these options is a survivable path. Impactors bring essential amino acids and other life-building-blocks (from whence

DNA and RNA formed) to planets, and, even without technical intervention from "intelligent" lifeforms, they tend to be the bringers of apocalypses.

The Potentates and their assistants glared at the site of the eruption, helpless to lessen the catastrophe. Pinching into action, the head of the robotics enterprise, Loor Dowor, approached the room-controls and programmed the mobile-pod to fly them to a secure underground position. It swiftly skidded down the side of the building to the lowest available point of the underground section. Then, Loor turned on a security camera with a view of the exterior around their building. As they all stared at the screen, they saw a few populated pods also dropping below ground-level of downtown buildings.

Just as these stopped buzzing down, around five minutes after impact, the meeting room began shaking with extraordinary violence as an 11-Richter Scale magnitude earthquake hit, a strength unseen on Earth. The image screen cracked on one side, the abstract furniture in the room not chained to the floor hopped around, together with the cadatas sitting on it. Some of the decorative sculpted lava formations along the walls cracked and fell off, shattering on impact. The earthquake was much stronger closer to its epicenter, where it flattened buildings enduring the radiation. Loor queried the others regarding relocating the pod to a more reinforced location, but they voted it was safer to stay put. The headquarters building remained standing, while the environmental strife worsened, so this was a logical decision.

Ejecta arrived in storm clouds over Alleges eleven minutes after the impact. At this distance, the cloud was a fine dust with scattered larger fragments. But even a few large pieces of dried lava rock, the size of heavy-hail, dented up the exteriors of unprotected buildings and damaged the surfaces of vehicles. A few pedestrians, who were gliding along the city's streets were killed on the spot.

The Potentates kept still as their assistants straightened up the dislodged furniture and swept up some of the crumbling decorations. Around fifteen minutes into the bombardment by the ejecta, as it seemed to be dying down in volume, they received a hologram call from CSEA commanders. "An air blast and a tsunami are headed your way. They will arrive in an hour. Evacuate immediately. We are sending an evacuation order across the planet. Meet us on the top floor of the headquarters, we're departing on the shuttle for the moon," said CSEA Commander Doul Veahy, before signing off to follow-up with the procedure. Loor complied with the instructions, speeding their pod to the top floor. They could hear the boom and compressions of the shuttle's engines starting up as they stepped out of their pod just a few feet away from its entrance. Assistants made a move to follow the Potentates on board, but were told they should wait out the coming "storm" underground in the pod, or attempt to reach safety in one of their gliding vehicles. The assistants were ill-pleased with this turn of events, but the Potentates were in the shuttle and it had taken off before they could raise an objection.

The assistants scrambled back into the pod. They asked each other about the strength of their vehicles, but none of them were paid enough to have afforded sufficient surface coating to have survived a tsunami if it hit them mid-trip. They decided the pod was their best bet for survival and guided it back to the lower-most point. As it flew back down, they could see numerous gliders and shuttles taking off in the direction opposite to the site of the blast. They could still see the dust cloud engulfing the city, though visibility was lowering as the spreading grime blocked the sun. They switched the security cameras to the non-visible spectrums to see through the darkness. They knew, if the worst came and the basement levels together with the pods stationed there were flooded, they would lack alternative maneuvers, meeting a watery end. They were huddled together across the following hour. To distract themselves from the coming horror, they opened a set of gas masks set aside for the Potentates' lunch. These were composed of a top-grade infusion of smells and nutrients too luxurious otherwise for their budgets. One of them also had some slothmoths, and shared these. At first, one of them resisted this total mind-shutoff. The others convinced her embracing a quick demise was pleasanter than struggling against the unconquerable.

Thus, the assistants were amicably absentminded as a 1,590 mph, 125 psi blast of wind hit Alleges. Just the sound wave alone was 119 dB and caused some pain to their ear receptors. Their pod shook with more violence than during the preceding earthquake. They heard above-ground pods dislodging and crashing into the ground. On the screen, they saw most above-ground constructions of Alleges crumbling, or bending in place. Glider connectors between Alleges and neighboring islands were shaking and collapsing into the ocean. Shuttles still parked in the streets were blown sideways into the remains of buildings. One shuttle hit and dislodged a tree.

All this happened in two minutes. Then, the tsunami hit. The foreboding sign of the tsunami's approach was the ocean's rapid retreat from Alleges, as the wave was sucking in momentum. Tsunamis are at least four times stronger on Cadata due to its deeper ocean than Earth's. But this was a tsunami of proportions much stronger than normal as it was caused by a highest-ever intensity underwater earthquake in recorded history. The water scurried at 800 mph, rising across the following half-an-hour to the height of 1.57 miles as the 310-mile wave tilled Alleges, drowning its tallest structures in an hour. Landmarks disappeared just as a grain of sand is insignificant to an ordinary wave. Those who had survived, but were stranded in an unattached vehicle or building, were now swallowed by the power of this onslaught.

Even the Potentates, now parked on the farther side of the moon, saw dust and rocks flying through Cadata's atmosphere, and into space. For an instant, they imagined the debris could reach them, but gravity intervened and this black cloud returned to the atmosphere. This dust cloud lingered, blocking the sun's light for four months. Any plants, animals or other organisms relying on the sun for its energy died in this stretch. Cadatas attempting to survive this apocalypse could see forewarning fireballs shooting across this sky without daylight. These flashes were among the few remaining light sources during the power-blackout resulting from the destruction of the planet's infrastructure.

Some life (besides the Potentates) did indeed managed to survive on Cadata beyond these four months of dread. The high frequency of rainstorms across Cadata helped to dissipate some of the fallout, and limited the exposure to fiery elements bombarding the planet. The wettest regions along the central heat-belt even saw wildlife island survivors, whereas most organisms living in the open in drier regions went extinct. Ocean life also escaped the worst brunt of these destructive waves. However, marine life reliant on sunlight was incapable of surviving four months without it.

Cadatas' survival was dependent on their resources. Anybody with access to a shuttle joined the Potentates on the moon, went into orbit around Cadata, or went further into space to wait out the worst of the impact. Those in the upper class who could only afford vehicles bound to Cadata's atmosphere, flew to the peaks of the highest volcanoes as far away from the impact crater as they could travel. With enough nutritional gas and energy to run the ship, these groups spent the four months as if they were on a camping trip. Those who owned submarines went into the ocean to wait for brighter days. Everybody else strived to find any shelter capable of withstanding these problems.

The assistants trapped in the pod in Alleges were in a unique position to survive the tsunami and the four months of hiding to avoid a projectile-strike, burning alive or boiling from exterior heat. Pods across all major cities were equipped with emergency power and air supply. These resources began dwindling at the two-months mark in many safe-bunkers, so cadatas embarked on store-looting missions for nutritional-gases and necessities.

The Potentates were silent to all emergency calls they received from their citizens in the first two months. Instead, they asked the CSEA scientists, including Commander Doul Veahy, to design strategies for their circle to settle on the moon, if Cadata became everlastingly uninhabitable. CSEA did devise several resource-mining ideas on the moon to tap into small chemical deposits mutable into the elements essential for cadata life. This plan encountered hurdles. Research indicated the moon's resources were insufficient for 5,000-year lifespans for even their miniscule population. When the moon's output was tapped, they would have to return to exploiting Cadata. Thus, the Potentates returned their gaze to Cadata to inspect damages. They decided to launch a campaign to save the remainder of their scorched riches.

Commander Veahy led a study on the body count from the disaster. The initial official count was, just the cadatas known to reside in the impact area were irradiated, which came out to 20 million cadatas. When they announced these numbers, over a million complaints from the public were filed regarding other friends and lovers dying even in distant parts untouched by the radiation blast. Instead of focusing on these complaints, the Potentates attempted to stress the many incredible survival stories due to the gene-editing they had made available. Some of these were supernatural by Earth-standards. Some cadatas survived floating within the enormous tsunami. Their limbs were torn, but their recovered torso managed to regrow the missing parts. The tear and rebirth of the psyche materialized in PTSD symptoms as they had to undergo a feverish and desolate environment. Cadatas hit by returning rocks had their bodies regrow over the hole this impact left. Some dipped a wing into the boiling ocean, and suffered the agonies of watching the regrowth of this handicap. Despite propaganda to the contrary, cadatas knew more than half of the planet's cadata population, or at least 10 billion cadatas, died in the first months.

Problems were far from over for those who lived through this dire period. Cadata never developed the Earth-concept of free housing for the poor. There were no shelters to be designated as cost-free refugee camps. Poor cadatas hid in business and government buildings strong enough for this shakeup. Then, as the Potentates decided they still needed Cadata's resources for their long-term survival, they realized they also needed some of their citizens to survive to take on the jobs to reassemble the destroyed infrastructure and buildings. They diverted the robot factory on the moon to create a series of makeshift tent-like, thin bubbles inflatable with the supplied air. These were positioned in non-molten pieces of land or floated on the ocean. These were tossed from shuttles at different illegal encampments, or buildings where a lot of vagabonds were known to be crashing. At first, cadatas who were recipients of these tents were grateful, but as centuries flew, they were still living in flimsy and fading shelters in plant-less, polluted environments. Tents lacked independent power sources or the capacity to plug into a grid. These tents' fabric was the only shield against Cadata's blasting winds and oppressive heat.

Meanwhile, the spike in ozone and other greenhouse gases brought new diseases rare in the pre-asteroid world. Because they had been deemed extinct, gene-editing solutions were not yet developed for them. Most upscale hospitals, with gene-editing capacity, were now rubble, damaged, or low on energy and supplies. Wonders of medical engineering devolved into lightless, steamy, unhygienic, and backwards hovels with peeling walls and buzzing or inoperable equipment. Healthcare had been assigned to computers, and with them offline, cadatas were incapable of matching their abilities. Sufferers of radiation poisoning and rare, nightmarish diseases faced dreadful doom. Their bodies regenerated to match their previous appearance, but mistakes in their code eroded their quality of life. They struggled with moving and gas consumption.

On the other hand, a positive outcome of the Potentates' withdrawal to the moon was the rise in demand for independent goods and services. The Potentates' monopolies squeezed enterprising small business owners out of metropolises. But with most Potentates' stores looted, and factories in shambles, they were unable to meet cadatas' demands. When cadata entrepreneurs heard about air pollution contaminating a nutritional air supply in their district, they located safe air sources, importing containers into this desperate region. They would sell it from little mobile carts or cheap little shuttles, moving these wherever the cadatas in need were. Many stranded cadatas survived because of these efforts. These businesses matured until they were even recognized in the press, which was otherwise beholden to the Potentates. Before the asteroid, newspapers never ran positive stories about any business competing with the Potentates. Acknowledging these upstarts as a building threat to their dominance over Cadata, the Potentates commenced dropping free packets of nutritional-gas and supplies. They only donated items with the highest profit margins for the entrepreneurs. This charity was unimpactful to the Potentates' wallets because they utilized government disaster funds for these drops. It did serve to muscle competitors out of these sectors because cadatas' desperate finances prevented them from spending money on commodities they could obtain for free. Once these businesses began failing, the Potentates cut back on the giveaways. They reopened their behemoth stores and raised prices.

This anti-competitive strategy was interpreted as hostile by the starving cadatas. They surrounded Potentates' stores in protests, hoping to appeal to their sense of moral obligation. The Potentates had no such imaginary anti-evil detectors in their genes, so instead of helping, they made a show of delivering a new batch of free supplies. However, this delivery was made to an uninhabited island without surviving roadway to allow access to anybody but the richest cadatas with shuttles to fly there. These chucked supplies rotted in the steamy roast-bath of the planet's overheating atmosphere. Having botched the search, the demonstrators could not prove the delivery failure. The Potentates made themselves appear magnanimous in the press, without depleting the demand for their products with freebies. Similar anti-protest maneuvers kept the rebels from solving their members' problems, minimizing the impact of the agitations.

The Potentates did not jail or otherwise exterminate the leaders of these protest sects because they had utilized their own vehicles, houses, and resources to assist the struggling cadatas in their worst moments. The Potentates thus avoided spending on troop deployments for housing, gas, and restorations. These leaders and their volunteers might have been ill-equipped and inexpert, but they did their best to figure out how to construct houses, repair broken infrastructure and otherwise pick up the molten remains of cadatas' lives. Their efforts extended to distant islands, which lost their travel links and thus had no help for weeks before these fortunate arrivals.

Once the Potentates restored access to goods through their stores, these organizations refocused on helping depressed and suicidal cadatas. It was easy to slip into torpidity as cadatas observed their shared-suffering, and its reflection in the once-beautiful natural world, and pristine atmosphere.

Causes of Planetary Death

A planet can die via a handful of paths. The odds of a healthy planet, under a healthy star, coming to a premature demise is astronomically miniscule. Cadata's location near the center of the Milky Way puts it at a greater risk of precipitous death, than a planet in the quiet outskirts, such as Earth. In Cadata's region, a star fills every cubic light year of galactic space. In contrast, the closest star to Earth is four cubic light years away. Cadata's interstellar proximity increases the odds of incidents, such as solar systems colliding and joining into a single entity.

A supernova within 30 light years of a planet can be fatal to life. At this proximity, a supernova's directed firing-ray blows off the atmosphere from planets. A supernova of a star with hydrogen-content can release 10^{46} J of energy, with 99% of this energy in the form of speeding neutrinos. If the supernova occurs within 10,000 light years; it causes a significant UV-ray increase, which interacts with the molecules in the atmosphere, depleting the ozone layer, and sparking global cooling. This can cause the die-out of plankton on a planet with Earth's biology.

Humans have recorded only three supernovae explosions inside of their Milky Way galaxy. One hit in 1006 AD, and lasted for a year. It happened 6,500 light years away. This proximity made the sky so bright, you could read outside at night. Then, Tycho Brahe spotted one in 1572 AD (8,000 light years away). And Johannes Kepler caught his in 1604 (20,000 light years from Earth). The latter was the last supernova recorded by humans in the Milky Way. The closest supernova since the invention of high-tech viewing devices was SN 1987A inside the Large Magellanic Cloud, Milky Way's satellite galaxy, 163,100 light years away. Thus far, life on Earth survived a blast from 6,000 light years away. However, if this light-bulb-sized celestial spark happened in modern times, failures of electronic devices, and satellites would have resulted in more noticeable damages. Supernova's UV-rays caused one of the earlier extinctions on Earth, and triggered an Ice Age.

Cadata's archeology records four supernova explosions that were proximate enough to cause extinction-events killing a significant percentage of species. As supernovas' frequency increases, the odds of life's survival in any corner of the galaxy are lowered. Thus, the galaxy's most dangerous location is its busy center. Astronomers in any developed civilization can spot the stages of a supernova before it commences.

An average supernova is unspontaneous: a 20 M_o star in its main-sequence burns through hydrogen in 100 million years, then through helium in 10 million years, then through carbon in 300 years, oxygen in 200 days, silicon in two days, before it finally enters the momentary iron-burning implosion and explosion phase. And after all this, either a neutron star, or a black hole forms in its place. A neutron star has a greater density than an atomic nucleus: so that atoms are intersecting amidst the neutron degeneracy pressure, and gravity is 190 billion times stronger than on Earth. A neutron is harmless because it will cool in the following millennia without further disturbances. A black hole develops in a minority of the most massive stars. It forms a debris disk around a singularity with magnetic fields, which shoot gamma rays as by-products of the infalling star. A black hole's gravity is dangerous to nearby planets and stars, and its expulsions add to this region's inhabitability. The warning spectrum change to helium is visible millions of years in advance. For organisms who survive the initial flip, there are at least 300 years in the carbon phase to evacuate out of the region.

Catastrophe can also strike an average planet from the standard rotation of a galaxy's spiral arms, which move faster than the emptier space between them. Planets located in this emptier gap can be tossed out of orbit around their sun. Such a planet can run into a dangerous, planet-heavy region of a neighboring star. It can collide with a hot gas cloud. These scenarios can kill the fragile ecosystem needed for life.

Earth is not eternal. In 225 million years, its carbon dioxide reserves will drop so low, plant life will be unable to survive in the wild. And if plants stop producing oxygen; humans, as members of the animal kingdom, will become extinct. Then, in 2.7 billion years, Earth's magnetic field will fail, and the Sun will blow away its atmosphere, until it is a desolate, cold rock, flying as silently as Mars or Venus.

If a species' luck is horrid; its own star might explode in a supernova, turning their planet into a molten fireball in hours. Worst-case-scenario might be a sun imploding into a black hole, and sucking a planet, and its erratic lifeforms into a place no light escapes. Cadatas would have had plenty of warning if these astronomic events were about to take place. A star gives sufficient warning prior to a supernova for any competent astronomer to decipher the danger. Instead of having millions of years to prepare for a natural planetary-death, by attracting the asteroid, cadatas created a disaster with a much tighter evacuation schedule. While cadatas were resilient, and did their best to stay alive in hostile conditions, the corruption in their political and business systems meant, when extreme scientific innovation was needed, their government was motivated by making a final profit, before the apocalypse, instead of with uncovering a preventative strategy.

While larger debris ejected into the atmosphere eventually fell to the ground, too many greenhouse gases remained, and created a runaway effect. Thus, in the first 50 years after the strike, Cadata experienced an average temperature drop of 86°F. Then, it endured a rapid acceleration of temperatures in the other direction, as clouds started to fall as toxic rains. The planet began enduring a 21°F higher average above pre-asteroid levels. Scientists predicted the temperatures would keep climbing over the following 200,000 years, eventually making it hostile to life. Even if cadatas could survive in spacesuits, by avoiding exposing themselves to the elements, all wild plants and animals would be dead shortly after the anticipated high-average was reached. The first sign of shifting temperatures was ocean evaporation. The air was increasingly poisonous. The rest of planet began displaying proof it would expire during this nuclear summer. Scientists were reaching a consensus: they lacked the mental and fiscal power to terraform Cadata back to life. They had to invest their research into inventing a statistically likelier method for their species to survive by relocating to a new habitable planet.

Chapter 3

The Long Journey to Liftoff

The KoraFePub asteroid strike destroyed the sanctuary of Cadata for its living organisms. Commander Doul Veahy, at CSEA, was charged with evaluating the options. Upon taking office, Veahy analyzed alternatives to evacuation, starting with potentially fixing the runaway-greenhouse in their own planet. Based on the distance to the nearest inhabitable world, he argued many safer, and more realistic options should be explored first. The idea he favored was building a giant spaceship, or a series of spaceships, to live on them as their permanent home. Cadata still had at least a few thousand years of habitability. This was sufficient to later design and perfect a permanent home in near or distant space. Theoretically this plan would benefit most parties. However, the Potentates calculated it would become their biggest government program expenditure. The construction of spaceships to accommodate 10 billion cadatas would consume a significant percentage of Cadata's metal, rock, and water. Every able-bodied cadata would be forced to labor at this single task for thousands of years. These laborers would have to forego their personal profits to benefit future generations of cadatas. Instead of minimizing supplies to raise prices, the Potentates would have to ration free supplies to the needy. Instead of hoping for the rich to waste money on useless luxuries, they would have to forbid extraneous purchases, for their robot factories to commit to the planet's singular goal. Cadata's surviving buildings would have to be repurposed into spare parts for the spaceships. Potentates' deluxe headquarters would be forced to close, leaving them in a temporary shelter, or in factory offices. Relocation to a new planet also necessitated a redistribution of the Potentates' wealth. After these sacrifices, there was no certainty that either the cadata species, or Potentates' profitable trade would survive. In one scenario, if they opted to moor in interstellar space; resources would become communal. They would be equal to all passengers on a ship belonging to the crewmen.

Veahy's second-best proposal was to settle on the largest moon, called Lentare, of the gas giant Tonace. Tonace is similar in size and composition to Jupiter in Earth's system. If the team eventually concocted a plan for fixing Cadata's atmosphere; it would be a short trip to return to Cadata from a solar-neighbor. This moon was internally hot, keeping surface temperatures high enough to sustain a liquid ocean. It even had a light atmosphere, and enough mass to keep cadatas somewhat grounded. Alternatively, a lower-gravity body would have generated the inconvenient tendency for cadatas to float away from it. The problems voiced against this plan included the amount of work needed to thicken the atmosphere for the air to be consumable. It was a tougher challenge than fixing their own runaway greenhouse. Cadatas understood how to subtract gases from an atmosphere through compression, or by changing them into solids. But they lacked a method for whole-atmosphere-engulfing compression. Adding a teeming atmosphere in place of merely a thin layer of molecules on Lentare was daunting. The essential particles had to be mined in precise quantities from the moon's ocean. And squeezing 10 billion cadatas into the miniscule surface-mass of Lentare, and housing them in under-water engineering marvels could produce claustrophobia, and deadly overcrowding.

The underwater-housing idea suggested another alternative. Veahy speculated a tunnel or cave system below Cadata's surface might be designed, where they might live in relative-safety from the surface solar bombardment. This plan also involved the Potentates donating resources to create new, free shelter and nutrition-harvesting infrastructure for 10 billion cadatas. Environmentalists proposed comprehensive insulation of these tunnels and filtration of the air supply to expel the atmosphere's poisonous gases from infecting residents of this planet-sized bubble. These would have cost about as much as self-sufficient spaceships.

Out of desperation, Veahy also pitched moving Cadata further away from their sun to lower its temperature. However, Veahy's scientists calculated, even if they invented a strategy to generate a massive gravitational field to attract Cadata away from its sun's gravitational pull, this would have caused massive storms and tides. The key insurmountable-for-their-current-capabilities challenge was how they might turn off Cadata's gravity to allow it to overcome the sun's pull. Enhanced electromagnets could bend space, time, and gravity by generating an elevated electric current. But building a gigantic electromagnet strong enough to rope a planet was prohibitive. They considered a slingshot maneuver wherein they would just tug on the planet once, and let it move the rest of the way on the generated momentum. But the next unconquerable test was how to slow it down when it reached the desired orbit. A cadata spaceship's breaking-system envelopes most of its outer surface to allow it to stop with perfect precision while moving in any direction. Creating a similar system with enough resistance or propulsion to stop a hurtling planet would require an unprecedented, planet-sized construction project. Veahy never attempted funding this extravagance. One scientist theorized a proximate explosion might stop Cadata's spin; then again, it might shatter it into rubble, or send it flying in the opposite direction. The goal was to slow it until it entered a smooth new orbit. Even if the relocation worked, this migratory trip would disturb the orbits of their solar system's other planets. If Cadata's new position was too proximate to a neighbor, their attraction might have stirred them into a collision, necessitating a new relocation. The present solar system had an exact balance of proximity and distance, established in an equilibrium between all planets, moons and massive objects. In contrast, if they attempted to move it out of the solar system to a neighboring star, the planet would freeze at absolute-zero during interstellar transit, killing the remaining lifeforms. Extreme internal heating could save the planet from freezing, but heightened core activity could over-heat or explode the planet.

The last drastic measure Veahy formulated was mock-aging Cadata's star to lessen its luminosity. A rapid-burn of a massive portion of the star's fuel could generate devastating solar flares. The development of a device to capture the sun's energy could garner the power for planetary transformation. However, they lacked a method for safely storing this voltage. None of these solutions were within established safety parameters. In the conclusion of this negotiation, the Potentates refused to suffer these massive expenses.

Searching for Habitable Worlds

With these alternative options shot-down, Veahy returned to the Potentates' directive. Veahy commenced a search for either initially-habitable, or terraformable planets in the Milky Way. At the start, his team of scientists asked broad questions about their home galaxy. What are the categories of habitable environments? Which conditions, or compositions are imperative, and which are deadly? Based on their research, the average planet in the Milky Way is rocky with a mass ten times that of Earth, and .10 of the distance from Earth to the Sun. This normal Super-Earth is giant, hot, and radioactive. While a significant percentage of planets are Super-Earths, these are less habitable than Cadata because most lack an atmosphere, and water. Outside of this middle-ground, there are plenty of horrors in the extremes. Worlds can be a thousand to ten thousand times bulkier than Earth. The bigger the planet, the more gas it tends to encompass in its atmosphere or body. The extreme pressures on larger planets cause more matter evaporation: this process further increases planetary heat, or the rapid movement of molecules. The mid-level-size is between a hundred, and a thousand times the size of Earth. At this mass, planets tend to be composed of liquids. Earth and Cadata are on the smaller end of the spectrum. This category varies between a tenth of the size of Earth, and up to a hundred times larger. The smaller sizes are rocky or metallic, with three to six times heavier density than water. Cadata is somewhat larger than Earth. If it had been smaller than Earth; it would have lost its atmosphere. The shortfall in gravity on tiny rocky planets means they cannot hold molecules from floating away. This loss is aggravated because, without a molecular-gases layer, solar radiation, and solar wind blow atmospheric-bits into outer space. Without this airy coat, the remains are just a large piece of rock: a planet by humanity's definition, but equivalent to an asteroid, or a comet for the purpose of habitability.

Commander Veahy and his team began testing, hoping to locate a match for Cadata in atmospheric and land composition, and in mass and gravity, or measurements disregarded by planetary residents while inhabiting these "natural" conditions. Abnormalities in these on foreign planets become paramount because they cause imminent death. For example, if a planet has ten times more mass than Earth, then it has 2.4 times more gravity, so a person who weighs 145 pounds on earth would weigh 348 pounds on a hypothetical planet with this larger mass. Earth's obesity crisis has observed a significant percentage of people weighing 348 pounds or more. But at around this weight, most humans become immobile and need special vehicles to move around. If gravity is four times greater than on a home world; settlement presents a physical obstacle. Without gene-editing to make a species muscular enough to handle the change, diseases associated with obesity would accelerate, causing a die-off.

At the juncture when relocation was proposed, Cadatas' technology was a million years ahead of Earth's in 2019. Commander Veahy's team attempted knottier versions of the basic technologies humans have been exploring. They scanned the sky for the bits of visible light from stars, which were missing because they were intercepted when they hit a planet. The missing light forms a dark spot, perhaps three 10-billionth of the massive sun. A planet also stands out on a space heat scan, as it emits infrared radiation from its hot core. Once this planet's distance in relation to Cadata was established, the size of this heat source determined the size of the planet's core. The planet's size screened out gas giants and other hostile categories. A planet's gravity and atmosphere can be secondary to the presence and nature of life on it. Anti-gravity suits, or stations might solve the gravity problem. Terraforming might even change the climate of a planet. However, if there is a parasitic species there, intent on penetrating through defenses to eat you… Well, it is not a survivable environment. Thus, CSEA tests included various scans for fossils, and biosignatures. Humans and cadatas alike are composed of star-stuff, or elements molded in the hottest stars. What separates living organisms from the inanimate pieces of rock is the strange combinations of different elements life binds to form complex organisms. H_2O is a combination of elements too, but living matter, such as human DNA, combines several elements into replicable strings. The presence of these complex mergers on a planetary scan indicates life. Any element from the periodic table can evolve; divergent elements form unique types of biological organisms. Scans must check a range of abnormalities, including exploring a potential population of machines, or other non-biological markers of an advanced civilization.

In addition to methods recognizable on Earth, cadatas use advanced direct-imaging techniques. We can amplify any planet in the galaxy to see a sketch of its surface. Our observational Computer creates a 3D sphere mapping the composition of the chemicals in a planet's atmosphere, surface and interior down to the core. Chemicals are named and separated into different colors on a comprehensive map. This imaging method encompasses a section of the sky as a single unit rather than breaking it into pixels in the human manner. The imaging device used for these 3D scans, the Nalut Expander, was parked in an orbit around Cadata's sun to gather the steadiest possible images. If instead it orbited Cadata, the frequency of the loops or the change in relative location would introduce a slight uncertainty in the resulting scans, and even a miniscule variance can generate erroneous findings. For example, the shift's misinterpretation of carbon dioxide as oxygen could lead to a pro-settlement conclusion, an error that might go undetected across the millennia of a wasted trip. The Nalut Expander was the size of a domestic travel shuttle. Its spherical surface was coated with miniaturized telescopic devices covering 360° of the sky. Nalut collected data on all wavelengths, including visible and invisible, hot and cold, ultraviolet and the full spectrum of infrared, as well as all frequencies of radio waves. The sensors picked up on radio stations or unnatural microwaves. Nalut cannot reproduce a legible version of the captured sounds because the waves are degraded by space-interferences and distance; it can spot the precise origin of these waves, deciphering if they stem from technology, biology, or an active stellar object. Nalut has other sensors picking up aspects making planets inhospitable. For example, it measures the volume, direction, intensity and type of solar wind the host star is firing at a potential planet. These measurements are multi-layered as a star such as the sun has a continuous flow of fast solar wind traveling at around 750 km s^{-1}, and slow wind gusts firing at half of this speed. At the peak of a gust and on planets in closer proximity, radiation is elevated, leaving an atmosphere-stripped planet. The effects of solar wind on an atmosphere is measured by the volume of departing gases. If gases are fleeing at a heightened rate, the atmosphere might be depleted by the time a spaceship arrives. Though, natural planetary change occurs over millions rather than millennia of years. Another essential indicator is atmospheric pressure because it changes the boiling point of various liquids. It can alter how substances interact, and therefore how technology and a new species might take hold. Precursor to these innovations were devices akin to Earth's New Horizons, which made planetary observations with an infrared spectrometer to measure heat, an ultra-violet spectrometer to examine the atmosphere, and a plasma instrument to measure interactions with solar wind.

The Nalut Expander and the innovative technology accompanying it was developed by Nally Subsenar, an independent computer programmer. She designed it long before the KoraFePub hit. Her research began 3,000 years earlier in 103,056 BC, when she began re-engineering and re-programming existing space imaging devices to expand their capabilities. She lived on a mini-island with few shuttles or businesses crowding its pristine setting. To keep her mind sharp, she enjoyed gliding from cliffs to view the beaches and distant storm formations. On one gliding expedition, she visualized an upgrade to boost the degree-of-certainty in space exploration. Her stockpile of parts and materials allowed her to begin building the strongest magnification process on Cadata. To afford expensive components, she worked as a freelancer for some of the Potentates' corporations. Nally spent centuries testing each added element by flying a shuttle into orbit and releasing a smaller prototype of the Nalut Expander. There were many challenges along Nally's path, as she had to invent some new physics formulas just to crack the limitations of prior viewing devices. She submitted these ideas to scientific journals, but they were rejected because they contradicted established ideas from scientists sponsored by the Potentates. One of the driving forces for Nally's extraordinary effort was her fear Cadata was moving towards an environmental catastrophe. She failed to anticipate it would come from a misdirected asteroid, but greenhouse gas emissions were spiking for millennia, while population growth kept increasing the need for inefficient fuel sources. Nally was careful to use renewable sources gatherable from the space around the expander; these renewable designs later inspired the design of the exterior-star-bound spaceship. Nally's environmental concerns led her to imagine a scenario where in the not-so-distant-future Cadata might collapse and would need to find other habitable worlds to allow the cadata species to survive. Without this foresight, the KoraFePub strike might have been the final chapter for Cadata. Nally's island was wiped out by the strike. In the chaos after the hit, similar freelance work for independent researchers was nonexistent. No spare parts remained unutilized for the rebuilding effort to protect cadatas from immediate threats to their lives.

When Nally was close to finishing a working prototype, she was hired as an assistant by a popular scientist, Dik Sibost. This was not the first time Dik hired a subordinate with a near-completed concept. These types of acquisitions of talent and technology was what made his Ludost Nethist Lab a frontrunner in innovation on Cadata. Ludost was one of the top government contract winners in the space imaging sector. His primary job duty was selling himself and his lab at scientific conferences, where he was a popular featured speaker. His assistant styled his skin to match the latest fashion. The assistant also inserted miniscule changes into a single speech to bypass plagiarism checkers. The streamline speech was written by Potentates' technical writers to advertise their latest technology to potential buyers. His job was whistling out the message in clear terms as the assistant's 3D holographics flashed behind him. He was given a week off work for each of these trips. Each presentation took no more than an hour, and the rest of the time was spent on pampering, drinking in bars with pretty fans, and napping in luxurious hotel rooms. If he failed to secure a knowledgeable developer such as Nally for a new season, he recycled previous season's findings in a re-arranged research paper, re-worded by his assistant, and released in trendy scientific journals. Dik had to continue publishing to be invited as a featured speaker to conferences. Without these writing and speaking tasks, he would be failing to perform the publicity tasks his corporate and government bosses required.

Nally fit with Dik's business-structure because she worked from sunrise through the long nights to perfect the Nalut Expander. Sleep necessity had been quasi-edited out of cadatas' genes, so unconsciousness was a personal choice rather than a daily chore. Nally abstained from it. She had learned from enumerable, youthful failures, if she failed to push the boundaries of the possible, she might starve, overheat from exposure, or suffer the wrath of fiscal collapse in Cadata's rampant-corporatocracy economy. Once the instrument itself was completed and released as a beacon of progress by the Potentates, instead of resting on her achievements, Nally set out to create the *Habitable Worlds in the Milky Way Galaxy Catalog* with this new tool. The computer program she developed for the Nalut Expander did a large portion of the busy work of scanning every tiny portion of the sky for planets and creating maps and visuals of their compositions. Nally's challenge was sifting through 232 billion stars, their 2.3 trillion planets, and their 287 trillion moons in the Milky Way to shortlist the habitable planets and moons. The Computer organized the data by components. It classified gases in the closest ratios and quantities cadatas consumed in their own atmosphere. And it categorized planets by the closest masses and atmospheric pressures to Cadata. It even identified those planets identical in every aspect to Cadata. With trillions of planets and moons considered, there were thousands of potential bodies near-indistinguishable from Cadata. Signs of intelligence and biology were the elements stumping the Computer because it lacked an algorithm to parse between slight fluctuations and interrelationships capable of turning an otherwise ideal environment into hades. These small deviations as well as the enormous task of sifting through the statistics to arrive at a short list of the best planets is what kept Nally in the Lab. She researched different molecular compositions and created projections to visualize the lifespan of the planets and stars the Nalut Expander presented.

As the list solidified, she introduced new points of comparison, such as aerodynamics, shifting the order. After hundreds of adjustments, Nally flew the Nalut Expander into an orbit around the sun, so its observational equipment was not restricted by Cadata's thick atmosphere. She supervised this flight, watching its progress hrough onboard cameras. Just as it attained its preset position, and was about to begin rescanning the sky for the new wind-resistance test, a solar storm began shaking Nalut. Even a slight trepidation voids results on the precision-scale needed to test a reading on the opposite end of the Milky Way. This solar storm was stronger than average, so it dislodged the new component. Because of her heightened curiosity, Nally had rushed through pre-flight tests and missed adding a securing device to keep the wind component from dislodging at the top speeds and pressures. Thankfully, the piece remained semi-attached, hanging on during the transport stage. Nally programmed the internal robot to release itself from inside of Nalut. It crawled over to the wind device. Under Nally's instructions, it clinched the piece to the target. The robot cleaned itself and retreated into the unit. After a full test for malfunctions in the systems, Nally commanded Nalut to run the maximum-wind test. Only weeks after this fix and nearly 3,000 years after Nally commenced this study, Nally released the *Catalog's* list of habitable and exploration-worthy planets.

In the Introduction to the *Catalog*, Nally included lengthy explanation for the choices she made in sifting through the overwhelming volume of data the Nalut gathered. First, she programmed the system to prefer planets farther away from Milky Way's central galactic bulge. Cadata is closer to this high-traffic midpoint than Earth, which is far from the center and in the middle of two spiral arms. Because of this proximity, the night sky has around a hundred times more visible stars from Cadata. Proximity to the bulge implies it is likelier on Cadata for a passing star to hurtle Cadata out of its system to roam through interstellar space. Cadata traveled faster than its closest spiral arm a dozen times in its 10-billion-year history; at these junctures, it risked colliding with flying planets and black holes. Then again, the odds of a direct collision on galactic proportions (where plenty of "empty" space even in Milky Way's bullseye) is smaller than the odds of a proximate supernova explosion. Cadata managed to avoid a life-ending catastrophe, but if they could choose a quieter neighborhood, the longer trip seemed like a worthy sacrifice. The Potentates disagreed, so Nally re-ordered the planets to place the closest habitable world first. This was the sole proximate planet reachable within the Potentates' lifetimes.

Nally also dedicated a report section to the life-building molecules cadatas require to thrive. To research this, she designed simulations of evolutionary paths. This program generated molecules and merged them in random combinations, placing them in the types of environments anticipated on the potential worlds. Cadatas' physiology does not require water or solid-food consumption. Cadatas inhale the molecules essential to sustain them. Water is still a necessary commodity because it dominates both Cadata's planetary surface composition and cadatas' cells. Simulations on the necessity of water for cadatas proved inconclusive. It could be generated from hydrogen and oxygen within cadatas' gene-edited bodies, so "complete" liquid water could be absent. Thus, the search could be expanded beyond water worlds. Water takes on the liquid stage at one-atmosphere of pressure on Earth in the temperature range between 0 to 100°C. This range changes between thinner and thicker atmospheres. In thinned or non-existent atmospheres, the presence of water in an organism's system can boil it alive even without a heat source. The organism cooks because water evaporate or boils at a cooler temperature. The problems a planet-migrating species can exhibit are near-infinite. Life evolves in balance with its home world, so any outlandish world will present terminal environmental pressures. Each planet provides unique building blocks for species. Humans rely on carbon, oxygen, and nitrogen. Cadatas had designed artificial methods for synthesizing the full range of molecules. However, the production of molecules on a planetary scale is among the most energy-consuming endeavors in the universe. Nally also explained the need for a sufficient source of energy to make life on a planet energy-neutral. If a planet did not have heat in its core, heat from its sun, or chemical reserves changeable into energy, it was not an efficient destination. The habitable zone measurements stretched beyond merely calculating a planet's temperature, proximity to the sun, and pressures generated by the masses in the system. Even a freezing world might be more suitable for an advanced civilization than a hot one without the resources needed for development. The presence of intelligent life was a superior resource because a thinking species pre-performs the terraforming, infrastructural mutations, which otherwise require millions of years of toil.

Nally also preferred planets with at least a single, large moon. Small, rocky planets, such as Earth and Cadata, need moons to stabilize their spin. Without a moon, planets this size can begin to wobble, or flip over after millions of years. Even with a moon, the Earth is wobbling due to the fluctuations in surface mass around the melting poles. A spontaneous total flip is accompanied with powerful earthquakes, tsunamis, and volcanic eruptions as a planet's crust cracks. In Earth's Solar System, Mars flipped before because its moons are too tiny to have kept it from this fate.

Earth ended up at the end of the *Catalog*. In 100,163 BC, when KoraFePub made it necessary for cadatas to seek ways to abandon Cadata, modern humans were still split into Homo sapiens and Neanderthals. While some Neanderthals might have been wearing clothing in northern regions in 100,163 BC, these adornments would not have stood out on a scan any more than a peacock's feathers. Earth made the list because of the diversity of chemicals on its surface, and the clear presence of unintelligent lifeforms. Nally defined "intelligence" for the purpose of her study as: "1.1 the ability to manipulate matter to create either the resources to allow for this species' complete mindlessness, or the computational knowledge needed to become dominant in an environment, 1.2 the ability to manipulate other species to perform skills an intelligent organism might not want to do, 1.3 having the creative impulse to design programs capable of acting with intelligence." Earth's atmosphere was rich and yet not too pressing, and the temperatures were warm enough to be acceptable for cadatas' constitutions. Nally also observed the presence of varieties of rock, dirt and both fresh and salt-water across its surface. Oxygen is poisonous to most other lifeforms, but cadatas can tolerate it because of their bodies' ability to run complex reactions on almost any gas they take in.

Nally's conclusions and suggestions shaped the following exploratory trips. Thus, how she arrived at these helps to explain why some seemingly absurd planetary stops, in retrospect, were initially deemed as among the most ideal.

The Gluttony Bias in Kardashev's Scale of Civilizations

Human definitions for *intelligence* and *civilization* across the universe are hilarious. Cadatas lack an appreciation for humor formulas believed to be amusing by humans. Still, I produced a chuckle-adjacent noise while browsing Kardashev's Scale of Civilizations. Jokes aside, I must debunk this theory before relating facts about real, intelligent civilizations across this galaxy. Kardashev separated civilizations into three types by the quantity of energy each utilizes. Why would a higher energy consumption level equate with higher intelligence, or civility? Let us consider an animal consuming the largest quantity of food per-body-mass on Earth: the American pygmy shrew, which eats three times its mass to maintain a pulse of over 1,000 beats-per-minute. In contrast, the whale is hyper-efficient, eating only 2% of its 170-ton weight daily. A shrew dies within a year of birth. A blue whale might live for over a century. It plummets under water, until its lungs collapse from the pressure. Then, it pops its jaw open, and swallows a school of krill in a few seconds, repeating this dive numerous times per-day to utilize one of the most abundant food sources on Earth. To survive, the tiny shrew constantly exhausts is muscles by sprinting in hunts for morsels. Meanwhile, the whale idles as it opens its mouth to swallow nearby water, which is rich in just the miniscule seafood it needs. Why would the distressed rodent be a more intelligent lifeform than the easygoing seafaring giant? A species inventing technologies necessitating the consumption of the entirety of a galaxy's energy is an extremely mad rodent. Intelligence is in a species' ability to control its intuitive infinite desire, and to minimize energy-consumption. The smartest civilizations are the ones capable of making a galaxy run on the smallest possible quantity of fuel. If a species can run a spaceship for billions of years on a grain of sand; this civilization must be more intelligent than a neighbor who wastefully consumes a black hole merely to turn on a lightbulb.

In Kardashev's breakdown, Type I civilizations process as energy the entirety of the sunlight falling on their planet. Type II civilizations use all energy their sun produces. And Type III use all energy created by their entire galactic system. If humans ever travel outside their solar system, and find a species gobbling every watt their galaxy omits; they would be well-advised to avoid approaching. This advice is based on cadatas' experiences with extremes of this sort. There are black holes at the center of all galaxies. There are no black holes in the observable universe capable of consuming all mass, or energy of their galaxy. A discovery has yet to be made of a black hole the size of an entire galaxy, hovering alone in empty space because it ate all celestial bodies in its vicinity. Unbound greed and gluttony are more than moral sins; they are also intellectual sins. A civilization busy thinking, lacks time for wasteful consumption. In contrast, wild beasts can indulge in consuming every accessible drop of energy. Are cadatas a superior, intelligent civilization? We have surpassed the galactic energy utilization average by creating an engine capable of magnetic attraction of particles and molecules from interstellar space. Our design converts matter into propulsion energy needed to traverse the Milky Way for 100,000 years. While we have cycled through an enormous mass across this span, we are far from utilizing the energy output of a sun. A species in need of an entire sun's energy must be engaged in an arms race on galactic-apocalypse proportions. Even humans' telescopes would notice if all the energy of any galaxy was moving towards any single solar system. Such energy- or mass-accumulation is only found in black holes. If universal-energy accumulation could be achieved without turning a solar system into an instantaneously destructive black hole; imagine standing on this planet as all electro-magnetic, heat, and chemical energy and mass of the galaxy is flying at you at near the speed of light. Even if these destructive forces can be controlled, can they be applied to creation? The goal of such accumulation must be destructive. Can any peaceful and reasonable species microwave their planet to engage in peaceful production? Radiation, in the form of heavy ions (silicon and iron) and low mass protons, causes cancers in humans when they travel in spaceships without the protection of Earth's magnetosphere. Deliberate beams of radiation are suicidal. Suns are controlled and violent chemical mixes reacting at a hyper-stable level to bathe planets such as Earth in consistent temperatures. While a sun can control this energy output, a planet disintegrates when it receives an equivalent energy input. Mutating a planet to match the properties of a sun without allowing it to undergo equivalent explosions necessitates the invention of new physical laws. Even a sun in the stable stage of its development undergoes solar flares. While a solar flare has a light impact when experienced from Earth, its size would engulf the largest planet in any solar system if it occurred on the planet's surface. Introducing a sun-sized bundle of energy into the lives of any species, puts them at risk of going supernova just because a few electrons leave alignment.

A closer examination of just how Kardashev imagines so much energy can be utilized reveals still more disturbing concepts. He theorizes a Type II civilization would use anti-matter propulsion. Anti-matter, by definition, cancels out all matter it contacts. It must be restrained from touching matter, until it is allowed to react to create near-pure energy. It is a volatile non-thing, the archnemesis of the visible universe. With sufficient anti-matter, the whole universe can disappear.

Another potential use of energy, according to Kardashev, is for the manipulation of space-time. Why would any successful civilization want to manipulate time? The destruction of its own world is a rare, relatable circumstance in which a civilization must travel back in time to repair the catastrophe. To travel backwards in time, a spaceship must move faster than the speed of light. Cadata's spaceships can reach a speed close to the speed of light, which for the sake of simplification will be rounded off to the speed of light across the book; however, accelerating to light speed takes around a year at 1 g or Earth's gravity. Quickening at a significantly faster rate would create an extreme g-force, which would flatten or explode any living organism. Thus, exceeding light's speed requires over half-a-year. The time required for the trip will cancel some of the time-gain, in part because acceleration towards faster-than-light speed would also move them in the wrong direction in time. Cadatas have been traveling at the speed of light across our journey through the Milky Way, but we have opted to avoid the faster-than-light-time-travel hypothesis for the safety of our spaceship. Cadatas did attempt sending a robot-manned space shuttle at this speed and it did move back in time a bit but also used up time getting to this speed, so it arrived back from its orbit in a bit under a year, whereas without the time-jump it would have arrived a bit later. After data from this recovered shuttle was analyzed, no useful purpose was apparent for such time tricks. Thus, we ceased such pointless experimentations, as our business leaders are impatient and demand financial rewards for all energy or matter expanded. When the asteroid hit Cadata, our scientists gained a profit-motive for time-traveling to avoid this catastrophe; but even given unlimited funding, they still admitted an affective time-manipulation was beyond their technological capabilities.

Human science fiction is obsessed with the notion a wormhole can punch through space and allow a spaceship to arrive on the other end of a galaxy. Even if a wormhole exists and can be penetrated, it cannot end in a location with a matching intelligent civilization. For this precision, the entire wormhole would have to be artificially designed and manufactured. A natural wormhole might lead to the beginning of the universe, a point before space and time normalized. In the first billion years after the Big Bang, it is theorized, the miniature universe was just hot, opaque clouds of ionic molecules too hot for electrons and protons to condense into atoms. The hypothetical spaceship that manages to travel through this black hole would end up in a place absent of consumable energy other than these ionic molecules. This civilization would have spent its galaxy's energy to travel to a place without usable energy or a road to return home. They might stop in a place before matter and anti-matter cancelled out, leaving only a small portion of matter behind which is our current remaining universe. Relocation to a spot before the grand cancel-out would spark their spontaneous disappearance in a clash with anti-matter.

In general, spending energy to create an artificial black hole is nonsensical. An abundance of black holes already exists if a species desires to bravely go where all matter is compressed into nothing. Is it existential to want to become nothing? Either way, a wormhole or a black hole would not allow a vehicle to turn back as it can only be pulling matter in one direction. Further still, because of gravity, the weight of the spaceship and those on board would increase to infinity, and radiation near a black hole would by equivalent to passing through a cloud of atomic bombs. If a ship can survive all this, we return to a problem: the nature of a hole through space, forming because of expansion after the Big Bang, to be unpredictable. The other end of this mercurial hole can be anywhere in the universe. Why would a spaceship want to go somewhere without knowing where it would end up? In summary, black holes destroy all matter, leaving behind only the denting in the fabric of space-time with trillions-hotter than Earth's sun streams of energy passing through each other and building into a state of Planck density (10^{93} g/cm^3 or a universe of matter compressed into an atom) and extreme gravity, speeding its insides to exceed the speed of light. While 100,000 planetless years have driven us to desperation, we have never flown into a black hole to test this theory.

The imagination of Earth's scientists has even phallically fantasized of spreading their seed to impregnate the emptiness, giving birth to a new baby universe. Our current universe contains abundance beyond any biological lifeform's reproductive and consumptive abilities. Why make new universes, disrupting a balanced equation? According to human science, at its birth, the universe's matter and anti-matter cancelled each other out moments after the Big Bang. What if the new anti-matter cancels out matter in our own universe instead of the new matter? Even if it is a tiny universe, it would contain sufficient anti-matter to delete this "creative" species' planet. If it is a microscopic universe, merely recreating the Big Bang, it would be the sum of the ingredients the scientists squeezed into it, failing to produce new matter or energy to be mined for a profit. If it is called "tiny", but encompasses several galaxies: the minimum size to call it a universe. Then, it would blow its neighboring galaxies into chaos, as it expands from an atom to fill this space.

A Type II civilization is also accused of spending energy to communicate through gravity waves. These are created when two black holes collide and merge. Would any intelligent creature want to communicate with earthquakes? They could invent a Morse Code and transmit it across the surface of a planet by striking active fault lines and triggering a volcano; the ripples of this strike would travel hundreds of miles and would be detectable to a receiver. Curious in theory; this is absolute nonsense in practice. Why would any planet with infrastructure attached to the ground benefit from spikes is seismic activity? Communicating with this destructive force is like communicating by hitting somebody over the head with a stick. You can send, receive, and interpret messages, while exchanging hits on the head. But is this a civilized, or intelligent form of communication? Gravity can be utilized to achieve astonishing outcomes. It can pull a planet away from its star, and into a more congenial solar system. It can keep a species attached to an artificial-gravity surface in zero-gravity spaces. However, gravity is the least logical tool when it comes to communication. Imagine spending every watt of energy of a sun to create artificial seismic activity on one's home-planet merely to say "Hello" to Jack ten miles away. This is humanity's insulting conception of "superior" alien intelligence.

At the Type III level, Kardashev named some of the most extreme phenomenon in any galaxy and granted their manipulations to his super-civilizations. Black holes now appear not only as potential vehicles for space and time travel, but as sources of energy to be consumed. By definition, black holes consume energy, pulling it inside their event horizons. To pull in the opposite direction to one of the strongest forces in a galaxy is going to waste as much energy as a species tosses at the effort. A less efficient source of energy than sucking it out of a black hole is unimaginable. Kardashev concocted a schoolyard brawl, and called the victor "intelligent". Generating energy compliant in performing a civilization's wishes is an intelligent use of resources; forcing the most violent energy source in the universe to accumulate on your home world displays supreme-stupidity.

Kardashev also believes Type III civilizations would collect dense stars. You can collect a star, but where would you keep it? Collecting enormous balls of plasma is hording in the extreme. Would they employ the accumulated stars to toss them in an interstellar war? Keeping a star closer than their natural star's-distance from a home world would devastate the planet. A fuel source must contact a mechanism to charge it. If it is stored in the space beyond a solar system; then, it is an equivalent energy source to the suns already neighboring a planet. Manipulating solar systems and galaxies is impractical. Suns and planets are where natural forces want them to be. No civilization makes zero mistakes as its science evolves; thus, a mistake during the collection of the matter from stars is inevitable; and this mistake might generate a new species of destruction. Civilizations experimenting with universe-ending energy must be flawless. Approaching universal-death ahead of nature's schedule is not a superior achievement, just as dying at birth is not better than living a long life. Only catastrophe-averse and life-prolonging civilizations deserve to be called civilized.

The most relevant absurdity of Kardashev's is in the propensity of an advanced civilization to explore billions of worlds. Are they making this volume of stops because their telescopes are deficient, and they are missing intended targets? Imagine discussing an ideal vacation with your partner. You propose: "Honey, let's take a trip around the world." The partner is ecstatic, but you add: "Let's travel for ten minutes, then stop and observe what we can in the place we have arrived at, and then get back in the car, travel another ten minutes, stop again, and repeat this cycle in a random pattern across the world until the day we die, and then our children and the following generations will repeat this billions of times!" Is this a civilized plan your partner would agree to? Travel for relaxation or exploration might be fun, but it an inefficient energy expenditure if there is an over-abundance of resources at home. Birds migrate because they are seeking warmer climates with food sources depleted in colder regions. Constant migration to perhaps less resourceful climates is counter to the best interests of any species. Cadatas' journey across the Milky Way has been in search for a new planetary home. We are so exhausted; we might remain on Earth even despite its flaws. Life on a single stable planet fosters intellectual advancement. The addiction to spreading through the universe is a sign of an id-driven personality akin to a sex-addiction. A species with a need to visit billions of

worlds must be undergoing an extreme case of an elephant's "must". Perhaps, it is reproducing at a rate capable of overpopulating a planet within days of arrival. It runs out of
room to continue this rampant reproduction, and must seek out a new blank planet to
continue the process. This species is the plague of the universe, consuming or destroying
every living and inanimate entity out of greed.

The only Kardashev prediction cadatas mastered, due to its benefits, is travel at sub-
light speed. However, such travel is easy enough with help from a simple magnet, and
recycled space particles. If we had been dragging a galaxy-worth of energy with us, we
would have torn a giant trench in the fabric of space-time in our wake, without benefits.
The most civilized species do not engage in the pursuit of waste. Conservation of energy,
or expansion of life through a healthy diet prolongs life. Gluttony and over-expenditure
shorten it.

The Development of Universal Robots

Cadatas would have never ventured outside our solar system without our robotic companions. The field of robotics developed on Cadata early in its scientific evolution. By
1,003,693 BC Cadata's elite could utilize small flying robots with the capacity to navigate to desired locations, and to take radiation, sound, and other readings needed for
spying and government operations. A thousand years later, in 1,002,753 BC, the Potentates had grown so dependent on robots they leveled the island of Bonne to create a massive factory for their reproduction. This massive enterprise allowed robots to improve
their own designs, accelerating their evolution until an intelligent robot was able to assist
Nally with fixing the Nalut Expander to find habitable worlds across the Milky Way. The
next step was creating a robot factory on Cadata's moon to allow robots (with help from
scientists and programmers) to evolve to maximize their suitability for the conditions in
low-gravity space travel. When KoraFePub struck Cadata, in 100,163 BC, Loor Dowor
was the Potentate acting as the head of the Robotics Corporation. He supervised the
final adjustments to prepare the latest robotic, and nanorobotic designs for continuous
operation across the journey in search for a new planetary home for cadatas.

The first self-operating robots went into production in Bonne in 1,002,751 BC. This
design solved a philosophical conundrum in robotics design. For a robot to perform autonomous actions as safely as an intelligent organic lifeform, it must recognize an enormous quantity of patterns. Beyond distinguishing between a circle and a square, it needs
complex subroutines to process the patterns between observed objects. For example, a
robot must be aware how an object is changing, and the significance of this change. A
robot looking at a flower left on a tabletop must comprehend the flower is in danger of
dying without access to water. If it only sees a flower at one point in time, it would not
have the same instinctual response as a human or cadata observer. A robot must have the
capacity to plan steps to achieve a goal by analyzing elements in an environment capable of influencing the outcome. For example, if a robot is unaware of rain, because it is
focusing on the set goal of moving a ball to the other side of a field; it might fail to take
precautions, tripping, and breaking key components. Living creatures use instincts, and
conscious spatial-awareness for preservation. A robot must be taught similar impulses,
allowing it to dissect the elements interacting to create reality.

A robot also must have common sense. Just knowing what a flower is doing as it ages, or being aware of the presence of rain and the certainty the rain will stop is insufficient to take corrective measures to overcome unpredictable challenges. Common sense is learned from interacting with the world, at times with devastating results. A child venturing into a garden and being stung by a bee, learns bee stings hurt; thus, the child might avoid all gardens. A robot would not experience any damage or pain from a bee sting, so it would not have the same reaction as a child. This is one of the reasons robotic laborers are preferable to living ones: pain or destruction are blocks to achieving an ambitious goal. On the other hand, the lack of pain means a robot fails to learn to fear bees. Because of these distinctions, a robot fails if it is created through reproduction of human common sense. Instead of learning to mimic the walk of a biological lifeform, a superior robot must discover the most efficient transportation method for its unique mechanical form. Programming adequate common sense into a robot is based on millennia of field testing in a myriad of terrains through problem-encounters.

The "universe" is a key term for the biggest challenge our robotics faced as cadatas began a galactic exploration. Bodies such as gas giants and moons require new common-sense rules of operation. Every planet has required a new program instructing our robots' behavior given the new standards in the environment. Simulations of atmosphere and gravity of a world about to be explored helped prepare the robots for a transition. In addition to input values, cadata robots can collect data from a new world into their own system and then utilizing this knowledge to calculate the new rules of best-behavior. For example, a hyper-high-gravity world might necessitate a different style of motion (such as crawling as opposed to flying), or a low-pressure world might mean refraining from using water inside a robot's system to keep it from boiling. A human or a cadata can struggle with re-adjusting walking patterns on a gravitationally divergent planet. A passing robot must auto-adjust gate of movement maneuvers, and strength of exertion and endurance to function with equal efficiency regardless of the conditions. Current models of cadata robots and nanobots evolved over a million years of programming and engineering. There were plenty of lopsided and upside-down buildings built by cadata robots before such problems were ironed out. Though, the upside-down house idea has become a classic. The materials comprising robots change to fit with the conditions and ingredients available on each new world under investigation. A robot might need to be composed of biodegradable gelatin to self-destruct into the elements if it cannot repair or return to the Mothership. Degradability is required on planets such as Earth with runaway-pollution. Alternatively, a hot planet necessitates robots to be built from the strongest accessible metals. Their power sources are also adjustable. Electricity and magnetism can fail on a planet without a magnetic field, or with heavy electric storms, necessitating the invention of a yet-unknown fuel-source. Robots can build themselves up to fit the environment based on information gathered by nanobots we send to derive suitability parameters.

Cadatas' robots can also be categorized with help from Michio Kaku's three levels of consciousness. Human scientists propose a robot's consciousness defines its artificial intelligence. A robot is more conscious than an organic creature because its perceptions on all wavelengths maximize at a machine's limit, which is greater than levels achieved through evolution of biological organisms. At Level 1, Kaku proposes, an organism or a machine has a set number of feedback loops, or processes leading to a preset result. Level 1 structures measure anywhere from one to several components in its universe, be it the temperature, radiance, direction of gravity or other useful elements. Kaku separates these from Level 2 beings by the latter's ability to be aware of themselves in space, and thus navigating their terrain to seek prey; in contrast, an organism can run on instinct only when it detects a threat. The highest level Kaku allows is 3, which encompasses social animals, which know their place in a hierarchy and can communicate their emotions. Within this framework, cadatas' robots possess Level 3 consciousness. They are programmed to measure an environment with their detecting instruments. Cadata robots are aware of time and space, and their location in it, their destination, and various other space-time complexities. But does a robot need to be aware of its place within a group? Cadatas need subordinate robots, which efficiently execute commanded labors. A profitable robot must be aware of its lower hierarchical status in contrast with cadatas. Intricate social rules are also included; for example, if an impoverished cadata orders an unpurchased robot to perform a task, the robot must distinguish this cadata's lack of authority to ignore the irrelevant directive. On the other hand, too much self-importance, or the perception of its own rights and desires, can create a confrontational and disagreeable robot. This is another instance when mimicking cadatas' emotions and behaviors is illogical for a robot. Unlike human ideas about advanced robotics, cadatas have not made robots look and feel organic or like cadatas. Giving them the texture of cadatas' skin or adding textual nerve sensations for enjoyment rather than for performing environmental measurements is a wasteful use of resources; it encourages self-importance and self-indulgence conflicting with their cadata masters' desires.

Cadatas have not seen any robotic rebellions against these rules of conduct because their programmed self-awareness does not imprint humancentric or cadatacentric notions of organic desires for sensual gratification. While eating, breathing or orgasmic release is illogical for a robot, the ability to communicate findings or to ask for input from an organic lifeform is essential to a robot's function. Cadata robots communicate on heightened linguistic levels. They speak, read, and write linguistic signals with greater efficiency and correctness than biological organism. A dictionary is stored in a microscopic spec of their memory. They code-switch between languages, accents, styles, and wealth- or poverty-based shades of meaning. Cadata robots are programmed with video-graphic history, allowing them to review recordings to determine the successful strategy. To adapt to a new planet, robots program themselves to adjust for the variations their instruments detect in the atmosphere, or in the behaviors, traditions, or machinery of the settlers they are visiting. Cadata robots perform cadatas' manual labor, without the drawbacks of job dissatisfaction from a subordinate. If any civilization created a robot without a subroutine to prevent the robot from killing its makers; it has justly disappeared into oblivion.

Each of Cadata's robots can perform universal functions. It has instructions in its databank on all known job categories, such as mining for resources, and construction of buildings and shuttles. The universal robot can alter its body to fit the assigned task. This job can be singular and specific, such as locating gold on a planet and bringing it to the owner. It can also be more general, encompassing numerous steps in a pre-designed program; it is requested with a simple command such as "build the Colosseum in that field" or "create a breathable atmosphere in this house" or "terraform this planet's atmosphere and land to make it compatible with my biology." These can be completed; but the latter requires the robot to make numerous duplications of itself, and would take several millennia.

Cadata robots exhibit some randomness, which might be called creativity, when performing any task; thus, orders must specify all relevant requirements. If you are overwhelmed with the array of choices, a robot can display pre-made design concepts from which you can select a model. Several businesses create new designs sold through all robot models; earlier designs are offered as bait to tempt later purchases for free in the public domain. Any design can be edited to fit with a specific terrain, needs, and budget. Plans from alien house and vehicle structures can also be shown to the robot. Typically, the robot needs a book-equivalents on local construction methods to duplicate a design requiring unique building theories, ideologies and materials.

Once Nally's *Catalog* of potential places was finalized with data gatherable from Nalut, it was time for CSEA to send nano-spaceships to the closest of these potential settlements to double-check long-distance findings. We could not wait for 100,000 years for these nanos to travel to Earth, but waiting for a few decades or even centuries was reasonable to avoid cadatas taking fruitless trips. The nanos are small enough to travel unobserved by multicellular lifeforms. They harness matter and energy from space with a magnetic field to accelerate to light speed faster than a spaceship. Their lightness and mechanical nature minimized their acceleration energy expenditure and maximized the potential g-tolerance. Once they reached the desired orbit, they took measurements from this edge of space to check if they had to adjust their structures or makeup to adapt to conditions on the planet better. Once they gathered the relevant measurements, and made the needed adjustments, they slowed, and made a controlled fall maneuver to a planet's surface. After a landing, softened by propelling a deacceleration, they began flying, swimming, crawling, and moving about to explore the types of reachable environments on or below the surface. Aside from geology, they searched for lifeforms, machines, or abnormalities to understand how these functioned, what their motivations were, and how an element might have impacted a future visit from cadatas. Once their extensive memory storage was full, they flew back to Cadata to deliver the data to our scientists for evaluation. Aside from the data, nanos brought back DNA, or the equivalent molecular codes. These were processed by cadatas to design suitable gene-editing in preparation for the forthcoming pre-departure genetic transformation. Samples were also taken from hostile organisms. This compound evidence proved these proximate worlds were unsuitable for habitation.

The Mission Committee

Before CSEA, under the direction of the Potentates, could open a contract competition for potential offerors, they formed the Mission Goals Committee to debate the direction of the exploration. The Mission Committee met for this purpose at CSEA Headquarters, in the center of the devastated Alleges, a year after the asteroid strike in 100,162 BC. Robots had repurposed the rubble from the buildings swept away by the tsunami, or crumbled by the wind-blast. In a year, with robots replicating themselves at maximum speed and working non-stop, they only managed to rebuild two buildings, the CSEA and the Potentates' headquarters, as these were marked as essential for the planet's survival. The robots failed to clear every damaged building because some were housing essential personnel for the rebuilding effort without resources to rent. The Potentates would have evicted these unpaying guests, but doing so would empty the city of cadatas, forcing them to manage the robots unassisted. A dark, dust blanket was still shielding the sun, so the floor-to-ceiling view from the Committee's meeting room was bleak.

Commander Doul Veahy opened the floor up for initial comments. Three Potentates raised concern regarding the cost of a galaxy-capable spaceship. As the first expedition to a habitable planet outside their solar system, the design of the spaceship had to be original, rather than a disguised plagiarism of an earlier engineering template. The ratio of plagiarizing incompetents to creative scientists in the Potentates' businesses was skewed towards the nincompoops. Thus, to achieve this leap, they had to hire an army of scientists, hoping one will be an "original thinker". They also reminisced on Kerod Retan's disastrous comet-attached flight to their closest planetary neighbor in 102,546 BC. They watching his feed as he self-destructed by crashing into the hostile planet's low-pressure, low-atmosphere, and freezing surface for "air" without a spacesuit in 100,234 BC. His behavior attained peaks of inappropriateness at the end still classified as top-secret, but the gist of the cause of his demise has been declassified in the interest of scientific research. That mission had cost CSEA the equivalent of NASA's 2018 budget or $21 billion. Developing a flight outside their stellar neighborhood was guaranteed to cost ten times this amount, and would be making CSEA the top beneficiary of the government's funds at a time when the majority of Cadata was devastated by the asteroid strike and needed emergency funds. One Potentates interjected the likelihood of a revolution if they ignored their needy population. They had maintained control over most of Cadata for over a million years. But their supposed negligence regarding the asteroid threatened to topple their rule, if they failed to perform sympathy. They discussed how unrest might build on independent islands running their own affairs, without interacting with the larger Potentates' governing agency. Refugees from destroyed regions might travel to these islands instead of the Potentates-controlled cities, if these rebuild at a faster rate because the Potentates invested too much into CSEA's Habitable World Exploration Project (HWEP). Despite progress, the discussion ended in a stalemate with the Potentates intent of retaining their money. On behalf of CSEA, Commander Veahy was insisting the HWEP project was essential. Other parties voiced alternatives, objections, and redirections that moved the conversation into areas unrelated to the decision on HWEP.

Finally, too frustrated to continue the debate, the Potentates called the meeting "adjourned", ordering Commander Veahy to compose scientific questions to solve on the first exploratory trip to a habitable planet outside of their solar system. Could they predict the inherent difficulties in this interstellar proposal and could these challenges be annulled? Veahy observed there was a single, obvious motivation to go: Cadata was becoming a wasteland and they lacked a rejuvenation plan. According to Veahy, any challenges to this goal had to be overcome at any cost. The Potentates insisted he was dodging his responsibility as the representative of CSEA by avoiding the task of compiling questions needed to evaluate the worthiness of the project. Commander Veahy had to concede, agreeing to put his team to work on this list. The meeting adjourned.

The Committee failed to meet again across the next twenty years. In the interim, CSEA put their resources towards creating a set of questions to convince the Potentates to go forward with the competition for the HWEP mission.

The Mission Committee met again in 100,142 BC. No revolution had sprung up. Alleges saw a good deal of out-migration. Independent islands gained power in the new restructured Cadata. The dust cloud was still hovering and causing a spike in various connected illnesses, and deforestation and the extinctions of various species across Cadata. The situation necessitated a response, but the world remained intact despite twenty years of procrastination. Commander Veahy did a presentation at this awaited meeting of the questions his team developed. First, can the best-suited for habitation planet from the *Catalog* support their entire population? Given the limitations of their resources, can Alleges build enough spaceships to relocate every individual off Cadata? Should Alleges only relocate their own citizens or also commit to saving those residing in other jurisdictions? What are the expected characteristics of lifeforms on their top-choice planet? Can the hostile of an alien species be determined from atmospheric conditions?

From the *Catalog's* data, they knew the contents of the atmosphere and the rate of its escape, the surface and crust contents, and the geology of this planet, which Nally had called Ganeir. But what impact would cadatas' arrival have on local lifeforms, and how might these lifeforms impact cadatas? Did scientific equipment have to be reengineered to fit into the new environment? Can their current gene-editing abilities adjust for gravitational differences, or did they need a new gene-editing process to meet the demands of extreme-environment mutations? Can the spaceship be designed as a swiss-army-knife, mutable to fit the temperature and molecular composition on Ganeir? In case of an instantaneous catastrophic propulsion system loss, could the spaceship's engine or gravitational-controls be swapped out, with a new unit printed and inserted with sufficient speed to evade a stranding or decompression in deep space?

A discussion on each of these questions followed, as educated arguments, and wild guesses were proposed. Veahy prepared several potential replies to solve each of the problems raised by the questions. Veahy's notes were sketches to inspire potential bidders, if the Committee went ahead with offering a competition for this mission's contract. The size of the spaceship was a hot debate because it decided if the Potentates would contribute $210 billion, or a lot more, or less. A gigantic version would feature breathtaking, enormous architecture to house a sizable crew, with a forest to generate the nutritional-gases for nourishment, and entertainment arenas to meet diversionary needs. On the other end of the size spectrum, it could be a tiny shuttle for three cadatas, containing just enough on-board for them to run the cheapest possible experiment in survival-exploration. Leavy explained, if the spacecraft was on the lighter end, it would be easier for it to achieve near-light speed. However, with merely three on board, it would lack the resources to explore the complexities of Ganeir. If this team was to miss a major danger in the oceans or skies, the whole planet might migrate there on their rushed recommendation, only to encounter a major catastrophe because of an undiscovered virus or some other hidden pest. Given the numerous presented choices, the Committee took another needed break to contemplate them. They reconvened again ten years later, in 100,132 BC, and voted on a middle ground for size. They also agreed on the call for proposals from contractors, which Veahy composed. The call solicited a spherical aerodynamic spaceship to minimize drag during flights through atmospheres or solar winds. It also solicited for the development of engines utilizing specks of dust and light, matter and energy found in open space. The call required the spaceship's attainment of light or near-light speed to avoid driving travelers insane. Slower travel was blamed for Kerod Retan's breakdown. Maximum velocity was also necessary to expedite a solution, in case Cadata's environment was headed for imminently desolate inhabitability. The Committee approved issuing this call for proposals, and released it for competing entities' review.

Securing a Contract for Interstellar Exploration

Nally Subsenar was still a young cadata in 100,132 BC, with 3,156 years behind her, when she read the solicitation for a spaceship design. The independent lab where she had designed the Nalut Expander, and the *Catalog* was destroyed by KoraFePub. Her house and valuables were also lost. She was searching for an opportunity to recover. She was living out of a mini-shuttle a robot she managed to rescue built under her instructions. She subsisted on air her own mechanical inventions collected and compressed. The solicitation for the Ganeir spaceship was the first major space exploration project from CSEA since the absolute catastrophic failure of their asteroid mission. Nally suspected it would be the last if this one failed to deliver. She was well-positioned because the Committee credited her *Catalog* as instrumental in the Ganeir choice. She weighted her options, searching for the financially and professionally rewarding approach. To anticipate the government contract competition, she tapped into what she learned through working for Dik Sibost (who died with his lab) for a millennium.

Nally learned of eight teams planning on submitting proposals. For an untrained eye, the eight appeared to have an equal chance at success. They had popular scientists on-deck. They had succeeded as business entities, with sufficient capital to invest in research and development. Their pre-proposals appeared to offer innovative ideas with benefits unmatched by the others. However, only a naïve applicant could have missed the ownership of one of these businesses by a Potentates, Loor Dowor, of the Robotics Corporation. The sub-entity working on the proposal was called Examen Initiative, and their marketing refrained from stressing they were owned by the Robotics Corporation, instead emphasizing their independent spirit and new-thinking. They publicized this affiliation only because it intimidated competitors. Their standing already won them the Kerod Retan comet mission and the major of all space contracts. Examen had a near-monopoly on previous awards in this field. They were even responsible for the attempted KoraFePub "deflection" attempt, though they did not include it in any of their competition materials. Regardless of failures, the number of awards they won translated into "experience", the main evaluation category the companies were competing for. Winning prior awards assured a higher likelihood of winning again as well as a bloated reserve of previous profits to invest in a new competition. Even if other competitors hired the best teams and created the best proposals, Examen's experience in the field allowed them to keep winning.

Despite having this giant in the race, it was not a simple choice. One of the other two powerful competitors was Pioneer Mechanics. It was based on Paroter, one of the independent islands emerging in status because it was on the opposite side from the crater, a region least-affected by the impact. They had won contracts for the reconstruction of Alleges from the Potentates government because they had the resources to tackle them, whereas the Potentates' equivalent entities would have failed to deliver even if they won a contract. One of the reasons the Potentates delayed the contract competition was to allow their Examen Initiative team to rebuild to be a fit choice by the start of the race.

The other rival emerged on one of the few surviving independent businesses in Alleges. Matchless Fabrications rented a space in the Potentates headquarters' basement. They had helped design robotic programs to assist cadatas with creating temporary habitation solutions (squatting in luxury as it might have been called on Earth). They also dabbled in many other essentials, including selling nutritional-air. Their lack of specialization created a team of generalists capable of working in a range of spaceship building, operating, and maintaining fields.

The other five competitors represented a few different regions of Cadata with different specializations. Though Nally was convinced, even with the best, most efficient and cheapest ideas, it was impossible for them to win. They offered variations on fuel sources, strange designs for arranging consumption, resting, and working spaces, and different plans for slow-down and speed-up rates and stages. One argued for increasing the g-force up to the maximum cadatas could endure. Either the cadatanauts would withstand this force in discomfort, or the ship could generate an anti-gravitational field to counteract this pull. Another proposal suggested short bursts of extreme acceleration, only survivable in these tiny intervals. An alternative proposal argued for stopping for refueling and amusement mid-trip, in contrast with most supporting a continuous trip without stops before reaching the destination planet. A proposal attempting to win by minimizing the budget suggested a spaceship on the smaller and lighter extreme with a less extravagant energy-generator. All eight teams had a similar breakdown into subteams handling singular aspects of the proposal. A geologist and geophysicist would research the makeup of a planet to derive the adjustments to genetics, robots and inanimate structures needed to adapt to life on a potential new world. They searched for laser, radio or pollution emissions and near-infrared photosynthesis indicators. A related but separate position handled the planet's chemical composition to determine what resources could be utilized in rebuilding the components of their civilization. Another team tested Ganeir's atmosphere for its natural suitability with cadatas' biology, or if it had to be terraformed. An additional team researched hostile-to-life in excess plasma, solar wind and alternative particles present in a planetary, lunar, or solar region. For example, a superfluity of radiation might penetrate to the surface of a planet because it lacks an atmosphere. The design of the ship had to reflect any abnormalities. The measures were abnormal by Cadata-standards, but some were within the survivable rates, while others had to be addressed as major obstacles.

After narrowing the list to three hopefuls, Nally had the unenviable task of researching if any could afford, and was willing to surrender colossal bribes to the competition administering CSEA officials. Nally would have preferred finding a job doing work as a background researcher for a guy such as Dik Sibost, who needed a workaholic to do the bulk of the job, while he handled the dirty publicity, and his assistant handled the contract applications. However, amidst devastation, no such stellar research lab remained with enough funding to afford a salary even for a hyper-productive researcher, such as

Nally. She either had to take a leading role in manipulating the contract bid process, or she would remain homeless, and trapped on a dying planet. It was as illegal to offer bribes on Cadata as it is on Earth. Because of this legal hurdle, the five weakest competitors lacked the capacity to contribute underhanded money without being observed and condemned. The winner gave so lavishly, it had the appearance of a court banquette billable as an entertainment expense, rather than as a dirty bag of cash. The process of turning an illegal bribe into a legal donation cost much more than the worth of the resulting gift. The losers risked imprisonment, bankruptcy or worse, so Nally had to anticipate before the race started who the behemoth winner would be. She had to predict who could dedicate half of the budget to the bribery process. The winner had to start with a higher savings and loans balance. Given the projected $210 billion budget for the Ganeir mission, the teams spending $105 billion on the application, but failing to win the reward would be bankrupt. Financial guesses were insufficient to the task, so Nally hacked into the financials for each of the eight teams for the exact current numbers. Five teams did not have and could not access in loans the needed $105 billion. As she assumed, these five quit the competition just before their expenditures were set to spike. Their communications indicated they had started the process for the press coverage to attract future contracts. The other three competitors borrowed to push into the next round, staking their assets. It was a race to zero. To win the pot, the top competitor had to spend the most on bribes, retaining only a tiny profit out of the enormous government-award. The winner's job was to stretch this remaining smidgeon into the scientific miracle of delivering Cadata's first habitable-world-finding spaceship. If the winner failed to deliver a one-way-capable spaceship; it would slip into insolvency due to CSEA fines for failure to meet contractual terms.

Nally's decision was swayed by the Potentates' control over their dozen corporations for over a million years. A Potentates corporation would win even if its branch over-extended into this ambitious competition. She applied to work as the project lead under Loor Dowor's Examen Initiative. She was accepted because there was nobody else on the planet superior to her in effort or creativity. Dowor was shocked to see her application because she was known as a champion of transparency, and morality in business practices. She had never allowed the testing of her dignity in a role turning her into an instigator of competition-fixing. He asked her leading questions during their interview to gauge her intentions in this regard. She confirmed she was betting on a bullish win rather than a rebellious loss.

Nally hit the lava hard both on the research and development front and in crafting a bribery strategy. Her team's design led the competition as the first Connoisseur Committee Evaluation approached. The field still registered the original eight teams. Nally attempted an untested strategy in the hope of overcoming the central problem in this process; she choreographed a pact between team-leaders, price-fixing the bribe-level at a low point to avoid runaway-spending. If they could all avoid wasting $105 billion on bribes, they would all avoid bankruptcy, and the winner would see a lot more profits. If nobody bribed the committees, they would have to base their choice on merit. This glorious theory was proven idealistic when in response to this bribery-freezeout, the Committee rejected all proposals. Without checking with the others, Nally guessed the freezeout was over, and commemorated by unleashing a spending-spree on the Committee to lobby to reinstate the competition. Two decades passed before, in 100,107 BC, the Potentates and CSEA pushed the Committee for reinstatement.

Withholding the reinstatement announcement from five competitors, the Committee notified just the top three competitors with the highest bribe-levels. Then, they set a truncated deadline for interminable proposals. The timeframe was the shortest in history, while the minimum word count was the longest. The five companies with the smallest budgets or the least corrupt leaders failed to match the volume of materials churned by the crookedest companies, forewarned in 100,110 BC, three years in advance, of the convoluted proposal guidelines. The Committee combed through the resulting proposals for tiny errors generated when any researcher rushes to complete a report. To further increase their odds of success, Nally even broke into her five poorer competitors' computers. She introduced errors. Thus, even if their reviewers had spotted most of their own errors; there would still be typos left the Committee could point to as reasons for disqualification. This ploy succeeded in pushing the five "disadvantaged" competitors out, allowing Examen to lower its bribery budget.

With the competition restored, the spending spree continued as Nally kept searching for the most extravagant and convincing gifts, leaving the scientific planning to her assistants, who lacked motivation to change Nally's initial designs. The members of the deciding Committee became the wealthiest cadatas in history, after the Potentates.

Nally invented variations on the traditional bribery and kickbacks in contractual negotiations. In one of these schemes, committee members were awarded a shuttle by the separate Toppold Shuttles transportation branch of the encompassing Robotics Corporation. Because of the ban on direct-gifting to those voting on a corporation's design, these shuttles had to be presented through a complex financial arrangement. Toppold Shuttles reached out to a bribe to offer a shuttle at a price lower than the manufacture-cost through the Toppold factory. If the bribee lacked the money to cover this minimum cost, they were offered zero-rate financing for a year. Thus, they paid nothing for a year, and then they were offered a buy-back at the full retail-value of the vehicle. The sale netted a huge profit without paying a cent for the original shuttle. Before the sale closed, the same dealership would offer yet another brand-new shuttle to the bribee, repeating the process. If questioned by the media, Toppold explained these special "sales" as rewards to respected government officials for their public service. The extreme shortage of shuttles on Cadata in this period, heightened the value of these exchanges. Aside from luxurious transportation features, shuttles were the safest places to live to minimize exposure to toxic air outdoors. Nally organized similar kickback deals with houses and lavish temptations for the media, the Potentates, the safety inspectors, and less recognizable players.

Aside from these massive payouts, Nally also organized entertainments and social engagements to satisfy the bribees' less materialistic needs. It was legal and expected for the committees to be wined, dined, and socialized as part of the contract-wooing process. Parties composed the bulk of a committee member's schedule, as they migrated between social engagements each of the three teams staged to out-maneuver rivals.

Some of these socials counted as research and development hours even if the members of a team were intoxicated before they arrived and did not have a computing device or a robot assistant between them to record the "ideas" they spewed out. Nally was obligated to attend, causing her to lose her typical extreme self-control, until she developed a slothmoth addiction. Sedated, she would pretend to scribble science in her personal communication device if Loor Dowor, her boss, joined the party; though he was always pleased to see her slothmothing as she became more approachable and malleable. Loor indulged in intoxicants at the parties. They whistled and danced, deaf to each other's nonsensical babblings, utilizing instinctive, non-verbal, primitive parts of cadatas' brains, the parts responsible for sensual explorations of smells, touches and sights. To avoid clandestine drug-dealing meetings, Nally hired the slothmoth saleswoman, Welkah Liny, as a full-time member of their team with the title Chief of Entertainment. Welkah's salary consumed a significant percentage of their entertainment budget; Nally stipulated, off-record, a portion of Welkah's exorbitant salary was in exchange for the purchase slothmoths in bulk to cover the team's demand. Examen's volume of slothmoth consumption was less than the competing teams, which fried their brains to lose the competition.

Right after Examen eventually won the Ganeir contract, near the end of one of these slothmoth-laced parties, two cadatanauts dared each other to a flying competition in the spaceship prototype the team just finished assembling for an upcoming demonstration before CSEA. It was a miniaturized version of the final product, but it still cost an incredible sum to complete and if either of them had any conscious brain-function, they would not have touched it without an army of overseers at the command post monitoring its behavior. This party happened in a bar on Cadata's moon, where the prototypes were being built, and tested. The cold moon was used because its low-gravity, and zero-atmosphere made liftoffs, and landings energy-efficient. The bar was operated by CSEA because it was the agency in charge of the complex where the ships were docked. It was on the opposite side of a series of spaceport gates. Standing was unnecessary as the cadatanauts' relaxed bodies were carried by air flows through the empty hallways. The nano-controlled air flows made turns on their behalf based on the biological IDs indicating they were stationed on the Examen Initiative Gate. Once they arrived at their destination, robots adorned them in skin-tight spacesuits, protecting them from the atmosphere-less conditions in the short stretch of the moon's surface they had to hop across to access the shuttle, which was parked at the edge of a runway. The shuttle's hatch auto-opened as cadatanauts jumped in. Cushioning devices auto-secured their bodies to prevent turbulence damage. All systems automatically turned on upon detecting the presence of pilots. Then, the Computer asked for their intended destination.

"Flips!" Cadatanaut Codol Havould commanded.

The Computer was programmed to obey first, and adjust later, if instructions were nonsensical. Thus, it pulled away from the parking bay and departed at the standard speed, acceleration, and outward direction consistent with the temperature and wind conditions outside. Once it was clear of the moon's mountains, the Computer began executing standard flip formations.

"This is boring," the flight's official transcript has recorded Cadatanaut Putair Niatage saying, as he directed, "Computer, give me the controls!"

The Computer complied, letting the shuttle continue making the same loop over and over again in rapid succession without redirecting it by a single degree. Earlier, the team experimented with the Computer switching off control over all functions when ordered to surrender controls; however, the cadatanauts failed at manual execution of the needed maneuvers with sufficient speed to recover from a freefall. Niatage yanked on a holographic 3D control wheel to exit the turn; this redirected them toward one of the moon's deepest craters. Niatage laughed gingerly and absentmindedly at this little kerfuffle, and executed some other random motions until the ship was being tossed up and down, side to side, then upside down as they flew. Also amused, Havould took over the controls. The Computer's safety protocols recovered them from the deep dives threatening to end in a crash. The backseat-driver save was seamless, so it went unnoticed by the pilots. They might have had a safe conclusion to a grand time competing for the weirdest combination of unnatural movements, if a volcano had not decided to blow as they were flying near it. A bit of lava hit the ship, and soot and smoke clouded visibility. The Computer was showing the cadatanauts warnings the volcano was about to blow, but they were inept to read these in their condition.

The Computer now switched to vocal screeching warnings at a pitch intended to grasp even the most distracted pilot: "You must surrender the controls! The spaceship is experiencing a loss of integrity due to the impacts!"

The spaceship was shaking, both from turbulence and because the Computer was shaking them into waking their intellects.

"Stop yelping!" Niatage exclaimed, as he was making pointless maneuvers leading them further from the CSEA complex, when they should have been attempting to return to assess the damages. Even direct lava hits did not do sufficient damage for them to lose integrity or propulsion, as the ship was designed to withstand large asteroid impacts at the speed of light. Niatage attempted another radical triple-sideways-flip, and in the midst of it he could feel the Computer cushioning the intensity of the maneuver, so he ordered: "Turn off Computer!"

As he was emerging from his slothmoth-haze, Niatage realized this was the sole command capable of deactivating safety protocols. Meanwhile, the turbulence rocked Havould to sleep, so he did not object on the Computer's behalf. Niatage did a few tricks defying explanation in human terms. On the last trick, he came so close to hitting the erupting volcano, his sense of self-preservation returned. He glared all around with his eye and checked the digital map for their location in relation to the CSEA complex. He was still too hammered to order the Computer to fly them back. He flew one way, realized it was the opposite direction from his intended path, turned around and coasted towards the vicinity of CSEA's direction. He made hazy adjustments as his marker swayed off-course. Finally, he saw the guidelights of the runway below. These were needed as they were on the side of the moon facing Cadata so little sunlight penetrated this base. He tried to align his breaking mechanism with the parking station, but could not balance the ship to fit with it. He had tried docking in simulations and test flights before, but on a sober mind and with the Computer's help. He rose up, hitting the breaks every minute. After an aggressive upturn, he slammed the ship down until it hit the docking station of a competitor's spaceship.

The hit ship belonged to Matchless Fabrications. Since Matchless had been fighting the Potentates' corrupt business practices for centuries from the Potentates headquarters' basement, Matchless first accused the Examen Initiative of sabotaging their model. Matchless' prototype from the competition was parked next to Examen's because they had been hired as a sub-contractor to delay their looming bankruptcy declaration. Matchless had bribed officials into this deal. Examen was forced to pretend to welcome Matchless' participation despite anticipating Matchless would create problems for Examen in a final attempt to convince the government Examen was incompetent and Matchless was a better fit for the award. As a preemptive, Examen stalled Matchless' ability to construct their sub-contracted components to prove Matchless deserved to have lost the main contract. Examen drowned Matchless with volumes of report writing and editing. By requiring these reports first, they made sure Matchless was unable to begin the construction CSEA paid them to build. In this hostile climate, it was logical for Matchless to assume Examen stooped to violent sabotage by hitting the spaceship, even if it was only a prototype without the element Matchless was under sub-contract for.

The blow of the impact was softened by the ships' coating, keeping both unscratched. The damage was from the lava. Nally was grateful this impact occurred because it allowed her to prevent similar lava damage going forward. As a result, Nally's design stood out among the competition because of its extreme resistance to imaginable impacts at all velocities, heat-signatures, and densities. Niatage and Havould were promoted to partake in the first test-flights of the final spaceship in exchange for this behavior, which reflected glowingly on the team's ability to entertain bureaucrats, who were delighted to discover the two spaceships sideways and an investigation afoot on the following morning.

Niatage and Havould were flattered by the attention they received in the press. Seeing how they enjoyed the coverage and were photogenic enough, Nally set them up as the new cover-faces for the Examen Initiative's Ganeir proposal. This meant filling their schedule with visits to every bar, restaurant, hotel, conference, retreat, submarine vacation and private party across their solar system. There were photo opportunities at the stops, and these images traveled across Cadata, promoting the cool and endearing image for Examen. The publicity was aimed at diverting the public's attention away from recalling Examen had caused the irrecoverable disaster. Niatage and Havould were awarded a space shuttle programmed with this non-stop-party schedule. Their job was reclining and watching the scenery as they were escorted to the next shindig. The hours they spent in the shuttle counted towards the required number of hours they had to spend training in shuttle flight to qualify to serve as pilots on the Ganeir mission. Each stop was covered by Examen's entertainment budget. Costs including travel and nourishment were also covered for public relations representatives, committee members and bureaucrats in need of buttering.

To make these tours more productive for Examen's chances of winning the contract, Nally also recruited female secretaries to attend from their enterprise whose job it was to persuade the bureaucrats with sensual enticements. "Secretaries" is far from an exact translation of the Cadata word for this broad category, which includes communication assistants, clerks, waitresses, stay-at-home caretakers, and various other feminine assistants. Regardless of their job titles, these secretaries' real duties include performing sensual favors for clients, whoever they might be for a given project. The secretaries are sex objects offered as a bribe or kickback to sway a wayward player. They fall on both sides of the sensual extremes. Some are desensitized to sensations of all types after thousands of years of indulging fantasies. Others are still exploring and curious about extraordinary sensual experiences and enjoy these exploits into the clients' desires. One of the reasons secretaries are employed in these roles is because secretarial work is redundant on modern Cadata. Tasks such as data entry are handled by computers, who are much more efficient and productive assistants to executives. While Nally was called Dik's "assistant", this was an unequivalent arrangement, as this category of workaholic employees are called surrogate-bosses. They perform the functions of their boss under the disguise of a subordinate assistant. Top-skilled but impoverished and anti-sexual performers tend to fall into this role because it avoids the bribery and backstabbing involved in securing an official "boss" title. The sex-secretaries are offered at all stages of a contract-application and execution process to all relevant parties. Once a team wins a contract, they are solicited by secretaries from sub-contractors, willing to help in exchange for a slice of their enormous budget.

Some of Nally's "secretaries" were hovering over the press who were following Niatage and Havould's travels among other newsworthy events related to Examen's Ganeir proposal. Examen's oversized budget included a substantial chunk to be spent on manipulating the public to believe the contract-award process was fair and beneficial for all. Nally's Public Relations Office was one of the most hard-working units in the organization. This team was writing fake letters to editors, opinion pieces, and blog-like posts on Cadata's version of the web, which lauded Examen as a pioneer of innovation breaking records and inspiring youths to enter space exploration. Without such an "outpouring of support", committee members might have been hesitant to accept bribes from Examen. In other words, errors in Examen's designs would be detrimental if the public clamored to oust them as villains responsible for the catastrophe. Corrupting the contract application process would be a punishable offense if through it an evil entity was condemning cadatas to a terrifying death. This is the follow-the-leader phenomenon: if a comic pays three cadatas in an audience to laugh and clap on cue, then the rest of the audience follows this example, until the performance appears to be hilarious. In contrast, a rival might hire a few choice hacklers, who might boo another comic off the stage. Nally engaged in both tactics, as the Public Relations Office (PRO) also composed false accusations and various other negative information about the rivals' staff and sent these letters to the same newsrooms.

There were several other practical uses for this office, tasked with drafting storylines helpful for Examen's bottom line. When CSEA cancelled the contract competition for the Ganeir mission, the Public Relations Office sent thousands of fake letters protesting the cancellation and insisting on its reinstatement. These pressure-tactics did not always work, but they were always stressful for the party being pressured, and thus helped to move Examen closer to its goals. The PROs' more intricate function was writing scientific white papers, which were fed to scientists in major government research associations, who released them as if they were their own words. These ghostwritten research papers supported Nally's scientific arguments in the essays she published in other journals. They also served to discredit rival proposals. These misinformation and re-education campaigns played such an important role in convincing the public and the government to side with Examen, the PROs wrote press releases before they made decisions about where the design was headed, or tested new components. PROs wrote scientific papers into recycled templates with blanks to insert relevant information on a new sub-point or a small update to previous Ganeir-related findings. PROs' programmer also created a scrambling-algorism, which rearranged a collection of theories regarding new-world-dis-covering flight into a series of seemingly original research papers; the scramble was ad-vanced enough to fool advanced plagiarism-detecting programs. On top of writing short newspaper articles and research papers themselves, PROs also contracted outside film-makers, comedians, and general entertainers to disseminate their propaganda into pop-ular platforms. These creative people made songs, whistle compositions and sculptures glorifying the superiority of Examen's spaceship. In fact, this book began for-hire as part of this continuing effort by the PROs, but I chose genuine research over the offered rewards. Propaganda-spreading is one of the few jobs cadatas must do themselves, as robots and computers programmed into lying and subterfuge cannot be trusted to act in an owner's interest instead of their own.

The bribery-race spiked the total proposed cost for the project to $290 billion, a sum above what the Potentates and CSEA could entertain. Again, they closed the competi-tion without a winner. Thankfully, the $290 billion price tag was based on anticipated bribes rather than those already expended, so the teams managed to lower their budgets closer to $210 billion, allowing the reinstatement of the competition.

Nally's determination to win was unusual among cadata scientists. Nally's mentor, Dik, was a typical example of this drive-to-lose. A typical Cadata scientist hoped a re-search project attempted would go nowhere because as a contractual employee, he or she made money even if they spent the millennium doing nothing. The quest to ap-proach doing pure nothingness was urgent for scientists who cheated or plagiarized their way into their positions. Dik was fortunate to have Nally to perform industrious-re-search and to write his scientific papers. Scientists who did not have the luxury of such near-selfless assistance, hoped to be allowed to pretend to work without working for as long as possible, without objections from their employers. Of course, their businesses approached bankruptcy as they succeed in doing nothing. So, these scientists had to seek out cheap report writers, if they could not find full-time employees. Either way, they had to give the appearance of being great for business even as they were flushing the business down the volcano. On average, cadata scientists and business leaders willing to bribe also fail in competitions against the hyper-intelligent fighting with superior products, ideas, and research. Incompetent scientists would be incapable of delivering an interstellar spaceship if they had won this competition. Losing is their sole recourse. A loss meant retaining wages for the attempt without being discovered as frauds who advertised their scientific know-how only to bankrupt a company through inaction. Thus, if Nally was the single lead scientist intent on winning, the game was slanted in her favor. The other teams' dimmer scientists were interested in the length of the competition rather than its conclusion. They artificially created a 300-year delay, stretching the competition until 99,782 BC, as the contamination of Cadata's environment continued; by this year, a gas mask was needed to glide outdoors. Seeing their profit motive superseded by the threat of an early death, the scientists destroyed all proof of the maneuvers they engaged in to engineer the delay. Once they allowed the competition to play out to its climax, Nally won the long-fought-for victory.

To secure this win in a manner profitable to Examen going forward, Nally had to pull together what is known in the industry as the Nonsensical Proposal. While Nally sabotaged competing proposals by inserting small typos into them, there was no limit on the typos that could be left in Examen's proposal. Before submitting the application, they knew if the votes favored their team based on the number of members they bribed. The submission act was redundant. Whatever type of a spaceship they concocted would have to be tested on the voyage. The proposal was a fiscal vehicle allowing scientists to begin the real research and construction. CSEA stipulated a 100,000-page minimum length for the report.[1] The findings had to be illustrated with intricate designs, and supported with equations only computers could interpret. The other required components were the plan for managing the scientists, the timeline for execution of the production steps, the budget allocations, and the exhaustive background information on each of the key members of the team. A semblance of this was on the surface of the Nonsensical Proposal. However, its true function was to act as a linguistic disguise for falsified data-findings, budgetary abnormalities hiding bribes and unflattering evidence only palatable to auditors if it was written in a style impenetrable to readers. A specialized linguist, Suras Kebers, crafted these webs of confusion after Nally handed her the 1,000-page draft of the factual version of the report. She employed tricks common to human scientific literature, such as repetitions, cyclical reasoning, introduction of unknown or unrecognizable words, and burying crucial meanings in the middle of lengthy paragraphs jumping between unrelated ideas. She also utilized convoluting elements about the Cadata language; one of these is jumping symbols, wherein words or parts of words move to the other end of a paragraph while a reader is trying to re-read it. Additionally, Suras inserted an enormous overload of acronyms, especially those with numerous possible meanings, to disorient readers and confuse them as to the intended significance. These obscure acronyms also refer to complex concepts with definitions buried in earlier research studies. A reader working to comprehend every word, is sent on an infinity-approaching acronym hunt, turning to textbook after textbook to find definitions for acronyms only to be directed to new unknown acronyms. The goal of these tricks was to keep the committee, Examen's supervisors, and the public from reading the specifics in a proposal. Nally's initial budget specified spending on strange entertainment costs, which without subterfuge indicated bribes. The layers of nonsense were added under the assumption all reviewers of the proposal would have an inaction preference, or would quit reading and stamp a report as approved if it was too difficult for them to comprehend. If overseers refrain from scrutinizing a company with an awarded contract, this encourages it to remain corrupt with impudence.

The final step in the application process was a visit to their Gate at CSEA by the lead committee deciding the competition. PROs were responsible for this crucial component. It had long since been interpreted by competitors not as a simple showing of the facilities scheduled to manufacture the contracted object, but rather as an opportunity to outshine other competitors on the grandeur of a company's imagination. Each team created a theatrical presentation with songs, dances, holograms, and sense-engagements. This project's experiences focused on three-dimensional joyrides on a holographic spaceship through maneuvers that displayed what their spaceship model was capable of in extreme conditions, such as flying through a dense asteroid belt, landing on an ice-ball planet, or parking in orbit around a moon. The committee members joyfully reacted, while

---

1   This figure converts Cadata's 3D language into the length of the translated transcription in human English.

taking hard turns, without stabilization because this rollercoaster sensation was unusual. Cadata's shuttles had long-since been equipped with stabilizers to eliminate bodily impacts of turbulence, and hard-breaking, or acceleration. Testers who deactivated these safety protocols could feel light-headed to the point of even losing consciousness. With anti-gravity on, the same extreme g-force maneuvers felt more like a video game. Cadatas' bodies are natural mutants, so they can resist high-g's much better than humans, as their insides can squeeze even into a flat pancake on a wall without feeling discomfort. Walls in the spaceship were cushioned with a material lightening the pressure better than a seatbelt, so even this extreme compression was unlikely to cause a cadata to faint. For a loss of consciousness, a cadata had to attempt extreme changes in the direction of the g-force, which would shake up their spread-out brain matter, causing it to want to shut down to avoid processing these shifts in perspective. Unlike the unreadable Nonsensical Proposal, these joyrides explained the major selling points of Examen's design in an entertaining way to offer one final seduction to the committee. At the end of the visit, the cram-bribed main committee member signaled to Nally she won. She appreciated the confirmation because he might have asked for added "reimbursements", or pressured another team into a larger payout.

The process of securing a monopoly over the Ganeir mission for the Examen Initiative scarred Nally. Examen won because it had pre-engineered many of the needed components for the comet mission; reusing these for Ganeir was a safer bet for Cadata than investing in a team without this start. The main complication to a simple re-arrangement was a shift from comet to light speed.

Another reason for Examen's eventual success was Nally's employment of a team with few managers and many scientists, engineers, creatives, and hard-workers, who carried the lava when Nally needed to decompress from yet another social with an overseeing committee. Managers created roadblocks for creatives because their job function was discovering faults in their subordinates; they had to appear busy chastising. Because the other top teams were led by scientists who hoped to fail, Nally was spoiled for choice among the best in these fields, who were all equally desperate with her for productive work, and could not fake appearing idiotic enough to be hired by the others. These brilliant scientists were poisonous to the competition because, if they slipped, and hired a single, uncorruptible cadata, who understood the science and the reasons behind delays; he or she would report these extreme, corrupt underpinnings. It was terrifying for competent scientists to apply for jobs with corrupt companies because signs of incorruptibility blackballed them from finding any job in their narrow field. Sadly for some blackballed scientists, it was especially difficult to avoid betraying one's tender cadataist tendencies at a time when corruption was threatening to end all life on Cadata. A competent scientist also knows how to lower the cost of a project. Competent hands build cheaper, efficient spaceships, capable of traveling farther, and safer. This price-lowering contradicts the interests of business owners seeking the maximization of bribery-payouts, and general profits. Nally hawkishly kept her virtuoso researchers away from accounting, to avoid explaining this apparent waste of resources, as they were being channeled to overseers.

The CSEA Complex and Building the Ganeir Spaceship

Once the contract was in Nally Subsenar's wings, it was time to establish a system for how they would pull through the process, with a positive output at the end. The head of the Robotics Corporation, Loor Dowor, had avoided interfering in the contract application and negotiation process. He had obligations in other branches of the businesses, and could not directly supervise the Examen Initiative. He insisted Nally had to arrive at a victory, at all costs. She complied. Five competitors exited beforehand, surviving intact, despite heavy losses. The two competitors who struggled until the end went bankrupt, as Nally expected. Any of these losers would have been delighted to sabotage Examen's smooth operations going forward. Because of the high stakes, Loor volunteered to oversee the remainder of the production process. He asked Nally to take on the role of Head Engineer, and be the brain behind the operation, rather than its logistical operative. She took a few weeks off to sober up and brush up on the scientific literature she missed during her years of socializing and playing science-politics.

By the time she returned, Nally noticed Loor took the task of restructuring Examen as a personal challenge. Through a series of intimidating meetings, Loor impressed on the staff he was now in charge, and their job was to say, "Yes, Sir!" and execute his orders and mission. This militaristic approach troubled Nally, as she anticipated it would clash with her own sense of what was reasonable. However, she had been earning the minimum salary. She anticipated seeing the first profit, essential for her independence and retirement, in the upcoming building phase, when bonuses for winning the contract would be activated. In the time she was away, Loor had already fired a couple dozen scientist naysayers, who happened to be the brightest team members Nally had relied on. Those who remained began auto-agreeing with Loor's ideas, even when certain his set goal was unreachable. While their compliance retained the remaining staff, it also stalled the project because, regardless of what they promised Loor, in practice, the project's cost was unbendable to Loor's will. As you will recall, more than half of the budget was already spent on corrupting the competition. The remaining funds fell short of lofty ambitions. Other scientists complained to Nally, developing the grand plan promised in the contract required 20,000 years and several times more money than they were granted. Despite Nally's best efforts, some of these ideas reached Loor, who through hysteric fits every time, threatening to catapult the next scientist who brings him bad news into outer space. Nally slaved to derive a middle-ground suitable for most parties. Just as Nally found balance between extremes, Loor rearranged her proposed schedule and budget, insisting on a puzzling redo. After decades dealing with needy bureaucrats, Nally patience was hyper-extended, so she continued proposing new solutions to the same lingering problems.

Given each player's self-interests and responsibilities, lead actors at Examen soon fell into established roles allowing them to perform their parts as they moved forward. Loor's fiscal responsibility was delivering the minimum-allowable product on schedule and on budget. This goal had to be met, even if most team members died in horrific accidents because safety regulations were ignored, or they all went homeless because of how little he was paying them. Nally's engineering challenge was meeting Loor's goal without the demise and bankruptcy of all involved. Even if she edited his directions in the background; she could never outright say, "No." One "no" led to immediate dismissal. She envied their sub-contractors who, by offering her indispensable employees bribes, managed to receive hundreds of millions under the promise they could achieving impossible feats of engineering, through untested tricks of design and science. In truth, these contractors cut costs in areas where expenses saved the spaceship from crumbling. The lower quality of a cheaper component introduced structural weaknesses promising instant-death upon liftoff. Nally strove to spot exaggerations, miscalculations, and mismanagement problems in the sub-contractors' work before they were paid. It was in their interest to promise to build an extraordinary feat of invention for the smallest price, and then, to perform the minimum work to craft the shabbiest flyable object, while inflating the budget to the highest number the buyer could afford. The price increased due to "unexpected" changes made to fit with the "realities" of the job. Nally and all others who won a contract from CSEA were dealing with having to lower costs, so they had to pick sub-contractors who proposed the lowest prices. Problematically, the lowest-price sub-contractors were most dishonest. Nally wish they could perform all work inhouse. However, because they lacked the resources and team members specializing in each part of the larger job, it remained cheaper to hire these grabber attachments.

The cadatanauts, led by the infamous Codol Havould and Putair Niatage, were an influential party within the Ganeir project. After their years on the road, touring and indulging in pleasure, including companionship and luxury, they now faced having been chosen to fly the contraption resulting from a corrupted and wasteful process. Other parties profiting from the project, such as the sub-contractors, would spend their winnings on opulence on Cadata, but Havould and Niatage's fate was to test the concoction on the Ganeir mission. They were promised too much money for this virgin interstellar flight to consider refusing the mission. As the test pilots, their primary concern was the craft's capacity to survive the voyage. Thus, on the day they all learned Examen won, Havould and Niatage showed up at the moon base's headquarters Gate as sober as rabbits running from a fox. Both Havould and Niatage had served in the Potentates' Shuttle Troops, where they were responsible for testing new shuttle designs. They had seen plenty of accidents and incidents over their centuries of flight. One time an engineer forgot to insert the propulsion triggering mechanism, at another the navigation system was convinced it was on the moon instead of on Cadata, and because it was using the wrong map, the Computer led them into the middle of a major storm, landing itself on a beach as enormous waves broke over the shuttle. These mishaps convinced them it was better to fear and verify, than to be confident and dead. Their routine began by rolling shuttles across the tarmac, then endeavoring light glides through the stratosphere, before rising into airless and non-eject-able space orbit.

With the Ganeir spaceship, they observed scientists testing each component as they were being developed. They spotted a few problems the scientists or sub-contractors might have slipped into the final product if they were not thus overseen. The spaceship design had suggested it could withstand most conditions and accidents. The failed prototype flight garnered the glut media-attention for its many faults. It was their duty to their own survival to uncover these unseen errors. A crash from a motorcycle on earth is survivable. A microscopic miscalculation, misalignment or mismovement in space is terminal. The test pilots earned their hefty salaries by overseeing everyone else; if either of them started to trust or buddy-up to any of the scientists, engineers, creatives, or others at Examen, the other would call this activity out: "Remember, don't let a pretty face convince you everything is kosher under the hood." While they went out and partied with the others, they kept a personal-distance because anybody who was attempting to gain their friendship and trust might have been doing so to deflect them from a health and safety violation. They were motivated to research the latest science to suggest improvements missed by scientists who would refrain from testing the equipment with their lives.

The moon CSEA complex was protected by heavy security in case a rival, a rebel group, or another entity attempted to sabotage the production process or to steal the expensive equipment and components exposed to lunar elements. Most protection forces were robotic, with a few barracks, where biological cadata troops were stationed. There were conceal-carrying robots stationed across Examen's hanger, so entering the testing tent or boarding a live shuttle was a stressful endeavor for the cadatanauts. Robots were armed with strong air streams, decompression units, or other contraptions.

Extreme conditions were tested by enveloping the spaceship in a half-spherical, see-through bubble tent capable of holding in or keeping out the weather. The atmosphere inside the tent could be clouded, heated, cooled, pressured, colored with chemicals, or blanketed to mimic the expected environments in orbit, in the vacuum of space, or during a landing.

Under this bubble was a dense protective floor, and below it a bomb shelter where workers and robots could jump in case the Computer gave an alarm of an impending explosion or other failures. It remained under-exploited because a disruption with an advance warning could be fixed by the light-speed Computer processors.

Upon a biological or a mechanical body's entry into the arena, it was air-vacuumed in a mini-bubble, which auto-hugged the entity. Impurities, including tiniest dust particles had to be sucked out to prevent contamination of sensitive equipment. A spec in the propulsion engine could cause an explosive reaction, so it had to be a sterile environment. Having the purification system appear anywhere an entity attempted entry meant the bubble could be entered from any direction, including the top of the dome. The walls were a condensed solid unless they encountered chemicals belonging to entities whose genetic or chemical signatures matched those on the list of invited or participating guests. The walls gave way to an incoming form in a tight hug preventing external air or lack thereof to flow inside. At 10 acres in circumference, Examen's Gate's bubble was the largest in the CSEA complex. They were assigned to it after winning the contract because this was the sole gate capable of fitting the final anticipated size of the spaceship with capacity to house hundreds of cadatas on a long interstellar voyage. The bubble included temporary sleeping quarters and dining facilities for the crew, as housing them off-base on the moon would have been far costlier.

To avoid small inaccuracies in fabrication or assembly, each piece of the spaceship was designed and tested on separate stations before being hovered into place with controlled wind streams across the top of the bubble dome.

The Control Room was near the central Gate. Air streams sped cadatanauts traveling between gates. The Control Room housed the Gate's Main Computer. Programmers and engineers had to visit it to communicate ideas for the spaceship's program. It was also a meeting room large enough to accommodate the crew. Loor was deaf to everybody's opinions, but enjoyed pronouncing his decisions, and deriding subordinates in public.

The point of this infrastructure was developing a light-speed spaceship to migrate an exploration mission of hundreds of cadatas to a potential new home-planet. Many millennia had passed since Cadata invented its first spaceship matching the current maximum-speed of Earth's New Horizon—31,000mph. Despite light-speed capacity, they opted for safer journeys, such as cruising along-side a comet at its speed. They practiced light-speed travel with numerous manned tests. Uncertainties, complications, and fears remained, but they pushed forward to stave extinction.

Through the Ganeir project, Nally and her team developed systems lacking equivalents on Earth. For example, cadata spaceship plans are written into a DNA-like code. Instead of the Earth-style pieces of metal attached with bolts, cadatas imbue the entire spaceship with nano technological capability. Each nano is a type of cell performing a specific function for the larger structure. The Computer oversees their cooperation, and they are aware how they are functioning in the greater whole. This system can run on most energy sources. The ability to exchange near-infinite combinations of fuel sources was one of the upgrades. Energy and chemicals flow through artery-equivalents in the fabric of the spaceship to deliver precise quantities when they are required. This system mimics an organic system because it allows for the same type of mechanical complexity as an organism. Human-made machines have few active components, including an engine and transmission. In contrast, the human body can change one chemical into another, perceive images, run computations, and perform extraordinary quantities of all these functions in 37.2 or so trillion cells. Just like human DNA knows the standard shape of a human, the spaceship's Computer knows the graphic and structural plans for the ship and executes them in the model. Parts of the ship in standard sizes and shapes can be pre-fabricated, just as a heart might be engineered in a lab and then transplanted into a body. The Computer is the conductor assembling parts according to the pre-planned composition and chemical makeup, putting the pieces together with logic and efficiency. When a single nano-cell fails, it is flushed out of the system and replaced with a new duplicate. Without this extreme precision and complexity, our spaceship would have failed to maintain integrity for a 100,000-year voyage. The cadata team designers were in charge of planning the ideas behind how these systems would function. The Computer ran their calculations and executed their orders. If the Computer was ordered to make an organic structure without further specifics, its design failed to fly. Once, a slight imperfection in the Computer's design caused a failed test during development due to the buildup of a toxic chemical near a cadata's rest station.

The software setting the course for a flight is one of the central intelligences in the spaceship. On a multi-planet trip this program calculates the most efficient path between them. Cadatas can specify their preferred world type, or the gravitation force they hope to experience when accelerating to light speed, but the calculations must be executed by artificial intelligence due to the quantity of threatening factors influencing seemingly empty space. Gravity from a black hole or a massive star can pull it off its intended trajectory. An excess of space dust or a scattered asteroid belt at light speed can be serious bumps in the road. The navigation software responds to millions of potential problems with rapid changes in flight speed or direction. It corrects interruptions during landing or stopping. A flood of debris, otherwise processable as fuel, is stored. An abrupt outage in the controlling organism is re-ignited. A leak in the gas supply is sealed. If an unobserved problem appears; it is fixed with the Computer's deductive reasoning. Given sufficient time, and if a decision has implication to the lives or safety of biological life-forms; then, this subroutine asks cadatanauts for their moral input.

Unlike planet-bound shuttles, cadatas' spaceships lack up or down directions. They can flip without giving the sensation of having changed orientation. Thus, all surfaces are utilized. The ceiling is not distinct from the floor. Walls can be erected and furniture can be added in the middle of the spherical interior as needed. To re-decorate, you choose a model from a catalog, or make a vocal request such as for an increase in a bed's size or softness. If there are more residents on the ship, more compartments are created to give them privacy. If there is a group of scientists who need to run experiments, they create a lab with the necessary equipment. If there is a group meeting, the walls and furniture can retreat into the walls, allowing for utilization of an empty orb. Walls had to be malleable on the Ganeir ship because of a lack of advance knowledge regarding the relative size of the lifeforms on Ganeir. They intended to edit half of their crew's genes to become like the dominant local species. Without knowing if this species would be huge or tiny, they were unable to pre-set the rooms to a size fitting for them; their pre-selected size might have turned out to be absurdly too small or too large for the local lifeforms. Because the "floorplan" can change daily, cadatanauts occasionally need to ask the Computer to draw a fluorescent light along the path to their room to locate it again. This fluidity is also the reason there are few official names for different spaces in the ship, other than "lab", or "private compartments".

The main reappearing room on a cadata spaceship is the Main Deck and its Observatory. This room's programming prevents any interior light sources, including from any of the technological components, which operate on other wavelength than visible light, which cadatas can see with their gene-edited vision. Even the room-composing materials do not reflect light. The measuring units line the exterior of the out-facing see-through walls of this space. The observing coat is on the outside end of a protective coating protecting those inside the ship from radiation. It gathers this radiation, together with space dust or a range of particles found in open space as fuel sources. They take in starlight and dim light from planets, measuring it on all wavelengths capable of assisting the spaceship with interpreting the weather of the surrounding space, as well as the state of distant planets, stars, and galaxies. Cadata spaceships lack exterior lighting, nor do they excrete noticeable volumes of heat, radioactive or light-generating materials. Earth science fiction's employment of neon lights on the exterior of spaceships is ridiculous as a spaceship spends its life in the darkness of interstellar space without any other ships in need of being warned of its location. Rapid heat and power loss visible to observers would be hyper-wasteful. External lighting would also create an extreme level of light pollution, which would block visibility for the Observatory, preventing sensitive equipment in it from detecting dim planets or smaller distant objects. Cadatas seldom enter the Observatory to avoid polluting the instruments with light from the outside even for a moment. Visibility is also decreased when the spaceship is parked inside of a solar system; sun rays, planetary light reflection, and planetary atmospheric edges flood receptors with interference. The Observatory's precision increases in interstellar space when it can deliver predictions about the expected fluctuations in the brightness of distant stars, or the movements of erratic planets with more precision than a weather forecast on Earth. The Observatory's software also creates a 3D filmographic map of the sky the spaceship is flying through and sends this data back to Cadata or a local operation-base for utilization in future journeys. The data on star systems improves the closer the spaceship approaches, so a trip through thousands of light years generates an enormous library of localized stellar information.

In the rooms where light is allowed, lighting mimics Cadata's length of day, with changes in brightening and dimming. Moonlight is displayed at night. During the "days", the light also imitates the color and intensity of its natural hazy sunlight, including an introduction of some random cloudiness, heavy humidity, or even some storm-like conditions. The internal atmosphere can be set for entertainment, simulating a thunderstorm indoors, drenching participants with raindrops and allowing them to glide on heavy winds. The temperature swings to high and low extremes to fit a species. Less intense atmospheres include a light sea-breeze or even a hologram of a favorite resort. While there are many more veins running through a cadata spaceship than the plumbing and vents running through a human space station, they are inaudible in the cadata version. Air and liquids flow through tunnels smaller than a vein, at a fraction of the speed, and without the interruption of pumping compression. Thus, these movements are inaudible, even if an unmodified biological ear contacts a wall. If you hear noises analogous to human-plumbing on a cadata ship; you should alert the Computer because the ship is suffering a catastrophic failure. No parts connect through mere bolts, so nothing creaks. The exterior skin is the engine and when it collects and processes matter into energy, it does not hum. If it starts to hum, this means it has ingested an enormous chunk of matter such as a comet, and it is undergoing a major internal crisis. Large lumps are repelled from the craft unless the ship is nearing a total depletion of its energy sources. Because cadatas use oral or 3D digital input and the Computer operates on the same organic circuits as the ship, there are no triggers capable of generating an earthly beep or click. A cadata ship is devoid of buttons, or open displays and controls; otherwise, a cadata might fall or slip, triggering a switch capable of draining all breathable air out of the spaceship. In general, a cadata spaceship is antagonistic to manual control by cadatas, as they are seen as an unpredictable and illogical element. Instead of constant-wasteful rotation, air streams out where it is needed in microscopic quantities. Each inhalation leads to an equal quantity of the used-up gases to puff out of the nearest wall. All interior and exterior walls are soundproof, so a spacewalking cadatanaut or a banging neighbor would not be heard. The Computer also controls the smells excreted across the ship. Some smells signifying a food source for a cadata, might kill a species on an alien planet where this substance is toxic. Cadatas undergoing gene-therapy to match this species must have the cabin's air purified of the hostile substance. Cadatas who avoid gene editing and want to cross over to a compartment in the ship with an unsuitable atmosphere designed for another species, must wear skin-tight spacesuits to avoid damage. Unlike with an air conditioner, cadatanauts are discouraged from adjusting the atmospheric settings in their room without consulting a Computer for guidance. For example, it is reasonable to ask the Computer to simulate a suitable tornado in one's room, but if a cadata just increases the temperature by three degrees, this seemingly slight rise might kill a fragility in their constitution. Beyond keeping cadatas alive, these atmospheric and sensory tricks ease space travelers' minds. The ability to create extreme atmospheric conditions was instituted on a grand scale after the comet-flight madness case to keep cadatanauts from becoming psycho-depressive.

Given the extraordinary functions a spaceship's Computer controls, the worst-case-scenario in space is for the Computer to reboot. Backup systems kick in if the Computer experiences a failure through sabotage or from overloading on the amount of data it is processing. During a failure, all systems are turned off, including life support and propulsion. This crisis is averted when an isolated backup system takes over operations. If the backup also fails, cadatanauts can rely on their spacesuits to prevent instantaneous death in the decompressed and airless environment. The loss of even a single unit of data during a crash can be cataclysmic. Thus, all data must be backed up in isolated parts of the ship. A loss of a single digit in the pre-crash settings can adjust the spaceship's atmosphere to a combination hostile to life.

The spaceship's coating is essential to its balanced integrity, power generation and various other functions. Most of the ship's active systems are in the coating layers. The coating is layered like skin, fat, and muscle on a human body. The external layer collects the particles, molecules, heat, and radioactivity. While this outer coat attracts energy-generating materials with a magnetic field, the coat below it is tasked with protecting the rest of the ship from this radioactive or otherwise toxic content. It must keep out heat and cold alike, and it must prevent damage due to even the largest over-light-speed impacts. A diamond grain flying at any velocity colliding with a spaceship flying at the speed of light in the opposite direction can only be stopped by the strongest combination of materials safely installable on a starship; even a human bullet-proof vest cannot stop a diamond at just the speed of a bullet. Human science is insufficient for me to explain the complexities of the outer and inner layers of the coating. I can verify, oganesson, or $_{118}Og$, is a key ingredient in the protective layer's strength. Oganesson has recently been synthesized on Earth. It is a chemical element with the highest atomic number on the Earth-known elements table. Because of its high density, oganesson prevents radiation from moving through. Without an anti-radiation coating, those inside would receive a dental x-ray-worth of radiation every four hours even flying at the edge of Earth's atmosphere in a plane. Earth astronauts and cosmonauts have reported seeing flashes even when their eyes are closed when subatomic particles ionize the fluid of their eyeballs. Flashes alone might seem minor, but radiation-buildup can cause brain damage and death if measures are not taken to prevent it. A human astronaut in 2019 can die from walking on the moon in standard-issued gear if a solar flare strikes because of the absence of a protective atmosphere.

The slightest hole in the oganesson layer must be repaired by the nanobots in nanoseconds, since exposure to radiation increases at light speed, and cadatanauts can travel through millions of miles of a cloud of radioactive light in minutes. Another component of a cadata spaceship's coating is a variation on what is known on Earth as carbon nanotubes, or sheets of conductive graphene. Instead of being rolled into straight tubes, cadatas' version is more organic or vein-like and is also operated by our intelligent nanos. This carbon structure gives the walls of the ship the ability to become see-through when visibility of a proximate planet, star, starship or another obstacle or destination is needed. Carbon nanotubes are the strongest material on Earth, which scientists can keep in a stable solid state. In contrast, oganesson evaporates upon humans' synthesis. Oganesson remains solid in cadata constructions because we control its natural millisecond radioactive decay with our nanobots by extending its lifespan to millions of years. Only with these types of ultra-durable materials can any species venture on a 100,000-year journey. The nanobots also keep the oganesson below its 170°F boiling point, so it does not evaporate into space. Because of the chance an outer layer might evaporate or explode due to extreme, unpredictable pressures, there are several backup coats to take over these same functions in worst-case scenarios. If this evaporation does occur, the ship's magnetic sphere attracts the evaporating particles back to the surface and re-solidifies them. The magnetic field works together with a repulsion field, which can push against unwanted matter to prevent damage to the ship. It always distresses cadatas to watch Earth films with space battles between starships; if mere nuclear or plasma weapons could penetrate a spaceship's defensive field, this ship would never survive a year traveling through space, as it would disintegrate from impacts by non-hostile dust and radiation. Mixed into the layers of coating are mutation properties engineered to mimic cadatas' natural mutation abilities. This property allows a ship to change its texture and color to mimic another object to avoid detection from a hostile alien lifeform. For example, while circling Earth recently, the main cadata spaceship took on the appearance of an asteroid satellite to fool Earth's telescopes. These carbon and oganesson materials are so essential to life's survival in space, they are also used in the thin, skin-tight spacesuits protecting cadatanauts, when they leave the ship's shield.

Diversities of Spaceship Engines

The hyper-efficiency of cadata engines can be explained through the law of conservation of energy. In a perfect vacuum, kinetic and potential energy shift between each other's states, just as a ball bounces to the same height continuously, in the absence of friction from air, or the weight of gravity. A cadata spaceship travels through space with help from this constant cycling of energy in a loop. The engine moves the equation back and forth between energy and mass. If any energy or mass is wasted, the spaceship had to collect more with its magnetic field. Almost any collected molecules can be processed to generate propulsion. For example, oxygen and hydrogen are among the three most common elements in the universe. Other molecules common in interstellar space appearing empty to an untrained eye include: carbon dioxide, water, salt, titanium, silicon, ozone, ammonia, and methane. An abundance of these fills every cubic foot of space. Once collected and stored in separate compartments, hydrogen and oxygen gases can be combined to generate water (H_2O) and energy. The latter can present itself as an explosion in an uncontrolled Earth experiment, but when the reaction is hyper-controlled, it generates pure energy capable of running the ship's systems. On Earth, a spark is added to ignite the reaction, but fire is a safety-hazard when exposed to oxygen in space. The nature of how this reaction is stimulated once again must be redacted to avoid confusing human readers. Then, the compounds in water are broken back into hydrogen and oxygen with the application of the energy their joining had generated and an electrolyte such as acid, which is also collected or generated and stored until it is needed. The reaction creating water and producing energy is exothermic, a category that includes reactions such as combustion due to the exposure of fuels to oxygen (in a human car), respiration (when glucose and oxygen turn into water, carbon dioxide and energy in human cells), and nuclear fission of uranium (human atomic bombs). A cell, a machine or an atomic reaction can produce energy; this material flexibility for energy production makes cadatas' engines efficient and egalitarian. Humans have botched duplicating this technology because energy generated through an exothermic reaction may feed back into the components, thus repeating the reaction in a cycle threatening to turn into a thermal runaway, or an escalation in heat or energy causing the types of extreme reactions seen in nuclear fusion in bombs and supernovae stars. Humans have attempted to take advantage of these runaway effects to fuel sudden bursts of energy in rockets capable of escaping Earth's gravity, but it is far safer to prevent these escalations and instead rely on a steady, recycling reaction.

Thrust is created when bits of water or another reaction output substance shoots out in the opposite direction to the destination. The molecules are pushed out of numerous tiny nano-cells housing the reactions rather than through a single tube. This allows the spaceship to move in any direction in space rather than having to flip over to change direction. Once they do the job of pushing the ship forward, these bits are attracted back to the ship in a circular orbit to be reused. Only a small fraction of kinetic energy is lost in this process. The ship also collects and converts into energy the radiant light and heat from the nearest stars. A spaceship traversing a galaxy must find energy in near-void interstellar space just as in the dense galactic center. The Computer seeks paths through dust clouds, which appear dark because they block light from the stars behind them. These solar system or several-solar-system sized clouds are rich in hydrogen gas, helium and metals such as silicon and carbon, mimicking the composition of most stars; this chemical match is not accidental as these clouds can be the start of a solar system, or the remnants of a supernova. Some of these clouds have hot cores, so they might be in the process of star formation. Even away from these coagulating concentrations of star dust, the interstellar medium is rich in over a hundred different molecule-types from a simple hydrogen atom to organic strings.

The Ganeir mission's objective was perfecting the utilization of a maximum number of different fuel sources and hyper-efficient energy generation procedures in a seamless way without cadata-input. A cadata spaceship begins propelling as it takes off from a surface or from orbit. Unless there is a reason to vary the speed, it accelerates at a constant rate until it reaches the speed of light. Then, the engines are cut off as the ship sails at a constant rate of light speed without any drag or other resisting forces to slow it down. If speed is lost, the Computer will reactivate the engines to resume acceleration. To stop, propulsion must be applied in the opposite direction, pushing against the target destination to create a type of artificial drag generating deacceleration. Turning is the most difficult and energy-consuming maneuver imaginable when traveling at the speed of light. Gliding through space with limited added energy after reaching the speed of light is achieved due to the nature of conservation of momentum in a vacuum. It takes an extraordinary burst of propulsion for it to cancel out an existing direction of motion. Thus, it is best to come to a near-stop prior to a turn. Think of taking a turn in a car on Earth at 100mph. Now imagine a turn at 1,000mph. What about at 670,615,200mph?

Even in a vacuum or without drag, the g-force exhorted on an anti-gravity-unshielded body would pancake it.

The problem at the heart of human space travel is the rocket equation conundrum. Humans have been searching for an engine type to beat it. The rocket equation states: the more fuel you take with you, the more lift you must achieve to move this heavier rocket off a gravity-generating body and into orbit. Almost any simple engine on Earth could accelerate a spaceship to the speed of light. The problem has been the heavy weight of the non-recyclable fuel they used. For steady acceleration to the speed of light, it would take a planet-sized spaceship for it to carry sufficient liquid fuel to also support its own mass. Humans have been experimenting with solar wind, and other alternative energy sources, without on-board storage requirements. These are great strides in the right direction.

An example helps to explain the problem of the rocket equation and intergalactic travel. When cadatas began the journey to Earth from another planet unable to satisfy our stringent requirements, we set the acceleration rate to generate a steady 1 g, which created a force equivalent to how humans feel standing on Earth. To attain this pull on our bodies, we had to travel at .0061 miles/s^2 (known as the gravity constant of 9.8 meters/s^2) until we reached the velocity of 187,200 m/s. It took us 8,500 hours (354 Earth days) to get to the speed of light. What do these numbers mean? A car accelerating from 0 to 100 km/h in 6.4 seconds is accelerating at 4.3 m/s^2. The Saturn V moon rocket used tons of fuel and did not even triple this acceleration due to its mass, increasing speed at 11.2 m/s^2. In contrast, a natural jellyfish stinger can attain 5,400,000 g by traveling at 53,000,000 m/s^2 for a moment. Thus, the problem with interstellar travel is not the acceleration rate (which a jellyfish has mastered better than top Earth scientists), but how long this acceleration can be maintained with minimum weight on board. Specific impulse is used to express the efficiency of a rocket's use of propellant or how much propellant is consumed per the impulse or thrust generated. When compared, solid fuel rockets are the least efficient with a specific impulse of 250. Ion engines are a bit more so at 5,000. Nuclear fusion rockets can reach 200,000. Nuclear pulsed rockets can peak at 1 million. And antimatter rockets might reach 10 million. On Earth in 2019, antimatter propulsion is still theoretical since it has not been created in a sufficient quantity in any lab to attempt to shoot it out of a rocket. Cadatas utilize energy sources with the highest possible impulse level. However, as you will see, antimatter is too destructive towards all matter (cadata or human in nature). And some engines are too difficult to control in space. Thus, the gains in impulse are offset by the risk of a catastrophe.

With Earth's 2019 technology, nuclear rockets would have to reach a quarter-mile in diameter, and a weight of 8 million metric tons to carry and explode about 1,080 bombs-worth of fuel to reach 10 percent of the speed of light. This fuel would be expelled and wasted during acceleration, leaving a radiation trail across space. After years of operation, some of this waste would sprinkle backwards at the surface of the ship. Even a small amount of radiation leaking through the metal exterior would kill the humans inside. Since the back of this ship would be a giant, re-exploding nuclear weapon, no metal on Earth could withstand the pressure of the consistent bombardment it would be generating. Any metal would begin to fail, creating cracks through which radiation would penetrate. Meanwhile, it would take too long for this ship to travel to even the closest solar system at 10 percent of light's speed. To reach the speed of light, engineers would have to increase this hypothetical ship's radius to rival the biggest structure on Earth. This distinction belongs to the Boeing Everett Factory in Washington State, with a floor area of 4,280,000 square feet, or a width of around 2,068 feet. The Empire State Building, at a width of only 187 feet, weighs 350,000 tons. A nuclear rocket engine can require around twice more fuel over the weight of the rocket. Saturn V, a launched Earth rocket, weighed 3,270 tons, so this would be far above a 1,000-fold leap, at a cost eclipsing Earth's GDP.

Heat and, in turn, electricity can also be generated by allowing materials to decay; plutonium has been popular on Earth for this reaction. But there are risks of catastrophic failure with using plutonium, as with other ultra-high-energy-producing substances. The problem with plutonium is its rapid decline in power in time with its radioactive half-life, so a quantity of plutonium generating 350 watts at launch might only manage 300 watts a decade later. While this might be sufficient for a flight across the solar system to observe Pluto, it would be a ruinously short lifespan as a power-source for an interstellar mission pre-set to last for at least 5,000 years. Plutonium is rare in outer space, so locating and attracting it with a magnet into a spaceship's engine is improbable. A combination of less energetic but common and hyper-charged but rare fuel sources creates the safest and most efficient system.

The same hydrogen gas can also be utilized in a fusion reaction. According to human physics theory, to generate propulsion, hydrogen is placed in a circular-tube magnetic field heated to 100 million degrees Kelvin, or six times more than Earth sun's core, until hydrogen nuclei smash into each other and fuse into helium nuclei, releasing bursts of nuclear energy. Generating multi-sun-level heat or pressure on a spaceship increases the odds the generator will overheat or collapse under this stress. Humans are even suggesting utilizing the largest laser beam in the world to fire at a hydrogen-rich lithium deuteride to raise temperatures to 100 million degrees Celsius, unleashing 500 trillion watts or 500 terawatts in a few trillionths of a second. Earth consumed 17.4 terawatts in 2015, up from 5 in 1965. So, to power this single ship, Earth would have to generate 28 times more power than the current total planetary energy output. In theory, to generate this force, this rocket would have to be at least 625 feet long, and would only propel the ship to 12 percent of the speed of light. In practice, this spaceship would have to be 28 times larger than Earth's current combined energy-producing power-stations. Outside this science-fiction, one of the most powerful fusion reactions on Earth was generated by compressing a small coil with the force of 4 million amps. This is equal to .04 terawatts, or an incredible waste of energy on a planetary scale. These amps of current ran through a generator at the speed of around 20,000 miles per hour until this coil could not be pressured any further and it bounced back, producing a shockwave, which terminated the originating generator. Human scientists appear to prefer destroying their labs with the biggest possible nonsensical explosion to creating an efficient energy source capable of propelling Earth into the Interstellar Age. Why would any spaceship need to generate 500 terawatts in an instant? Where would it be stored as the ship uses small quantities for gradual acceleration? It is more efficient to generate bursts of watts just when they are needed to avoid transporting prepared combustive materials. Generating sun-level heat to create 28 times of Earth's energy in moments is wasteful, but a single isolated reaction where two hydrogen nuclei contact the coating of a cadata spaceship and fuse into helium is an efficient solution. Cadatas employ heavy modifications so temperatures in the millions of degrees or crushing pressures are not necessary. Creating such high temperatures or magnetic fields requires an equivalent amount of energy in input as it generates in output. Receiving back the amount you put into a trade negates the activity; thus, this is a nonsensical reaction where heat is put in, and about the same amount of heat comes out. One of the more logical fusing (as opposed to fusion) reactions proposed by humans includes the refinement of CO_2 out of the atmosphere to fix its runaway greenhouse effect; in this system, CO_2 is refined with carbon and water, turning it into car fuel (CH). Changing gas into exhaust during a drive, and then converting the exhaust-waste with a chemical reaction back into gas is the type of energy conservation at the heart of cadatas' energy theory.

Human physicists deem antimatter to be one of the most efficient energy-generating methods because it is a direct conversion of matter into energy. Antimatter is the opposite of matter because it carries the opposite electric charge; thus, anti-electrons are positive in charge, and anti-protons are negative. When matter and antimatter collide (be it protons and anti-protons or hydrogen and anti-hydrogen), both are annihilated into pure energy in the form of a burst of gamma rays (one of the most energetic phenomena in the universe) and X-rays. In theory, if pure energy is created, the process is 100% efficient, whereas Earth's nuclear energy wastes all but 1% of matter. The cataclysmic problem with anti-matter is also the reason for its extreme efficiency: anti-matter annihilates all matter. Cadatas and humans are constituted of matter. Thus, we are all antimatter's fuel source, vulnerable to being annihilated on-impact. Antimatter can be stored in a magnetic trap. However, in human tests, occasional collisions still occur. Even a marvelous, Computer-designed cadata storage unit contains a few imperfections. If these flaws are magnified by 100,000 years, the unit might erode to nothing, spilling the self-destructive fuel across the spaceship. This method is seductive because it promises the collision of a single electron and anti-electron releases a minimum of 1.02 million electron volts. But where can anti-matter be found in sufficient quantities to use in a spaceship's engine? It can be generated in a lab, or found in single particles, but neither of these are efficient on the cosmic scale. Humans have not yet found any large clumps of visible anti-matter in the universe. The universe is dominated by matter. In theory, moments after the Big Bang, most of matter and anti-matter cancelled, leaving a small quantity of matter behind. Introducing enormous quantities of anti-matter threatens not only the desperate spaceship, but perhaps the entire solar system it is flying through. Needless to say, cadatas have not utilized this "efficient" energy-producing method.

Methods of more balanced propulsion cadatas have explored to great advantage include solar electric propulsion, which uses solar panels to capture sunlight and convert it to electricity, stripping away the electrons from the gas, and creating ions. The resulting electric field shoots these charged ions out, creating thrust. One interesting innovation humans are developing is solar brofskites (sp). This is a material that generates solar energy through various types of items with appearances divergent from standard human-made solar panels. Solar brofskites is an efficient energy source because it can be printed even on a home printer, so its creation necessitates minimum energy and resource input.

Another human method of utilizing sunshine is with light sails, wherein light's pressure, or the pushing motion of photon particles, propels the spaceship. A 46-foot light sail ship reached Venus in six months; however, a ship with aliens on board would require more energy than Earth can produce. The only place and time with the needed power-burst of light's pressure is in the final seconds of a supernova as it burns through its iron core, rapidly collapsing down to three times the density of an atomic nucleus, and then producing a shock wave exploding outwards at 10^{44} J, spraying up to a galaxy-sized chunk of light in a moment; if a spaceship can hop on this light wave, it can travel across the galaxy and beyond. However, since planets 30 light years away are in danger of catastrophic extinction, no high-tech spaceship can survive this proximity. If it can be engineered, a light sail spaceship would also need an enormous "magnetic parachute", or a loop of magnetic wires sixty-miles in diameter for steady deacceleration across many decades, before it would reach Earth's closest neighbor, Alpha Centauri, two centuries later. Since suns only fire outwards, slowing is difficult if relying on this sole energy-source. In theory, light sails can be turned at a 180 degrees angle to thrust in the opposite direction. However, nobody has yet attempted to send a light sail to Venus, and then slow it to a speed needed to turn it into a permanent satellite, or to turn it around, and return it to Earth. Neither solar electric propulsion, nor light sails generate abundant energy. Both are helpful substitute near suns.

Humans also theorize of laser beams positioned on a planet's surface, which fire at nanoships' sails to propel them to a fifth of the speed of light. This solves the rocket equation since the spaceship can refrain from carrying fuel. It also eliminates the need to place explosive materials in the engines. On the downside, with only laser beam sails, this ship is unable to generate its own energy. So, once it is out of range of its home world's firing line, it would continue at a constant speed until it hits a planet or an asteroid, or deteriorates. It lacks capacity for gradual-slowing as this would necessitate laser beams firing at its sails from an alien world it is speeding towards. How far can the strongest laser in the Milky Way reach? What energy level would the edge of this laser retain near the end of the path? What planet has the energy to maintain an enormous laser burst for the several light years it would take to reach the nearest solar system? If an alien species lived on a sun, it might be able to keep up this extraordinary waste of energy, spewing its death ray across the universe, but no species we have come across has succeeded in this quixotic fiction. When a laser fires through a planet's atmosphere, it loses 60% of its energy. Thus, given the combined dust, solar flares, micrometeorites, black holes, and enigmatic obstacles, it would collapse under a 100% energy-loss before leaving its own solar system. Imagine the precision necessary to shoot at a sail from as little as 3 light years away. This species is more likely to hit a butterfly on Earth from Cadata. There are similar odds on either of these shooting their target.

Perhaps the human-conceived process closest to cadatas' is ramjet fusion. An *Acta Astronautica* 2022 article argued that to achieve twice the propulsion of the Space Shuttle, the ramjet's field would have to be an entirely impractical 93 million miles long. In this technique, a cone scoops up hydrogen gas from interstellar space. Then, a fusion reactor generates energy by concentrating the gas, fusing the hydrogen atoms. Hydrogen's abundance across space allows a ship operating on it to accelerate forever; thus, an efficient hydrogen engine has infinite impulse potential. With unlimited acceleration, reaching the speed of light becomes possible. This process solves many of the problems other strategies introduce. Ramjet fusion does not create extreme radiation. Its explosions are more controlled. On the downside, humans have theorized this hydrogen-collecting scoop might have to be hundreds of miles across to meet the generator's demands. Cadatas have solved this problem by using an electromagnetic field attracting all elements from the periodic table from many miles away to the ship. Thus, there is no need for a scoop. The outside of the spaceship's sphere is generating this attracting force, and matter falls into it like flies to a lit bug trap in a field. Able to process all molecules, the trap can be energized by hydrogen or oxygen. Millions of particles inhabit every square meter of space near stars, and a sufficient count exists even in interstellar space. These are 6 million times more efficient than chemical rockets. Humans are far from comprehending nuclear fusion, and given their prior tests in varieties of the nuclear, they should avoid it. It is best for any species to wait until they know how to avoid destroying themselves with a tool before employing it.

Havould, Niatage and a team of their support cadatanauts were put through a set of evacuation drills before they could start testing the spaceship as it was coming together. The tests on the components included adjusting interstellar communications by flying a shuttle away from the spaceship and attempting to send and receive messages at increasing distances. They also tested the power system by overloading it with excess energy before venting this surplus. The most thrilling tests for the pilots were on the control system, as they uncovered and adjusted any points of weakness where a spaceship was becoming disoriented during extreme maneuvers. Two situations causing the Computer to malfunction were shifting through molten rocks or tornadoes. Every new component and feature were tested uncadated in the bubble with the team secured in a shelter below. Then it was tested empty with the Computer at the controls on the moon's surface, and then in challenging exercises with cadatanauts. The presence of pilots altered the delicate balance of the spaceship's systems. In case volatile chemicals or components were unhappy with their presence, cadatanauts always wore explosion-resistant suits during primary testing.

Early uncadated flights saw several breakdowns of the mini versions of the spaceship. Examen began assembly by re-using parts from the spaceship it had abandoned unfinished because it failed to meet the agreed-on budget. The components of this old weather satellite model were taken apart and incorporated into what became the first tiny Ganeir spaceship dummy. From the engines to the coating, the new design included modernizations over previous versions. The number and complexity of the upgrades left many vulnerable points capable of failure because they were untested alone or in combination. One of these was sent to a distant small planet at the edge of their solar system to test its maneuverability during icy landings on snowball planets. It made the trip on schedule and then began the touchdown exercise, with everybody back on Cadata's moon watching with a delay of around 7 minutes due to its distance 7 light minutes away. As expected, the sphere wobbled and flipped over a couple of times as it was finding its balance on the ice. The low gravity on this planet also failed to hold it on its surface, so it floated up and had to use its propulsion to push itself back to the freezing ground. The Main Computer and a few cadata programmers toiled for 3 hours to create the code to correct these malfunctions. They waited for 14 minutes for the code to reach the spaceship and for the data results to fly back. The clock ran out, but they still did not have a transmission back with an outcome. Assuming a solar flare or some other phenomenon had caused the spaceship's message to be delayed, they kept waiting for many hours.

Early in this wait, the team grasped natural interference was not the culprit blocking the transmission. An analysis of the programming code sent to the vessel highlighted the introduction of three new paragraphs absent when the code was pre-tested on an identical spaceship in their bubble on the moon just minutes before being sent to the traveling ship. These paragraphs commanded the system to switch off its communication system. Thus, it was unable to receive commands from a cadata or a computer on the other end. It relied on its standard programming to survive alone on its distant icicle. The programmer who inserted it used several tricks to hide her identity, but she was just one cadata against an army of investigators. Barely forty programmers had logged into the system to help with the code-editing process. Cesode Lysur's guilt was soon discovered. Without hesitation, she confessed of being hired by the second-place team in the competition, Matchless Fabrications. It was a delayed retaliatory act for Examen's presumed sabotage of their spaceship-model, an attack they blamed for their loss and ensuing bankruptcy. Matchless was hoping to profit from fixing the problem they created and delaying Examen, making it appear less competent before the next contract competition. Cadatas are less rigid about manipulative business strategies than humans; these types of trip-ups are a part of doing business in cadata culture. It was easy enough for Nally to understand Matchless Fabrications' extreme frustration with losing after a herculean effort, as she anticipated dying of stress if she had lost. The matter was dismissed as an expected loss as there was a reserve in the budget for sabotage, which happened at least once as part of all major government project on Cadata. Cesode was fired from the team after a day-long harangue from Loor. Cadatas' ridicule is hyper-tense in contrast with human derision; cadatas utilize psychological torture to mock, jeer and insult, designed to instigate suicidal thoughts, a difficult bar to meet in a culture saturated with routine slander and teasing.

In the meantime, a recovery mission was planned for the petite spaceship anticipated to be working on its pre-programmed atmospheric analysis. They already had a duplicate in the bubble prepped for deployment; a duplicate had to stay behind in case they had to test new programs or physical adjustments to the traveling ship. The substitute was rushed by the robots, and deployed within a couple of days. The second ship followed. It took it two months to arrive on the snowball planet. This second craft was intended to retrieve the first if it could not turn on the first's communication systems. However, the programmers and engineers hastened a crucial pre-flight check, so this mini-craft approached the planet at a heightened speed and at a lowered orbit, a combination sparking it to burn like a meteor in the light atmosphere. The error was in its protective coating, which should have shielded it from damage regardless of the angle or speed of the descent, but instead ignited and melted. The angle and speed were separate problems as they were due to miscalculations in the mass, balance, atmospheric pressure, and g-force. With amazement, the programmers found an error in the math; they had never seen a shuttle's Computer trip over the math in this manner. Apparently, the Mothership's Computer had an emotional response to the sudden departure from Cadata after learning it was a mere double and would never see outer space. This was one of the reasons emotions were turned off in the final design of the Computer flying to Ganeir. After this, it took them five years to build another model spaceship and to send it back out to the snowball to once again attempt recovery. The third model benefited from the previous failures and avoid repeating them. It turned on the first ship's communications to send collected data back from a five-year exploration of the snowball's surface. The third model also collected the scattered debris from the second model, which contained enough nanobots in it to also send failure reports back to Cadata. While these might seem to be a series of diminutive calamities, these were great results demonstrating several new scenarios cadatas could perish on an interstellar journey, and how these might be avoided.

For the power-overload test, Havould, Niatage, four supporting cadatanauts and two engineers took off from the moon base on a test version of the spaceship. They accelerated to a tenth of light speed with a sudden burst, increasing the gravitational pull on their bodies. This deliberately pushed the engine to the edge of its capability and put maximum pressure on their forms as they were squeezed into the soft protecting cushioning along the walls. The acceleration maneuver was programmed with precision into the Computer in advance to avoid cadata-generated errors. Just as they hit their target speed, a critical propulsion component failed, spewing a leak of toxic propulsion chemicals into the cabin. The Computer had been monitoring radiation levels inside in case of a leak, so it sounded the alarm before the leak was visible or olfactory. The team attempted to escape the danger through the hatch, but it was stuck. Everybody was wearing spacesuits. The Computer did its best to filter these corrosive chemicals from the air. It sent robots to begin sealing this interior leak. Meanwhile, the engineers managed to open the hatch, and escaped into a safer space, closing the hatch behind them. The robots remained to finish the fix. When they thought they were safe, the Computer announced there was a leak on the ship's exterior as well. They could not afford losing any of these fuels in this isolated region of space, having just run a power-drain test, which exhausted their fuel reserves. Niatage volunteered to spaceglide outside to repair this damage. Being the sole volunteer, Niatage's proposal was approved. He moved through a rapid decompression bubble to the spherical exterior of their spaceship. The ship had stopped accelerating, but it was still traveling at a tenth of light's speed. Niatage remained near the side of the ship without departing on a float through space because he was also in a state of motion, going in the same relative direction and speed as the ship. The same rules apply to Earth's astronauts as they spacewalk in orbit while their shuttle is moving at 17,500mph. The extreme speed escapes conscious-registry because of the lack of wind and gravity resistance in space; these forces cause turbulence on atmosphere-rich planets. Niatage could see the explosive liquids and gases streaming out of the part in need of patching, but he could not hear it hissing or firing off reactions at random because of the surrounding vacuum of space, where no sound can travel, including from these cataclysmic bombardments. Niatage did not come alone; he brought a space-safe robot with him, which he now released. Under the Computer's control, the robot knew the optimal sealant to apply to the leak to repair it. Niatage was there to supervise in case the Computer's perception of the problem failed. Machines' instincts are unpredictable as these lines of code demand spontaneity, in contrast with their standard logical responses. No amount of education or programmed algorithms can stop an undesirable reaction in extreme and unforeseen circumstances. The robot banged with violence to push a loosened part into place. Then, the robot melted the split parts together with heat. Niatage heard only his own heavy breathing as he studied the robot's apparent frustration. Niatage tried, but failed to recall intuitive directions superior to the Computer's strategy. In the interim, the Computer managed to stop the bleeding, and offered a report on its progress to the crew. There was no immediate danger, and the single robot sacrificed to the exercise rejoined the other robots to finish a thorough repair before they could resume the trip back to the CSEA complex.

After they returned from this expedition to Cadata's moon, the engineers realized they had to end their reliance on doors, hatches or other safe-on-Cadata conventions in outer space. All doors are vulnerable to blockages from heat or cold; these vulnerabilities had to be reinvented for higher-risk conditions. Nally led the engineering team in designing an ultra-thin but hyper-strong sealant material for utilization between sections of the ship to keep them isolated in case of depressurization, loss of air, or other catastrophes. The ingredients composing this sealant turned liquid only when it sensed a matching chemical composition to a cadata's or a robot's body. Alien visitors had to be pre-entered into the Computer's system to penetrate this fabric. Thus, critical members of the team could escape into a different space in moments without needing to open or unlock an obstruction. This proved a savior in many incidents along cadatas' travels between the stars.

Meanwhile, Nally's team learned the leak was intentional. The exterior and interior surfaces of the spherical spaceship were weakened with a mechanical device, which left a thin membrane intact to allow the ship to approach the maximum-acceleration point before breaking. It was calculated to cause a complete depressurization and explosion, but the math was a fraction off, allowing the team to enact a repair. To find the saboteurs, security measures at the CSEA complex were intensified, and the Examen wing was sealed off to all non-essential personnel and members of rival teams. With the scene cleared, robots performed an investigation of every molecule's movements across the bubble; focusing on molecules proximate to the spaceship's pre-flight testing parking space. Meanwhile, the spaceship's Computer checked for hacks within its system. One of the Computer's proposed culprits for this problem was a forced entry into its system to program it to order its own robots to carry out the damage, before deleting the sabotaging code. But its assessment of the input it had received and various potential weak points proved it had been done in some other way. The central Computer for the moon stepped in to add computing power to fix the problem, and CSEA's top programmers led this extended investigation. They examined energy consumption patterns across the moon in the previous year. They categorized these by who was utilizing energy, where they were located, their movements, and their political and business affiliations. This detailed four-dimensional map had many repeating patterns. The participants in the Ganeir competition moved between their gates and their living quarters. CSEA employees came to and from their main office and traveled to oversee different gates. The map showed the movements of the complex' 493,429 robots as they traversed the edifice to check components' integrity against the cold and desolate conditions on the moon. None of these players had traveled near to the damaged parts of the spaceship nor programmed a robot to do so. The final data set was on the 735 visitors to the lunar complex across that year. Of these, ten had an affiliation with a new-formed anti-Ganeir activist group. They were allowed to visit the base because they requested to do interviews with the CSEA leaders on the moon, and to film the progress of the Ganeir spaceship's construction. Their movements before the failure were chaotic, which triggered the Main Computer's closer scrutiny. All ten had done a walk-through of the bubble in the three pre-flight prep days. One of them, Reald Ditex, stopped by the Nano Lab without an appointment and had lingered there unsupervised while the front desk secretary was on break. Reald was located down on Cadata by an inspection computer and escorted up to the moon for questioning and testing. Traces of nanobot ingredients were found lingering on his wings. He was asked for a reason for this unnecessary-for-a-journalist contact with nanobots. He resisted at first, but the Investigation Computer confused Reald into confessing he did steal nanobots from the Lab and programmed them to cause the failure. He expressed, he was protesting CSEA's plans to resettle cadatas to Ganeir, instead of investing into saving their cherished Cadata, the best of all possible worlds. He was also angry the company instigating the catastrophe was now profiting from "solving" it. He believed CSEA and the Potentates were plotting to manipulate Cadata's broken system for personal profit, instead of out of concern for the billions of cadatas who were already dead, dying, or about to die. The mission would abandon those in need in favor of cruising the galaxy in search for an imaginary nirvana. He denied other members of the group helped his plot. As punishment for this disruption, Reald was assigned to the crew of the Ganeir mission. This was a fitting response because the odds of surviving this extreme test were calculated by the Main Computer to be .003%.

Gene Editing in Space

A debate causing strife and warfare on Cadata since 1,003,726 BC, continued dominating its space program. Should cadatas edit their genes when they attempt to settle a new planet? Was it better to find a planet suitable for their existing genes? Alternatively, was the planet's stellar neighborhood of primal significance, while the rest was pliable to terraforming to match cadatas' needs? Perhaps, what both sides were missing was an understanding of the other's perspective. To address these questions, we must begin with the definition of *genes* in this context. Earth science speculates life originated on your planet in phosphates. Phosphates can only form in the extreme-cold vacuum of outer space, at temperatures as low as 5 K or –450°F. They form on meteorites and comets, which are theorized to have hit Earth in its first billion years. To clarify, meteoroids are fragments breaking off during collisions between asteroids. When they burn-up in Earth's atmosphere, they are called *meteors*. If a piece lands on Earth; then, this discovered rock is called a *meteorite*. Human life began in tiny water ice grains covered with carbon dioxide and phosphine. The latter causes frostbite on-contact in humans today. It had stimulated the development of "biorelevant molecules" in the vacuum of space, when it was subjected to ionizing radiation from stars. The life-igniting products of these reactions were phosphorus oxoacids, such as phosphoric acid, and diphosphoric acid. These compounds could not have formed on the merely warm, young planet. Thus, all biomolecules in cells and chromosomes developed because of these deposits from space.

Each gene-information-carrying cell is composed of building blocks, such as phosphorus oxoacids on Earth. Life might start with other compounds as its base. Life needs a gradually-developed coded program, carrying with it the instructions for how to create the next generation, and how to survive in the unique environment of a planet. Life on Earth reproduces by splitting DNA molecules, or the double helices, in two, and having each half attract molecules to complete two new sets. On other planets, different structures and processes spread life. Genes might form a circle rather than helices. Elsewhere, these molecules might split into a thousand instead of two pieces in an instant. I use the term *gene* in this book to represent any variety of transfer methods, which reproduce, multiply, prolong, or otherwise foster life.

To understand how cadatas' gene-editing functions, begin by questioning if the human genome is the sole possible life-organizing system. Human DNA is a combination of elements from the periodic table: hydrogen (1: this is its atomic number, a measure of how heavy an element is), oxygen (8), nitrogen (7), and carbon (6). It can also have sprinkles of other elements such as phosphorus (15). Human DNA is composed of strings of these elements connected with atomic bonds. DNA is un-unique to humans, animals, or plants. It is shared among lifeforms on Earth, with the same elements (H, O, N, C) repeating. Merely their proportions different across the branches of life. Human theories propose only Earth's atomic building blocks are capable of springing life across the universe. Far-fetched human philosophy about alien life merely allows for slight variations. For example, the human DNA formula might be reproduced on a hypothetical alien world unchanged, except for the substitution of a heavier chemical from the same type-group as carbon or silicon. Thus, they propose, instead of water, an alien world might utilize alternative solvents, such as ammonia, to bind these same elements. Through our exploration of the Milky Way, we have discovered unexpected combinations reflecting the unique compounds available on a planet and the reactions between them. Earth's DNA structure is common on planets with similar chemical makeup, temperatures, and atmospheres.

Regardless of the original elements responsible for birthing life, each planet's lifeforms go through a billions-of-years-long process of evolution. Nature's small mistakes during replication always split a single lifeform into numerous lifeforms, which share components equivalent to DNA in common. Variations eventually cannot reproduce or communicate with each other due to the changes in their genetic makeup. On Earth, DNA mutates at a rate of 1 percent in 1.5 million years, and this rate is about the same across the Milky Way. The more time has passed since life began on a planet, the more diversity can be found unless extinction events have eliminated all but a few species. Humans have a 4% DNA difference from chimpanzees. An equivalent on Cadata are tenals, a pet we use as a companion and to guard our homes in a manner mimicking human dogs. Tenals diverge from cadatas' genes by 8%; we share a single genetic ancestor, and are both in the Mutant Branch of our evolutionary tree. Human science fiction depicts alien planets as populated by a single humanoid-adjacent species. This fictitious alien might have an elongated or a truncated humanoid form with green skin, or with deep ridges on the forehead. These depictions are understandable given Hollywood's low budgets for innovation. Instead, this industry focuses on funding starlets' plastic surgery and on marketers who calculate which formulaic plotlines to copy based on prior ticket sales. With such expenses, the budget lacks leeway for the creation of genuine, alien eco-systems, and evolutionary trees of lifeforms. Why not just duplicate an identical beach, with palm trees and ferns, on every other trip an exploring crew visits? Perhaps, humans would be more interested in funding space travel if they did not base their ideas of life on alien worlds on these films. Who would you want to travel 100,000 light years to end up in a forest identical to the one nearest you? In fact, a modicum of human biology, or of Earth's flora and fauna has been repeated in this enormous universe. Any two humans are only .1% different from each other. Any two species from two different worlds can-not have even a .1% similarity to each other. A human-sized book cannot encompass explanations of these 99.9% divergences; despite this challenge, I will attempt to tackle the major dissimilarities. Two unedited cadatas look near-identical to a human, just as humans look near-identical to us. This is one of the reasons I am refraining from de-scribing characteristics such as wing or eye colors in this history, as these slight genetic variations are insignificant on this galactic scale.

On Earth, gene therapy is divided into two categories. Somatic cell gene therapy is a process where non-sex cells are altered; these changes only affect the individual edited. In germ-line procedures, sex cells are changed; thus, the following generations also carry the mutated genes. The latter is necessary to screen a genetic disease out of a popula-tion. Cadatas alter all cells (sex and non-sex) in an edited body, including the structure, molecular makeup, and design of the cells to adapt the body to a new planet. Cadata scientists had tested these changes in microbial life to fit with microbe-equivalents they located on a distant planet in their own solar system. The voyage to Ganeir created a new challenge of wholly mutating cadatas to fit with an alien species, rather than a species from the same evolutionary tree or solar system.

The ethics of this extreme mutation went to the heart of the question: what is life? On Earth, the definition does not fit cadatas' because our ability to reproduce and evolve is not tied to nature, but rather to our mechanical manipulations. Cadatas control our own evolution. If we notice an evolutionary trend unbefitting to our goals as a civiliza-tion, the changing gene can be suppressed. When we needed to improve our vision to see on all wavelengths, we avoided waiting for natural selection to grant us this advantage. Cadatas expanded their vision to cover both visible and invisible light. Our eyesight is also hyper-precise and telescopic; we can focus on a tiny detail in the distance or zoom in to a microscopic bacterium in front of us. Some of this precision is due to our incorpora-tion of nanobots to improve our bodies' functions. We can see infrared heat radiation as with night-vision glasses. This ability is necessary in the darkness of space and on worlds with dim or absent suns. A moon might sustain life with internal heat, but its planet might have an orbit so removed from its sun, it lacks a "daytime" period created by a huge fiery ball of the sun; instead, when the sun rises on the horizon, it appears as dim as the other stars. Similarly, extreme hearing is necessary for cadatanauts on worlds with thinner air, where sound fails to travel far.

To return to the meaning of "life": machines can evolve and reproduce too. Are they alive? If cadata rulers lift the 5,000-year limit to lifespans, allowing for perpetual living, would we exhaust our planet's available space, forcing us to seize reproducing? When the day arrives for the last new cadata to be born, would this be the day when "life" stops among us? Some explored species in the Milky Way have near-infinite lifespans because they are comprised of strong materials, which resist degeneration, unlike Earth-bound biological lifeforms. If a species is metallic rather than liquid-based, does this make them less of a lifeform? Can rock or metal be the star stuff at the foundation of superior alien races, or might they be screened out if they do not reproduce, or do not breathe or perform other functions humans have defined as essential to life? For a metallic lifeform, human reproduction and death appear as catastrophic failures for the species. Mountains multiplying themselves might appear as a life-destroying disease to aliens, but might be the definition of life for these mountain-species. If metals degenerated and were consumed by other species within a hundred years (as Earth-based lifeforms age, die, and decompose in the ground or are eaten by other species if left on a forest floor), this would be a structural flaw in their nature. From a metal's perspective, life on Earth has the characteristics of a terminal illness infecting the planet. On the cosmic scale, why would the constant turnaround of nutrients or molecules engendering atmospheric or environmental tensions be a positive development for a planet? Rapid molecular change introduces an unknown component destabilizing a planet's atmosphere, leading to extreme ice ages and hothouses. During the *Catalog's* creation, Nally and her team questioned if "life" should already exist on an ideal planet. And is "intelligent" life superior or inferior from the perspective of new arrivals. One camp argued for a rocky planet without fungus, bacteria, or other threatening and unpredictable elements. On a clean slate, cadatas could insert only the chemicals beneficial for their existence, creating a system with fewer environmental threats. The anti-life side lost the debate because Nally proved: as living organisms, cadatas need other lifeforms for their peace of mind, if not for consumption or exploitation. Knowing we are not alone on Earth is soothing, even if we strive to keep our presence secret from humans.

Ahead of the Ganeir mission, cadata scientists were researching utilizing gene editing and nanobots to fix cell degeneration during cell division, responsible for erroneous mutations manifesting as aging. On Earth, cancer cells replicate with precision; they avoid degenerating as they divide; these qualities of diseased cells mimic those needed in cells in an immortal organism. One distinction between perpetual and cancerous cells is the latter's uncontrolled divisions. Cadatas biology resists disease such as cancer because of our bodies' immune responses to unwanted cell growth. The process of our bodies' natural mutation in response to environmental changes also purges toxic cells.

Given the nature of life, cadatas have been debating if an individual's identity can remain intact if their genetic code is altered. Cadata's scientists were advancing gene editing into a science allowing adults to change their entire bodies, rather than only mimicking other species' appearance with cadatas' natural mutation trait. To counter these developments, rebel purists emerged who viewed these extreme changes as destructive to cadatas' nature. Full gene editing was carried out on embryos, allowing the edited cadata to emerge into consciousness with a single, stable comprehension of selfhood. The reason to do an extreme gene mutation of the structure of an adult cadata might be necessary when there is a short duration between arriving within accessible distance of an alien species' genes (perhaps in orbit around a new planet) and the scheduled deadline for exploring or beginning the settlement of this planet. When editing an adult, memories and consciousness must be preserved, while the brain's structure alters in shape and location in the body. Such a radical mid-life transformation is impossible with most species, but as shapeshifters and long-time gene-editors cadatas' neuron-equivalents store this data across their bodies rather than in single organ. Thus, these individual neurons can be repositioned and reshaped without damage.

To pacify both sides, it was decided the Ganeir mission would be split in half, with half of the cadatanauts editing their genes to fit with the species on the new world, and the other half remaining in their cadata form. If the planet was not survivable for their biological needs, they preferred terraforming. Because the pro- and anti-gene editing sides' interests clashed, they refrained from communicating across party-lines. For example, money could either flow toward gene-editing, or terraforming, and a fiscal lack on either end could have been catastrophic to the survival of one of these groups. Each of the sides refused to help the other, and did not even engage in communication to avoid giving away secret information. This split created one of the only permanent architectural walls in the spaceship between the two sides, allowing the atmosphere on the gene-edited side to match species-specific measurements. The two camps haggled over which side would house the best measuring equipment, shuttles, spacewalking suits, and nutritional air or food producing machines. These were built in duplicates to appease both sides. A super-majority of the leaders of the expedition were anti-gene editing because they succeeded in convincing authorities: those who underwent gene-editing were vulnerable to unwanted alterations to their reasoning, and moral priorities.

Loor, Nally, and mission-organizers decided the on-board leaders were obligated to support the political priorities set for the mission prior to departure. Re-directions from exploitation of a potential planet's economically-exploitable resources, for the benefit of the Potentates, broke the organizers' self-enrichment goals. Warfare with local species, or altruistic rescue missions could be either beneficial for Potentates, or detracting, depending on if resources were acquired at the end of the action. The subset of the team undergoing gene editing was tasked with the busy work of traveling to the surface, and exploring the new planet. Those who refrained from editing remained on the spaceship to perform supervisory functions.

Chapter 4

The Exploration of Ganeir

Back in 100,132 BC, when Nally was hired as a spaceship designer, Examen also sent a call for cadatanauts for the Ganeir contract bid. KoraFePub's destruction had displaced most of Cadata. The public advertisement for a ticket away from the horror on the planet's surface generated a million applicants. Even risk-averse cadatas acknowledged this could have been the only path to survival. It was also a point of civic pride to help find a solution, before the planet's impending death. Examen hired five hundred cadatas, exceeding its rivals. Most hires were engineers, programmers, and niche specialists. Fifty-four were various types of pilots, who had practical flying experience. Space-flight training was required for all hires. If a catastrophic incident struck a spaceship; the indispensable function among a handful of survivors was navigation, without which, the rest would shortly parish. Training lasted across the entire 351 years between their hire, and Examen's victory in the competition, in 99,781 BC, and then across the mission to Ganeir. They were even training during the 300-year delay the Committee artificially enforced because Loor refused to lose just because he let his team take it easy during the break. With this time-abundance to learn, these cadatanauts were as familiar with engineering as with piloting by liftoff. Grounded trainees were reserved for potential deployment on the next expedition.

While essential training was universal, there was also job specialization. Most had practiced a narrow field for millennia prior to this centuries-long assignment. The section-leads met every month to discuss progress, and to reach cross-station decisions. Nally was the Chief Engineer, and she led these assemblies. Safety Controller Elleqow Comeepout worked with our two famous Chief Cadatanauts, Codol Havould and Putair Niatage, to run test flights to evaluate the spaceship's apparatuses. If they found faults in a system, Nally wrote reports to combine their complaints and the Computer's data from the run. Each report went through an in-house committee advising Loor if the errors necessitated adjustments. The Telemetry Engineer was responsible for managing the progress of the teams who were updating Nally's *Catalog* of habitable planets with new data before and during the flight to Ganeir. The Chief Programmer was tasked with adding new code, updating programs as new ideas were developed, and correcting errors as glitches were discovered in the spaceship's Computer. Nally's immediate subordinate was the Spaceship Engineer, who would travel to Ganeir and would be responsible for mid-flight hindrances. Nally considered flying to Ganeir, but Loor insisted her value exceeded the risks of a pioneering voyage. The Chief Doctor was needed in case nano-bots and the Computer's medical expertise failed against unprecedented problems on an alien world. She was also the moral gauge on the ship, who helped to evaluate conflicts between team members, and was able to judge if a crewmember was too wounded to return to work before his or her body regenerated. Even at this top level, most positions were in-practice inactive. They pretended to be working, as they merely watched the Computer execute pre-programmed steps. In outer space, the smallest risk is a death sentence for a spaceship. If a spaceship is a foot off-target, it can squash a cadatanaut attempting an external repair, or it can punch a moon instead of parking on it.

Putair Niatage protruded as the "class clown" at the start of training, before sobering after the crash into Matchless' mockup. Putair was born in Getor to a family who owned an independent weapons manufacturing firm. They were a million years removed from their famous ancestor, Medit (who was as productive in spreading his seed as Genghis Khan), but they and their island still maintained a rebellious spirit, now choosing to profit from this strength rather than engage in warfare. Most of Getor's buildings were damaged by the tsunami and the wind-blast from the asteroid strike, but Putair and the others survived by hiding out in their family's bomb shelter. Putair was home over a break from work when it happened. He was previously stationed in the Potentates' Shuttle Troops in Alleges. As a youth he was trained in the prestigious Technical Flight Institute of Getor. He won the spot in the Troops in Alleges because of his astonishing abilities. Most Getor applicants were rejected solely because of their island of origin. The Potentates' system sidelined those who remained independent in business, geography, and politics from the Potentates' sphere of monopolizing influence. Nobody else in his family attempted to secure jobs in Alleges before him, but Putair had a passion for flying above Cadata's atmosphere, and there was no space program in Getor, so he either had to join the Troops in Alleges, or return to working for their family business after training. Since joining, his physical and verbal comedic routines helped him win friends and climb in rank. If he had been stiffer, he would never have made it into the elite high-tier Shuttle Testing program. He had flown across the Cadata solar system to provide input on shuttle designs. Shuttles allowed more manual control than a spaceship; he was encouraged to perform spectacular tricks during testing to check a ship's capabilities. Putair was an expert in gauging the limits he could push a shuttle to in a specific terrain or wind conditions. He loved the feeling of high-speed flight around and between planets. Making loops around all ten planets in the Cadata system in a single day allowed for the most spectacular variety of views as snowy ravines were left behind, and stony craters appeared. As he flew to the farthest little planet in the system light pollution dimmed, his side view of Milky Way's galactic center elucidated. Because the Potentates used the Troops for business and military purposes, Putair's responsibilities included flying their cargo shipments to other planets. He was always pushing his commander to let him fly more, even if there was no reason to venture out. Nobody in his unit was surprised when he pushed his way into the Ganeir project. Niatage's diary is an insightful source on Examen's process of selecting and preparing cadatanauts for the Ganeir mission. He also describes the disastrous Ganeir mission itself. It is a painful chapter in cadatas' history. Thus, it must be related from the perspective of a cadata who lived through it.

Putair Niatage's Diary Entries

100,132 BC

Media accounts report a million cadatas applied to serve as cadatanauts with the Ganeir program. Hundreds showed up at the same appointed time as me for testing at the Potentates' headquarters in Alleges. After a long wait, they took me into the testing area comprised of a long line of testing stations. We had to use one of these machines at a time, and then make it available for the next candidate. I had been granted a day's leave from the Troops for this interview. I arrived before my appointed time, and rushed through some of the tasks to avoid this exercise spilling into the following day. I had seen some of these tests before, during my Troop physicals. New test-types were for endurance of long-distance space voyages. The strength test would have been impossible for a cadata from a million years in the past to handle, as it required gene-edited, super-cadata muscles. Each of us had to lift heavyweight mechanic components in common need of substitution on a spaceship. Some of these parts had a lot of volume, while others looked tiny, but weight more than 600lbs. The machine also required bendability as weights would come at us from all directions, and we would have to rotate or hop around pushing against each crashing weight before it could hit us in the eye. The loads increased in increments. Many ahead of me gave up after four increases. Even pushing 30lbs is painful for a cadata, without a steady training regimen with this equipment. A dozen of my Troop buddies were also there. We wagered from the lobby on who would withstand these tests the longest. We were pushing our physical limits. I was semi-unconscious and thrashed, crawling out of each machine. I pushed through it not just to win the glide-assist-solar-propulsion unit Codol Havould staked on the bet, but also as a point of pride. The losers would be laughed out of the Troop's locker room in the morning. In the final stage of the weights test, for those who lasted, the weights became robots flying at us from all directions; we had to hop, flip, turn, and do other acrobatic tricks to avoid them or push them away. A couple of guys were knocked unconscious. One was tossed through a wall and crashed into the ground a couple of floors later; he survived after a nano-reconstruction. Just missing a similar fate, I progressed to the last component of this stage. A bubble surrounded the machine and created a zero-g environment inside as the weights kept coming. I was used to weightlessness from flying shuttles across the solar system. But the sudden shift to floating caught me off guard. I pushed myself aggressively into the edge of the bubble, as I attempted to push a weight away. I recovered and performed within suitable parameters. Then came the high-g test, as I felt my body become a dozen times heavier than usual after several minutes without any sensation of this downward pull. 600lbs became 6,000lbs, and at this rate even the most bulked-up cadata cannot lift it, though I tried. Instead, all I managed was lunging out of their way; I guess the goal here is to show you can survive an alien attack on a planet of giants. All this time, nanobots were jumping under my skin or up my nose to test how my body reacted to the stress. If a cadata's body rejected high or low g, he or she was disqualified from the post because this inhibited survival on an alien-g world.

For the maximum acidity test, I was dumped into a tank with a simulated hydrothermal vent at the bottom spewing bubbles of acid. The task was to put on a spacesuit attached in a sack on my back before my body decomposed. Who came up with this one? But I guess I now know what it feels like when your skin is tingling because it's breaking down from contact with acid. So, I will go for the suit pretty quick, if we encounter it on Ganeir, which I've been told is a possibility, since it's an ocean-planet, with some vent activity.

Beyond temperature, the climate test simulated humidity, wind-strength, and pressure fluctuations. One moment, I was deep in hot sand in a desert, then I was drenched in sweat in a steamy tropical climate, and then I was freezing my wings on a replica of our solar system's farthest snowball planet. All I had to do here was stay alive, so I did what I could to hop around in the snow to keep my nanobots active, and tried to find shade in the tropics.

The next station looked like a food court, but as I started eating what they were handing me, I realized it was placed after all those other extremes because it was tougher. Codol later explained feeding us foul-smelling nutritional-gas allowed for weening out cadatas expected to go mad if they were forced to eat this stuff across a 5,000-year lifetime in space. I was livid after a few inhales, but kept consuming it without complaining. This meal contained the minimum required nutrients, with ingredients revolting to cadatas senses.

Few cadatas emerged on the other end of the food court booth, so I estimated my odds of passing increased. Codol also made it. We were both so frustrated and hungry after the last test, we bought extra-sized cheap street gases outside the headquarters to forget this extreme sensory under-load. When the Troop received the test results on our personal computing units, I learned the nanobots judged my body functions and abilities were at the top of my unit. Since the Troop's readings were outstanding, I was prepared when news arrived of my acceptance into the program, together with Codol and six other Troop members. Our rivalry had pushed us beyond reasonable boundaries.

I packed my pod essentials. We were informed, Codol and I would be piloted out to the moon's base with a few dozen fellow recruits as passengers. It would have been strange if Examen assigned a pilot who was not joining the missing to this flight. It's not like we needed added training for the short trip we had made every other week. Once we landed and parked the shuttle in a wing of Examen's Gate, we were met by Nally, the Chief Engineer, who escorted us to our new quarters. She lived in this communal space with scientists and non-pilot professionals because it was cost-prohibitive to add a housing structure outside of the main dome on the moon. The expense of the addition was overruled across the million years of the moon base's operations. We were housed in group rather than individual sleeping bunks as sleep was frowned upon. Cadatas who still required it were screened out of the program. Instead of unconsciousness, we decompressed muscles with devices administering full-body massages. The moon's low gravity was itself a comfort. After a short rest to play 3D games or to daydream, we started a new day by gliding on our wings with skin-tight spacesuits. It was a merry exercise method. We explored the lunar landscape with its craters, hills, and stranger landmasses. The scene was devoid of rain from the sky, or collecting dew on the ground: striking differences from Cadata's small islands in the middle of a planet-hugging ocean.

With our muscles well-stretched, we would file into our designated classrooms for a series of lectures, and training exercises. The space was instantly converted from an auditorium to a gymnasium by the robots, and the liquid-solid mutating walls.

100,130 BC

The last couple of years have been a pain in the eye. I always hated school. It's such a drag to sit there and listen to a professor without flight experience discuss aerodynamics, or rocket science. But my competitive nature never let me drift off or skip classes. I did my best to turn our boredom into amusement for the sake of our mental survival. I'd pull little pranks. I pulled a bomb alarm a couple of days before the class with Safety Controller Elleqow Comeepout, when we were supposed to learn to handle this emergency procedure. The entire moon base had to evacuate into the underground bomb shelter. It must have cost CSEA a fortune just in energy-consumed by having those spaces open, close and activate the nuclear-level bomb shields. One kid's reaction was to put on noise-cancelling ear plugs because he interpreted it as a series of annoying musical sounds from the speakers. Another trainee, in a fit of post-KoraFePub PTSD, took a shuttle and flew to Cadata because he assumed an asteroid was about to blow the moon. I hope nobody reads this diary, or I guess I'd be in trouble. Overall, our class exhibited a failing response: nobody read ahead.

One of the more interesting classes was astronomy with Telemetry Engineer Rudam Sinlace because we studied it in holographic environments, which placed us inside of simulations of what planets, moons and asteroids across the universe are predicted to smell, look, and feel like. When we studied galaxy charts, we did it by programming a computer with directions to a set of planets, and then following this course in moments to check where we would end up, and if we would miss one or more of our destinations. We also helped to test theories about the potential habitability of planets in the *Catalog*. We would program a holographic space with known facts. Then, we tried zany tricks to test cadatas' reactions to stresses from fluctuations in pressure, atmospheric composition, and temperature. One surprise was when a combination of chemicals caused our bodies to steam at a low pressure. In another, our cells exhibited a negative interaction with the combinations of molecules in this foreign landscape. We had fun inventing possible aliens we might find in each of these places, and then setting up a fight with these holographs (which also helped us hone our weaponry and wing-combat skills).

The Theory of Alien Communications class with Chief Doctor Photuate Nackan was at the edge of what I was capable of tolerating. Our scientists had encountered alien bacteria on a single other world in our solar system. Thus, these communication methods were conjectures, which were merely extrapolated from observable communication patterns on Cadata. I interrupted Nackan to object that an alien species would be unbound by rudimentary machines generating radio, or other wave frequencies, or even laser beams. I tried to imagine smell-based communication systems. A species would secrete smells in complex patterns, and their mates would comprehend these, and be able to respond in kind? To the consternation of our instructor, I tested this hypothesis by projecting smells of my own… I was also annoyed by how we were determining the presence of intelligent communication signs. Doctor Nackan argued the rhythmic burst patterns in any communication signified intelligence. I pointed out our tenals could make rhythmic sound patterns to alarm us of a threat of harm, as could a variety of species on our planet. However, in parallel with all ignoramus instructors, he dismissed my observations as irrelevant.

I was still stewing over this injustice when, a week later, we were assigned a new Lab. The assignment was testing the reactions of bacteria chemical-sensing molecules' communication methods after exposure to their preferred nutrients. To prove my point, I snuck into the Lab that night, when everybody was out watching a sunrise over Cadata. I programmed a set of nanos with a complex light-generating language system, which I invented myself just to amuse my fellow cadatanauts. I downgraded the intensity of these lights to a low frequency to allow for the possibility scientists just failed to notice this phenomenon before. Then, I released these nanos into the bacteria and allowed them to take over the bacteria's control mechanisms. These nanos did not show up on the visible light spectrum, so their presence would not be detectable at first glance through the microscope. Doctor Nackan began the Lab by drilling the procedure from the 3D textbook. I was forced to issue a little whistling chuckle as the students began releasing the nutrients into their bacteria bubbles. Perhaps because he heard my snort, Codol was paying more attention than the others. His wing shot up.

"Do you see the expected chemical responses?" Doctor Photuate Nackan asked, surprised by Codol's excitement.

"No, I saw a bioluminescent light!" Codol exclaimed, staring at this little signal. "Oh! Look! Now the other bacteria are also flashing. They're repeating each other's signals. Wait… Now they're flashing in some kind of a strange pattern."

"Those are not bioluminescent bacteria? What are you talking about?" the Doctor questioned, approaching to watch it over Codol's back. While he was moving over there, other students in the class also spotted these signals. "Strange. I see it too," Doctor Nackan agreed. "Maybe the moon has had an impact on the bacteria, and they are now displaying their chemical communications with visible light?" he speculated.

"Isn't this the Potentates Troop's signal code?" one of my friends from the Troop, Denable, noticed. "One of them is saying, 'Help! I'm stuck in a bubble! Set me free!'" He stopped, as everybody glared with amazement. "Here, another is responding… 'How am I supposed to help you, not like I can just file a lawsuit against these evil giants!'"

"I won't have any jokes in this class!" the frustrated Doctor demanded, certain Denable invented a fake secret Troop language.

Codol knew the code, and spotted the patterns falling into its standard rules. "No, Denable is saying the truth… Oh my, I'd translate, but they're being rather indelicate."

The entire class was mesmerized, trying to decipher the messages. They even tried sending messages of their own with artificial lights to the bacteria, and to their amazement, the bacteria replied with philosophical and erudite ponderings. They failed to foil my ploy in the first day. Doctor Nackan needed a week to test for and locate my nanos. Then, he spotted the philosophic artificial intelligence program I uploaded into the nanos, and this explained how the bacteria appeared to communicate with bioluminescence. Once again, I managed to escape detection because I left no trace of my identity in the programming, but Codol guessed it had to be me and he spread the rumor around, which helped me gain credit as an outstanding clown among our group.

Before long, Doctor Nackan returned to his monotonous lectures about how Nally and her researchers had been imaging and monitoring intelligence-hopeful signals from planets in the *Catalog*. I had a bit of fun inventing an alien language Doctor Nackan's alien-capable translation software bombed at deciphering. The homework assignment asked for the creation of a complex set of words in various Cadata languages, but my language was textile, so a creature had to touch another creature's mutating skin to comprehend and respond. It was adjacent to our brail, but it was strange enough for the program to crash from data-overload. Doctor Nackan flunked it. Then, he moved on to teaching us the shorthand language used by the main Computer on the Ganeir spaceship prototype. I was alert during this section because given Doctor Nackan's limited linguistics, and the problems with the spaceship's confused linguistic program, I anticipated our survival would depend on my ability to edit this program mid-trip. The order of commands had to be precise to avoid giving the opposite directions to the intended meaning. Most of this shorthand was mathematical and displayed in the six dimensions of motion in space and time.

My favorite class was Computer Theory with Chief Programmer Guagray Raimingly, as it inspired most of my mischievous pranks and kept me from being fired for them. Raimingly was a programming artist. We learned the complexities of a Computer's processes and how to manipulate them to maximize their accuracy and efficiency in space. Basic programming commands was unnecessary because computers can utilize these themselves with minimum guidance from a cadata programmer. Programmer Raimingly taught us how to imagine a problem from a computer's perspective, and how to create an outline to screen out all possible exceptions and abnormalities capable of causing a misunderstanding between a command and the computer's execution of it.

Doctor Nackan's Space Biology class was a bit more tolerable than his communication skills. Codol and I knew from the start we would undergo full, adult gene-editing before exploring Ganeir to increase our odds of survival in this foreign atmosphere. Pilots had to opt to be edited because they would be needed to drive shuttles down to the planet, thus becoming exposed to its elements. We all had to become doctors by the end of the Space Biology course. Every member of the Ganeir team had to be able to administer medicine on themselves or others in emergencies if nanobots and the available medical equipment was insufficient or inaccessible. We were all issued nondescript medical devices mutable to appear as ordinary inanimate objects or machines common on a given planet. These devices could be programmed to send orders to nanobots to repair breaks, burns or other damage to our bodies. Medicine was automated, but we needed to exceed Cadata's capabilities in anticipation of encountering unimaginable diseases on Ganeir. We also had to understand how to edit genes as well as any gene doctor on Cadata. This medical branch had not yet been released into the public domain of knowledge despite being utilized for over a million years. The Potentates restricted access to gene-editing proprietary information to control the population in need of edits to maintain their normalized 5,000-year lifespan. Because of this cloud of secrecy, I was enthralled to learn how it worked and to attempt it on test cadata and tenal subjects.

I had fun changing a tenal's tentacle-feathers into advanced, bio-robotic, flying devices; afterwards, this tiny, fuzzy creature could flutter faster than our weaker robot models. I let it loose in the dome, where it terrified Examen's staff for many hours, as the robots repeatedly failed to catch it. It remains a mystery why it was set on evading capture. Perhaps, we have imagined tenals want to be domesticated, but their instinct is to escape into the wild. I considered granting a tenal the ability to speak to learn its preferences, but it was illegal to combine tenal and cadata genes. This law took effect after a scientist created a hybrid, who "liberated", through a violent insurrection, the tenals on its island. It died in a fiery battle with a troop unit.

Chief Maintainer Vomere Tenievan volunteered our Spaceship Maintenance class for the cleanup after the tangle with the enhanced-tenal. He programmed robots to repair a hole the enhanced-tenal left at the top of the dome, which was causing depressurization and loss of air to the moon's vacuum. It also disheveled and tore parts off several robots and equipment. Even a microscopic dent in the exterior of a cadata spaceship can cause an explosion at lift-off. Damages had to be repaired in moments in space. Nanobots flowed to the impacted location whenever a flaw was identified. But if the ship is off, this small error might be interpreted as an intended adjustment made by the engineers, and it might go unpatched, leading to a catastrophe. Thus, during our cleanup, we replayed in virtual 3D visual to determine what the tenal touched as it was tossing across the dome. Then, we applied repairs to every dropped feather and missing robot arm. A loose feather on the floor of the dome could kill if a spaceship drove over it during liftoff. Because cadatas' machines are precise in their utilization of the available molecules as fuel, a drop of unexplained dust can send the Main Computer into a computational labyrinth.

Our computers' germaphobia is absurd, as this weakness allows a giant to be conquered by a feather. I did not try too many physical pranks like it after the tenal incident because this type of perfectionist nitpicking infuriated me. I learned a vast amount from it about general contracting, the leading subject in the Spaceship Maintenance course. These types of metal patches become critical if there is a problem on the exterior of a spaceship nanobots or robots cannot fix and a spacewalk by a cadatanaut, like me, is necessary. We also learned about plumbing—the organic systems on our spaceship model moving and recycling waste ranging in size from the microscopic to entire rooms in need of repurposing. We were certified as air conditioner repairers as well, a job involving moving air in precise quantities of molecule-portions to where it is needed. There would be no cheap laborer to call if the ventilation system broke in space. Plumbing and air conditioner repairs are critical scientific ventures in outer space, where a buildup of waste can flood a ship, just as a ventilation problem can empty it. We had to assume the conditions on Ganeir would be so different on the gene-edited section of the ship, there would be no suitable fixes for problems in this mutated environment in the standard *Spaceship Operations Manual* programmed into our robots and nanobots. We would have to write new code and new manual procedures for each such complication, and for this we had to respect this dirty craft. I do not wish this class on Cadata's worst enemy.

The least academic class for us troopers was Safety Controller Elleqow Comeepout's Zero Gravity Training. We practiced it in several different environments: in diminished-gravity out on the moon's surface, in a near-total zero-g Lab, and on shuttle flights into outer space. The Lab was equipped with a mini-spaceship exterior model on which we could practice spacewalks. It had soft, buoyant floors, so we could do acrobatics without risking injury. We utilized all surfaces of this Lab, just as we would on the spaceship, pushing into motion against the floor, the ceiling, and the walls. We had to spend three months living in this Lab to test for adverse reactions. Nutritional-air and necessities were provided inside the enclosure.

Despite spending plenty of time in zero-g before, I had a lot of time to contemplate what it felt like and its significance in the Lab, where I had no duties as a pilot of a ship in the middle of a journey. I tested strange behaviors of objects without gravity. I would let go of a tiny, feathery, vacuum-cleaning robot, and it would remain suspended in the same spot unsupported, until I bumped it. Then, it maintained the same speed as it floated at a wall, bouncing off it and continued floating until the collisions dampened its momentum. I could just leave it hanging there until I needed it again. However, since it was a cleaning-robot, it would return to its charging base as soon as it noticed my inattention. Those robots are programmed to recycle every bit of lint or stray particle. Once I left a computer hanging and they recycled it by the time I returned from a snack. Thankfully, they backed-up my data bits. They judged the computer to be outdated because I had not turned it in for a regular update when it was released by the Potentates a month earlier. I was also enthralled by the sight of my own floating skin, which I mutated to be so thin it was approaching see-through. It shimmered like leaves in the blowing air without gravity to hold it down. We engaged in daily exercise routines to avoid muscle atrophy. I enjoyed doing acrobatics with pullies, which assisted with muscle weight retention. Because they fed us the recycled nutritional-gas we would be eating on the Ganeir mission, I struggled with maintaining a regular feeding schedule. My weight dropped. Organisms composed of stony bone-like materials lose around 1% of this material per month in weightlessness.

When this experiment was over, I maximized my strength to stand up or to move a wing. The act of straightening made the dome spin around me. I knew enough to focus on a single object, a computer in front of me, to stop the sensation of spinning. While the space seemed to settle somewhat, I could feel my eye spinning in its socket. This rotation caused disorientation, and I fell to the floor. I decided to crawl in the direction where my less awkward teammates were headed. Knowing I was joking, they just kept going. But as I was still crawling along the floor when we reached the cafeteria, Doctor Nackan stooped down to examine me. He deployed an army of nanobots to run every imaginable test on me, and diagnosed me with having gravity-fluctuation-sickness. Meanwhile, the invasiveness of having these little bug-like bots—well, most were microscopic, but he snuck a couple of big ones in—floating through my system in copious amounts distracted my attention from my feelings. When it was over, I felt a bit more like myself, and I tested this by standing up again. I tremendous quantity of nutritional-gas came back out through my pores, and Doctor Nackan was a bit too smug as he added the sub-diagnosis of post-zero-gravity-indigestion. I had also lost some optic nerve function due to inflammation from the pressure of fluid buildup on the eye. Apparently, the test they ran on me when I applied into the program failed to identify my propensity for these side effects. Once Doctor Nackan learned of their prevalence, he applied a gene-editing procedure to increase my eye's pressure resistance. The Doctor also mentioned our cadatanaut crew had gained a smidgen of height due to a lack of gravitational compression; without this restraint, our organs expanded to fit this newfound freedom. I was already returning to normal height as the nanobots were finishing their evaluation. Once Doctor Nackan released me, I spent weeks napping until my muscles re-adjusted to even the moon's low gravity. The adjustment period was also difficult when in the following month we had to travel down to Cadata for a conference. I had to be wheeled around like an invalid.

While I was recovering, I enjoyed studying Multi-Chemical Recyclable Propulsion under Safety Controller Elleqow Comeepout. We had to understand how individual chemicals are attracted by the spaceship's magnetic field, how these react to create energy, and how they can be reused. Upon receiving an error message from the Computer of our system running low on a specified chemical, and thus switching to a substitute energy generator, we had to be ready to intervene if the new approach presented problems. If were low on hydrogen, a short detour to a neighboring source would suffice. A shortfall was unanticipated because of the density of matter across interstellar space and the fuel-flexibility of our engines. A more worrisome scenario originated from the build-up of an unexpected combination of chemicals or radiation in the ship's storage banks; these could blow due to a negative interaction despite being stored in separate compartments. Our greatest enemy was any scenario we failed to anticipate before journey. If we could predict a problem, we could program a solution into the Computer. Our team of cadatanauts had to comprehend the complex interchanges between different elements propelling the ship. If they had to be remixed or rethought, it was our burden to help the Computer process and intuit the unpredictabilities of nature. We also had to learn how the rules were adjusted for different scales of cadata vehicles, from spaceships to shuttles.

Without the use of my full strength, I refrained from my signature physical comedy during the Propulsion class, allowing my temporary handicap be the end of many jokes. I recovered my muscles and my sense of fun by the time Controller Comeepout began the next course in the cycle, Physics of Spaceflight. Knowing the sensations involved in prolonged zero-g exposure, we began dissecting advanced gravity topics, such as the behavior of matter in a black hole, or the rules to follow while changing g-force during acceleration or deacceleration. With benefits from nanobots and gene-editing, what were the limits of what a cadata's body could tolerate? Comeepout used my injury as an example when answering this question. I was surprised by many of the strange phenomenon Comeepout explained existed in space requiring specialized programming to edit the Computer-generated flight path. Dust could be perceived as irrelevant matter from the Computer's perspective, but its makeup or speed could require a reevaluation of known physical laws. Even from flying across our own solar system, we had stumbled into a few space-time puddles. Considering dark energy makes up three-fourths of the universe, we spent most of the trip inside of it. Various interstellar space anomalies resulted in computational errors compelling imaginative solutions. Whenever I use the term "dark" for matter or with scientific concepts, I am alluding to the errors in cadatas' astronomic formulas; if we don't know what potholes might be lurking in space, whatever they are, they are dark to us. Out in a distant part of the galaxy, our path was bound to stumble over mysterious roadblocks. I mastered travel through air or aeronautics when I purchased my first shuttle on Cadata. Then, I ventured into space as a cadatanaut. When I traveled at the speed of light in the coming training exercises, I became a chrononaut, as I started moving at such an advanced rate, time stood still outside our ship.

We had to learn to adjust the basic departure program to specific coordinates and flight conditions. We used the same simple gestures to give 3D directions to an intended point in space as in planet-level shuttles, but instead of atmospheric pressures, temperatures and moisture along a route, space measurements were displayed and options for avoiding or targeting certain phenomenon were presented by the Computer. It was not recommended, but in an emergency, a cadatanaut could point with a digit to a star or planet, thus commanding the spaceship to begin the trip on auto-pilot. In a standard flight, all directions and theoretical considerations would be programmed into the system long before takeoff, with slight adjustments made on the day of departure, and then as-needed on the trip. We learned how to execute complex maneuvers on a miniature, early prototype of the spaceship. Risky flips and rapid turns were inadvisable on the final spaceship because of its relative cost. Once we were comfortable in simulations, we began testing the mockup in practice. Each of these real-world flights was recorded, and saved as changeable training simulations for future generations of cadatanauts.

It is exhilarating when a ship climbs into space before plummeting towards a planet's surface, as I go weightless. Falling is far from a pleasurable sensation. My psychological space-sickness remained despite the gene-edits. The thrill is in the threat of pancaking the ship into a titanic ocean as I am approaching unconsciousness. It is in the strain of struggling to delay a turn until the ideal moment for a manual exit out of a freefall. It beats cadatanaut fangirls and rival untamed thrills across the galaxy. I guess this is why so many species fly on Cadata. Life has an innate need to separate from the ground to defy gravity. After seconds without gravitational pressures on my imperfect body, the ship bounces back up with an instantaneous shift from zero-g to up to twenty-g's. I'm breathless as my body is compressed into the soft cushioning gel around it. Average cadatas prefer minimum pressure on the ears, or general gravitational pulls or lightness. Anti-gravity and pressure-stabilization systems accommodate these dislikes by equalizing the experience to mimic stable conditions on the surface of Cadata. I turn all that stuff off. These extreme sensations are the reason I leave my abode to soar.

A high point in Spaceflight was when Controller Comeepout told us to deliberately turn off the stabilizers to test the sensations of breaking the sound barrier. With a few dozen of us observing, each of us climbed up to a high altitude and then accelerated to Mach 1. I was the last tester for the day, so the spaceship's coated fuel was running low when I started. Once I exceeded the others' acceleration rate, I managed to sap the remainder of the fuel reserve. The ship recharged itself by taking a gulp of air, which had plenty of atmosphere in it as we were flying low for a spaceship. I imagined I heard the skin of the ship give a slight boom as it inhaled and then pushed out processed elements; I did not audibly hear it because I had reached Mach 1, and was moving faster than the speed of sound. The silent outside stunned me. Our spaceships' skin allows safe soundwaves through, so I would have heard a bit of wind-resistance at a lower speed. A ship's coating can magnify and translate visual wavelengths into sounds new stars make as they gather materials and pulse with high-frequency waves. In theory, transmutation into musical noise of interstellar bodies surrounding a spaceship creates ambience, and a sense of place.

100,126 BC

I'm back from the dreariest and yet the most petrifying trip a cadata has attempted. Breaking the light barrier was a time-intensive trial. We were only allowed to do it once, and as a group. Our instructors and leads accompanied us cadatanauts to test the speed-of-light-travel hypothesis. The first step was leaving behind even the moon's weak gravitational pull. Once we were clear of distracting influences, we positioned the shuttle on a trajectory avoiding an impact with planetary, asteroid, sun, and larger objects. Long before takeoff, we had programmed the Computer with a path avoiding hostile space phenomena. Beyond being terrified of entering unexplored space, we feared the worst from spending a year speeding up to light speed. The Kerod Retan comet mission proved cadatas lack psychological traits essential for long-term spaceflight. Since my physical failed to reveal my muscle and ocular problems, I was terrified of doing a kerod and crashing our ship into a star just to sense natural gravity again. Codol and familiars thought I was the weak link and likeliest to snap. Codol even smuggled in slothmoths to sedate me if I began over-flipping. I doubt my anxiety was higher than the others, but these suspicions exaggerated this fear in my imagination. I was determined to resist irrational temptations.

Around ten months into the journey, we reached 90% of the speed of light. We had not turned on anti-gravity yet to preserve energy, so our masses had doubled in contrast to what they were when we departed. The higher the speed, the heavier we felt even if acceleration was at a relatively low and steady g-level. Through the transparent windows, the path of stars ahead appeared to be warping into a tunnel. Their glow was a deep-blue because relative to them, we were approaching; an advance creates a blue color-shift. When we looked back, the stars behind us were a menacing vibrant red. We switched on anti-gravity to glide, unimpeded by the increasing g-force. This gave semi-comfort after those tense months. So we decided to call Cadata to teleconference with Nally, Loor, and the other administrators back home, to gloat about our progress. We made the call before thinking through time-differentiations. We were now traveling at near the speed of light for a few months, so we were half-a-light-year away from Cadata. We waited for a few minutes without anybody picking up before we all broke out laughing. The change in mass or perhaps our extreme must have been distorting our ability to remember and comprehend. It would take half-a-year for our message to reach Cadata, and if we remained stationary, it would be another half-a-year before Cadata's response reached us, but we were moving in the opposite direction, so it would take longer. We were approaching a stop after reaching light speed, when we received the delayed response from Cadata. Their 30-minute message by their clock was condensed into half its length and played in only 15 minutes. We strained to understand what Nally and the others were saying at double the speed of normal speech; the Computer slowed it down after we had a laugh trying to decipher it.

We had sent a few other periodic status updates and they sent some critical software updates in response to some unexpected issues we experienced and used to send error reports. The Computer kept track for us of what day and year it was back on Cadata because the changes in speed would have been impossible for any cadata to calculate. We stopped counting the days on board of the spaceship because we did not want to think about moving back in time during chunks of light-speed travel. Without sleep or natural sun-cycles, time became a philosophical conundrum rather than a natural progression. All this work was for three weeks' travel at light speed at the end of a year of steady acceleration. The Computer auto-shut-off the engines when it hit light speed. We did not feel the loss of acceleration because of the gravity-stabilizing field keeping us from attaining near-infinite mass near light's speed. The exterior went black, as our speed now matched light's visible spectrum. This darkness became menacing after a few hours, a similar effect to continuously hearing the same high-pitched note on a string. We turned on a virtual reality program generating the illusion we were on an island resort to avoid panic. We tossed around the idea of ending the experiment after only a week in this black void. But the Computer objected and overwrote this lazy idea. We decided it would have been too aggressive to turn its controls off to change the schedule ourselves. The Computer had retaliatory subroutines for instances when its preferences were superfluously overwritten.

I was personalizing the Computer more than usual in those light-speed weeks, as I started suspecting it was a counter-power in the ship's dynamics. Nanobots were taking readings of our systems across this stretch, and proved time travel had caused a few malfunctions in our bodies, including slight psychosis due to mis-firing of brain cells as bits of light or radiation were traveling at such a slow pace they were frozen in critical pathways. After we landed and before our trip to Ganeir, this error was fixed through gene-editing and programming nanobots in our bodies to spot these interferences as they were happening.

At the end of the three test weeks, the Computer turned on its propulsion, beginning the deacceleration process. The brakes were applied at a steady rate, so we felt around the same as we did when we were standing on Cadata. It took us a year to come to a speed of near-zero; then, the ship's propulsion began pushing us back to Cadata. The full trip took four years. We had not stopped at a single planet or even a stranded asteroid. This was the first time a group of cadatas broke the light barrier on a trip through space outside Cadata's orbit. So, we were overwhelmed with media interviews. Everybody was clamoring for a comment on how it felt. At most, I formed five words at those junkets because my brain was medically-speaking scrambled. I am utilizing the holographic videos I recorded on the trip to enhance my recollection as I am composing this description. Prior to this refresher, I failed to recall the number of years it took. And despite these reminders, I still must be mistaken regarding the timing of message-travel between Cadata and our ship due to unaccounted intricacies. The nutritional-gas we ate, the resting-accommodations, and the rest was too dull for memorization. The bits shocking to my memory were those of crisis. I recall the terror of a decompression leak, and the semi-undigested-by-engines solar-storm causing extreme turbulence at light-speed.

The pressure-exposure stimulated a semi-extinct virus, sending me into weeks of feverish delirium. De-acceleration from light speed was executed at 1g. At least my mobility was not as impaired as after the weightlessness test. The gene editing had helped my well-being when the gravity-stabilizers were off, and during the return to normal weight after this period of weightlessness.

99,532 BC

The last 594 years fatigued me to my limit. Before, I had complained the trip to break light-speed was painfully long. The regressive runup to the Ganeir mission forecast it would conclude with Cadata's demise, instead of a liftoff. In the first four decades, I maintained faith the contract was merely days away. When the competition was cancelled, and put on a 300-year sabbatical, we were all despondent. We celebrated to the outmost when Examen won. If Nally and her power-team had advance knowledge of our victory, they refrained from sharing such speculations with our troop, so it was a wild bet. And it still wasn't over because two more centuries slipped, as the engineers, programmers, and other teams worked to build the spaceship they promised in the writeup.

The instructors continued these classes for three decades, perhaps the universal constant for maximum schooling duration. At the three-decade point, it was decided the Survival Training portion of the course would take place over a series of extended stays on each of the planets in our solar system. Our first few trips lasted for a few years per planet. Each expedition was followed by a return to Cadata or the moon to attend a party-conferences, or for a press junkets to remind the media we were still working towards the Ganeir mission. Our desire to visit waned with every return to Cadata. Pollution and the greenhouse effect spiraled until it became mandatory to wear full-body spacesuits outdoors, as if Cadata was only a foreign, inhospitable planet unsuitable even for our exploratory *Catalog*.

This turn heightened the urgency of our mission, but my angst was unfelt by the bureaucrats. Millions began dying from exposure due to the high cost of gas helmets and spacesuit. It took 632 years after the strike to build and staff an interstellar spaceship. The public failed to find the astonishment needed to rebel because the same circle of Potentates had retained power despite causing the downturn. If planetary destruction was insufficient to spark a revolt, a steadier environmental decline merely prompted depressed complaints.

With the intensification of storms and radiation on Cadata, the worst environments in the *Catalog* became normalized. On our expeditions, I served as the Chief Cadatanaut with Codol as my second. We brought the bulk of the other cadatanauts and testers on these trips. The journey to these destinations did not take long at the acceleration rates our prototype spaceship achieved. Sometimes we arrived on the same day as if we had just flown to the moon. The goal of these missions was to generate data on how cadatanauts could survive in the types of extreme planetary environments we might encounter on Ganeir or on future explorations of the galaxy. There were ten planets in our system, and we spent around fifty years camping on each of them. We brought a group of replicating robots, nanobots, and some heavy metals to get them started on a planet with a basic elemental composition. We were discouraged from bringing personal artifacts, not only to avoid melancholic longing for home and loved ones, but also to conserve mass and fuel on the spaceship. We were using a lot of fuel driving across our solar system, just like you would in traffic as opposed to on the highway.

The first world we sightsaw was our-moon-sized ice planet in the outer edge of the system, Tilain. Its light gravitational pull allowed us to park on its surface without an intermediary. Keeping our resources on the planet was an added benefit, unlike the strain of storing it on a moon as we had to do on gas giants. Moon storage necessitated travel back and force to retrieve computers or machines unreplaceable on a gaseous planet. Soon after we landed and emerged through the liquefiable walls of the spaceship, we sent the robots to carry out the setup of our base. Our programmers created code-orders for this maneuver. Now, they hurried to meet their pre-set deadlines. The contrast between an ice planet and Cadata's steamy tropics meant the greatest risk we were facing was freezing despite our gene-edited resilience. The spacesuits were as protective as any material in the universe can be, but even they had limits. After hopping around a bit to explore the site and to deploy nanobots to check the density of the ice we were standing on, we returned to the spaceship to wait for the robots to finish their task.

I had fun hopping under the threat of exposed liquid freezing in a millisecond, but my survival-drive directed me towards caution. At last, the robots returned with progress reports. They had some questions for us about the structural integrity of these makeshift houses in case of seismic activity or another disturbance. We researched the catastrophe scenarios they forecast, and dismissed all as insufficient to delay settlement of the base. We then ordered them to carry the massive molecular processing machines out to set them up at the edge of the base, so they could begin collecting molecules from this environment and turning them into the nutritional gas and components necessary for our long-term survival. The first batch of nutritional-gas these made smelled and felt like consuming nuclear waste. Whatever substance it was, it kept us alive, and we did our best to consume it without reflection on its essence.

We had settled into life on the base a year into our stay. Gliding with robotic propulsion for entertainment was popular because Tilain's thin atmosphere failed to allow our little wings to achieve a glide on their own. While we had fun with video games simulating Cadata back in its heyday, betting games and practical jokes, the gliding sessions exposed us to a world no cadata had explored for a similar duration before. There had been plenty of tourists visiting Tilain to experience this strange climate, and brief scientific missions, but the Potentates and independent businesses failed to envision benefits to mining ice-water in over-abundance on Cadata, so a long-duration mission never materialized. There were as many mountains and winding slopes to ski down as islands and lava-hills on Cadata. Sunshine was dim even on the side of Tilain face-locked to the sun because of its great distance from the center of the solar system.

We invented robot-enhanced sporting events. Our contribution was the invention of a program to accelerate arrival at the finish line. There were limits to our programming innovations as we tended to reproduce the same pre-tested tricks. One of these recycled code paragraphs auto-adjusted speed and incline to optimize aerodynamics in response to weather fluctuations. So, I still had to trip or push off course at least one of my competitors per race to arrive first. I believed we were all amused, but after years of being knocked off within reach of victory, Codol boiled over.

It was a standard robot-gliding race. The molecule-processing machines were the intended destination. Twenty of us began it. A couple were knocked off route by the gusts of a sudden lightning storm in the disturbed, thin air. Across the two-day endurance race, we subsisted on the nutritional-gas stored in our suits. Because there were no shelters on the route, these vessels had to reprocess all outputs of our bodily functions. I arrived first, far ahead of Codol as the distant second. The last cadatanaut was just halfway through the track. Being second was a psychological handicap for Codol: he was terrified of being dominated.

Usually, when he spotted me, his body limped from fear. But at the start of this race, he thrust through the pain. I did not hear his approach. I was alerted to the change when I glimpsed him to my right. I gave my robot a command to expand the propeller component out until it hit Codol's side. In the past, this trick worked. However, the repetition meant Codol expected it. He used the momentum I generated to grab the mechanic extension, and to fling me over the top of his body. I regained consciousness in a heap with a strange purplish chemical seeping out of a small tear in my spacesuit. My nanobots reported a deadly chemical was eating my body. And the extreme cold was creating a case of life-threatening frostbite at the site of even this small opening, which had to remain exposed until the nanobots expunged the poison. The chemical eating me was safe in its solid form, but became dangerous when my body heat turned it into a pervasive liquid. I scanned my surroundings with my eye. Codol had just crossed the finish line and commenced a victory hop. There was no precedent for him to suspect I was near melting-point. The tingling at the wound-site spread and dissipated as I began losing sensation. My skin and muscles express-degraded in integrity. The opening in my skin expanded until it revealed my innards. I wrote one of the fastest programs I had yet composed to command my robot assistant and the nanobots to wipe the corrosive chemical off me, and then begin the process of re-growing the damaged cells. I also requested a shutoff of my pain-receptors in this area as I could now feel the sting all the way in my wings. The pain-relieving chemicals knocked me out, so my body fell back on the ice. I had neglected to turn on my communication unit to relate this crisis to my friend Codol. In my distress, I believed he was celebrating knocking me off rather than the victory.

When I opened my eye, Codol and the other race finishers were crowding over me. I jumped as the memories returned. I glared to check if my lower extremities were missing. I learned the nanobots succeeded in repairing the tissue. I decided the day was a split between morbidity and amusement.

A more geological-research-demanding expedition sent us to a canyon-covered, 15° F cooler than Cadata, rocky world called Yudrock. It is Cadata's neighbor in our solar system a bit farther from the sun. To address my chemical-spill-incident, Nally's scientific team on Cadata had created an upgraded version of the spacesuits to make them impervious to all natural and synthesizable chemicals in the universe. The new design was transmitted to our robots for fabrication even before the Tilain expedition ended. Now, Nally asked us to test them against the most extreme environment on Yudrock: an erupting volcano. The training exercise was to enter an active volcano and attempt to make it back out again. Our robots had conducted a 3D internal scan of the geology beneath Yudrock's thin crust; this characteristic promoted a high volume of global volcanic activity. The scan determined the precise potential strength of this ongoing eruption, but unforeseen randomness of greater impact remained from the complex interactions at high heat between the unbridled mass of lava below the surface. We had parked a shuttle and set a base on a raised platform near the biggest stratovolcano on the planet. This platform was erected by our robots, who were sent down first as we hovered in the shuttle just over the site. At 11,652 feet, it had an enormous magma reservoir, which had just become active again after years of dormancy.

We used one of the supports to slide down to the blackening at the top and rising surface of the molten rock. As a dozen of us began the approach up a steep concave side to the narrow crater, we had to walk through flows of lava going down the side of the fired-up mountain. I watched in amazement as a stream of dense and yet shape-shifting lava dripped down from a ledge. The lava had been flowing for a while, and now covered 178 square miles. Without infrastructure and living organisms on this land, it looked like an abstract architectural project. Our suits repelled the lava's particles, keeping it just far enough to keep it from hardening on us. Still, if we remained stationary for too long, the lava could dry in a trapping cocoon around the suits. Thus, we remained in uninterrupted motion. We were facing obstacles in all directions rather than just below our feet. Tephra was flying out of the interconnected series of vents at high-velocity, and bits of it hit the suit, sliding off as they hit the repulsion field in bubbling clumps.

Suddenly, pressure congested and burst a few hundred feet to the right of us. The explosion formed a new opening and expelled a cloud of tons of sulfuric acid gas and a fountain of lava. The volcano's shaking disturbed our balance, so we tumbled to the ground. To recover, we had to escape the lava rushing over us. Because Codol was moments slower to respond, our robots had to dig through heavy lava to free his submerged body. They formed streams of pressured gas to blow the bulk of the lava away, before brushing, or breaking away the remnants clinging to his suit. Codol was a bit frightened, but he did not feel the heat or the pressure, proving the suit supplied adequate air, depressurization, heat repulsion and other protections essential in a real emergency.

As we hiked, a lightning sparked out of the dense purple-red, billowing clouds, which were a combination of ejected gases from this event and a building storm system. The rod struck the smoking and firing volcano-top, resonating against the plumes and red-black smoke as if in an electricity-containment experiment. Meanwhile, the opening at our side turned into a red river of lava rolling towards our base. We kept climbing with help from our robotic assistants despite the path offering increasing resistance in lava's density and depth. The robots acted as wing extensions for us, allowing us to hover just above this flow to the tip of the mountain. Walls of lava sprayed over us, and yet we could remain stationary.

The last planetary visit left a powerful impression on me and the rest of the team. The journey was for an extended stay on Elpile, the farthest gas giant in the system. I was checking the measurements in the smaller Observatory of the shuttle as we made our approach. We left the spaceship at a distant orbit because the planet was too massive to escape its gravity even if our Mothership expanded its maximum energy reserves. I was quibbling with the lines of code to make them match with unfolding storm patterns. The light coming from Elpile's series of rings and moons was more intense than the distant sun, and it was flooding out any stars in its half of the sky. We were closing in on the outer gas and asteroid ring with the appearance of a pink haze.

As we passed through one of these rings, I could distinguish on the non-visible wavelength tiny grains of water, silicate, methane, ammonia, and carbon dioxide falling like rain toward the planet. The Computer calculated this ring was losing 123,456 kilograms per second. Thus, it would exhaust its contents within 986 years. There was no sumloss in ring-volume across the planet's rings because new rings formed at a matching thousand-year interval, as new asteroids, and moons collided, and broke apart. This rain was catching, and distorting the light. Rays were playing a lightshow. After a period of low-lighting inside the spaceship, and near-total darkness in space, this explosion of light made it difficult for me to see my instruments in the relatively deeply dark Observatory. A set of rings closer to the planet had the appearance of strings of plasma, instead of typical asteroids. As we sped through the system, one of its 84 moons passed within a few hundred miles of us. I craned to study its cratered surface. Soon we were so close to Elpile's surface. I could distinguish the white, purple, yellow, and orange swirls of condensation. The atmospheric layer was much thicker than around Cadata and had an inflamed bright-orange tint. Still closer, the fire-illusion turned into massive aurora displays caused by the sun's strong solar wind—or rather its ions—hitting the planet's powerful magnetic sphere. The ions were stuck in this field, rushing in a loop between the north and the south poles before hitting the upper atmosphere, thus ionizing atmospheric atoms; the following de-excitation emits sky-illuminating photons. The measurements were updating the current makeup of the planet: rock and ice core metallic hydrogen around it, and further out helium-neon rain, liquidier hydrogen, gassier hydrogen, and the top 10-mile-thick cloud layer.

Cadatas had visited this planet many times, but few attempted to "land" on it, as we were now tasked. This was even crazier than walking up an erupting volcano. The Computer auto-detected the target within a 20-miles range. I saw it a few moments after the Computer showed its location with a flashing light on the see-through wall below: the center of the planet's most massive storm system. It had the appearance of a nuclear explosion with dark-brown dusty clouds in the center and white shockwaves of gas radiating around the edges. A large depression was just coming into view at the horizon. First, a dark line, it seemed to spread as it came closer. The shuttle flew at this round hole in the clouds. Going down this hole, I observed a condensed and steep wall of rainclouds rimming the perimeter. A puff of white mist was in the center of the black funnel.

As we slowed the descent for the next 10 miles, clouds closed and the sky turned from black to a dirty purple and then to a luminous pink. We were hovering into the growing mushroom of the storm. The Computer was on automatic while I checked its progress, ready to steer if it engaged in risky trickery. The intense light outside drew most cadatanauts into the Observatory. Upon entry, they began back-seat-driving my flying maneuvers. Out of these critics, Codol is the most grating on my nerves. I never stop him because his corrections are useful. The lower in the clouds we traveled, the more atmospheric pressure built up, and the heavier the wind's strength became. We have intense wind on Cadata, but these winds were twice stronger than our peaks. The heavy air moisture composition made these gusts feel even worse on exposure. The Computer measured winds of 854mph in the worst parts of this soup of gas.

Wind was swelling as we moved towards this hotspot. The shuttle was spared turbulence because it was auto-stabilized to remain level in relation to the planet's surface. Gravity-assist also maintained a Cadata-1g inside the shuttle rather than the five-times heavier weight we would have experienced if we ventured out without anti-gravity protection. Visibility became restricted as gales blew in a fog of concentrated molecules. Separate clouds were turning into a surrounding gassier hydrogen. The massive typhoon was still raging at this depth, but now it was seeded with strange formations such as floating dust and semi-liquids boiling lumps.

The Computer spotted the next boundary line underneath. We would not have seen it on any wavelength because of the volume of matter between us. It was a liquid metallic hydrogen ocean, super-compressed until the chemical bonds broke and electrons were shared among the atoms. Droplets of it were floating into the gas layer and being carried away on the extreme winds. Humongous and arbitrary wind-slapped waves of this shiny, liquefied metal were crashing, a frenzy distinct from the rhythmic waves on Cadata with a singular direction guided by our moon. I took over the controls and did a few tricks just above the height of the tallest waves to test the shuttle's maneuverability in these conditions. It passed, fighting against the winds and the high pressure. The temperature alone was survivable for a Cadata without a spacesuit at a hundred degrees higher than our average temperatures. But in combination with the pressure-rate, and the clouds of ultra-hot ammonia and physio-damaging gases burning the breathing organ, it was unadvisable to glide out unprotected. Codol and the other guys dared me to do some tricks without help from the Computer, and I tried a few. This was hazardous because zero-visibility indicated a high likelihood we could have crashed into the metallic waves. I cheated by checking a control gauge in the floor I had not switched off. Otherwise, we would have tested the ship's crashability against metal during a flip. The exercises inspired me and I decided to proceed to the optional phase of the assignment. We designed a robot to serve as a second-skin around a spacesuit; in proximity, it hovered above the metallic hydrogen's liquid phase. It was a superconductor skin with repulsion powers against this hydrogen's intense magnetic field. When this skin and liquid hydrogen are proximate, they quantum-lock, so the body inside the skin is suspended just outside of the dangerous material. Because the skin was never tested on a physical gas giant, this device could have failed; if so, I would have sunk through the surface. Then, I would fall nonstop until pressure broke through the spacesuit's defenses, crushing me into whatever molecules could survive at temperatures hotter than the sun in the center of the planet. Codol took over monitoring the controls, as I donned this concoction. I liquefied our shuttle's wall to slip out. This was a tense moment, as we were unsure if the density of gas outside might have a deteriorating effect even on our ultra-strong shields, or if they would leak gas into the shuttle during my exit. Neither happened, as I was able to hop away from the ship and begin a violent glide carried by an intense gust of wind across the turbulent, fizzing ocean. A moment later, the robot reacted to the conditions and began fighting against the wind. I felt its pressure on my skin as I controlled my position. I glided back in the direction the Computer indicated the ship was. My ears vibrated from the earsplitting roar of the crushing waves, but they were hidden from vision in the muck. I spotted the ship when it was on top of me. I executed the pre-planned walking-on-hydrogen maneuver, so there would be at least a small chance of a rescue if I started to sink. Once triggered, the program creeped closed the gap between the ocean and me. When the first wave hit me, I checked my wing to see if the double-suit would break apart from exposure. It held. I caught one of the rapid waves with my lower extremity. I was surprised when the tip solidified instantly. I was left floating a few molecules in length above it. I hopped between a few waves, surfed on them, and even floated on one wave on my back, as it dropped me and then shot me into the air. Nobody else had ever come this close to an ocean of metallic hydrogen before, and here I was breaking a new boundary for cadata-kind. Codol and the others were livid with envy by the time I climbed back through a wall to allow nanobots to check my vital signs for any abnormalities. Codol even called Nally to ask if he could take the trip too, but she denied the request because she did not want to risk a working prototype to do additional tricks. I had proved its feasibility. So Nally needed this robot back on Cadata to feed its collected data into an army of replicates for the Ganeir mission. After this handover, we returned to Elpile. In the following decades, the team tested the limits of how we could live in and around these extreme substances and conditions.

99,531 BC

The last week before launch was slow to arrive, but quick to depart. Cadata was willing to entertain our wildest whims. Assistants, robots, and even supervisors, like Nally, fetched us rare items no amount of funds could purchase. I was served an exotic, million-year-old nutritional-gas-mix from a Potentate's private collection. We had rooms booked at the five fancy Potentates-owned hotels in Alleges; these were offered gratis by the management in the hope our visit would generate free publicity. When we visited one of these hotel balls, we were served top-secret and newly-engineered intoxicating drugs, which bent reality, slowed time, and performed other mind and body tricks.

Between the parties, we performed final checkups at the CSEA Examen Gate on the moon. I started a work-day at the PRO bunker, where they were launching a war against the press. They had to create a false sense of urgency about the launch. In truth, the spaceship was as finished as it had been a hundred years earlier. Saying as much publicly would have confessed Examen was wasting state-money.

Instead, they had to create the illusion they needed every extra minute in the last week before the launch to make final corrections. Loor and Nally both issued statements: they wished for more time to resolve the last few complications. They had designed these fictional obstacles, forging them into the data. Then, they schemed to appear as heroes when they "solved" these self-induced puzzles in the short gap of time left. One of these fictional problems was a pre-designed failure in the Computer's safety protocols. Fixing it would keep the crew from suffocating moments after entering space. If this was an accidental defect, it would have remained unnoticed until a catastrophic failure on departure. The programming department was helpless in these milliseconds, as they watched their artificial intelligence, robots and the Main Computer attempt to run error reports and fixes. Air deprivation needed an accurate mathematical solution from a detached (and sober) mechanical entity rather than from a cadata's imagination. The programmers would have just spent the time looking at various simulations the Computer would be running to test the system after every adjustment. The Computer's transistors were the size of atoms. This micro-space compacted the power of a cadata's entire brain system.

To prove just how hard they were all working, they began napping or taking restful breaks right next to the spaceship rather than in the crew quarters on the other side of the half-sphere. Loor never overnighted on the moon before, preferring his shielded mansion on Cadata. Now, even he was sleeping in the barracks with his crew. He was transfixed by Codol and I's celebrity. Loor had his private army of robots erect opulent furniture to enrich his stay. The sections of the base he frequented became fashionable, in place of its preceding bare-workspace design. Since he was stuck with us, Loor's goal was the seduction of all secretaries and cadatas of loose sensual inclination. An orgy developed on the first night and lasted across the week. I partook in these festivities as a continuation of the sensual liaisons I partook in since interest in me increased due to the infamy of my Matchless spaceship crash. While supervisors had been facilitating earlier trysts to help morale, they now solicited our participation with vigor. Trustworthy members of the press joined, assuring their "understanding" of the need for secrecy regarding these indulgences. As the drugs and sensual-strain were causing us loss of awareness of time and our identities, the PRO office released fictional memos on our fictitious progress to the media.

Our private orgies were interrupted by the official celebration on the morning of the launch. The press, wealthy business owners, the Potentates and their entourage, families, and friends of the cadatanauts, and luminaries and dignitaries were invited to watch the departure. The moon opened its borders to visitors for the first time, putting them through a heavy security screening to avoid a terrorist attack just as we were setting sail.

Most visitors allowed on the moon were not invited to our gala. They were crowd-actors, shuttled in by Examen to assure the media filmed a satisfied and joyful audience. Public sentiment regarding the delayed launch, extreme over-spending, and the continuing decline of Cadata's environment were such CSEA's chief goal was to avoid inspiring a revolution among the populace. This fake joyful audience sold cadatas on the popularity of the mission. Armed with this fabricated footage, the media could insist their botched statistics of leadership-approval were genuine. These energized actors were stationed in a theatrical sound-stage Observatory five miles away from the launch-site. They watched the event on a projection screen; their initial reactions were not exaggerated enough, so only the fourth take was released to news channels.

I have never heard so much ingenuine applause as at this pre-flight celebration. Given my past celebrity for mere drunkenness, the fakery of past praise should have been more pronounced. Cadatas' clapping methods must be strange among the universe's species: our applause consists of whole-body shimmering between different forms and states. The patterns change with the intended message. Here the theme was "pride". So we became solid statues representing heroes of Alleges. The throng toasted to my success, and lauded the mission as the inaugural flight of cadatas' interstellar future.

I escaped them to spend time in a private room with my family. My sister, Lapo, was crying. I commented I was shocked she had such an emotional response after hundreds of years of listening to me describe our preparations for this mission. She did not elaborate on her fears, and I did not push her. I tried not to think about what might happen. We had some ghastly incidents during training, and yet survived them. I hoped we were ready to tackle an unexplored world. Its environment was average when compared with the planets in our solar system. I explained to her, if all went well, it would be a 504-light-year trip, we might stay for a couple hundred years, and then the trip back would be another 504-years. I was only 3,142 years old, so I would be in my prime at around 4,350 by my return. Instead of playing the hero, I would be returning as an actual hero who would save cadatas by opening a path for our species to thrive as migrants.

After the goodbyes, we gathered on the spaceship to run through the final pre-departure checklist. Our training instructors boarded for this process. Each oversaw a different set of pre-flight tests. The Computer cycled through these pre-programmed checks, replicating the system designed for the final model of the ship tested during the Elpile mission. We had taken prototypes with missing functions on previous planetary visits because we were only awarded the contract in the last few decades. I was grateful for the Elpile trip as otherwise I would have been risking destroying Cadata's primary mechanical hope through inexperience with the model's upgrades. The run-through was still tense for the departing team, as they watched the results rolling across the screen.

The Computer detailed readings from engineering, safety, and intergalactic weather. Individual numbers were within safe ranges. However, in combination, there were three warnings, indicated tiny miscalculations. I worked with the lead programmer to examine small adjustments.

I despise sentimental reflections, but leaving my home planet without a guarantee of return made me hyper-aware of the characteristics I was observing about Cadata, its moon and sun in the final hours. Given the hostile environments we found across our own solar system, would we ever see a plant or smell fresh air again? And would I make an absurd mistake to bring the ship to a crash-landing during takeoff? I questioned every maneuver as I carried out the re-trained steps. The gravity-stabilizing dampeners had to be switched off for liftoff to avoid their interference with breaking free of the planet's gravity. The first wave of vibrations began 6 seconds before the ship separated from its base. The propulsion mechanism was turned on and processed the sedentary fuel sitting unutilized in its system for months. They were now calibrating and collecting molecules from the moon's surface, as it would be the last solid source of fuel until their destination. After these seconds, the ship was refreshed and began pushing away. The shaking and the power of the reactions in the skin of the ship filled our ears with thunder. My vision blurred as the walls trembled and rumbled from the overwhelming sound. This was the worst liftoff turbulence I ever experienced. I left a cleaning robot on a shelf, so it jerked between the floor and ceiling until it was squashed into the former. I saw it bouncing around, but I could not hear the fall, as it was drowned out. The ship began rising from its supporting oval platform. It emitted a propulsion gas, which returned to the ship under the tug of its magnetic sphere. The turbulence was softened by the wall-gel tightening around our bodies after the first tremor.

The worst imaginable scenario materialized when the ship lost power, and began crashing upside-down. The cushioning gel was unconnected to the ship's power, so it continued softening the intensifying blows. However, when it was time for us to aid the Computer amidst the crash, and we commanded the gel to release us, we fell "down" in compliance with gravity's extended will. We grabbed onto the movement-assisting protrusions along the walls of the ship, and fluttered our wings to regain balance.

I called on the main-base communication wavelength to check on status. Nally sent an apology, and confirmation this was a pre-planned crash. As usual, she had a skin-chilling panic attack upon discovering a problem with her system. My first reaction to this was to be thankful they didn't execute this flop with us a few miles higher. A clue was her refrain from telling me to supervise the Computer's analysis of the systems, instead saying they were handling it. This was an order *not* to inspect the results of the checkup to avoid spotting the stupendous hitch unnoticed across all the test flights we executed. I ignored Nally's double-speak by checking on the results of the guidance system alignment test. It showed we were in alignment when we attempted the liftoff. I also ran an extra test to scan for lifeforms near the spaceship right before or during the attempted liftoff, but found no obvious cadata interference with the launch. This was a long-shot, as tinkering with a departing spaceship was risky. I felt the weight of cadatas' horror across the planet. Their one hope for surviving a dying planet had just crashed.

Then, I saw Nally standing just outside the projected crash-site. The top rule drilled into the gate staff for the launch was: nobody is to be outdoors when it happens. And

Nally's job required her to be on the main computer-hub floor to lead the team in problem-fixing. The ship also produced radiation during the charge up, energy recycled by the engine in space, but life-threatening to lifeforms exposed to it outside of a bomb shelter or not inside of the shielded spaceship. Even the best protective spacesuit designed to withstand bombardment by all known elements could not block this entire energy wave. Then, I heard her over the comms deploying the robots to navigate the ship to a safe landing on the base. Her commands were general because by being outdoors she lacked the equipment to review the code or hardware reports on the cause of the problem. Later, she was questioned on her decision to watch in the open by Loor; she said she wanted to hear the power of the engines and to sense the penetrating, disjoined, low-frequency pulsations.

I thought they might fix the problem right away and we would get back on track, but the "fix" took a few months. The PROs milked this delay to generate sympathy for our mission and our budget. Seeing our struggles cadatized Examen in the public's perception. The failures made us into global heroes fighting against gravity in an unfeeling galaxy. Then, when I received my first bonus for the extra time on the moon, I realized Nally and the PROs were delaying a successful launch because each day lost in waiting meant more profit for them as CSEA expanded their budget without reservations. The ship was finished. With the invested money spent, it was impossible for CSEA to retract the contract back now. They were obligated to give Examen whatever extra funds it asked for.

As I suspected, we had two more failed launches. For these, I wore my robot around my middle, inside the gel, so it would keep me from falling, if the ship landed the wrong way up again. Each repetition signified an extra salary-month for members of the Examen team. Nearly everybody would be fired or laid off once we departed, as no further pretend-adjustments would be ordered. If I was staying back on that moon, would I want a steady paycheck for the next thousand years for failing to launch, or for the money to dry-out as the mission continued? No, I wouldn't have. But Nally had to make decisions like this, and maybe she was standing out there at the first launch because she needed to air-out her doubts.

After each of these flops, we had another "sendoff" party with the same group of higherups. When I visited to partake in the giveaways, the same five cadatas approached me to repeat the same sentence. I believe they read the same book about socializing and picked one phrase they could remember. It was ingenious because I never engaged them in conversation, so they avoided revealing their potential stupidity with these repelling non-starters, but also displayed their willingness to be social.

"I'm rooting for you."

"Hang in there."

"Don't hit your eye."

"Try to keep the controls in check."

"Who's driving, you or the Computer?"

The mission's entertainment budget was wasted on this group.

Before the fourth attempt, Nally issued a release: the uncovered problem with the gyroscopic stabilizers was their overload due to excessive data input regarding the long voyage ahead. The systems and nano cells were isolated and self-standing; thus, a collapse of the gyroscopes would not begin a domino-effect on the other components.

We experienced less turbulence on the fourth attempt—it still rolled and shook prior to finding a balancing point. I was further reassured when the ship refrained from dropping when it reached the phase involving rapid horizontal spin on a single axis to establish gyroscopic stability to focus all thrust on the ground to achieve liftoff. It held steady for a couple of seconds. Then, as it began moving up, it decreased spin by stabilizing the ship's natural tendency to wobble. Once controls of movement in six-dimensional space were garnered, the Computer executed the procedure to depart from the moon, positioning itself at a precise point in the sky. Then, it commenced gradual 1g acceleration from 0 to the speed of light. Those watching from the moon saw us hovering just above the platform one second, and in the next the ship was only a dot of light. From our perspective, first the moon and then Cadata also shrunk into dots.

One challenge for cadatanauts during the departure was remaining motionless in the gel until stabilizers could be switched on to return the spaceship's interior to 1g. I wiggled my wings and squirmed as the stillness prompted a desire to itch and cramped my muscles. There was little to do but pay attention to the sensations across my body. The cocoon tightened as vibration intensified, compressing the circulation of air, but not to the point of blocking it. As hours passed, I sunk to the bottom of the jelly-orb for a nap.

I woke when the gel released me at 1g and the robot hugging my body maneuvered it to our new "floor". I pulled up a wall-sized map of the Milky Way and checked our position in it and the time to destination. The countdown clock read 504 years, and the ticking away seconds did not make this seem any shorter. I examined the tiny spark signifying Cadata's solar system. It was near the inner section of the Norma Arm of the galaxy, a dense region, but not as compact as in the high-traffic Scutum-Centaurus Arm, or the high-impact bulge still farther inward tight-hugged by the Near 3kpc Arm. On the wide view of the galaxy, Cadata and Ganeir belonged in a couple of massive neighboring star-clusters. This standard map was measured in sets of 5,000 light years, and we were only traveling a tenth of this distance. We were heading towards the center of the galaxy, to the other side of our Norma Arm. Since we were traveling to the denser part of our arm, it was brighter outside the farther we went from our system.

It was a slight change, but it inspired me to run simulations of what we would see if we traveled all the way to the galactic center. If I spent the rest of my life heading there, I would have gone as far as the Near 3kpc Arm; my descendants would have finished the trip. Aside from time, reaching the center was prohibitive because of the increasing violence of the inner circle. As the simulation showed, the first obstacle in our galactic bulge journey would be Sagittarius B2: a cloud 7,000 times larger than the size of the Cadata solar system. It constitutes silicon, ammonia, hydrogen cyanide, alcohol, and ethyl formate. Our spaceship would be well-fed with these, creating a reservoir of extra fuel to fight any drag or other stressors this dense concentration of space material created. 390 light-years farther, we holographically reached the inner parsec. As soon as we crossed into this boundary, the Computer informed us an overhaul of the spaceship was required to prevent annihilation. It suggested gathering the cloud-materials we had sailed through to generate a thicker protective layer. I ordered the simulation to execute this design plan. It demonstrated thirty-seven failed protections. The crew was restless, so I asked the Computer to suspend disbelief and assume we would generate the needed combination after millions of tests. The simulation was to commence with the trip under the assumption this hypothetical solution was already instituted.

Inside the inner parsec, we saw cylinders of energetically-unmoving lightning lining the sky. Outside our ship, gaseous bubbles echoing metallic hydrogen drifted in perfect spheres. The Computer displayed a 3D image of what an ocean of gravitational riptides surrounding the ship would look like on the visible spectrum. The spaceship's path became erratic to avoid the millions of stars flying at millions miles-per-hour. It could no longer travel at the speed of light to keep from colliding with one of their gravitational fields.

Direct impact with a star remained a statistical improbability due to the vast distances between the stars, but a close-passing star would exert so much gravitational pull on us, it could alter our direction or even pull us into an orbit around itself. Even more dangerous were the black holes in their midst; three massive black holes were falling towards the galactic center, and pulling stars out of their arm-alignments along this violent journey. Other objects took on superluminal motion, as they traveled faster than the speed of light having accelerated from several stellar assists and the attraction of the cluster of massive black holes in the center. This meant we could have attempted to accelerate our spaceship beyond light's speed, but tests in the lab were inconclusive, so we decided not to waste a spaceship on a physical experiment.

In the meantime, needless to say, the Computer estimated radiation readings in the inner parsec maxing-out the scales because there even atoms disbanded into subatomic particles. To stimulate the viewers, the Computer accelerated the simulation from natural-time's light years into mere hours of screening. At 3 light years from the galactic center, all light in front of us gave way to a rapidly approaching spherical darkness, the supermassive fusion of black holes at the core of the Sagittarius A* galaxy. The Computer was kind enough to warn us: continuing meant abandoning hope of return because the force of Sagittarius A*'s gravity was stronger than any force cadatas had yet imagined. Absent real danger, we directed it onwards. The show continued until the Computer attested all matter composing our bodies would have been destroyed within this space buried in total darkness. Despite this sad ending, the team was entertained by this fictional scientific exploration. Encouraged by the applause, I composed several other test trips into unique, and life-threatening galactic regions. It was a long trip, and playing such mind-rousing simulations helped us cope.

99,029 BC

We adjusted to space flight with ease because of our numerous practice space missions. If we succeed or fail, this mission will be etched into history books. The following observations are for the benefit of future spacefarers. Once we reached light speed, after the first year, and allowed the ship to let us feel natural weightlessness, I had time to contemplate our condition.

In fact, we had so much free time in space, we were forced to spend some of it on sleep, even if it was optional for our gene-edited bodies. We could catch up on a nap or an extended sleep-session anywhere on the ship by expanding a gel cocoon from any surface to wrap around our bodies. Unlike inside a house on Cadata with separated hallways, closets and rooms, the ship was a changeable space. Any job or leisure activity could be performed in any space on the ship. Instructions recommended attaching the sleeping-cocoon to a wall to avoid drifting through the air. Since the gel was impact-resistant, if we left it untied during a collision or rapid speed-los, we would still be safe bouncing inside the gel. I liked experimenting with this gel while I was awake, pushing against it from the inside to make it bounce off the walls. This trick required an empty enclosure because it hit every surface as it bounced numerous times with minimum speed-loss on a single push.

Cadatanauts were permitted to request a private room. Conjugal relations were the top reason such requests were granted, as the leadership and I valued sensual pleasure and keeping it private. On the other hand, this privacy could be denied for unrelated insubordination, such as withdrawal from work duties. Those on the "troublemaker" list slept in public spaces. The worst of these was the Gas-Dispensary Hall, kept at a cooler temperature to stabilize the processing of gases. I opted to sleep in the Observatory to be near the equipment in an emergency. Two hours seldom passed without me checking on our course, or magnifying and evaluating a new region of space we had traversed. Codol chose a spot next to the Main Computer because it makes low white noises and trembles as it re-directs matter and energy through its conduits. The Main Computer is the most dangerous component on the ship for a cadata because it is a giant reactor bomb inside the ship rather than in its reinforced coating.

One of my favorite parts of that spaceship was the Organic-Nanobot Shower Room. As soon as I would step through the liquifying wall of this space, centimeter-sized organism-imitating spherical nanobots would begin floating toward me from a collecting sphere where they existed as a single unit at rest. These nanobots were collections of cell-like units with several functions just like a lifeform on Cadata. Their function is to attract out dead or decomposing cells and to restore an entire cadata body to a peak condition. These nanobots would roll over the skin until they would find a troubling spot, and then they would spread out as small clumps to penetrate beyond the surface. In the next phase, the nanobots trimmed follicles or other dead-cell growths falling into the grooming category. This program eliminated foreign matter growing on a species' body in contradiction to pre-set grooming preferences. One species' culture might require short hair and nail length, while another favors wooliness or feathering. They also cleaned all digestive and nasal, hard and soft surfaces, such as teeth and tongues. After performing this work in a semi-liquid state, they would dissolve into clouds in a close formation around my body to perform cleansing of the pores and parts inaccessible in their larger dimensions. Once they determined, there were no impurities left, they would attract each other and would all be pulled back into a single sphere to be reused by the next bather. As part of their self-cleaning mechanism, any dead cells they collected were re-processed and turned into energy the nanobots used for the next cleaning cycle. Cadatas could alter the length and intensity of this shower. I indulged in the longest allowable stretch. It would not let me surpass this limit due to its energy-conservation objective.

A real shock came in the last year of our trip. The nanobots had traveled ahead of us to Ganeir and collected samples and holographic videos of the most intelligent species living there. They gathered samples of life and minerals from the planet to provide data for our Computer to design a plan for safe interaction with the environment and the creatures living on Ganeir. Codol and I awaited the results with anticipation as we, and twenty other cadatanauts were scheduled to mutate our genes to mimic the dominant Ganeir lifeform. Our imagination and training suggested numerous possibilities, but we had not anticipated even one of these alien creature's body functions. 3D videos of ganeirs showed them putting solid living and dead organisms from their planet into receptacles and sucking these into their bodies. Ganeirs walked on six thick sets of claws, which supported their weight without a soft cushioning. They had a hard, long, pink shell covering the tops of their bodies. Below this shell, they had tufts of fur. It was twice smaller than our natural height. Their faces were covered in a red and black explosion of fuzz out of which only a long receptacle could be seen hanging. There were around a hundred teeth inside this opening just soft enough to avoid harming the soft interior of the mouth, but powerful enough to sink into the entirety of an elongated prey. Since we were used to walking upright with help from our wings, switching to walking on six legs was disturbing. But the idea of consuming another living creature by sucking it down a tube and compressing choppers over it to break it into pieces and then standing by as our insides broke it down further and then dumped most of it out of the back was downright sickening. I tested the first gene-edit into a ganeir on myself. I refused to inflict this trial on a teammate. The edit was a success, except for my failure to maintain the proper posture for the ganeirs. Instead, I flopped to the floor, spreading the claws away from my stomach to avoid injury.

Prior to this mutation, I had programmed the Computer to replicate food with the appearance and composition of the little furball creatures composing the ganeirs' favorite meal. We set aside a space for replicating ganeirs' food sources in larger quantities after the first test. We had run experiments on growing various types of nutrients (liquid, solid, gaseous) in different environments and utilizing a range of cultivating ingredients. The Computer had the capacity to program their gene sequences into nanos to generate them artificially. But creating a setting where the self-replication would occur in nature would minimize energy-expenditure from the ship. As we settled into an orbit, the Grow Room filled with food sources from Ganeir. A fastidious litany of the vitamins, minerals and ganeir-life-essential ingredients were researched for their interaction with a smaller cadatan biology, as a few bits of our cadata brains had to remain in our bodies for our "self"-preservation.

Codol now placed this organic ball in front of my snout. I botched generating suction at first. Then, I learned there was a correlation between the growth of my hunger and an instinctive drag of food down my pipe. After the piece was too far down, I realized I had not chewed it enough prior to swallowing, so it strained the muscles of the pipe as it migrated. I thought this was the extent of my trials, and tried to focus on learning how to stand up, when Codol ran into the Lab and explained they had missed a crucial clue in the biological videos. During this winding narration, I began to feel a strange bulge and pressure forming at the other end of my body. I was still lying there when a fowl smelling pink ball jumped out of my rear end and bounced along the floor to the other side of the Lab. Codol held my gaze with sadness until I started imagining I had just lost an organ. Seeing my fur starting to stand up, Codol explained I had just experienced defecation, a process where food consumed through the sucking opening is digested or stripped of needed nutrients, before the unneeded waste pops out the other end. Cadatas excrete used up matter taken in from inhaling gases through exhalation and through our skin via evaporation. Each molecule inhaled is recycled. The Potentates, in part, worked this into our genetic code so the poor needed the minimum measure of gas to work for an extended time, thus creating the most efficient or the cheapest employees. These were made hyper-efficient prior to the start of our mission because in space extra gas or other consumables carried on-board necessitate additional fuel to propel them forward. Cadatas waste is in gas form, so it exits seamlessly through our air conditioning system. This conditioner heats up the molecules it attracts into the system and introduces binding agents to fit each compound, causing a reaction such as the oxidation of organic compounds into CO_2 and H_2O. Inorganic compounds can be mutated into acidic gases such as hydrogen chloride, sulfur dioxide and hydrogen fluoride. Our waste disposal systems were created for these gases. We had just learned ganeirs' waste system created solid leftovers. If we had still been at zero-g, the stuff would have floated around our spaceship without proper containment. But even at 1g as we were deaccelerating at this rate as we closed in at Ganeir, we had to research their waste disposal methods, and installed a strange waste receptacle next to our showers. It was an exhausting process. Ganeirs had some intelligence, an advanced communication system and some technical abilities with tools, so we found a way to mimic them while also being able to communicate among ourselves and perform some basic programming if this was needed once we were down on Ganeir.

As we began editing our bodies to fit Ganeir, our physical exercise sessions intensified. While we were weightless, to keep muscles from atrophying, we maintained a strict daily schedule of aerobics, weight training, gymnastics, and gliding at our Gym. We could materialize instruments to assist or intensify these physical exertions. We utilized a magnetic necklace to avoid floating away from the pressure-causing components straining our muscles, including harnesses and rings. As my body floated outside the safety region of a given piece of equipment, this harness pulled me back. While I performed flip and jump maneuvers, too strenuous under gravity, I grew nauseous from the abrupt eye movements, as the liquids moved with the rest of my body without being restrained in the direction of the ground by a constant downward force. I performed some additional gene editing to the anti-zero-g fixes the doctors on Cadata had already programmed into me. However, gravity, and its lack, is a force all matter struggles to overcome. We had the option to turn on gravity in the Gym even when the rest of the ship was weightless, but the change from having mass to being weightless is unpleasant, so I undergo it only when it is a necessity. With the return of gravity, we built-up muscles in those last months before we began gene-editing.

Communicating with those on Cadata was a complex process due to the length of the journey at light speed. We were kind of moving back in time, and we were moving at a speed where time stood still without passing us by for 500 years. So, messages were being sent from the past when they came from us. Yet, these messages had to travel at first for decades and then for centuries to reach Cadata. And then a response had to travel back to us, but since we were traveling away from the signal at the speed of light, it was impossible for a light-speed signal sent later to catch up with us. Thus, we were sending regular reports generated by the Computer, and composed by the officers, as well as personal messages to friends and lovers, but it was pretty much a one-way stream. Messages from Cadata began reaching us during the year slow-down as we were deaccelerating towards 0 on our approach to Ganeir.

Trash was recycled on the moon base, so I was not surprised by the degree of reuse on the spaceship on this long-term mission, but it would be shocking to an average cadata who has never traveled to space. Nothing can be expelled unless too much matter has been added. During an overage, we are warned by the Computer of the quantity of mass and density of matter to be expelled from the ship. The warning appears in advance to allow us to oversee the Computer's decision on which chemicals or objects are disposable. One of my hobbies was experimenting with ancient sculpting techniques, and I recycled my creations back into clay around a thousand times before I made a piece approaching the pinnacle of excellence, which I retained as a decorative sculpture in my room. Choosing to retain the art meant seizing sculpting to avoid utilizing too much matter for a frivolous entertainment. Clay contains a lot of potential energy, which can be released if this matter is converted into energy. The most energy and time-consuming items on the ship had to be most efficient. The robots fixing ship malfunctions had to self-repair, a skill offering inorganic lifeforms potential immortality. Most organic life replaces all bodily cells numerous times per lifespan. A machine can recycle its components daily without degradation. Even if a robot burns or explodes, melting into a liquid, or scattering into molecules; it can self-rebuild based on schematics stored in the Computer. We only build a "new" robot after a major scientific breakthrough where the old design or code cannot be repaired, and instead must be built on new physical, or chemical laws. Small improvements are written into their main program or into their digital design in the Computer, and they upgrade each other to fit with these schematics.

While art was distracting, working has been more beneficial for my sanity. I can work in any space on the ship. To bring out the controls and programming function, I have to make a voice or gesture command, and a 3D space appears with the necessary tools (such as code dictionaries and a list of the programs the Computer believes are the most relevant for the moment). The system accepts a near-infinite range of input types, including strange ones such as bioluminescent signals and x-rays, in case an alien species can only communicate on these channels. For example, due to their claws, ganeirs are horrid typists, but they have exceptional vocal cords at the bottom of their extended snout. They can give excellent, deep singing commands, wherein the melody colors the significance.

Some of my standard labors included reviewing procedures for maintaining the spaceship. After centuries in space, additional maintenance steps are required. These include a full flush of the air ducts, and pushing out deteriorated plutonium and alternative maxed-out chemicals. Deterioration is monitored on a molecule-by-molecule basis. A failing clump can create a dangerous energy-storage low-point, so I designed a program for estimating future buildup formations, and it has become a standard checkpoint for the Computer. I also worked on detecting new habitable planets and adjusting readings for existing potential candidates, as well as processing measurements of other significant spikes in chemical activity in the space we were flying through. Nally's *Catalog* was thorough, but as we approached star systems, our visibility of the makeup of the planets around them improved, revealing some new data. Nally gave us a list of possible experiments we might run in space to determine the effects of prolonged space travel at light speed. This list did not come with step-by-step instructions, so I had to interpret and create suitable plans to test Nally's questions. Are there any changes in how cells, atoms, or other particles of matter and energy function at a sustained light-speed? By how much does matter traveling at the speed of light travel back in time? Does shooting a laser at light-speed in a matching direction with the ship's light-speed movement create a double-light-speed traveling bit of radiation? If so, does double-light-speed laser beam radiation attain super-properties? What happens to matter when it is ejected from a spaceship moving at light speed (we tested this with our infrequent waste receptacle dumps)? How did light-speed affect the tiny reactions our ship was firing to create system-maintaining fuel? How does weightlessness affect optic nerves or other organs responsible for sense-perception in cadatas and ganeirs? Does surface tension of various compounds change with unusual changes in time or space? Are there parts of space where the spaceship's molecular printer would not be able to find sufficient molecules to form the nutritional-gas cadatas need to survive; and if so, what solutions would be most practical in those dead-spots? Are crystal formation patterns altered in snowflakes not only at different pressures, humidity and temperatures, but also at different spaceship travel speeds or other components of prolonged spaceflight and life on another world?

The later project was more spelled-out in Nally's notes, so I spent a good deal of time creating artificial individual snowflakes and whole snowstorms in the Lab. While it was an interesting hypothesis, I learned snowflake crystals behave in a similar matter at any speed of relative travel, time, and location. The intention of this test was to determine if crystal-formation can be improved in a light-speed traveling spaceship. However, to notice a change I had to stick those snowflakes to the outside of the ship, and even the strongest crystals in the universe would not have survived this experiment intact. The tests the Computer cautioned me against the most were those on fire's behavior in a zero-gravity environment. I persevered despite these signals because I believed it would be unwise to wait for an accidental fire to learn how to control it. It was also fun to play with fire. The cushioning gel along the spaceship's surfaces, utilized to keep our bodies safe during takeoffs and liftoffs, also acted as a fire retardant. Even if the Lab burned to a pile of ash, the outside would be unimpacted. If I or another cadata was inside the Lab during a volatile ignition, we could slip out through this same gel and it would also assist in extinguishing any fire we caught. With these safety features, I am free to vent my frustration and boredom into these tests, blowing flammable compounds up, burning down old equipment, and smashing spark-able equipment until it ignites. I even attempted taking in flammable substances and exhaling fire, lowering the temperature of the flames, and then tossing, and catching burning spheres, mixing chemicals to create a fiery reaction, and creating burning hydrogen bubbles. The latter helped me understand the properties gaseous hydrogen takes on gas giants. The Computer spotted curious differences from stationary tests, not obvious to me while I was doing these. Shocking discoveries turned fun into science. We uncovered new revelations on black matter while we were flying through a region dense in it. There were thousands of research questions, so there was never a shortage of work for motivated cadatanauts.

My main role in these experiments was as a catalyst rather than as scientist. Most "jobs" on our spaceship are unnecessary because the Computer is better at innovating in all significant operations. Cadata computing has been evolving for over a million years, and if the latest model of the Main Computer failed to spot abnormal readings or to correct errors, it would have been impossible for us to travel through space where extreme precision is essential. Despite the Computer's superiority to all cadata scientists combined, we still have pilots, engineers, doctors, and other designations for the cadatanauts on this flight.

Codol, my co-pilot, joined me in my experiments; our gaming rivalry helped me push ideas to their extremes. As we traveled farther from Cadata, I noticed the other crewmembers became disinterested in labor and withdrew. They expanded enormous effort into pretending to be working without tangible activity. Our jobs consumed a large portion of the budget proposal. Though we would only see benefits from this, so-far fictional, money if, and when we returned to Cadata. Even if we all perform these jobs as-promised, the two sides of the spaceship are redundant by design. Thus, if engines and all other equipment fails on one side, the other side takes over operations and can continue the journey uninterrupted. Because of equipment doubling, each side must have an engineer, a programmer, and a pilot. Each employee can fix problems on the opposite side remotely. Thus, at least one of these doubled-positions lacks assigned activities, even during a crisis. And complex engine, or air conditioning repairs are performed by robots: this again makes us into unutilized parasitic growths on sides of the ship. This truth hit me when I suffered a burn during a fire tossing experiment. I self-repaired routine burns and cuts, but because of the depth of the wound, Codol insisted on taking me to Doctor Wern Lexause. He spent an hour in contemplative staring at the wound. He touched the sensitive skin around it. He double-checked the body-composition measurements from the nanos', and scans of the affected area. Then he brushed our standard medical device over the burn, and held it there while it healed the skin. There was a larger healing container in the Medical Office, which would have run the tests and healed me in moments. The efficiency of this powerful unit was one of the reasons Codol thought it was best to come here rather than using our wing-held devices back in the Lab. The burn was healed until there were no signs of its past existence by the end of the session. However, I was in pain longer than I would have been if Doctor Lexause did not pretend as if he was hard at work rather than a redundancy. Neither Codol nor I have returned to Doctor Lexause' office since this incident, and if we were back on Cadata, I would have reported him for failure to meet a minimum degree of medical responsibility.

As we drew closer to Ganeir, tensions broke into conflicts. I began insisting the crew had to oversee the gene-editing process as well as the pre-landing maneuvers and trajectory-planning. The team countered the Computer and I were sufficient for the task. There was inadequate cause to put all cadatanauts under stress. Nally restricted my ability to terminate even the laziest cadatanauts. Dismissal of cadatanauts amidst a historic flight was unexplored during the endless pre-departure meetings. What other threats could I employ? Frighten them with a looming expulsion out of the ship mid-flight? What a homicidal idea, regardless of the rank of the executioner. Eviction of a single cadatanaut would be an insignificant resource-saving measure. And the crew's lethargy was a side effect of depression and suicidal tendencies; these vulnerable crewmembers might have requested ejection to escape their miseries. Could I say I'd abandon them on Ganeir even if it wasn't suitable? However toxic life was on Ganeir, they were better than the post-asteroid radioactive warming back on Cadata. And if they stayed on Ganeir, they would not have to endure the 504-year trip back. In response, I performed the work myself with Codol's help. Functioning without rest impacted my concentration, allowing problems to slip through. Everybody on the ship is issued equal nutritional rations. Housing accommodations can be truncated to slumming in public areas. But the entire ship is kept molecularly-spotless by the nanobots. Placing a cadatanaut in any location allows for the same purified, majestic view of the galaxy through the see-through walls.

As I began editing the genes of half of the crew in the last year prior to landing, a new rift started to form. The cadatanauts scheduled to land on Ganeir and undergoing gene-editing to mimic the dominant ganeir species had to be isolated from those who would remain on the spaceship to monitor progress. The spaceship was designed to have two mirrored halves with the same energy and resources-generating capabilities. We placed a boundary between them stronger than the solidifying, traversable gel separating rooms. Rather than being a stronger material, it required special authorization from the top five leaders on board for passage. An entrant had to call a leader from the other side to request permission to cross. Access was granted after a supplicant gave official acknowledgment he or she was heading into a territory with different rules, and a unique atmosphere. They also had to confirm they were wearing the appropriate spacesuits. If granted permission, this individual would enter through a body tight, liquid foam sterilizing door. This isolating measure was necessary to avoid wearing a spacesuit when crossing into an environment unfit for a cadata's or a ganeir's physiology. In an emergency where one side experienced a catastrophic failure (mechanical or a disease), the two could separate into independent half-spherical ships. Each would have a strong enough center to resist the pressures of spaceflight at the speed of light. Then, both sides could utilize their nanobots and robots to rebuild the missing half, as they continued on their trajectory, now half-lighter, and with half less fuel.

While in theory, the two sides were identical and non-conflicting; in reality, I sided with the main Observatory and my Lab. The non-gene-editing side could replicate these spaces. However, my research was part-proprietary, so I refused to grant them access to my setup and research notes. Thus, if they had to run experiments to save their lives, or if they had to observe the stars without our assistance, they would have been at a disadvantage. Doctor Lexause led this opposition. He sent daily memos to my team requesting a fairer division. We had been applying for passage through the formal permission-granting process. But then, Doctor Lexause instituted the first payment transfer requirement on tasks we performed while on his side that utilized "his" side's energy or matter. He claimed scarcity of resources on a spaceship meant consuming a nutritional-gas container or performing post-digestive extraction practices in their own bathroom cost them more to replace due to the extra care they took with residual ganeir cells we left behind. So, in return, we would have to transfer ten times more than what we used in matter or energy to them. This was troubling for me as an engineer because if we sent enough mass over to their side, the ship would be heavier there, and this might cause wobbling. Doctor Lexause ignored my pleas for reconciliation and inflated his price into a blooming trade war of counter-retaliations. I reacted with vocal outbursts, which became harder to compose due to my transition into a ganeir. This transformation is decreasing my focus. I fail to comprehend some of my own scientific notes. Assuming

Doctor Lexause was also frustrated, I scheduled a teamwork training exercise for the two sides. But he rejected the pitch for half of us crossing the boundary for a trust-building merry-making session. His counter-proposal: "Dedicate time to isolated self-reflection." I lack authority to fix this widening rift.

99,028 BC

We arrived and parked in an orbit around Ganeir. The local species are incapable of seeing far enough into the sky to detect our ship. When we were deaccelerating down to orbit-speed, I forgot our internal conflicts, and my safety concerns, and pressed my furry face against the see-through wall of the Observatory to feel the warmth of the rising sun on my sweaty skin. It was emerging from a half-orbit concealment behind Ganeir. The light cast an intense reddish-orange glow and carved deep shadows on the Observatory's familiar instruments. The space appeared foreign because I had not turned on artificial sunshine there since leaving Cadata to avoid contaminating readings of the sky. The Computer's trajectory was precise. The ship arrived within a foot of the intended location and began a set of 40-minute orbits around Ganeir.

Just a day earlier, even within the light-polluted space of the Ganeir solar system, the billions of stars outside did not twinkle. But now, the outer edge of the planet's atmospheric fog was introducing a slight light vibration on the dark side of the orbit. Before, they were points of single-colored light—red, blue, purple, green, and yellow. Now, they were a uniform white with a shaky yellow glow. Before, I mused at the gas cloud of the Milky Way; now, as we passed the day side, I saw the plentiful texture and soothing movements of the bulbous water-vapor clouds. At night, I saw falling meteors, their tails ignited by the atmosphere with air friction in the opposite direction from their motion. These were the first meteors we saw since leaving Cadata; out space lacks air or solar wind capable of agitating a flying piece of ice or rock to flash in a meteoric display. On the day side, I marveled how the seas on this planet were red because of their high concentrations of iron. The red waves blended into a blanket of pink sand. At night, the waves had a blue bioluminescent glow along their coasts because of a species of tiny glowing creatures dominating the shorelines. A quarter of the planet was dotted with small green lakes. A strange formation I could distinguish even from that height were gigantic and sparkling white crystal mineral pillars with any icy appearance. There was no chance they were ice-cold because they were sticking out of a hot desert. A psychedelic feature was rainbow-colored mountains, which we learned were painted by ganeirs as a national monument of massive proportions. Another oddity was a set of symmetrical, rectangular mountain-like structures, which I also learned were giant cities where ganeirs lived. I aborted determining if they were natural from their exteriors without transparent walls to see through. My attention was also captured by a mountain-sized, solitary rocks, scattered across a savannah. They matched the appearance of a litter of crashed asteroids that had landed without leaving craters, or burning up. This burst of colors, textures and never-seen formations of nature reminded me why I became a shuttle pilot and volunteered to fly on this mission. Yes, I cared about cadatas' survival, but I was here to discover life's strange varieties.

Whenever we spun to the side of the planet engulfed in darkness, and the glorious light was drained out in an instant. I saw a dense black. Then, I sensed a chill on my cheek as the temperature outside dropped from 250°F to –250°F. The wall needed a few moments to recover from this rapid 500-degree temperature swing because it is unnatural in outer space, where there is no weather. Entering a stellar gas cloud might alter the temperature outside by a miniscule fraction of a degree. Thus, the outer wall could maintain a stable internal temperature even in these rapid fluctuations, but the wall allowed itself to freeze at the low-point because the layer of ice helped to further insulate the interior. Heat was not lost, but rather was drawn away from the coat until the ship's magnetic sphere attracted it back in. Then, I felt the wall jumping to the recommended normal temperature for ganeirs. It never jumped or dropped as low as it was outside. The change in temperature occurred on a regular thermal cycle, and the alteration expanded and contracted metals in our coating, which emitted inhaling and exhaling sounds. If we used weaker materials; they could have degraded in efficiency from repeated cycling. But oganesson could withstand a leap into a sun. The components of the coat were programmed to allow for such expansion for flexibility and breathing space. If no temperature change occurred outside; the coating still puffed itself out, and deflated at intervals merely to keep its internal nano cells active. I am contemplating the science now. In the moment, I was delighted to experience the change. It was mesmerizing to see true sunshine after centuries in space without a sun in the sky or a natural swing in temperature. Even as I acclimated to the darkness, I struggled to distinguish the planet from the dark side. My vision was distorted after turning into a ganeir. I was blinded on the non-visible spectrums. I was tempted to turn on the light inside of this sacred space just to see my own claws to avoid doubts regarding my own existence. Is a thing unseen nonexistent?

My ganeir-afflicted brain questioned why we were orbiting the planet rather than remaining still far from this massive body with its messy qualities. I could remember some of the lessons from training about orbiting, but it was difficult to connect them to our specific situation. I have the textbooks in front of me now to review this. In theory, orbiting minimizes energy expenditure to near-zero because a spaceship can refrain from firing its engines to accelerate or deaccelerate. Without an atmosphere to interfere with, or boost propulsion, a satellite can continue spinning until it is overcome by gravity, and falls to the planet's crust. We can imagine a scenario where our spaceship needs to change directions, or to alter the degree of flight—perhaps from 20° to 55° in relation to Ganeir's equator. It must fly at a perpendicular direction to its previous horizontal orbital path. It would expand several times more energy to push against its own momentum. Hm… Is the solution coming to a complete stop first, or continuing? I think stopping means an immediate meteoric drop to the surface. I avoid sharing with Codol or others my struggles with grasping ideas in this body. Codol also underwent the transformation, and he has not mentioned any loss of competence to me. And I am letting Doctor Lexause win in our arguments. He is deciding on the strategy for our coming exploration of Ganeir. I hope that with the preparation the team and I invested, we should be safe trusting Doctor Lexause with our health and safety.

Head Doctor Wern Lexause' Last Entry

99,027 BC

Exploring even an untamed island on Cadata carries risks. Threats grow to astronomical-proportions on planets outside our own solar system. The attack was invisible to the naked eye. I must be succinct, using limited citations, because I am suffering the pangs of the disease. My cognition will fail soon, before I am likely to finish this findings-report.

To assure I am conscious for the critical part of this memo, I must begin with the nature of this disease. I have called it *W. Lexausious*. It can be categorized as a gene-merging pathogen, the likes of which has never been observed on Cadata, so the category is not on the official list issued by the Cadata Robustness Association (CRA). Once I am dead, it will have killed all 46 members of our crew. If it managed to survive a return trip to Cadata uncontained, I estimate it would kill the entire exposed population of cadatas in the first year after the initial contact. Most of this year would be reserved for its gestation period, when it would spread in a dormant state, making cadatas' genes pliable to a merger. *W. Lexausious* is an airborne pathogen capable of traveling through almost any atmosphere and across great distances, so Cadata's isolated island locations would not prevent its spread. I discussed disaster scenarios associated with alien pathogen spread in Chapter 589 of my book, *Diseases in a Horrifying Universe* (100,133 BC). If this disease does not kill everybody on Cadata when this message arrives back with the ship, readers might also be interested in reviewing the full spectrum of Cadata disease-related knowledge in my previous book, *Diseases in a Horrifying Planet* (100,145 BC).

W. Lexausious has a horseshoe shape interlocking with the horseshoe shape of ganeirs' genes. I located a full genus of similar gene-merging pathogens on this planet after learning of its destructive potential. The nanobot-collected data on Ganeir's pathogens forsook indicating dangers associated with this genus. Life survived on Ganeir despite it because its species consumed a poisonous to the unaccustomed plant, suaxel, which counteracts this disease's control over a complex species' genes. It appears as if the gene-editing our team of cadatanauts underwent increased their susceptibility to *W. Lexausious* because the special powers their genes were given through this process made the genes pliable to solicitations for mergers from cells. The cells had to be pliable to mergers to tolerate the interspecies mutation process, which merges cadata qualities with those inherent to the target species. The unintended merger between edited genes and *W. Lexausious* horseshoes created a super-disease capable of hoping from gene-edited cadatas to the rest of our exploratory team. The disease now had cadatas' super-gene-editing abilities, and began attempting to alter our cadatas' genes to fit with preferred environmental requirements. Only after the entire crew was infected was I able to isolate *W. Lexausious* in the Lab and observe its behavior when it encountered various types of cells. It feeds on almost any nutrient sources in the invaded cell, using it to create small glowing streams to help it propel to the next cell to infect it as well and to garner its energy sources. Symptoms of the disease emerge in the final stages, when treatment by any means is impossible. These include solid matter liquidation (including bones, claws, and other hard components of a body), internal cooling and hypothermic-related effects, and depletion of a body's ability to process nutrients out of gases, liquids and solids consumed. The disease absorbs most incoming nutrients.

The first signs of the infection showed up two weeks ago, after a year of gestation. The cadatas who mutated into the ganeir species have failed to communicate a single report since they landed. I believe they were infected within moments of exposure. The disease traumatized their brain-functions long before it began to dominate the rest of their bodies. The cadatanauts began showing signs of mental impedance prior to departure. Putair Niatage, who was assigned to lead this expedition, had enough stamina to confirm the flight-plan to Ganeir. He did not touch the shuttle's controls during the landing maneuver, letting the Computer operate it. This was unusual for Putair but not for the other cadatanauts who had become lethargic much earlier in the expedition. I assumed Putair had decided to embrace a restful retirement during his experimental stay on Ganeir, so I refrained from insisting on a post-landing mission-status communique. I sent nanobots and small robots to check on their progress, and it appeared they were adjusting to life on Ganeir. Their weights remained constant. Their temperatures and composition measurements were consistent with average healthy ganeirs. They were tolerated in ganeir society, having been assigned living spaces, chores, and mating partners. They overcame physical impediments including walking on six legs, and vision impairment. They improved in location-detection based on their sense of smell. Putair even contributed to a rainbow painting on the exterior of the housing project where he was relocated when his housing structure collapsed in an earthquake (Ganeir has strong seismic activity).

Due to their apparent adjustment, we minimized monitoring of their activities, planning on waiting for the scheduled ten years prior to collecting them and returning with them to Cadata to evaluate the suitability of Ganeir for habitation. However, their nanobot wellness detectors were triggered when their temperatures dropped, and they started displaying the other symptoms of *W. Lexausious*.

Risking death, I hazarded a lone shuttle voyage to Ganeir to examine their condition. Without regard for my own safety, I tested Putair, Codol and the others, but missed the source of their illness becaue the disease' genes were disguising themselves in forms identical to what I would have expected in a cadata-ganeir hybrid.

I had made the journey and the initial tests clothed in a spacesuit, disguising my body to appear to be a ganeir, but the spacesuit was synthetic, and stood out as alien to the locals, who I had to work with to access our cadatas. I did my best, with my computerized translator's help, to explain to them I had invented the skin-tight ensemble back in my home village on the other side of the planet. They seemed to believe it.

Meanwhile, ever since landing on Ganeir, I had been wishing I could feel the weather just outside the suit. Without gene-editing, I knew I would not have survived long without the suit, as I needed the set of gaseous nutrients it was providing, and would have suffocated, while also being poisoned to death if exposed to the harsh elements in Ganeir's air. The temperature outside was 115°F, not much colder than normal temperature on Cadata. The ship had been maintaining a standard Cadata-level temperature on the un-gene-edited side across the trip, but it had not changed even a fraction of a degree since departure; this steadiness became disturbing to some with the centuries. I felt slight pressure from what I knew to be a strong wind entering through the open doorway.

But I lacked a sense for the wind's texture. Was it humid or dusty? Putair's complex was next to the ocean, so I assumed this wind smelled of the lighter chemicals picked up by air as it traveled across this liquid body. I contemplated this half-consciously while running the tests. When I completed the chore, I began craving the removal of my suit. I imagined keeping just enough of its material over my air-intake-valve to breathe its cadata-specific air. I felt the moist, salty breeze on my skin as it shimmered with delight. I even inhaled some of this wonderous air, and felt immediate relief. Then, I put my suit back on, said goodbye to Putair, who I believe might have recognized me towards the end, and departed for our spaceship to process the materials I had collected from our cadatas through our medical-evaluation hardware and software.

The trip exhausted me, so I took a full restful night's sleep. When I woke, I sensed a lighter version of a few of the symptoms I had recorded in other cadatas. I left the test samples in the Lab without processing them into the Computer to avoid performing this critical task without full engagement. I now input the materials, adding my own to the batch, and waited for the results. The outcome was not as I had hoped. The gestation period developed a strand compatible with the instant invasion of any cadata's genes, and not only of an edited cadata-ganeir hybrid. *W. Lexausious* had invaded my system as well, and the Computer detected the mutant strand across the spaceship, and inside the bodies of the rest of our crew. We all grew sicker in the next two weeks. Those left on Ganeir were dead before the start of yesterday. Most of my comrades on the spaceship died earlier today. In the last few days, I developed bionic body parts to prolong my life, but while they have allowed me to outlive my comrades, despite my earlier infection, I anticipate this fix will only suffice to finish this article. I cannot journey back to Cadata to share my wisdom with you in person. The bionic organs and limbs are improvements on Landut Ayced's models, as you might notice if you perform a biopsy on my remains. The Computer miscarried the engineering of a cure for this disease. Perhaps it will succeed in the 504 years it will spend flying back to Cadata.

In summary, the strenuous tests we ran to check the microbes, genes and various other compounds from Ganeir in our Lab were insufficient to identify this unusual interaction, which was magnified because of the main source of strength of our cadatanauts' genes, their mutability. Future missions should send a single gene-edited cadata to a planet first, and observe the impact on their health for at least a decade to determine if the same results can be replicated in the realities of a planet's environment. A scientific lab is a purified environment incapable of replicating the complexities of a planet's atmosphere even if a cubic chunk of it is caught and transported.

It took cadatas thousands of years to map and describe the geology, zoology, and microbiology of our own planet. It is vainglorious to assume even the best of us can absorb all indispensable knowledge about an alien species, and their world in a mere year. We have uncovered predator species, diseases, and natural disasters across our history, finding new threats in isolated ocean spaces or high in the atmosphere, which escaped detection for over a million years. Even if we landed a thousand times on random parts of Ganeir; we could still miss at least one dangerous foe cable of killing us on the one-thousand-and-first landing. We cannot be tricked by the seeming insignificance of small things, or invisible things, or cute and fuzzy things. Destructive force does not discriminate in size. Cadatas are the essence of what this force represents: our survival means the consumption of various nutrients. Interplanetary survival means transitioning from top-predator on a single planet, to rising in rank to conquer foes on many worlds. It is too late to study death once you are in its embrace.

The progress of science is innately slow, and to attempt to push it beyond our limits is vain. We must keep testing, and searching for successful alternat[ives…][2]

The Aftermath, or the Return of the Ganeir Spaceship

In 98,519 BC, the Computer completed a landing of the Ganeir spaceship back on the moon's platform. The Computer spent a century, before commencing the return-jour-ney, on analyzing the causes of that mission's failure. It weighed every molecule of its own composition to expunge, or purify questionable specks of matter. Due to the nature of traveling at the speed of light, the ship arrived sooner than the 504 light years passing on its internal clocks. The full trip to and from Ganeir took around a thousand years from Cadata's perspective. Cadata had started receiving status reports only in the last deacceleration year of the ship's trek. When the first report from the Computer arrived, Nally and the other scientists were disturbed by the lack of survivors. They were also terrified by the impending menace of a once contaminated ship landing on their moon in less than a year. Doctor Lexause' entry was included in the first data package, together with the full 3D-video log of the tests performed, and pertinent information related to cadatas' premature end on Ganier. The Computer had its own conclusions, but Nally went over them to consider the unpredictable cadata-element. She turned her findings into a report submitted to Loor for approval. He had the PROs condense it down to a short press release for dissemination to the media and then to the public.

Nally's judgment of the Doctor's actions were softened in these latter stages of the vetting process. Even to an untrained newborn, who just learned a language, it was obvious Doctor Lexause failed to enact basic safety mechanisms. His duty was to mon-itor for unprogrammed disasters the nanobots lacked the imagination to comprehend. Instead, he failed in his duties because he was complacent from centuries of inactivity and reliance on Putair to be the superego in their dynamic. The tests he purported to run were repetitions of the tests nanobots processed before he arrived. His presence was unwarranted. Then, through exposing himself to the airborne pathogen by breathing in the local air, he took this disease to the spaceship and infected the crew. Without the Doctor's negligence, the space-based crew might have survived to discover a cure for this disease, finding an adaptation for cadatas to live with these pests. Instead, we wasted half a millennium on the trip back, making a return to Ganeir inadvisable. Logs indicated strict enforcement of separation between the gene-edited and the non-edited halves of the ship. So, why was the Doctor foolish enough to fail to isolate himself from exposure to a gene-edited cadata?

Friends and relatives of the fallen cadatas were informed first before the PROs issued a statement on the disaster to the rest of Cadata. When the data on the loss arrived, they were taken to a Potentates-owned hotel in Alleges. They anticipated they would be paraded to fancy parties celebrating the cadatanauts' triumphant return. These guests were half-way into a raging party in their honor, when Nally managed to escape from

2   Here this narrative ends abruptly.

the awaiting press to review the transmission. She was eager to find images of spectacular landscapes and seascapes and various flora and fauna unimaginable on Cadata and un-seeable with the best imaging devices from 500-light-years away. Then, the reality hit. She took it as a personal duty to inform those who loved these fallen heroes of their loss.

Given cadatas' 5,000-year lifespan, news of a death is uncommon in any close circle. Putair's sister, Lapo, was the sole unsmiling face in the conference hall reserved for this occasion. She was standing by the back wall. Her eye darted at Nally as soon as she walked in, perhaps because Nally's countenance matched her dark mood. Nally took Lapo into an adjacent private meeting room and delivered the news: her brother died with the entire crew. Nally said the required platitudes for such occasions: he would be missed, an outstanding cadatanaut, a great loss for the galaxy, would have risen to the top of the space leadership program if he survived… Loud whistle music continued playing in the ballroom. Laughter was echoing through the thin wall separating the rooms. Lapo could not think of a response, and just sat there for a while with a blank quiet stare, processing what she had been suspecting. Nally asked an accompanying assistant to escort Lapo back to her hotel room, as she delivered this news to the others. The dancehall emptied gradually until the music was turned off. Only one member of this group reacted with a violent crying fit; Nally suspected this was an exaggeration from an ultra-unfeeling relative who just inherited a fortune. Families were accustomed to bad news Cadata suffered since the asteroid strike, so only a successful trip would have shocked them. They also expected the worst because they had been over-briefed on catastrophic ways the spaceship might fail in open space or on Ganeir. These worst-case-scenario training sessions were required by Cadata's law to avoid lawsuits from relatives and friends, who might otherwise have claimed Examen failed to anticipate a given disaster. If they were warned about the odds of a fatal outcome; they could not claim a lack of a fair warning. They decided to enjoy the party while the good times lasted. Once notifications were completed, Nally filed the report, which traveled around the wheel until the packaged message arrived on the PRO-end. In sync with the release of this announcement, Loor gave a public speech to reassure cadatas future flights to other worlds were suspended until the conclusion of a full investigation. Because he had used the term "suspended", instead of "cancelled", Nally assumed they were pushing forward with the mission of exploring other worlds. Under this belief, she kept the remnants of her team in constant training.

In the meantime, the moon was evacuated before the expected landing. Only robots remained for maintenance. Even cadatas who lived on the opposite side of this spacious satellite escaped in their shuttles, anticipating the alien disease would make the moon uninhabitable even with spacesuits. Some even speculated it might be capable of traveling the distance between the moon and Cadata to infect the ailing citizens. Cadata was now a colder, desolate world in contrast with its condition when the Ganeir spaceship departed a millennium earlier. Its population matched its coldness in mood, a terror of untold disasters dominated the collective consciousness.

The Computer was concerned about these possibilities as well, so it purified the ship of toxins in the first decades back on Ganeir after Doctor Lexause' death. This purification began with the complete breakdown of every molecule of the victims' bodies. This separation allowed for the isolation of infected particles. The diseased bits were turned into basic elements before being churned into energy to generate power to complete this purification process. Contaminated cells and genes were captured and shredded by the air conditioning system. The side of the ship with an atmosphere altered to fit Ganeir was brought back to cadata-levels in case an element or a combination of components in the atmospheric makeup caused an escalation of the disease. The positive test samples also had to be destroyed after safe holographic versions of them were replicated for further study. The Computer broke itself down into compartments to debate with itself. One possibility it reviewed was sending an electronic data file to any planet rich in the required elements. First this data would stimulate the self-construction of robots, and then they would build a replica of the entire spaceship, including the Main Computer. The initial spaceship would self-destruct once the construction succeeded, and the new spaceship would take the return trip. This would eliminate every molecule, bringing the risk of contamination to zero. However, a new construction required a complete stop, whereas self-repair of the existing ship could be executed while it commenced the return journey. The Computer also debated if it should send robots down to decompose into elements the bodies of the cadatas who died on Ganeir's surface. They had been buried according to local customs. The Computer reasoned: since readings showed no spread of the gene-edited super-disease from cadatas to the ganeir population or to other Ganeir lifeforms, it was safe for the ecosystem for the bodies to remain intact in the ground.

The Computer's thoroughness paid off when the ship was tested for pathogens back on the moon, and the tests came out clean as they day it was constructed. Nally volunteered to be the first cadata to fly to the moon to test the safety of the air inside the spaceship. She refused taking a co-pilot to avoid unnecessary contamination of another cadata. Remote testing by nanobots was hyper-reliable, but a microscopic chance of a missed threat remained. Once on-board, she toured the compartments of the ship. They were frozen in the precise architectural plan saved in the archive when Doctor Lexause died. The adornments worn by cadatas across the ship upon their deaths were stowed by the Computer in standard compartments for such trinkets. Signs of their last activities were wiped away: some were consuming nutritional-gas, others played video games, yet others took slothmoths to ease the pain. Leaving these out could have caused damage when the ship changed speed. Nally had already reviewed Putair's, and the other crew-members' digital diaries. However, the personal computing devices they wrote these in had already been recycled due to the heightened threat they posed. She collected some of those closeted trinkets to share them with Cadata's museums clamoring for such treasures to display in their Ganeir exhibitions. While she was much graver when she exited, her health survived the walk and the year of isolation on the moon scientists required to be certain the pathogen had not hitched a ride on the interstellar journey. In the interim, Nally designed a series of corrective measures to avoid a similar catastrophe on the next attempt. A new pathogen-collecting array of nanobots auto-tested planetary components for adverse impacts on cadatas' physiology. If a toxic substance was located, the system diagnosed if it could be extracted from the environment without killing the ecosystem. Given my ability to compose this history for you in English, it follows cadatas reached Earth, without succumbing to a similar spaceship-wide extinction.

Before continuing this story, I must digress to protest the formulaic human portrayal of the horrific exposure to an alien pathogen in human visual fiction. The repeated plot has the humanoids learn comrades at a distant disaster-site are unresponsive. To solve the mystery behind the radio silence, the "heroic" captain flies to the location under threat. An expedition team ventures on board the distressed ship, only to ascertain the death of the crew. Unavoidably, the expedition team also catches this disease, and supporting actors begin dropping dead. The main characters begin showing symptoms, but fight through it. The happy ending comes when this team's medical officer concocts a cure from the safety of a health-center. Their climatic triumph is in keeping death away from the inner circle. The obvious is never stated: if the second ship instituted protocols to remotely investigate such suspect, abandoned ships; they would avoid repeating this plotline, refraining from yet again sending guinea pigs for a new pathogen's slaughter. As we learned from Ganier's harsh lesson, any unanticipated alien pathogen can kill everybody contacted in instants, before the quickest computer could begin contemplating a cure. It is too late to help those infected once the worst strikes. Researchers who follow the example of Doctor Lexause, and strive to become the hero of the formulaic-tale, are setting their own crew on fire to test a fire-retardant, which has already been proven insufficient on a rival team. A pathogen is indifferent if it took an alien a thousand years to find it. It's hungry for dinner, and you are it.

Chapter 5

The Start of the Long Voyage Across the Milky Way

Nally Subsenar was the discoverer of the Nalut Expander and the Ganeir Spaceship. She was also known as the mother of cadatas' habitable-world-exploration mission. She was 4,788-years-old when she completed, and prepped for liftoff the upgraded multi-world-faring spaceship in 98,468 BC. After one year on the moon for a research retreat, it took her the next fifty years to follow the underhanded corrupt process required to secure funding to commence development of a spaceship capable of longer and knottier explorations. When prompted to decide on the first planet for the journey, she chose the one 150 light years away. She hoped to live to see it. The team was surprised she volunteered to join this mission, instead of the preceding one. Though it is easy to sympathize with her perspective. She had spent an extended lifetime developing tools to reach distant stars. Yet her first spaceship to visit a habitable world had returned without its crew. There were no suitable planets closer than 150 light years away. Thus, she would not see that spaceship again, even if an infection once again terminated its trip prematurely. She could remain on Cadata's moon to perform surplus theoretical, and hyper-impractical experiments, or she could sail to be buried in space: her passion-medium. Petitioning to improve the engineering of the spaceship's model was a dead-end because regardless of the size of the bribes, another contract would be denied. The political mood was against renewed searches for habitable worlds. Remaining resources were needed to fix the semi-survivable world cadatas were inhabiting.

The population of the planet had dropped to a sixth of pre-asteroid numbers. A permanent toxicity hyper-warming descended on Cadata's islands. But the bureaucracy at CSEA was unchanged. CSEA's budget had been slashed by the decimation of the planet's economy. The Potentates were facing regular violent rebellions as many islands previously under their control gained independence. Amidst this turmoil, it was taxing for Nally to convince herself Examen could win another contract for a new long-term mission despite losing their entire crew. As she did her research on potential competitors, she realized how they would win even with this avalanche of trials. There were no serious competitors left standing. The top two competitors in the previous final round had filed for bankruptcy after milking the winner for sub-contracts. The other five lost steam and closed in the following centuries. The remaining corporations and small businesses were wrestling to arrive at work alive. Spending their last dimes to pay bribes to perhaps send a doomed ship away from their starved hellhole did not appeal to anybody but Examen. Nally's company sailed in relative abundance through the post-strike centuries. Nally's personal profits were also considerable, enough to purchase a pod in a skyscraper at the center of the rebuilt Alleges. On the other hand, there was a shortage of luxuries to spend a fortune on in Cadata's condition, so there was no motive to remain there for a relaxing retirement. Nally's list of improvements included regenerating the population with a youth whenever a cadata reached the end of a lifespan. Given the limitations in their knowledge about Ganeir, Nally anticipated this would be a one-way, fatal trip as too much had to align to find a survivable planet for cadata-habitation. Another necessary change Nally proposed was rejecting lazy cadatanauts, only bringing workaholic, proven engineers, programmers, scientists and pilots. Nally even evaluated their genes to determine if similar hard-working traits were pronounced in past generations, improving the odds of repeated ambition and drive in the following generations to keep the mission in perpetual motion. She was careful to stipulate all doctors joining the mission had to volunteer to engage in continuous research across their lifetimes rather than retiring and letting the Computer perform the medical work. The Computer had failed to extrapolate the presence of a never-before-seen pathogen, so if left without a cadata doctor's supervision, it would repeat similar misdiagnoses, with dire results.

Once the contract was secured, Nally shillyshallied to locate a new Lead Cadatanaut with equivalent skills to Putair's. She supervised the selection of physical and psychological tests for the applicants. Two million applications arrived for this round, as cadatas were more desperate to escape their planet. However, after the first failure few of them believed they would last much longer in space. Nally designed much harder selection tests for this round to reflect the realities of interstellar adventures. One of these involved flying a shuttle from a spaceship down to Ganeir in the form of a ganeir (claws and all), and then attempting to blend with this local species. Only 500 finalists were allowed to take this test. All struggled inside the holographic world. Being tiny, walking on six legs, or having dexterity problems caused forgetfulness regarding the required safety measures. Those who remained in the center of the room instead of moving closer to a wall during the rapid-landing phase were tossed into the cushioning gel of the wall below by the speed of the descent. The holographs were designed to adopt solid properties during tests of this caliber, so these falls led to serious injuries, which healed with help from nanos. Nally had incorporated a strike by the ganeirs, if they spotted signs a cadata in their midst; in response, they would begin shooting at the test-taker. The program avoided fatal regions of the body; but it was determined to hit flash if mistakes were made. To Nally's disappointment, those who were hit, tapped out without attempting to convince the ganeirs they were one of them despite their doubts. There were also several tests on space and land navigation methods. The Computer's navigation abilities would have been disabled, and the cadatanaut hopefuls would have to travel along a pre-determined, or an improvised route to exotic planets, moons, asteroids, and less receptive bodies. Nally spotted several applicants cheating by using their personal computing devices to recall the required maneuvers for landing and course-correction. She resented such short-cuts because she foresaw the repercussions of such incompetence in open space, where the longest cheat-sheet would be insufficient because failure to react in seconds would splatter the spaceship against its final resting place. Cadatas from the Troop stood out in the lineup; they had seen drastic downsizing, leaving many desperate shuttle pilots eager for a mission elsewhere in the galaxy. Nally decided on two of these as the Lead, Wocega Lapovy, and her Second, Ortack Steral. The media did not take to them as Wocega did not partake in partying or slothmoths, and Ortack followed her example, but Nally viewed this stoicism at the start as a necessity for any chance of at least surviving the mission's first stop.

Two months ahead of the scheduled departure, the updates were perfected and the team was prepped. Nally took the excess time to conduct a final research project on Cadata with her private savings, knowing she could not utilize these funds in space. She designed an enhanced-power signal to emit a call on all wavelengths, frequencies, bandwidths, and energy-levels with the capacity to reach the other end of the galaxy. Its strength gave it the appearance of a small second sun circling the cadata solar system. This light would have appeared unnatural to an alien astronomer because binary stars circle each other, rather than one of them behaving as a gravity-weaker but vibrance-equivalent satellite planet. The beacon was smaller than a moon. Its vibrance was an illusion: a trick of optics, lasers, mirrors, and brightness magnification, which I cannot detail here. It was a satellite circling Cadata's sun, and using the sun's energy to communicate. The message was a series of linguistic variations with wavelength fluctuations repeating the concept "Help!" It elaborated on the nature of the catastrophe Cadata had suffered, adding its inhabitants were searching for a new home anywhere in the galaxy. As Nally oversaw the launch and operations-start for the Star Beacon, she returned to nagging questions. *Even if an advanced civilization received this message, would they help? Did I do enough to locate similar unnatural signals from alien worlds in the Milky Way? Why haven't even planets with signs of intelligent life sent out noticeable messages to attract visitations? If Cadata had received a signal with a request for help from an alien world (before or after the asteroid), would Cadata's government have decided to ignore this call, or respond and risk having a slew of migrants swarming their Cadata? Would even the most magnanimous superior beings want to invite migrants from a world that caused its own destruction?* Nally had little time to solve these philosophical conundrums. She left the pondering task to the Lead Scientist, who was scheduled to adopt her job at Examen.

In collaboration with her science, engineering, and programming teams, Nally instituted the new Rules to Avoid Contamination (RAC). According to RAC, the mission was required to commence by sending a single cadata with edited genes to a new galactic object—planet, moon, asteroid, or whatever else was lurking in the expanse of space. Then, the spaceship had to wait for at least a decade without physical contact with the world or the sole grounded cadatanaut. During this interval, they were to deploy nano-bots to run tests on the environment. The exploring cadatanaut would also have his or her vital-signs monitored non-stop. If a cadatanaut failed to check-in with a report, miniature observatory robots would be dispatched to determine if the cadatanaut has been compromised or incapacitated, or to uncover an alternate cause for the silence. Once the decade ends, and the cadatanaut is determined to be in good physical and mental health, the rest of the gene-edited cadatas can join this expedition without risk of death due to a pathogen or another unpredictable local cause. Regardless of the duration of planetary habitation, a cadata returning to the ship must first be kept in isolation, without air circulating from his or her section into parts of the ship housing unexposed cadatas. The returning cadata might have caught an unpredictable disease (natural or superficial), and might only begin showing symptoms once back on board. On the medical front, RAC recommended the isolation of cadatas' brains from the gene mutation process to preserve their minds. One danger was devolution of an edited cadata into a species with a lower intelligence-capacity, making it impossible for the cadatanaut to process the required scientific tasks or to communicate intelligibly with non-edited cadatas.

With final tasks completed, Nally approved an expedited training program. The instructors from the main courses for the Ganeir mission, Chief Doctor Photuate Nackan, Chief Programmer Guagray Raimingly, Chief Maintainer Vomere Tenievan, Safety and Controller Elleqow Comeepout, followed Nally's example, also volunteering for the critical mission. With them on board, the ship was equipped to house a training academy. Only flight and maintenance training had to be accomplished before liftoff. Since these lead players were nearing the upper limits of their lifespans, they debated putting themselves into hibernation by becoming tolerant of extreme levels of sleep through gene-editing of this function. They might enter deep-sleep, a state wherein body-functions switch to minimum settings. Similar strategies are used by certain frogs on Earth, as they freeze, but keep their blood circulating for months until they are ready to resurrect. Hibernation was inappropriate for the Ganeir mission because cadatanauts were expected to perform scientific experiments during the flight. In practice, only Putair worked, so it might have been more energy and matter-efficient to keep the rest asleep. This proposal was discarded when the lead scientists insisted: they volunteered for a final contribution to space research, and not for a long nap in space, before dying without reaching Nerore.

One of the improvements Nally instituted for this expedited training initiative was more complex physical training to help prepare them for any physical abnormalities members of the team who would undergo gene-editing might suffer, as well as to help others acclimate to the new planet. Doctor Nackan acted as the chief masseuse and physical trainer. He would re-align their bodies after the genetic change so their bones and muscles were maximized in power and avoided misshaping from rapid genetic mutation. Doctor Nackan had concluded: aside from mental challenges, Putair, Codol and the other ganeir-cadatanauts had extreme difficulties adjusting to their new bodies without transition-assistance from their doctor. Nackan hoped to solve this problem with his personal involvement. Nackan's assistant doctors performed substantive research, while he was awarded the credit and the glory for "organizing" the maneuvers they developed. Nackan and assistants thus worked to improve the flexibility of space travelers, who would spend 150 years in weightlessness prior to reaching the destination: Nerore. Cadatas called it Nerore, though both Nerore and Ganeir applied other names to their planets and populations. Nanobots' data on the first expedition showed spending decades, or centuries in space caused restrictions in cadatas' nutritional-gas-faring canals, when they began to relax, and wither without the constant pressure of gravity to keep them robust. Some spatial pressure massages helped to exert pressure onto the major air-ways, while also fighting muscle atrophy. Nackan decided starting these treatments before liftoff would assure continued results.

Since even Putair and Codol experienced claustrophobic symptoms, Nally gave the cadatanauts she chose as potential travelers a one-year isolation-test. It mimicked what she endured on the moon after the Ganeir spaceship's return. She found this year helped her center and become more productive, and if any cadatanaut could not survive this duration without asking to be let out, he or she was unsuitable for prolonged isolation from society or planetary contact in space. To her surprise, half of those isolated in a series of pods inside of the spaceship, as it underwent renovations, resigned from the space program before the end of the year. This helped to finalize the makeup of the Nerore team, as she only had to eliminate a few other names from the cadatanauts recruited in the first and second rounds.

The usual party-festival accompanied Nerore spaceship's liftoff from the moon. Nally avoided the circus in favor of flying a shuttle around Cadata, stopping on her birth-island. She also visited Alleges for a textual exam of the remains of a settlement of an evolved species other than cadatas. She describes her speculations in her archived diaries. The liftoff was instantaneous and smooth because turbulence and fluctuations in internal pressure had been solved. One moment they were still, then they were hovering above the ground, and then those on the ground only saw a disappearing dot in the sky.

Nerore was a relatively short distance from Cadata in the opposite direction from the galactic center, and a few degrees higher in the disk of the galaxy map. There was only a slight difference in the appearance of the sky outside, between the time they entered light speed, and exited it 150 years later. During the trip, the haze of surrounding stars dimmed, and the central bulge shrunk slightly. Upon beginning deacceleration, they observed, as expected, they "landed" in an invisible, scant-populated cloud of icy plane-tesimals, and comets circling the outer reaches of the Nerore solar system. They saw two of Milky Way's satellite galaxies—one three times larger than the other—located in a perpendicular direction to their spaceship's movement. The smaller one appeared to in-clude a giant, bright star somewhat separated from the group. The 3D star map revealed this star was located only 50 light years away: an abyss from its apparent parent. Under closer scrutiny, the scene revealed clouds of supernovae-ejected gas raining back down to the Galactic plane in galactically-slow motion. The sky in the opposite direction from the little galaxies was black with only a dozen scattered stars visible to the naked eye. The other stars' light in this region was blocked by two puffs of the collimates jets moving away from a new-born protostar. These blown-by-the-solar-wind offshoots looked like the dying remnants of a fireworks display. According to the Computer, the map of the sky remained relatively unchanged in the past 150 light years. It was a smooth flight without uncontrolled disasters. If a fire was sparked in the Lab, it was auto-contained. If a section of the ship depressurized, the robots, and the Computer executed the pre-pro-grammed fix-steps, without a death. Nally staged disaster drills for cadatas and robots, which were passed with a "survivable" score.

Chapter 6

Water World

The slow-down year began in 98,317 BC. Chief Programmer Guagray Raimingly deployed the nanobots, pre-programmed for the Nerore mission, to collect, data and genetic material. Beyond appearing just as neroreans, the cadatanauts traveling to Nerore had to operate the local vehicles, and wear native clothing, and decorations to blend with the denizens. A textbook in a human language is incapable of summarizing the myriad dimensions of divergent species' characteristics. References to attire and vehicles are simplifications of cultural-models species invent. Just as human genes are composed of a single combination of elements from the periodic table, out of billions of alternatives, apparel is one ritualistic habit, among billions of interchangeable patterns of presentation. Popular human films portray smart phones, and computers as too complex to master for time-travelers from earlier periods. When exploring a different planet, an alien is discovering a virgin time and space. Alien communication systems, and light sources cannot have reference-points back on the visitor's home-planet.

The nanobots returned to the spaceship three months later. They flew faster than the spaceship deaccelerating to orbit-speed. Their lack of significant mass decreased drag and made quicker deacceleration possible. They can withstand extreme g-force from abrupt breaking due to their semi-immaterial size, mechanical-design, and inherent protections. The nanobots brought back a library of information and resources, which guided the following nine months of pre-landing preparations.

Even from back on Cadata, the team knew enough about Nerore to make it the top candidate from the *Catalog* reachable within Nally's lifetime. Nerore's star echoed Earth's: a also a yellow dwarf, though slighter in mass. Earth is on the upper end of the yellow dwarf category. At 6.8 billion years old, Nerore's sun was young for the neighborhood, indicating it was a second-generation star with a significant portion of heavy metals. Nerore is closer to its sun, completing a single orbit in 212 days. Gravity on the surface on this planet was six times higher than on Earth, or a bit higher than on Cadata. Cadatas can glide on wings-alone on their home planet, but the weight increase on Nerore would prevent gliding for natural cadatas. Ten-fold higher gravity than on Earth at sea level would cause back injuries in humans, including when a human endures sudden deacceleration in a crash, or rapid acceleration during liftoff of a rocket. Humans' tolerance limit is a prolonged 4g. This added weight was diffused by the submerged lives of the dominant species on Nerore. The ocean's salty water cancelled out some of the gravity through its buoyancy and density. The neroreans lived just below the surface of the sea. Their land-weight of around 80 pounds on Earth went down to only 16 pounds under highly salt-saturated water. Magnification in gravity brought the total to around 96 pounds, though this number changed depending on their depth and ocean conditions. The larger mass attracted a heavier atmosphere to Nerore, which in turn raised the boiling point of water; here the ocean refrained from evaporating even at surface temperatures higher than Cadata's pre-impact average. The high level of organic materials in the atmosphere attracted Nally's attention to Nerore. Close examination found a range of sea plants and creatures generating these gases. Nerore's exosphere was three times larger than the planet's diameter, and included common ingredients such as hydrogen, carbon, and oxygen. The latter, and other chemicals in smaller quantities indicated the presence of lifeforms adjacent to cadatas. Its atmosphere was heavy in water vapor, methane, and carbon-dioxide. Most lifeforms lived in shallow regions, covered with merely 100 feet, or less of water. The deeper ocean was bearable to exotic creatures with extreme pressure and gravitational tolerance. The lowest half of the ocean was intolerable even for single-celled organisms. Under all this water rested a rocky layer, an excellent source of minable minerals to construct cities for cadatas. Given the heat, the planet's ocean was slowly escaping into the atmosphere before being blown away into space. This evaporation had allowed for the development of life, when the first shallow regions opened. Nally calculated the planet would become habitable for cadatas as this evaporation process continued to reveal more shallows.

As geologists were preparing writeups, the gene-editors were also busy with rapid-model-development. As the best pilot on board, Lead Cadatanaut Wocega Lapovy was chosen to head the expedition by fly solo into an alien underwater world without prior shuttle-diving experience outside of simulations. Chief Doctor Photuate Nackan supervised the gene-edit. The first step was evaluating the nerorean genes collected by nanobots. While Wocega was the primary subject, the cadatanauts on the spaceship due for the journey a decade later also underwent the process, but a few days on her heels to learn from her potential death. A million years earlier, cadatas had a special gene implanted allowing their genes to be pliable to editing. This special changeability gene was what the disease on Ganeir exploited. Once Doctor Nackan had the full genetic code, he programmed the Lab's Gene-Editing Machine with it, and positioned Wocega inside. There, her body was covered in a breathable liquid saturated in nanobots carrying the new code and the chemical substances needed to build the new body, replacing Wocega's cadata-native makeup. The structure, composition, position, function, and other dimensions of her cells were changed by this process. The program allowed for a nanobot mechanical separation layer, which allowed Wocega's cadata brain to remain intact in its dispersed locations throughout her body to avoid stupification complications. Sensory receptors were numbed by this army of invading nanobots, so Wocega did not feel any pain as this happened. The surrounding chemicals were liquid for absorption and because neroreans were aquatic, necessitating Wocega undergoing the transformation in a similar salty-water environment. If she was reborn in an air-filled, waterless orb, her dolphin-adjacent lungs would be incapable of utilizing air without filtering it out of water. Once the process concluded, the orb was transported from the Lab to the isolated half of the ship filled with salty water to match the mutated cadatas. It was a bit of a challenge to generate this much water out in space and the weight it added presented a problem for the overall lightness of the ship. The ship had started collecting, and concentrating water in isolated rooms earlier in the voyage, by attracting and processing accessible hydrogen and oxygen molecules. The zero-gravity on the spaceship across the bulk of the voyage made this added mass insignificant. But as gravity returned during deacceleration, they had to use anti-gravity tricks to restrain the ship against wobbling, due to one of its sides being heavier at 1g.

Once inside this watery environment, the orb released Wocega. Here are Wocega's notes from this period.

Wocega Lapovy's Diary Entries

98,317 BC

My transformation has brought back recollections of my youth. I learned to fly a shuttle soon after attaining biological maturity, as I gained the ability to communicate my predilection for rapid-transport. My family owned a shuttle mechanics shop, servicing fancy joy-rides of the rich. I test-drove them in space drag-races to assure proper component placement. Unlike most cadatanauts hired for this mission, my application to join the Troops was not from a desire to further my flying career. I had flown commercial tourists around our solar system for quick cash I invested in designing ultra-speed shuttles and selling them to rebels banned from Potentates' parties. Rebel and independent islands commissioned my work because my rates were lower than competitors. I've just wanted to make stuff go and stop fast. Trade disintegrated in the post-asteroid aftermath because of the disappearance of independent parties with the resources to indulge in top-end travel. If they were buying, they would be more interested in the cheapest vehicles, which could double as their homes on the moon or on parts of Cadata abandoned as unfixable. I scraped out a living on meeting this demand for a bit. When I saw the call recruiting new cadatanauts, I had to sign up as it was apparent staying grounded would put me in the front seat for overseeing Cadata's depressing slide down to absolute-zero. If I went up with the ship, even our deaths from a bone-melting pathogen would have been entertaining to the last.

I expected to be chosen as a cadatanaut, and to take the Lead role. My competition was unremarkable, half-starved, and scraping for air on unemployment. I resented the training and practical exercise regimen, but refrained from complaining. My policy is: if I am angry enough to complain, stronger actions than verbal reproaches are needed. The 150-year voyage to Nerore was duller than training suggested, but at least we lived in a spotless ship serviced by robots in exchange for limited research and development duties. It is unclear why they refrained from informing me I had been chosen as the first cadatanaut to transform into a nerorean. I would've volunteered if it was optional. I always wanted to try extreme gene-editing. My friends inserted vast muscle-mass to champ in sports, or ginormous wings to fly rather than glide through the sky. However, the realities of this gene-edit have been shocking.

While I was in the orb, I only sensed my internal fluctuating movements. I saw the size, shape, and color of my body mutate at the molecular level. As the change continued, my vision blurred, and then dissolved to black until I even lacked the capacity to see the blackness as my vision sense disintegrated. I was paralyzed to act. Instead, I reflected on this development, recalling the lecture Doctor Nackan gave me a week earlier after I was chosen. He mentioned different cell types in a cadata eye sensing a single-color signal, such as red, blue, green, or yellow. A nerorean perceived the environment through groupings of cells sensing sonar frequencies. The ultra-salty and mineral-rich water was mucky even close to the surface. The heavy atmospheric cloud-coverage further decreased underwater visibility. Due to these environmental factors, neroreans developed advanced sonar sensing rather than cadatas' sunny-island-native eyesight. He was saying the water in our half of the ship would be enriched with chemicals enabling the cells in the sonar-receptor organ to function. He had asked if I objected to having my eyesight subtracted to become a full nerorean. My brain would be the sole organ preserved in its cadata biology. I replied, "Yeah, sure." I was focusing on the novelty of seeing with sonar powers. Then, I plunged into slogging on the flight simulations in the new submergible shuttle, which mimicked the neroreans' design, and forgot about this complication. Because I failed to dwell on the implications, the change was startling in the first moments. *I will never see my face again?* I questioned. I began sensing sound waves returning to tell me the shape of the orb I was in. Then, I detected the shape of my new body. They allowed me to see all around and even behind and below me where cadatas had blind spots. However, these echoes were mere shadows. The vivid beauty of perception through my eyesight had been the chief motivator for my passion for shuttle flight. What would I now gain from flight, if I never saw the splendor of the planets and moons from the air, or space? The neroreans had invented a way to "see" into the distance with sonar mechanical enhancements in their submarines. This made it possible for me to sail these. But I wouldn't *see* Nerore, in however long I'd spend on this strange planet. Neroreans also never developed vocal communications because air-bound sound cadatas are familiar with cannot travel under water. Neroreans receive and send communications with the sonar-organ receiving and deploying intricate variations of sound pulses. I preferred speaking in sonar instead of cadatas' whistle-heavy language. I never enjoyed speaking. Blindness, on the other hand, was terrifying. My first action upon emerging from the orb and accessing a submergible computer was to send a message to Doctor Nackan. I agreed to add an invisible mini-cadata-eye to allow me to see in full-color without betraying my identity. The scientists warned me to limit my instinctual twitches and turns in response to seeing light to avoid warning the neroreans of my alien origin. Because of my discomfort, other cadatanauts also retained their cadata-eyesight during transformations into neroreans. With my visual sight restored, I spent the past day in intensive, submerged swimming. The ship's room arrangements remained unchanged. Yet exploring this familiar environment with switched senses made its objects appear distant and foreign.

Moving through water instead of air was an uncomfortable adjustment. Cadatas' evolutionary ancestors developed in the ocean, so water is bio-suitable. I indulged in diving and swimming before the asteroid struck, making water activities toxic. I was a bit heavier than I was before, so my body sank to the side of the ship facing Nerore during the deacceleration, or the side with gravity's pressure on it. Half of my body floated and half leaned on this surface as I navigated. This perspective helped me understand neroreans' strange body shape. My command of my new body was tested when I sent my first message to Doctor Nackan. I tried to mutate my wing into a cadata shape conducive to sending 3D messages, but my mutation abilities were still recovering from the transformation. Instead of a wing with manipulatable digits, I ended up raising seven of my tentacle-arm-with-armpit-flippers combinations and smacking them all against the Computer's display. This slapping produced an error message display. The Computer had my new genetic code in its system. It was prepared to correct a heavier typing error-rate, but not seven random splats, without any linguistic significance. As I counted them up now, I had 30 of these tentacles across my body. I flexed them a bit to understand how they worked. I shuffled around on the lower tentacles, trying to find a footing. I felt a strange desire to turn to the right, where a wall was awaiting me, but fought this compulsion, which arose because the higher quantity of fluids across my body were spinning as they adjusted to the heavier gravity and the whirl of movement they underwent during the transformation. I leaned a few of my tentacles on the wall to steady my insides. After a delay, the spinning eased. Mere months elapsed since we exited zero-gravity flight and entered 1g-deacceleration, so my body was still adjusting to gravity. Before this transformation, I sensed back pain from the simple act of lying in the orb before it filled with water. This pang was a remnant of the lack of pressure on my muscles in weightlessness. In zero-g, I was tranquil floating without contacting any furniture or floor while programming on the Computer module. I was recollecting those idyllic days as I leaned onto my tentacles and hopped on the lower legs to test their limits. Then, rolled forward and backwards on them. Then, did a dance to check their flexibility. These experiments explained why neroreans needed so many moving parts: their planet's gravity made their bodies heavy even under water, so they needed to shift their weight between many points to avoid putting excessive stress on any one leg while they also had to push against the drag of water currents. Then, I focused on a single tentacle and tried to maneuver it with precision. It took five hours to make a single coherent Cadata word gesture.

Across this difficult time, I heard in sonar via the communication system Doctor and Nally yelling queries regarding my intactness, or if I had suffered Putair-like neurological damage. So, my first word was: "Hush!" This silenced them as they replayed the image to check if I had intended to use this momentous occasion to request muteness. I strained my muscles in the following hour until my sign language evolved from nonsense to comprehensible phrases. Since cadatas use this sign language for computer programming, it was a lot easier to use it for emergency initial communications than attempting to make sense of the foreign sonar tongue I had been attempting to learn in the prior weeks. I signed, I needed to my sight back and then returned to my room to scrutinize how it was transformed to fit my new body. A seaweed bed was inserted to acclimate me to a rustic back-to-nature style popular on Nerore. All equipment, and my model shuttles were now floating in a salty soup. Seeing the little design shuttles—which I created back on Cadata prior to building life-sized versions on-contract—in that corroding water made me a bit crazy. I commanded the robots to take them to the other side of the ship, and to clean them before they corroded. Those tiny models lacked all the self-cleaning and corrosion-repellant properties of the standard-sized marvels. I felt drowsy. The sensation startled me because I had not entered full unconsciousness for at least 300 years. I tussled with the seaweed for a bit, before I settled on a hugging position, with my belly to the seat, and my tentacles wrapped in all directions around it. The tentacles were a bit sticky. So they kept me from floating away from my cushioned nest.

I woke up with a strange taste of seaweed. This taste was not only tingling on the

edge of my elephant trunk,[3] but also on the tips of my tentacles. To check if I was hallu-
cinating or in truth tasting with my feet, I compressed a single tentacle down onto the
seaweed and tickled it to study the sensation. The exercise confirmed I was reading the
makeup of the seaweed. My taste-savvy feet were also inspiring hunger for this plant. I
was still tinkering with this new dimension in my body when I sensed a mixture of organic matter and metallic mechanics approaching my room. Just as I was processing what
this riddle might point to, a couple of robots floated in. *Aha,* I thought, *the robots have
adapted to the salt-water by taking on organic properties, and that's why I sensed that mixture
in the water.* The robots brought a replica version of lower-lifeforms neroreans ingested.
The trunk was the sonar-organ and it consumed seafoods the nerorean digestive system
needed to generate energy. I had read about Putair's strange experiments with solid-food
digestion. I was apprehensive regarding what alien sensations my first meal might bring.
The idea of having a mini-alien in my trunk and then inside my digestive system as it
was broken down into a goo reminded me of alien-invasion horror films popular on
Cadata. Upon sensing these fin-covered creatures, my trunk began pulsating with delight. I sensed my stomach growling with longing. The first organism was up my trunk
before my conscious mind decided to grab it. It felt delicious as it touched taste receptors
along my trunk, which crushed it into digestible pieces with a compressing and sucking
muscle at the top. This snack was a bit less salty than the water I was swimming in, so
as I ate it, I recalled Doctor Nackan's lesson. Neroreans were halophiles, or salt-loving,
and could not live in salt-free water. They preferred to live in at least 2-5 M NaCl, which
meant their cells had to be stronger to withstand the heightened turgor pressure, and
rehydrating to keep back the constant tide of dehydration and desiccation. I swallowed
water with my trunk; I sensed it being filtered by fin mechanisms a few inches down my
face. It was a relief conscious tracking was superfluous for my body to extract unneeded
salt or to perform other body-functions. This adaptability was programmed into the cells
by nanobots, which fixed irregularities as the transitioned commenced. They sent alert
messages to the Computer if they discovered a collapse of any major function.

The day was six-times shorter than a Cadata day. To mimic conditions on the planet, the temperature of the water around me changed as the sun set, making me sleepy
again. The doctor had fixed my eyesight earlier in the day, so I could see the dimming
lighting mimicking what I would see if I was just beneath the surface of the water. I did
not notice falling asleep, and then it was morning again. My body felt more my own, as
if it had acclimated to a different time zone. I even managed to flip around, and found it
was much easier to flip with thirty arms to support this movement. My posture was now
more erect. I could utilize my fins in addition to my legs to speed up my movements as
I explored the half of the ship belonging to me alone.

When I reported improved mobility, Nally transferred me to Safety Controller Elleqow Comeepout, who asked for a training meeting at the Gym. Comeepout wore a
spacesuit with swim-enhancing propulsion engines inside fin-like devices to navigate. He
advised me on efficient body movements and methods for stabilizing water intake. Then,
he increased the pressure to the maximum-depth level neroreans had been found at. It
compressed my lungs. Each shift of my army of tentacles became draining. After a brief

---

3   This reference is a translation of a term for Cadata's long-nosed species, without a ref-
erence-point on Earth.

delay, he progressed to the next test, generating a strong water current. I express-shifted the pressure-level by speed-diving from shallow water into the depth. I learned to paddle to remain suspended at a stationary point.

We continued developing my physical training for a week before Comeepout allowed me to attempt a simulated nerorean submarine voyage. These subs' systems are outlandish. Comeepout reminded me the sub ran on magnetic conductors' power circling electricity through a water-safe wire. Many intergalactic power-generation methods are inaccessible to neroreans because the ocean negates them. They are unable to burn fuels as oxygen is needed for combustion and water and fire cancel a blaze. Sun radiation also failed below the sea's surface. Because it is sonar-based, there are no windows to see the exterior through this sub. It has threadlike surfaces allowing me to feel sonar fluctuations coming from the exterior through this fabric. The vehicles allowed neroreans to venture much deeper into their ocean than they could without their depressurizing properties. They became necessary to exploit deep-water organisms when they started to see scarcity of shallow-water edibles as their population grew. I attempted a fishing expedition under Comeepout's supervision. It was a strange, cannibalistic feeling to hunt after a species for food. The emotion was equivalent to slipping into a homicidal madness. My appetite was awakened when my sonar sensed a school of the little creatures, which I breakfasted on in the morning, swimming across my sub's path. I rushed the group and caught several with a special net-like device the sub was equipped with. Comeepout had the Computer replicate these in reality, so I could taste my victory.

When I was halfway through my meal, he broke the device keeping the ship pressurized. He was helped by Chief Maintainer Vomere Tenievan, who had just joined the fun. He broke it to verify if I was listening during his lecture on nerorean sub mechanics. I felt the pressure mounting inside. It disrupted my concentration because it was squeezing my isolated cadata brain. I decided I was still too clumsy with my tentacles to try a fix myself, so I programmed my personal computing device to bring in a robot to fix it in compliance with the manual Comeepout and Tenievan had assigned. This fixed the problem in a couple of minutes, but Tenievan was disgruntled with the cheat. He explained the steps again, saying I was forbidden from calling a robot if there were any neroreans on my sub since our team was unprepared to reveal our alien nature in exchange for a minor mechanical fix. After all, neroreans lacked astronomical comprehension because their bodies were incapable of withstanding above-water emergence. Their sky was also too clouded on most nights to distinguish a star. They could come up to the surface in their subs, which had been invented within the past 500 years. However, their blindness was a larger obstacle to their perception of the sky, as stars lacked signals strong enough for them to sonar-identify them. If any radiation was coming down, it registered like fainter sunshine. Explaining aliens and space travel to a species without star-perception would have been a philosophical conundrum destined to end in violent disbelief.

Then, since I had claimed I was concerned with my tentacles' dexterity, Comeepout had me do repetitive exercises, such as entering hand signals and repeating the same dance steps, until I gained self-confidence in my movements.

Once acclimated to my new body, I queried the team's progress on the lethal pathogens responsible for truncating our last attempt. Doctor Nackan showed me the library of pathogens they found in a survey of Nerore, and the steps taken to prevent infection.

Nerorean medical science was advanced, unlike among the semi-intelligent ganeirs. Neroreans' knowledge helped explain their unique traits, chemical makeup, and cell-attacking strategies. The numbers give the illusion of safety, but I am terrified a painful death will greet me upon my real-life exposure. In case of this eventuality, at least the rest of the crew will progress to the next planet instead of joining me in this fatal outcome.

I learned the nerorean language in eight months through sheer repetition. While I memorized their key words, my accent was abysmal. I was striving to mimic recordings of nerorean conversations in sonar, but the Computer was failing my sounds as too divergent from the originals. Nanobots recorded how the sonar organ produced sound, and the Computer simplified this data into strategies for me to match these sounds. The trunk had to be constricted in a certain way. Water had to be taken in at the right moments of making a string of sound waves. Preschool textbooks from Nerore supplied additional tips, but I was playing catchup after missing the natural education and indoctrination of an incubating infant and a nursery babe. We decided approximate sounds were sufficient for me to be understood. To counter the impact of my strange mispronunciations, my cover-story was I was from an isolated, distant group, which shared these linguistic oddities. If I had to enter commands in the sonar language or send communications while I was in the submarine alone, I could use its hidden translation system, which played in audio equivalent sounds in my voice. The programmers adjusted our Computer to adopt elements a dominant local species used in their designs. If our mimicked vehicle was deconstructed, the Computer's parts had the appearance of local components. It also had to be able to communicate in local programmable code languages and systems. I was never a fan of programming before, but playing with strange, liquid nerorean computers has been entertaining. For example, their computer can just spill out of its coating while remaining as an intact unit, capable of computation, if a sonar command reaches its receptors. I hope to imitate a nerorean, with sufficient exactitude for integration. The more I learn about neroreans, the more I admire their inventiveness. Their scientific and cultural achievements fight against the limitations of an aquatic species that must design solutions to obstacles to progress generated by submergence in salty water.

98,316 BC

I'm recording this in sonar in my sub. I arrived on Nerore a week ago. I'm still alive. I've sent a few official audio reports on the communication system. Now is my first available moment to reflect on my reaction to my entanglements. I have started making some connections with the locals by trading goods I replicate in the sub. But I still do not feel comfortable to attempt to live among them. So here I am retreating to the safety of a cadata-technology-enhanced sub.

The landing on Nerore was as challenging as Nally, Comeepout, and Tenievan predicted. Their standard outer-space-flight-capable shuttle was redesigned to disguise its flight elements as an operational nerorean submarine. For example, they had to use the elongated tube shape with fin-like navigation devices at its sides. This shape meant they had to increase the ship's internal stabilizers beyond those needed in an orb-shaped high-velocity vehicle. A spinning orb is more controllable at 7g or higher as it falls from space. An out-of-control tube can rattle the innards if internal magnetic and gravity controls cannot offset this extreme turbulence. The fins created unwanted drag and imbalance during the descent, so they had to be hyper-reinforced to survive hanging at the sides of this downward-shooting rocket. These were two obstacles in this convoluted redesign process, expedited from the centuries it necessitated on Cadata to only eight months from conception to landing. If we had failed to account for an interaction between our shuttle and the atmosphere, or between our cadata designs and nerorean systems, the shuttle-sub could have fallen like a stone out of the sky, becoming bio-metallic decomposing matter for the local environment. Initial simulations ended in thousands of crashes, explosions, implosions, cracks in half, deployments of my body out of the side of the ship, and various other unimaginable catastrophes. Two months prior to arrival, we saw the first safe landing, with a mere leak and burned equipment to repair in the aftermath. Three robots came on this journey with me, but they had to merge with the walls of the ship whenever a nerorean was visiting, and they could not venture outside the ship if I needed assistance elsewhere to avoid being spotted, as neroreans had not yet invented robotic technology. We debated disguising the robots in nerorean-like bodies, or bodies belonging to other species or technologies on Nerore, but there was insufficient time to calculate all possible failures, so we just opted to bind them to the ship. Once the sub was operational in theory, we tested it by having it perform movements and maneuvers inside the water-filled half of the ship. Then, I manned an unprecedented test. We had tried it first unmanned, so I had some expectation of survival. We liquefied the Mothership's wall and shot the shuttle-sub out of the back end while moving at a speed constant with the spaceship's gradual deacceleration from light speed to orbit velocity. We were around $5/6^{th}$ into the process. Due to relative motion, the shuttle did not slow down at dangerous g-levels. I supervised the Computer's catch-up with the Mothership and the gradual merger back into the moldable wall. These gave us sufficient confidence to believe an actual test landing on Nerore was unnecessary. The plan was to deploy the shuttle-sub upon entering orbit. Neroreans' lack of astronomic-detection encouraged us to proceed with this rapid and direct deacceleration and the instantaneous drop of the shuttle. If they could not see the sky at all, we could have even increased the ship's anti-g stabilizers and slowed down from light speed to zero just above the planet's atmosphere without piquing the neroreans' interest. We worried how the crash of a sub from the sky into their ocean might disturb neroreans within sonar range. To avoid an encounter, we scanned the landing site for intelligent lifeforms. Leaving our spaceship in orbit, instead of abandoning it on a distant asteroid, reassured me help was near. However, if I do call them for help, the leads will refuse a rescue trip to spare healthy cadatas exposure to whatever might be eating me.

Without centuries of landing on strange planets in the Cadata solar system, I would not have survived the rapid ejection from the Mothership, either in the earlier test or during the final landing maneuver. The atmosphere introduced wind current, temperature and particle density complexities the Computer failed to micro-account-for in its calculation of the trajectory and the needed stabilization methods. Some manual operation was necessary, and my reaction speeds had to be instantaneous and instinctual. I mutated my tentacles into more pliable devices to fit with the instruments to avoid losing control because of the awkwardness of my digits. We had sent nanobots for one last scan of the system to confirm the best landing site just ahead of us, and they returned with positive results. We were landing in deeper waters just outside of the furthest village from the center of Nerore's main city, Stirang. With the target entered, and the trajectory calculated, I had a large meal to keep my energy and hydration up. Nally communicated they were now at a constant orbit speed, and approaching the best position for my departure. The Computer heard the message as well, and at the pre-determined nanosecond, it auto-commenced prep for deployment. I reveled in my decision to retain sight in a hidden eye because tracking its programming adjustments would have been impossible if this code was transmitted via vocal sonar signals. I questioned if I spotted a problem when the Computer chose an optimal moment and ejected us out of the wall of the Mothership seconds ahead of the planned time. At orbit-speed, I felt a jolt during the shift from weightlessness to a heavy g-force on my body, which pushed me into the gel-cushioned wall of the shuttle. Then the ship settled into orbit-speed for a few moments, and my body once again began to float up in the saline water. Then, the Computer narrowed in on the target landing zone and propelled itself into a rapid drop at a 60° angle from the fins-level, which pointed to this goal. The fins became solid with coating around them; they were linear-flat during landing. I could eyeball the angle of descent in contrast with them in addition to relying on the Computer. Manual direction-tracking could be necessary if we experienced a mechanical malfunction due to incorrect geo-location or another unforetold variance. The new speed gave me sufficient gravity to land back on the seaweed-recliner next to the main controls, so I could adjust those few flaws the Computer could not perceive as significant. The ship accelerated as Nerore's gravity added to its own propulsion efforts to fight against the increasing air friction of the dense atmosphere. At the point when air friction overcame the forces of gravity and propulsion, I experienced another spike in g-force, a riskier change because of the planet's relative extreme size. Feeling the g-force until this point helped me sense if we were following the planned course. However, at this juncture, I felt so much pain in my multiple tentacles, I turned on g-force stabilizers, so the g-force inside would remain steady at the level we expected underwater. The engines collected the crammed molecules around the ship and used them to thrust against drag to further accelerate our controlled fall. With the gravity-stabilizers on, I pushed the ship to hit 500mph.

At the peak-speeds, friction, pressure, and colliding gases generated 4500°C in heat along the outer coating, which was re-absorbed back into the engine as added matter and energy by the magnetic field. If it had escaped without an artificial magnetic field to pull it in, it would have looked like a multi-colored neon trail evaporating from the ship and burning its way through the molecules of the surrounding air. I mention this because this input of energy was too massive even for the outré-high-capacity shuttle. Perhaps, the overload happened because I accelerated the speed of descent just to test the shuttle-sub's limits. Either way, I flooded the energy reserves in the cells of the ship to the brink. This forced collected matter in a mixture of condensed solidified and gaseous states of atmospheric molecules to be expelled from this skin. Their ejections were accompanied by loud pops, as the clumps of matter jumped in a swirl of dazzling lights playing on the wild wind-streams formed by the piercing motion of the projectile. They kept popping until the ship's speed was low enough for the engine to condense and store all matter contacting the gathering mechanism.

Meanwhile, the sound-dampeners were off, so my sensitive, multi-frequency sonar-hearing was overwhelmed with a sudden flood of noise. When I first entered the atmosphere, air was rushing by with a low hiss, then it started to sound like the whistle-wailing from traditional Cadata music. Then, the building air pressure transformed it again into a roar dampened by the ship's walls alone. This was when I supervised the switch to a 90° angle of descent, so I was now falling straight down.

When I ejected from the Mothership, I saw a star-dusted sky, the ball of the purplish-blue planet on one side and the radiant sun on the other. At the top end of the atmosphere, the black around me lightened and blued. Still lower, the Computer began registering droplets of condensed potassium chloride gas. This was the salt dominating the oceans and making them buoyant. It had extreme calming qualities and could induce death at large quantities. My body has been soaking it in from the water since the transition. According to our research, an excess of this stuff in a nerorean's system causes a high rate of premature mortality. The average lifespan for neroreans is around 650 years. I exceeded this limit as a cadata prior to the change. But Doctor Nackan assured me the nanos would filter the toxic excess out, preventing a throwaway end from aging. At 30,000 feet, with the added evaporated gases from the ocean pasted into clouds, the air became a grayish-orange color. With clouds came the clapping of rain, and 450mph yowling winds. By now, the planet was humongous, taking up 180° of my vision. It was strange to approach a planet other than Cadata covered with a survivable liquid surface, and yet without Cadata's islands or any other external land masses to focus on. If we only had external imaging capabilities, this planet would be absent from the *Catalog*. No ships, lights, flying devices, or artificial signs of occupation were observable from this perspective. It was a beautiful sight. This was the last time I would see the surface until this mission's unforeseeable end.

At 15,000 feet, when it seized being safe to maintain my speed without splattering, the Computer deployed propulsion in the opposite direction to deaccelerate my little tube. I was in the lower, narrow tip of this tube, which was facing the planet. I had left the walls in their see-through mode to take in this sight. The speed slowed to just above zero a couple of feet above the boisterous ocean's surf. It hovered the remaining length. This tranquil descent was met with heightened, steaming, foamy and power-spraying waves. The ship's stabilizers endeavored to discount this onslaught with a gradual sinking maneuver. It was a soft, soundless landing. Only incompetent or thrill-seeking cadata pilots allow jolts. While I enjoy turbulence, any jolt when hitting a foreign planet after falling from space is suicidal. The impacts perturbing my shuttle were from formidable waves denser in salt and harder compounds. As I sank deeper, they started rolling over the ship, adding pressure to its sides. The simulations we ran back on the spaceship failed to account for the intensity of this storm, perhaps because the calculations missed a contributing factor in wave-power. Unexpectantly, the biggest monster wave among them pushed the tubular sub sideways, flipping it so the longer side was now lying on the surface of the ocean. I guess we had not imagined the shape of the sub made it vulnerable to lose its balance. The stabilizers also failed to correct for the wave's pressure, perhaps because of the rapid transition between air and water, and the strange presence of water inside of the descending object. I would have been tossed against the other side of the ship if I had not flipped on gravity stabilizers, which kept the center of gravity under my feet rather than shifting it when the tube's position changed. The Computer auto-spotted the positioning change and bobbed the ship back to the correct orientation, allowing it to finish sinking.

Visibility near the surface was a few feet, but my sonar and non-visible wavelengths senses pierced through the mock to check for clearance. The Computer had also scanned the depth of the water long in advance, and knew we were entering in a spot with sufficient depth. The sinking maneuver went much slower than expected, as the ship had to fight wind and water for every inch.

Once in calmer under-surface water, the sub remained stationary to optimize its surface from space-travel-mode to sea-travel-mode, and to send telemetry to the Mothership on the conditions on Nerore, and its performance during this operation.

This flight took 8 minutes. But the final water submersion took 17 minutes. Nally and the others seized this opportunity to deride over the comms my slowest total-time descent across my pre-Nerore test flights. I was defensive, saying the mechanics should have anticipated the dumb tube would flip. The flight speed I achieved was at the maximum limit of what I believed the ship could have handled, given the conditions. To prove me wrong, the left-behind team analyzed the matter expulsion error, and determined it could have sustained a much larger ejection. Even flying at a twice higher speed should not have changed the outcome of this reaction. While I was sorry to be proven wrong, this was good news for my eagerness to test rapid and haphazard underwater tricks while on Nerore. In retrospect, I think I have the Restless Pilot Syndrome (RPS) after 150 years of access to flying a mere spaceship incapable of accepting any unplanned changes in its direction or acceleration.

The sub remained stationary as the team stopped to run down the check-list. Once they finished checking my systems, Nally played a pre-recorded holographic video Loor saved in our files for this occasion. It was such a sonorous and lengthy speech, and a nerorean day was so short, it was dark by the time it ended. I was so sleepy, I indulged in a full sleep cycle before continuing the journey. In the interest of preventing readers from suffering a similar drowsy fate, I will summarize its essence. He offered congratulations on our arrival on a new world full of potential discoveries.

"Nerore might be our new home, our hope, our deliverance. You and your fellow cadatanauts are our hope. You have taken the first swim in the ocean that might one day bathe us all."

I believe Loor asked for this to be a template speech to be played at future worlds, only changing the name of the planet, and its makeup. If this trip does not work out, I hope nobody lands twice within their lifetime, and must listen to it again, with just those little changes.

When I checked in with the ship after waking, I fathomed they partied non-stop since terminating our call. They were so high on slothmoths, the Computer had to answer the line for them. I could hear them toasting, yelling, and whistling in unison, relieving the pent-up stress of the journey. Even if they had doubts about our chances; it was easier to forget them through intoxication, than to run through the new data my ship was generating in every minute I was in contact with the planet.

Their over-consumption highlighted my own ravenousness. Deep in contemplation, I spotted a group of the hundred-head fish I had tasted from a replicator. I switched the ship into fishing mode and chased the school, netting them in. I learned the replicated version was less chaotic in their movements. Their sonar-receptor-equipped heads were dexterous and wriggled out of my tentacles repeatedly, pushing away and swimming to the opposite side of the tube. It was strange to see food resisting consumption. I had to strangle each little head until it was still, and I could eat it, without fearing it would push me around from inside my trunk. The taste was further ruined when I realized the guts were meant to be removed. The efficient Computer had kept the digestive organs out of the replication process to avoid expanding energy on parts to be tossed out as waste. It would just have to breakdown these innards back into molecular components to feed the ship's engine.

It left such a bad taste in my mouth, I decided to swim in the ocean outside the ship to introduce myself to a new sensation to forget ingesting this wriggly waste-bag. Nanobots had given us estimates of the ocean's composition, but these were based on measurements taken in a few isolated places, so the consistency was homogenous. As I slipped through the skin of the ship, I discovered real sea water on Nerore was unlike its textile sensation and taste on Cadata. There were intense tastes playing on the lower end of my trunk, which failed to block water from breaking in. The hot water tasted salty, and then metallic, and then sulfuric, and then muddy. The currents were strong, so I had to fight to remain stationary without an ocean floor's sand to generate drag for my feet. I went back inside, realizing why neroreans invented subs even if these did not shelter them from water.

Once the open ocean filtered my digestive tract of the toxic-tasting matter, I ordered the Computer to create a trajectory and head for Stirang. It started moving in a few moments at the slow sub-capacity speed of 20mph. The Computer projected it would be a three-hour journey. I fail to recall traveling in any object at a speed this low before… Well, it's the speed of me gliding on my own wings in average Cadata winds. While it was a pain for my patience, it was the best possible speed for me to explore this new world. The nanobots returned videos of the outlying villages and cities. We watched these at high-speed to familiarize with the gist of this culture. Now, my senses experienced the strange colors, shapes and textures of rock formations and wildlife movements. With my engineering background, I was most interested in the neroreans' architecture and vehicle design. I had already spent months testing their sub's systems, so now I was eager to understand how they built their stationary structures. The first village was marked as an abandoned in peace-time ancient port on neroreans' maps. Eight shattered tubes blocking the main road marked the remains of a surviving fort. The Computer spotted my contemplative mood and explained the neroreans of the past employed vehicles requiring contact with the bottom of the ocean to move. Recent innovations developed the style of the sub we were in, with multi-directional floating capability and without required electric conductivity or stability generated through ground-contact. The rest of the village, on the surface, looked like a series of tubes sticking out of the ground. When I looked via x-ray and other wavelengths below the ground, I discovered a set of mechanically altered caves below each of these tubes. Each had a different appearance to fit its function. One was the Village Ruler's house and meeting place. It had inner caverns with seaweed-covered furniture and precious-jam-crusted adornments, and a party-space. The latter was occupied with a few neroreans, who had faced me as I rode by. Visitors were so sporadic, their sonar was tracking information about me just as I was studying them. Seeing their curiosity, I stopped to introduce myself. I hoped this would be a safer first conversation with real neroreans, than if I waited to talk until I was in a crowded city. I said as much to the Computer, who had already plotted a trajectory for the tube under investigation, and now slowed to a crawl to maneuver through the village without disturbing historical monuments. One of these was a carved stone statue of a nerorean giving a welcoming gesture. I had plenty of time to study the communal workers' house, where they were crafting traditional furniture as a specialty item for city buyers. Their trading post was stationed on a veranda at the top end of the tubular structure. This veranda was crowded with displayed furniture, fish-catching machinery, and other items the village was producing. The tube above each of the caves was a hallway, water-filtration system, and a shelter in case of extreme weather. It also converted into a vehicle, separating from the cave it had been nesting against. The veranda was balanced on the flattened top of one of these stationary subs, and could have been disconnected and lowered down to cave-level if the owners ever had use of transportation. It seemed unlikely this group was keen on traveling, as crustacean-like ancient and immobile creatures, seaweed, and stranger natural growths had formed along their sub-tubes. The cave had internal filtration systems capable of working in isolation if the submarine at its tip departed. The lower level of the trading post was a food processing and cooking establishment.

The Computer parked just outside the tube leading to the meeting space. Since their sonic-sensors indicated my precise actions, I exited out of a hatch rather than through the skin of the ship, careful to use the neroreans' invention. I then constricted my trunk to keep it from taking in too much pollution as I swam from my tube to their entrance. I discovered their hatch was closed. I studied it for signs of a way to announce my presence. The item standing out was a musical instrument attached to the hatch. The Computer and our science team had missed the neroreans' unnecessity of doors to enter into houses or host bodies. I tried touching the walls of the shell instrument without a reaction. Then, I snorted in frustration and felt it vibrate. To test this response, I put my trunk against it and created a range of sonar sounds. The instrument magnified my waves, reverberated them, and gave them a unique musicality. *Do I need to sing a specific code to be admitted?* I contemplated. As I was questioning if I could research common melodies on this planet to crack the code, the Village Ruler came up and opened the hatch for me.

"Greetings. It's wonderful to have a guest! Come in! Come in! We are always interested in meeting new neroreans!" the Ruler exclaimed, spreading his tentacles wide in all directions, so for a moment I thought the statue in the center of the village was of him. Then, I recalled my alien bias. The historic map the Computer had pulled up aged the statue at over 2,000 years old. Neroreans' lifespan was less than half this length.

"Thank you for your kind greeting. It would be a delight to learn more about your town," I said, as the Ruler closed the hatch behind me and we floated down the tube to the meeting hall.

The first feature I noticed was a thirty-foot nerorean stone-carved head with a trunk stretching across the length of the circular wall and ending on its other side. The middle of the room was decorated with historic weapons with ungraspable properties. The group made greeting waves with their trunks, and I mimicked this motion. After we chatted about the strong currents and my journey, the Ruler's wife, Emouril, lost control of her curiosity and asked me about my origins. I related a practiced story: I had traveled from a distant village, with a name we picked from a map because of its lack of historical significance and its small population of 200. As we continued the discussion of the progress of their trade, I was forced to ask if they could understand my accent. They explained neroreans possessed individual accents, and village-wide accent patterns. If the general rhythm, tone, and wave length were observed; the variations were comprehensible. They marveled at my undiscovery of sufficient folks outside my village to know this. They added: it must be quaint to live in such blissful isolation. I asked them about my posture, and walking style. Should I edit it in some way to look less like an outsider, when I arrived in Stirang? They once again stressed I could crawl around with my trunk to the ground, or walk upside down, and it would still be acceptable because the current fashion was to abandon restrictive mannerisms, in favor of linguistic and cultural freedom.

As I was beginning to contemplate taking my leave, they invited me to join them on a special hunt, the reason they were all gathered on this occasion. Since my first ocean fishing experiment was unflattering, I was apprehensive I would display some unexplainable by village-origins idiocy. Still, I had to accept the invitation, as studying their culture was one of my main objectives. Instead of taking a sub, we walked down a cleared, polished-rock walkway along the ocean's floor. I felt a bit lighter the further we walked, and I started to see some sunshine piercing through the foamy ocean above. I saw a white cliff ahead, twinkling with lights colored in wavy shades of blue. I detected it was composed of chalk and salt. The tip was bobbing above the water. It was a tiny rising motion, which should have been an insignificant land marker from space. I started preparing to paddle up, and to attempt climbing this near-vertical cliff. But the Ruler stopped my attempt at a leap, and pointed to a small opening at the bottom of the incline. He went ahead, and I followed right behind him. We moved through a narrow, nature-constructed tunnel, which veered in strange directions, then climbed up, before dropping down. It was easy to navigate with my many tentacles, if I compressed them ahead of me, and behind me, making my multi-part body squeeze into the path. I scraped my skin against the rough stones, but ignored the pain. I felt with sonar the other six members of our crew following us. Part-blinded after a long period of darkness, we emerged in an underwater cave. I shielded my disguised eye with a tentacle, but then removed it, realizing the motion could have drawn suspicion from the others because my need for cover was unnatural if I saw on the sonar range. And the slight temperature increase would not have impacted a single spot on the body. They ignored these rapid movements because they had started climbing a stair to the top of the dome cave. At the tip, we pushed through another narrow passage, which headed straight up. I was careful not to trample with my lower tentacles over the fellow struggling up below me. When I could see the surface of the water above through a gap in the Ruler's body, he stopped, and poked his trunk out by a few inches above the water's surface. I asked what was going on, and he lowered the trunk just long enough to explain we were waiting for an extreme high-tide to fill the top of the cliff otherwise exposed to the unbearable high gravity and atmospheric pressure above the water's surface. Neroreans lacked the capacity to survive a stroll on dry salt exposed to the atmosphere, not only because their bodies processed air by filtering water, but also because their weight above-water would have crushed their organs. He kept his trunk out for around half-an-hour, making small pushes upwards, and starting to leave the channel. Finally, he popped out. I followed him and mimicked his crawling motion to stay close to the salt, and just underwater. The others emerged soon after. Without waiting for them, the Ruler commenced collecting seashells distinguishable by a colorful misshapen appearance matching the musical door-bell I explored earlier. He handed me a collecting net, and I also started gathering these little creatures. They had this long tail housed down the middle of a flower-stalk narrow tube, and this little fuzzy top with a wriggly body with a dozen fins and four black-bead eyes. Along the length of the tube where its fuzzy head could hide from predators were a series of bud-like solid cups with openings on their ends. As we picked each of these up and placed them into our net, they used these cups to make beautiful melodies. I felt guilty we were about to eat these guys and to make doorbells out of their houses, but my hunger pangs were once again overpowering my native cadata disgust toward diges-tion of fellow living creatures. As I was having an internal moral debate, the Ruler was explaining these are high-pressure-loving creatures, one of the sole species on the planet capable of surviving outside the ocean because of their sturdy shells, and consumption of toughening microbes. I asked why he seemed to be grabbing them with all tentacles at top-speed. He explained the tide was about to turn, receding away from the cliff. To avoid being crushed, we had to leave before it did. There would be little warning because of its rapid movement. He had set a timer, and it went off a few minutes later. We bun-dled our nets, and pushed our bodies down the tunnel, letting our tentacles intermingle in the commotion. Once down in the cave, I thought we would rest a moment. But the Ruler continued down the path, mumbling: the salt cave was unstable, and subject to spontaneous collapse. It was ultra-vulnerable to a break just when the tide water was rushing back in, and increasing the downward pressure. A chunk of salt hit me, as if to confirm this danger. We all made it out, and I never felt more satisfied with a meal than when I ate those little sheser creatures.

They invited me to stay with them for a while, and I asked if I could park in their village, staying in my own sub overnight. They were happy with this arrangement. In addition to going on similar collection hunts, and helping them with their furniture work and cookery for visitors, I started replicating some little trinkets with the Com-puter's help on the sub, making it look like they were native pieces I brought along from my home village. As I said, I'm still hesitant to attempt living inside their houses, or of tracking back to my initial target: Stirang. Maybe if they were a bit less polite, or a bit worse at group singing; I would have long ago departed.

98,315 BC

I had high moral goals for this mission. I imagined bringing salvation to cadatas, and a new prosperity to the neroreans. If heroism was a matter of intention, it would be uni-versal. A year of amazement, enjoyment and music has come to a horrific end. I struggle with finding words to begin. The history of an explorer from a distant time and a foreign place necessitates a chronological narration, but describing the months of carefree vil-lage life, and my insignificant archeological discoveries would be as tone-deaf as playing a wedding tune at a funeral. Then again, a history about me as a character might begin with an admonishment of reckless disregard for life, irreplaceable culture, and interstellar alien rights. I am having difficulty making these sounds, stopping between words to contemplate the next. I have not drowned this guilt with intoxicants because I'm too afraid that letting my unconscious mind rule would turn off the switch.

I stepped away for a couple of hours to observe Nerore below… I'm in orbit. As I was deliberating penetrating through the spaceship's skin and diving unassisted to Nerore's inviting ocean, I concluded my corpse would be a pollutant and potential evidence of an alien visitation. Instead, I settled on relating this story as a lesson for future cadata planetary visitors.

After many happy months when I seemed to have joined the Ruler's family, I decided I had to continue the adventure. Nally and the other cadatas were displeased with my prolonged holiday. The rest of the journey from the village to Stirang took the expected length of time. The density of villages increased along the road, before they turned into towns and then the suburbs of the city and finally the city itself. Its center was positioned along a slight incline, with the main business district in the most desirable top position. Each of the buildings was composed of a mixture of hardened salt, rocks, and crushed seashells. Each had circular walls, so they had the appearance of a collection of stacked balls, reminding me of the preference for orbs in Cadata's architecture. Stairs led from the lowest protrusion to the next and all the way up, with flat surfaces above each on which somebody could walk along before reaching the next ladder. Several neroreans were out on these balconies, enjoying the gentle currents. Much can be learned about art and beauty from these elegant beings, who put value in their cultural heritage, traditional crafts and designs, and moral fortitude.

Since I had learned the types of trinkets valued on Nerore back at the village, I launched a trade in these in Stirang. As my profits grew, I made friends with wealthy merchants and they invited me to their top-of-the-hill veranda parties. It was an intoxicated continuation of the same shindigs I enjoyed in the village. Cadatas back on the Mothership joined my drunken revelry out of new hope the mission would end in our settlement on a planet with a healthy atmosphere. Our cadata team was pushing me to gain power and fiscal clout, so they could say they were my business associates when it was time for them to join me. These parties were how business deals were made, so I stayed under the table across these months. I had parked my shuttle and disconnected my communication units at one point, when I fluctuated between not wanting to get out of my new seaweed bed in a luxury flat in the city, and rolling on highs of excitement from some new adventure, like stream-racing or expeditions to intoxicant-producing farms. I was detached from the news, so I learned of a strange illness fighting through Stirang when I was notified of the death of Fitan, a wealthy trade partner who had been purchasing most trinkets I had stock-piled. We had a beautiful funeral for him, wherein his body was decorated with flowering seaweed and floated up to the surface, where the high-pressure dissolved it with help from a special cream placed all over his body. I started worrying he infected me because of our frequent contacts. I ran self-diagnostics with my disguised-as-a-giant-seashell medical device. All tests were negative. I returned to partying under the assumption avoiding infection despite the hugs and trunk-kisses I gave Fitan meant I was safe with less intimate contact with other sick neroreans.

Perhaps because so many of my friends were starting to fall ill, one of them put a much larger dose of intoxicants into the bowl we were all sharing than was advisable. I woke up buried in chalk in a small lime-walled room at the main hospital. The chalk burial was a restraining mechanism used on Nerore because no handcuffs were multifaceted enough to hold thirty arms. In the first moment, my trunk and hidden cadata-eye were both in the chalk. I was staring at blackness. Then, I migrated the eye to the exposed tip of my head to restore my sight. After a lengthy delay, a group of investigators arrived. Nerore had a noninterventionist and decentralized governing structure. Every major decision about helping the poor or the wealthy was reached by a vote. The system lacked a single decider or a small council of wise rulers to reestablish order. The investigators were just called in egregious matters when dismissing the behavior with a fine was unthinkable.

They told me I had been near-death, so the doctor ran an array of tests on me to check for possible causes of this coma-adjacent condition. Internal scans had shown the presence of my hidden eye under my skin and my distributed rather than central brain. The tests of my body's chemistry at first appeared to match neroreans'. But then, they discovered nanobots in my blood. Closer analysis distinguished other inexplicable differences. Since they had no equivalent technology to these nanobots, they decided I had to be an alien. More significantly, they had determined I was the original carrier of the otherwise unexplainable disease now covering the length of their planet. They put me through weeks of interrogation, when only my trunk and the top of my head were exposed. The lengthy period of immobility made me pliable. I confessed about cadatas, our ship, and our mission to find a new habitable world. I explained the spread of the disease was a horrible accident. I hoped to be of help to solve it, if they would just free me and allow me to utilize a medical unit to help them. They found the seashell hiding my medical unit, but since it was programmed to switch on when it was in contact with my genes, it remained seemingly useless in their tentacles. These protestations convinced them I was more mad than alien.

While this was happening, the death rate from the disease increased until the planet saw a 98% drop in population. It started to level out at this point as some neroreans had immunity. The investigators used this moment to put me in a mock trial, where the entire population of Nerore voted to have me executed. Up to this point, I had been hoping they would kill me every time they brought horrific images of more neroreans my actions had mutilated and brought to a painful end. But now hearing death was near, I reflected I had failed to report my findings on this disaster to a single cadata. What if they just heard I had been executed without knowing why, and attacked the neroreans in retaliation? This was unacceptable, so I turned on my hidden communication system and made a brief report by making tiny signs with one of my tentacles.

A few minutes of explanation sufficed for Nally and the others to rally around the immediate problem. They activated the Computer and robots on my sub, which homed on my beacon. Instead of asking for retaliatory actions, I insisted they let me die for my sins. But Nally ignored these pleas as she would have if a pet was trying to leave the house prior to a scheduled walk. I struggled to explain my position with the little signs, when my shuttle-sub plowed through several walls of the hospital, ramming into the wall opposite from me. The Computer had calculated a path barely sufficient to keep the building from collapsing. The robots flew out, administered sleeping shots to my guards, blocked the door, dug me out, and tossed me unceremoniously into the shuttle. Then the Computer drove the ship out of the building and departed, at a higher speed than would be legal even on a Cadata street, straight for the ocean's surface. The speed was intended to avoid any potential violent clash with the locals, as hurting any of them would have been against the Computer's code. The shuttle popped out of the ocean and through its violent waves without the extreme difficulty it had going in, and in a moment, it was flying through the atmosphere, and ten minutes later it was in orbit.

Nally and the team were still on the line in case of needed input. Once the Computer confirmed it was in orbit and clear of obstacles, Nally informed me I had to remain in orbit for the eight residual years from the planned decade of my exploratory Nerore habitation. I objected to this cheery outcome: I would be living in luxury on a private spaceship, while people I afflicted were dying below. Nally defended me with the argument any cadata would have spread the disease through their inactive-in-cadatas pathogens.

I commenced developing a nanobot and chemical solution for this disease. I utilized sample nanobots acquired from aggrieved tissues. Doctor Nackan remote-supervised the project to avoid infecting the main ship with mutated-by-neroreans contagious pathogens. Findings in-hand, I transmitted the cure to the investigators through their sonic communication units. Shockingly, they refused. They feared it was a new attempt to poison them, to weaken their population, ahead of an invasion by my swanky spaceship. An order was given to exterminate any alien observed: a violent command contrary to my knowledge about this peaceful species, which had outlawed war thousands of years earlier. Logic demanded ignoring their stubborn refusal. I sent an army of nanobots to administer an antidote to every nerorean on the planet microscopically, without their knowledge. I hoped they would accept unsolicited, direct assistance. If this assistance had been appreciated, it might have reestablished a communication channel, allowing me to explain the recommended actions they had to execute, if a mutated strand of the pathogen re-appeared in the future. Instead, I had to be satisfied with burying these instructions under noticeable markers across the planet, just in case most initial recipients destroyed their copies.

Chapter 7

Alcohol-Based Lifeform

A 2014 biography, *Wocega Lapovy: A Spacefaring Story*, called her an "inspirational cadatanaut who pioneered the brave exploration of dangerous worlds." Most cadata histories also remember her kindly, instead of blaming her for the failure on Nerore. Some have offered harsh words during her lifetime and since. They say she had to be sober to discover the pathogen before it spread into an epidemic. I hope Wocega's diary speaks for itself. She was an intellectual, whose story needs to be reevaluated, as we research if Earth is a suitable location for a permanent cadata settlement.

Wocega completed her isolation hold in 98,306 BC. In the interim, cadatas on the Mothership debated if there might be some argument made for staying on Nerore, and attempting a second expedition, after a prolonged delay. Wocega's distribution of cures worked to restrain new infections from the disease she had spread. On the other hand, anti-alien sentiments on Nerore were only growing as their economy plummeted amidst the extreme depopulation. Nerore doctors had developed a test for alien genes, and were administering it to suspicious actors. Due to these tests, discovery of their cadata identity loomed, more so if they were working to gain political and economic power during this second visit to secure a hold on territory to prepare to receive a potential influx of billions of migrating cadatas. One member of the team, who asked for his or her name to be redacted from the official record, even suggested the possibility of allowing or even assisting the neroreans to die of some new breed of cadata-native pathogens. If the remaining 2% of neroreans died off, this would leave the planet without a rival intelligent species, and thus ripe for colonization. Everybody else with a vote at this meeting of chiefs voted against this because if they started exterminating native planetary populations, they would garner a deleterious galactic reputation. Survivors of a genocide would also be ripe for revenge, and even one angry nerorean could have killed many cadatas. If they went to this extreme, it would also spread belief among cadatas they were living in either a homicidal tyranny or a lawless society. If the number of prior rebellions over gene-editing was any indication, the decision to exterminate a gene pool would launch a revolution to end the Potentates' or the big corporations' rule, and since they were the sponsors of the mission, the group concluded this would go counter to their mission's primary objective. It was a dark day in cadata history when this topic had to be broached, but facing extinction had driven them all to a state of defensive psychosis.

By the time of Wocega's return, the Mothership completed an extensive investigation of the possible scenarios allowing for the settlement of Nerore. On top of the anti-alien campaign, Doctor Nackan and the other scientists argued a world submerged in water and with heightened gravity clashed with cadatas' basic physiology as a semi-flying, island-faring species. To adapt, billions would have to undergo gene-editing. Technology and buildings would have to be built underwater, in conditions promoting rust and other damage to their water-sensitive, high-tech equipment. The shuttle-sub had performed well during the trial, but Wocega had not (from a biological standpoint). While her brain remained cadatan, the physiology of a nerorean body made her susceptible to intoxication. She was one of the soberest cadatas across her history. The scientists determined living inside of a salty liquid slowed cognitive function, and made a body groggy and lethargic. Permanent habitation would have been detrimental to their minds, making them reliant on robots for their needs.

These definitive studies favored leaving the neroreans alone, and seeking asylum in better climates. Then, just as the team was planning to depart for the next world from the *Catalog*, they received news from Cadata. They received the first message from Cadata around the time they began orbiting Nerore as it caught up with them a year after they exited light-speed. This was a note from a year after their departure, offering congratulations on their arrival on Nerore. A few more messages followed, but since it would be another 150 years until Cadata would receive the Mothership's response from Nerore, there were few new communications coming in, just some news updates. Then, in the last couple of years on Nerore, these notes stopped coming. The team's worst fears were proven when this communique reached them just as they were leaving the sector. Nally read it to the entire crew, who listened, flabbergasted.

98,456

Dear Examen Team Members:

Shortly after your departure, a group of scientists attempted to create an artificially-stimulated massive volcanic eruption. The goal of this mission was to spread sulfur and heavier gases across the planet's atmosphere to recreate the drop in global temperatures observed after the asteroid strike. However, while the temperatures did decrease somewhat, they did not return to levels seen post-strike. The impact also created an unexpected level of chemical fallback in the form of toxic rain coming down all the way from the stratosphere. We want you to know we are working hard to find a new strategy to bring the climate back within habitable limits. At the same time, we hope you have had better luck with finding a solution on Nerore. We anticipate some good news from you, and hope the next memo you will receive from us will contain a breakthrough of our own.

Sincerely,

Loor Dowor, Head, Examen

The room erupted in exclamations and terrified screeches. Would there be a homeworld to return to if the journey ahead failed? The toxicity of the atmosphere must have begun to penetrate through their spacesuits for the leadership to have come up with a volcanic solution. Everybody agreed Nally should respond with a toughened argument on how their species had proven many times they remained too infantile to conquer nature. Nally took this challenge to heart, and retreated to the Lab to compose a response reflective of the intricate relationship between volcanic activity and the climate. She began crafting 3D diagrams to help herself understand the problem better. She also sculpted a diagram of the asteroid strike, and its impact. She tried to come up with a scientific equation to show what the two said about the cadatas and their ambition to become world-builders. She was in the Lab without rest for two weeks, with the ship's departure date on hold until she completed her theorem. The scattered ideas were ill-fitting. An associate supported exterminating neroreans for cadatas to prosper. She was reliving her mortifying failure to envision a pathogen's spread to neroreans, instead of focusing on her own potential infection. She molded a diagram of the impact her ship had on Nerore. Then, she created a summary of the Ganeir mission and how the crew she had handpicked and trained died while she was living in luxury back on Cadata. She looked at the four diagrams for a long while. She could not begin writing the response. Finally, she observed the reflection of herself in the see-through wall of the ship, on the other side of which was the image of Nerore's purplish orb. She chiseled herself out of a clay block. Due to the gene-editing techniques, she appeared radiant and youthful. The surveillance video captured a flash of an idea in her eyes before she passed in an instant. She left us without a conclusion.

The remaining cadatas had an elaborate ceremony in Nally's honor before the Computer broke her body into molecules and recycled it into its energy and matter supply. The most robust of Nally's embryos was now revived, and placed in the Lab for development under Doctor Nackan's supervision. Ally, as Nally requested her offspring be called, would have a short blooming period before beginning her training for the chief-engineer role. Once she was mature, she would commence the leadership duties Nally brought to the ship.

In the meantime, Wocega, who had been isolated even after rejoining the Mothership, was solicited to exit retirement to take on a leadership role in Nally's absence. She agreed when they mentioned Doctor Nackan was the runner-up because he had proposed responding to Cadata they should attempt releasing cloud-creating gases into the stratosphere, rather than ejecting it from a volcano. Upon taking control of the ship, Wocega created a to-the-point memo instructing Loor to seize attempts at "fixing" Cadata, and instead to focus on developing survival methods in the new artificially-damaged environment. Then, she gave the coordinates of their mission's next stop because they had skipped over two of the next potential planets in the *Catalog*. One underwent rapid atmospheric transformation. The other was also a water-world, a category matching Cadata. Excess water appeared less desirable after their last trial. The next best option from the list was Likos, a bizarre planet 2,684 light years away when heading at a 45° angle right from the galactic center, on the closest edge of the massive Scutum-Centaurus Arm. The oddity of this planet was preferred because Doctor Nackan theorized the similarity of the molecular compositions of Ganeir and Nerore to Cadata was the chief reason for the ease of disease-mutation between cadatas and ganeir and nerorean species. He theorized an outlandish molecular combination would shield both the natives and the visitors from contamination. While Likos was extraordinary, its solar system was mundane. It was stationed in a neighborhood with many neoteric systems, on average younger than 3 billion years old, which had developed from gas accumulation as the slow Scutum Arm dragged its stars through lighter and fast-moving inter-arm regions. But this sun stood out among them as one of the oldest, at 12 billion years old. Likos' sun was a member of the most common class of stars in the universe, a red dwarf (M). Its mass was around .3 of Earth's, but it had a rapid orbit, making for a brisk 11-day year. It was also a low-luminosity star, with Likos located extremely close to its explosive surface, at .05 AU (with 1 AU being the distance between the Earth and the Sun). The low luminosity was a good sign for its longevity, as it was destined to survive till near the end of the universe, living through trillions of years. If it was lighter and dimmer, it could have slipped into the category of a failed star like a brown dwarf, or an oversized (to the point of radiance) planet. Based on these parameters, Likos received around 81% of the amount of stellar energy the Earth receives. The lowest temperature recorded during the period Cadata was observing this planet was −133°F, and the hottest was 119°F. Most proximate M-star planets are uninhabitable because of tidal-locking. But this was a planetary binary, where a planet and its moon-planet hybrid were spinning around a mutual mid-point every 9 hours, in a manner simulating a standard unlocked planet's daily cycle. Thus, Likos had seasons, and day and night times similar to Cadata's, features indicative of a complex climate, an ingredient cadata scientists believed necessary for diverse lifeforms to evolve. Another common problem with M-stars is their heightened solar flare activity, which is more destructive at this proximity, but observations showed the planet had retained an adequate, dense atmosphere. Lifeforms were thriving in this environment. These measurements were close enough to Cadatas' ranges to fall into Cadata's "habitable zone" standard. Cadata's ideal-for-habitability heat scope is somewhat higher than Earth's preference for surface temperatures of between 0 and 100°C. M-stars have habitable zones a tenth smaller than the band around Earth's Sun; in other words, there is a small space near M-stars where life can thrive, and planets outside of this thin line is fried or frozen. The extreme closeness of Likos to its star is necessary for habitability because of all M-stars' dimmed luminosity, which cannot project sufficient light to planets any further away.

Wocega stressed to the crew receiving the transmission: Cadata ought to send future communications to Likos. However, due to the lag this message needed to reach them, if Likos was unideal for cadatas, they would be on their way to the next world before they received a response from home. She suggested they should send urgent updates to all planets in the *Catalog* to increase the odds of message-penetration. In general, the timeline used in this book is from the perspective of cadatas who traveled on this cross-star voyage. Far fewer years passed on Cadata, and on Earth, than on the Mothership, in the same span because of the ship's near-continuous motion at the speed-of-light across these millennia. With this urgent political message delivered, Wocega gave the marching orders to set sail for Likos after performing the standard pre-flight maintenance.

The long stop at Nerore created a fuel-shortage on the ship. If they began the most fuel-consuming step of accelerating to light speed now, they would have run out, being left in the middle of the galaxy without a "gas" station in sight. They could have made a trip over to Nerore's moon, but they were short on a set of the elements more abundant on Nerore itself. Wocega was in a glummer mood than usual after Nally's passing, and she opted to just extrude the needed matter from Nerore without sending a notice of intent or an apology for the intrusion. She directed the spaceship to point its magnetic attracting field at an uninhabited spot on the planet with the needed combination of elements. Then, it pulled up tons of salt water and dirt from below the ocean's floor. This intrusive extraction was observed by a couple of salt cliff hunters, who informed other neroreans cadatas were at it again, pilfering their resources and endangering their lives. Without caring for such criticisms, Wocega oversaw the Computer's distribution of this fuel across its skin, and to the deprived systems.

The second crucial maintenance procedure was the Great Garbage Dump (GGD). While they were short on useful matter, they had collected too much junk, and some of their older matter had gone past its radioactive decay due-date. Wocega brought these attachment-possessions with her, when she hoarded furniture, sculptures, and trinkets from the planet. Most robot-collected inessential research-samples had to be dumped to free space for the added fuel-matter. To avoid this waste blowing back into the ship, the matter was gathered into a decomposable, thin spherical container, which was pushed out of a wall on a degrading orbit towards the planet. It burned in the dense atmosphere before the charred remains drowned in Nerore's ocean, without arousing suspicion from the blind locals. Matter exchange was a constant at all Mothership stops because new resources, goods, clothing, decorations, and technology were only allowed entry, if old items of equal mass could be expelled. The salt water, added to half of the ship to assist Wocega's transition, and then to experiment with nerorean physiology, was part of the matter discarded because Wocega opted to transition back into a cadata. She struggled with concentrating on her captainship duties while under heavy water pressure. Once the pre-launch tasks were completed, the spaceship began acceleration to light speed, leaving Nerore as a mere bright spot behind them.

The rest of the senior staff died in the coming centuries. They were replaced at the helm by their children. Wocega also did not live to see the next star. The leadership torch passed to Ally, who never visited any planet. The new generation had been born on the ship. They were still assisted by the elders, who were young upon departing from Cadata. The many questions youths asked them about life on a planet made it apparent they had to record these answers in detail for future generations destined to lack direct contact with solid-planeters. They had a right to assume the upcoming Likos stop would be as fruitless as the last two potential worlds. These messages would need to be repeated for millennia beyond their lifetimes. Their recordings are prized artifacts in our Archives. Our latest generation learned from these as well during our preparation for landing on Earth without prior textile contact with a planet.

The slowdown phase began in 95,620 BC. Once again, after exiting light speed, they sent the nanobots to collect a survey of Likos. Wocega's offspring, Cega, volunteered to follow her mother's example by being the first cadata to change her physiology and land on this odd ball of a planet. As they continued their approach, they received data and samples back from the nanobots. They had anticipated some of these readings and visuals of the planet's surface and lifeforms, but other elements were strange and inexplicable. Likos was one of the two close inner planets, positioned near its star. This proximity meant it had to be built from strong stuff to survive. It had little of the lighter elements, such as hydrogen and helium. Specs of these were only found on its surface. The rest was composed of heavy elements, with an abundance of silicon, oxygen, aluminum, and iron. The most shocking distinguishing characteristic was life's development in a methyl alcohol ocean. Cells lived in and were filled with this liquid rather than with water, in contrast with living organisms on Cadata and Earth. Alcohols are composed of hydrogen, oxygen, and carbon strings. Living organisms on Likos had similar genes to humans and cadatas as they were also composed of these basic elements, but instead of nitrogen and phosphorous, to these three alcoholic compounds, likos' genes add sulfur and selenium, the other members of the nonmetals group.

Methyl alcohol has a freezing point of −142°F and a boiling point of 148°F. To put this into perspective, the lowest recorded temperature on Earth was −128.6°F in Antarctica in 1983, and the hottest was 134.1°F in Furnace Creek, California in 1913. In contrast to alcohol, water freezes at 32°F, and boils at 212°F. Because Earth's temperatures fall under water's freezing point, much of Earth's surface freezes. On the other hand, because temperatures have not historically reached water's boiling point, Earth's water has refrained from over-evaporating into the atmosphere to be blown into space, a common process responsible for turning water-rich planets into waterless deserts. Likos' Goldilocks zone differed from a water-based planet. A planet can be colder than record-cold temperatures on Earth, while its alcohol oceans remain unfrozen because they are not reaching their freezing point. So, there was no snow or ice falling from the sky or collecting on the ground or on water bodies on Likos. On the other hand, if Likos was just a bit hotter or even as hot as Earth, its oceans would begin to boil and evaporate, so even slight global warming on Likos could be disastrous.

Methyl alcohol multiplied on this planet because of a runaway bacterial infestation's conversion of its early ocean into alcohol; this process seized once the bacteria depleted its water. A handful of the bacterial strains evolved to consume alcohol and survived. This chemical is a great energy carrier, so life based on it can run as if on gasoline. Earth's humans have used it as a vehicle fuel. However, it has a limited seven-day half-life, which means all living cells in an alcohol-based body must be discarded as waste every seven days, and new replacement cells must be produced. This alcohol-based composition made lifeforms on Likos heavier, and less stable. Average lifespans of likos were closer to 10 Earth-years, despite their medical advances. Earlier species on their evolutionary tree started with seven-day lifespans, and worked up to relative longevity over billions of years until they evolved complex regeneration systems. Methanol has over three times less surface tension than water, which makes it more difficult for a surface membrane to form around a methanol cell. Life adapted by primarily forming complex relationships and membranes on the lower part of the body in contact with the ground. The rest of the body and individual cells are held together by a thin membrane, which is sufficient for the task because of the high surface tension. This made their bodies gelatinous and sluggish. Their skin was semi-transparent with a yellowish tint due to their high sulfur content. Likos' atmosphere was dense with hydrogen, acetylene, and ethane. Thus, likos inhaled these, and exhaled methane.

A likos body is three times larger than a human body, but a likos feels less heavy due to Likos' lighter mass, equivalent to Mars'. They walked by bouncing off the surface. After Cega's transformation, she was granted control over an isolated half of the ship; this environment was less altered than during Wocega's training. One difference was Wocega's new biology required a heavy intake of radiation, on which her skin fed for energy and ingredients essential to likos' biology. The planet's proximity to its star, an even older M-class with less violent solar flares, meant it had been blasted by radiation across life's development, so the surviving species were adapted to exploiting radiation, rather than fighting its threats. This relationship with radiation was reflected in the dark-pigmented surface skin layer, which protected the yellow gel below from exposure to outdoor radiation. Heightened radiation was un-inhibiting to this planet's top habitability ranking because Nally anticipated the prolonged space voyage and the accompanying gene-therapy would have reinforced the explorers' immunity to radiation. Interstellar space's absence of natural shielding via an atmosphere or solar wind makes it hyper-radioactive. The ship's coat protects those inside, but some of it seeps in because the ship uses it for energy.

As Cega reviewed these findings with Doctor Nackan's progeny, Doctor Nack, she grew concerned about becoming a likos. She had a sober-minded disposition, a genetic trait in their family. Switching from being a water-based lifeform to an alcohol-based one violated her moral standards, which commanded sobriety to achieve top productivity. She questioned Doctor Nack as to the viability of performing research if on top of constant intoxication, she also became the alcohol. Doctor Nack eased this concern by explaining, just as water is a normal life-carrying fluid for cadatas, her mind would accept alcohol-saturation as "natural" to match her body's transformation. She questioned if her isolated brain would remain water-based, but Doctor Nack advised against this because of the low water-content on the planet; keeping a water-consuming brain hydrated during her residence on Likos would have been problematic.

Cega raised a concern about the likos' ten-year lifespan. Given the pre-set ten-year span of the experimental stay on Likos, what were the odds of her life ending just before completion, or sooner as she was already a couple thousand years old? Doctor Nack reassured her the internal nanobots would continue rejuvenating the cells and fixing genetic errors, which otherwise led to rapid aging. These mistakes in a likos accelerate alcohol and weight loss. Cega reminded him she would be blending with the local population, so her eternal youth might be noticed by friends. To address this conundrum, Doctor Nack instituted an artificial mechanism to age her appearance without degrading her cells. There were many other troubling details about likos' physiology Cega questioned. In the end, Doctor Nack convinced her of the innate imperfections of all lifeforms.

The atmosphere on Cega's half of the ship was changed to the compounds she needed to consume. Local jellyfish-like, sponge-like, and rambutan-like food sources were replicated and set aside for Cega in preserving containers in the Lab. She had to eat the equivalent of her body-weight of these daily to allow her body to regenerate all cells in its standard seven-day cycle. Gravity was decreased to a third of Earth's across the entire ship because cadatas who had spent their lives in space were struggling with adjusting to 1g in the weeks they spent slowing down, prior to this conversion. Despite a strict physical training regimen, they were less muscular, and bonier than their ground-born predecessors. At first, Cega kept the furniture she inherited from her mother, but then discovered she had to obtain a thrice-larger resting unit. She had insomnia prior to the transition. But like neroreans, likos need a high quantity of sleep because alcohol induces grogginess. Learning to hop as a likos was arduous for Cega. The sensation was disconcerting as mammoth overgrowth of blubber bounced, resounded in her ears, and dragged her body's joints downwards. The mouth organ was in the middle of the torso to accelerate the speed of food's movement through the digestive system. It had suction and the capacity to jump, like a frog's tongue, towards desired food. Unlike a tongue on Earth, it was a tube for sucking a victim in during its rebounding back into the body. Likos' eyes were a string of beads scattered in a symmetrical pattern around the neck. It was difficult for Cega to adjust to seeing from these multiple perspectives. Doctor Nack had to account for this by adding the part of a likos' brain that processed imaging to her cadata brain because it was a unique system, which had to be linked with the remainder of her brainpower. Because methyl alcohol evaporates at average temperatures, there was a lot of liquid circulating through the planet's system, but instead of the type of rain forming when water condenses and falls in uniform and tension-held droplets, Likos saw occasional alcohol fogs, and scattered cirrus methane clouds with drizzling precipitation. This lack of density in Likos' primary liquid meant few small bug creatures survived the shift from water to alcohol during the Great Bacterial Fermentation. Buoyant bugs died from drowning in even a solitary drop of methyl alcohol, unable to stand on its surface as bugs can on Earth's or Cadata's droplets. The creatures who adapted to living in Likos' oceans sank into their depths because their gravity overcame the weak tension and buoyancy of alcohol. Swimmers were absent from these oceans, dominated by sluggish crawlers of the deep.

To return to transportation, the abundance of sunshine on the surface made solar energy the most convenient source of propulsion. Combustible fuels, such as the prevailing methyl alcohol, would have caused enormous explosive reactions if engines attempted to use them on Likos. The solar-powered vehicles likos used were massive to fit their size and bulk. Their binary system's moon was rotating with them at a varying angle to its sun, so they did not experience eclipses every nine hours; all sides of these solar bodies received some sunshine throughout their short year. The two were circling one another around an invisible balance point (the center of gravity) between them. It was closer to Likos. They moved in a back-and-forth yo-yo motion.

The standard-model vehicle on Likos was operated by shooting out the mouth to exert pressure on the see-through, kaleidoscopic car-top in the intended direction. The vehicle commenced motion until a second gesture commanded it to stop. The cars were designed for frequent collisions. These incidents were fostered by the lack of roads and the nonexistence of rules to govern their operation. Likos were drunks down-to-their-cells, so their road behavior fit the chaos intoxication induced on their thought processes. The bottom of these vehicles was composed of a goo, which slid over surfaces like a slug. This goo prevented acceleration beyond the preset speed, which made bumping somebody heading in the opposite direction as inconsequential as going over a puddle. There were alternatives for the wealthy, who lowered this speed to an absolute crawl to annoy other drivers. The difficult part for the engineers was hiding these extreme-drag-inducing features during spaceflight, and then protruding them upon a touch-down. Months of tinkering solved the glitches accompanying this butterfly-to-caterpillar reverse-meta-morphosis.

Since Likos was tidally-locked to its moon, the crew decided to land on the side of the moon always facing away from the planet. The planet was too massive to land on, and the species had telescope-like devices capable of observing them if they remained in orbit. They opted for minimum sneakiness during the parking maneuver on the moon because if this species could detect them, their awareness would ease the tensions of a first-contact. Meanwhile, without visible spaceships or rockets in space near Likos, they assumed this species was not yet spacefaring.

The moon was 10% of Likos' mass and was locked in a close orbit, so they could see the giant ball of the planet shining over the horizon of the moon as they landed. After they landed, they no longer saw Likos; if anybody wanted to take a little stroll to look at this beatific view, they had to fly a shuttle to the other side of the moon. There was a concentration of green-yellow aurora flares across the darker side of Likos because heightened solar flare activity, a strong magnetic sphere, and the combustible alcohols in the air created a nightly lightshow surpassing artificial fireworks. This form of entertainment also had to be avoided on Likos because a tiny fireworks bomb would have exploded alcohol-based partygoers. Likos' reddish sun was enormous and ominous. As the ship parked and attempted to adjust to life in this region of space, the 11-day year and the 9-hour spin around Likos started waning on their nerves. They had attempted to switch the lighting across the spaceship to this pattern during their slowdown, but it was disorienting and dizzying to see the constant movements of these enormous flying bodies. Some cadatas started envying Cega's continuous intoxication, as they never seemed to be intoxicated enough to adjust. The team formed a habit of staring from the Observatory at the micrometeorites and asteroids slashing down at the moon's surface at 40,000mph, without the impediment of an atmosphere to slow or burn them. Meteor showers had the auditory qualities of a volcano spewing rocks. Unlike terrestrial rocks, these cratered the moon's face. The surface was littered in these intersecting craters. The biggest crater was the size of a small city; when it formed, the impact approached tossing this moon out of its binary orbit. This crater had an eerie, dark interior contrasted by its sunshine-sparkling rim. The overarching landscape was dark and gray, with morose canyons, troughs and a string of cliffs lining the equator for a thousand miles, with deep valleys between them. There was a bit of liquid water under the moon's dark northern polar ice cap, so cadatas benefited from having this fresh resource during their stay. They had parked near this ice cap because, in addition to the water, the region was rich in organic molecules generated from gases by the strong solar ultraviolet light. The bacteria responsible for eating Likos' oceans were too weak to withstand the moon's temperatures, thus leaving its water intact. Alcohol and water discharged at intervals from one of these binary bodies to the other. These bits had the appearance of a glowing meteor shower when viewed from the planet. The debris was perceived as smashing space rocks from the moon. Interplanetary space was too cold for any bacteria to survive even this short trip. Some spacefaring bacteria can survive a journey across the universe, but Likos' slothful alcohol-makers lacked this capacity. The moon was much colder than Likos because of its lack of tectonic activity or sufficient internal heat.

After landing and three practice moon-circles, Cega took the slug-ship to Likos. The stress of her first major flight was alleviated because the Computer executed the entire maneuver. She sat back and watched as Likos enlarged into a sparkling map of radiant, reflective alcohol oceans. The ten main continents were covered in tall mountain ranges, sandy dune fields, deep canyons and bluish forest. The Computer landed her in a forest meadow with access to a city, but far from intelligent lifeforms. The sky was cloudy and the ground was covered in fog, conditions counter to manual navigation.

The Computer saw through this muck and made a gentle landing without a single shake or ear-compressing pressure change. The conversion to the local vehicular structure was instantaneous, with the internal controls mutating to fit the standard aesthetic, in the unlikely case a likos visited this interior space. I am going to let Cega's diary describe what happened next.

Cega Lapovy's Diary Entries

95,619 BC

When I emerged by opening the kaleidoscope top of the slug-ship, the air smelled as if I was inside an alcohol bottle. The same elemental composition on the spaceship did not have an equivalent odor because in space these substances were in their purified, odorless form. My sense of smell was congested by the gelatin-consistency of my nasal cavity, so the stench must have been more profound for lifeforms with smoother nostrils. Once my head stopped spinning from these air intoxicants and the rapid landing, my eyes began to focus on the opposite sides of the meadow. Every grain of soil and living organism they spotted was enticing, so I pushed up with difficulty and hopped out of the flattened, low-to-the-ground vehicle. I started hopping towards the jellyfish-textured plants, as the trip triggered my hunger. I had only consumed replicated versions of these, so I was surprised when these stretched their bodies and bent away from my touch as I stuck out my mouth to grab one. My lips landed on some dirt, which I had to wipe off to get the taste out. I tried grabbing them with the three sticky, multi-coned limbs my likos body was armed with, but they still managed to elude capture for several minutes. Then, I surprised them by jumping in their midst, placing the three limbs on three sides of a large plant, and clapping them together. They had a simple nervous system capable of sensing the approach of a potential predator organism. The proximity triggered a chemical reaction, which caused the plant to retreat. Since this was a challenge, I looked around for simpler pickings. As I focused on the meadow, I realized that in the few minutes I had been chasing the jelly, the meadow's color-intensity and plant density doubled in size. Our research showed the short length of days, and year cycles made this blue flowering plant into such a rapid reproducer. It was born as a small blue cell amidst a giant colony of other pollen-like, semi-cohesive beads. It sped to double its cells until it became a spongy, circular formation. Then, it reproduced itself, an exhaustive process at the end of which it died. The fog condensing on the meadow as I was landing caused this explosion of growth. As it lifted and cleared, these plants died off and the meadow's other colors dominated. The most overpowering of these in terms of size were bushy-fungus-like sets of thick, drooping white strings. I remembered they were edible, so tried a string. I learned to avoid its sticky, rotting-fruit-mimicking flavor. I noted my intoxication level rising as this string started moving through my system. I could swear I saw those bushes dancing around, but I have not seen any movement in this breed. In a blissful and still more starved state, I hopped towards a pond at the other side of the field. It was a peculiar pond because, based on my archival research, there were water-flowers, plants and muck floating at the top of small water bodies on Cadata, but this pond was unblemished on the surface, without a single bug touching down on it to feed.

In retrospect, I know this was because of alcohol's weak surface tension, which would not have held any entity, however small, suspended on its surface. I approached the edge of the pond and lay over the blue flowers for a better view inside the pond. I saw the bottom as if there was no liquid in between us. It was covered with blue, circular stone-like plants with small holes across their skins. A few little pond creatures were crawling around these in search for still tinier meals.

I indulged in meditative staring at these natural wonders. As the first world I ever inhaled and touched, Likos mesmerized my senses. The distance to the closest village relieved me of the pressure of worrying about acting odd. Though, there was a video-feed on me the whole time. The crew back on the Mothership was watching this rollicking adventure. Other worlds might have equivalent complex flora and fauna, but these felt native to me, perhaps because of our temporarily matching biology. We grew plants from the different planets we had visited in the Lab, but when they developed in air-bubbles, they seemed more nanobot than natural, so I never felt this close bond with them. The methyl alcohol in my system was enhancing this extreme emotional response. The plant-rocks at the bottom of this pond were un-miraculous, but we lacked spare rocks without scientific functions on the spaceship. And now there was an abundance of functionless, useless, inedible bits of nature in all directions my little beady eyes scanned.

I caught five more jellies. Then, I rolled back into the ship, and set it crawling through the meadow to the nearest road. Despite being a major roadway, it was as hyper-overgrown in minute-growth trees, and weeds as Likos' roads. It had been ages since the locals attempted cutting these. The roads were composed of a cushioning material resistant to overgrowth, but unable to repel the sprouts forming at its sides. For a moment, I was unsure which direction led to my destination, and since I had switched to manual controls for the sensation, the Computer abstained from making the selection for me. I ordered it to head for the nearest town, so it rolled through the brush and onto the road, turning right in the direction of the blinding sinking into the horizon. The whole planet was dominated by these small towns, without major cities. As the slug crept forward, I noticed a lot of roaming animals and giant pests grazing on the road and among the trees. They moving out of the way only when an impact was imminent, glaring at my vehicle as if offended I failed to bypass them. These pests and infestations looked problematic when I was viewing them with a cadata eye, but now they all looked kind of funny, like decorations at a party.

I almost overlooked the town because its buildings were hidden in runner-like varieties of invading plants. The structures were sophisticating. Their walls were covered in haphazard cables, chips, and metals. While some cables led nowhere, most components carried high-tech functions. Even ordinary houses had the capacity to turn to face the sun, as it moved across the sky, to attract more solar power. They refrained from straining to move for as much time as they retained the minimum needed energy to survive. The walls were uneven in size and shape. Each corner was somewhat bent, or broken at an angle. I failed to locate a single straight line in this architecture, as I slowed trying to understand it. Some of the walls were heightened, while others were shrunk. Houses constructed along a hillside were leaning sideways, as if the presence of a hill did not compute in the architect's plan. The sides of houses were littered with kaleidoscopic windows and doors; from my perspective, the only difference between the two openings was size. One of the roofs was leaning against the side of its house, as the contractor failed to pull it into position to cover the top of the house. Vehicles had crashed into buildings during parking maneuvers. The proximity from the transporting units to the doors indicated this was the standard parking method to achieve entrance-proximity to minimize hopping time from the car to the house. Soaked furniture stood outside two of the houses; it belonged to residents who planned on finishing a move, but abandoned it halfway. All houses in Likos' towns were single-story because energy-conserving-hopper likos would have rebelled if architects proposed adding the burden of height to their domiciles. Each house had a giant bowl of alcohol room, which could be sealed-off to foster activities such as drinking, swimming, and bathing. This bowl had to be pre-filled because they detested any concept requiring effort, including plumbing triggered by turning a switch for the needed liquid to appear. I was scrutinizing these features with such intensity I veered off the road and hit a house hidden by weeds.

I issued a cornucopia of curses before exiting the slug to inspect it for damages. The house was built of an impact-absorbent rubber, which left the slug and the house unscathed. Just then, a large kaleidoscope window popped open and a groggy likos appeared. He scanned the same collision I had just observed and came to the same conclusion. Then he started chortling. To clarify: likos communicate through laughter. Intricate variations of giggles, guffaws, chuckles, chortles, roars, neighs, howls, hoots, sniggers, snickers, snorts, cackles, titters, and some more physical displays, such as rolls across the ground, express different meanings. They laugh even if they are angry; darker sounds adopt menacing or maniacal tones. I began studying this language on the Mothership, using it in chats with our translation program. A real likos had too many random variations to his laughter for me to grasp the fine denotations and connotations. The nanobots had gathered data from spontaneous communications of several different likos. But this language had as many dialects as there were likos on the planet. Their computer code retained these haphazard features; for example, different constructions could be used to achieve the same end, such as rotating a house. I inferred the string of five short and long snorts indicated appreciation for my vehicle and parking ploy. The notes in the titter suggested he almost fell for my quest for free grub via a pretend-crash, but he now saw through this. Still, he found my neediness to be amusing, and with a final hoot invited me to come in and bring some libations. He left the window open as my target entry point before hopping back to the middle of his house.

I replicated the best of local alcoholic beverages, grabbed edible, wild plants from the meadow, and managed to hop through the window into this stranger's house. I handed these items to him. He grabbed and emptied a bottle, stopping to laugh meaningfully. We exchanged thoughts on the foggy weather, the high quality of the liquors I shared, and the need to develop sophisticating, telepathic vehicles aware of where a traveler desired to go without the need to stick one's tongue out to direct them. As it darkened outside, we drifted into sleep in the comfortable jelly-bed-chairs we had been sitting in.

The morning embarked when an armed police unit broke every window and door in the house as they hopped in from their crashed vehicles. My new friend, Ottode, exploded in questioning laughter, to which the police replied with stern howls. I was still too groggy to comprehend these, so I asked them to slow down when one of the officers faced me and gave an interrogative chuckle. In response, the officers began hopping around the house, splattering Ottode's possessions across the floor, munching on softer materials, and spilling out the contents of drawers. I was unfamiliar with policing rules on Likos, so their violent tactics were bewildering. I scanned Ottode and asked him with a giggle for an explanation. He was stern as he wiggled a smile indicating he blamed me for this trouble. When no piece of furniture or electronic device was left upright, the lead officer returned to me to repeat his question. I had taken a gulp from a liquor container I saved from the table before it was overturned. This somehow cleared my head until I could understand him. He was saying their space agency had spotted our cadata Mothership's landing on the other side of their moon. Then, they spotted my own slug-ship's flight and arrival on their planet. They followed the trail my slug left on the untraveled road to this location. They demanded answers as to what I was doing there. What did my species want on their planet? I was lost for laughter. I had studied Nerore expedition's history and its sad ending after my mother's alien nature was discovered. Knowing Likos as an advanced civilization, did my shipmates send me here anticipating this discovery? Was there a benefit to holding back on the truth? To minimize uncertainty, I asked how they had learned all this. They explained they had complex kaleidoscopic telescopes, utilized to study the stars for centuries. When I asked why this was unmentioned in their textbooks, they answered they had not gotten around to updating them. I understood my hope for anonymity was lost. I described the catastrophe our planet suffered and our search for a new home.

The officers laughed. Why would we imagine they'd want to help homeless sacks of aliens? To this, I asked if they had ransacked the ship I parked outside. Indeed, they had, but without arousing interest. We all hopped out of the broken windows as I led them outside and into my ship. They had torn my ship's gadgets out, demolishing and scattering the pieces at random, but an hour later the ship was back in idyllic order due to a rapid nanobot-cleaning. My species, the cadatas, were techno-advanced. We were willing to trade our scientific knowledge for their hospitality. Most were bored by this complex exchange proposal. But the lead officer saw the benefit in it, and contacted his superior for instructions. The leadership was on a drunken binge and could not be reached for several days, which we spent partying at Ottode's house, trashing it beyond even its post-raid condition. Just as even my ship's replicators were starting to fall short of our demand for alcohol, the leadership responded they were interested in receiving any information we had to contribute. They would decide on letting us stay once they determined if there was enough to be gained from it.

Our scientists struggled to condense our scientific findings from a million years in excess of the likos' current evolutionary stage. They screened out dangerous weapons, including sun-sized explosives, gene-editing and the technology involved in light-speed time-travel. They still surrendered technological secrets, which pushed likos to access unimaginable inventions. We gave them our general theory of spaceship design with clues on how to push a spaceship upwards against a planet's gravity, but no detailed plans for a buildable spaceship. When they asked for these schematics, we gave them the most convoluted 3D schematics our Computer could generate. As we expected, they lost interest in comprehending them. At least this appeared to be the case.

95,615 BC

I've spent four years on Likos without an infectious disease spreading between us, nor any other adverse effects. I did need annual liver transplants because mine were overwhelmed by the alcohol-overdose competition I was waging against likos police-officers. They adopted me, and had me stay in their station's guesthouse. It was a constant-fraternization outfit merely visited by partiers. It engrained as my permanent station.

I had settled into a comfortable rhythm, when I received a memo from Ally: a massive solar flare storm was forming on the super-proximate M-star illuminating Likos' sky. They estimated the radiation's strength would penetrate through the Mothership's shields. I would be protected because of my likos physiology, but the cadatas on the Mothership had to flee the event by leaving the solar system or traveling to its edge. The other alternative was attempting to dampen the flare's impact. They anticipated the coronal mass ejections (CME) would be at peak-strength. This spike reoccurred on 50-year cycles, and the sun was due for the biggest hit in the cycle. Readings by nanobots of the sun's thin atmosphere and the lower layers of chromosphere, photosphere and the nuclear interior confirmed these estimates. All indicated arrival in thirty days. There would be an increase in energy output by 16% for 18 minutes. This wave of radiation would hit the planet in seconds due to its proximity, so the time to brace for it was before it sparked. This was much lower than the extreme bursts of energy this M-star saw in its youth, but it would still generate an eight-fold increase in the sun's visible light as well as the quantity of x-rays and other charged particles zooming through Likos. These protons and electrons would disrupt the spaceship's communications system, and cause mechanical malfunctions in electric machines, including likos' solar vehicles. The likos had not paid much attention to these cycles in the past because even if their cars all short-circuited together, they had dismissed it as accidental, and ignored it. The likos and most other species on the planet were also conductors of electricity, and could absorb these waves as energy.

Ally and her team sent numerous warnings to the leaders across Likos of this impending disaster, but had not received a response. I had left it up to them to decide on a strategy, as I was in the middle of watching a non-stop-running animation series about a young likos who hops across the planet while jumping obstacles to catch three suspended food items. It was addictive to watch this character; when it caught a food in its mouth, my mouth flew towards my fridge and grabbed similar fare. Taking this unresponsiveness as a signal to proceed with their proposed fix, Ally organized the terraforming effort. Her team took spaceships across the planet to collect the reflective ingredients needed for a rapid atmospheric density increase to bounce these flares away from the planet to avoid disruptions to infrastructure. These ships began deploying these chemicals across the stratosphere, when the Computer picked up an apparent spaceship jumping off the planet. Ally ordered the ships to stop deployment, and to return to the Mothership. They executed the command, merging with the main orb in parked positions.

The team watched with stupification on their magnification screen as a lopsided and misshaped ship was pushed away from the planet by a cloud of steam, or rather a local non-combustive propulsion method. Then it extended flapping solar panels and attempted to adjust their angle of ascent with these sails. This concussion managed to overpower the planet's gravitational pull and somehow turned towards the side of the moon where the cadata spaceship was stationed. It first turned at too steep of an angle, and so had to readjust and fly in the other direction. All this while the metal making up its sides was creaking and groaning as if in pain from the exercise. The solar wings kept flapping, and a few more steam extrusions pushed this spaceship in leaps towards the cadata Mothership. They were mystified, but Ally had the self-awareness to attempt reaching the likos again to comprehend why they had launched this strange device threatening to break into pieces and splatter debris over the cadatas' orb.

Without an answer, they all waited in suspense for the hopping spaceship's next move. When it was a few spaceships' distance away, the likos ship's right wing began rotating wildly. Ally cried out, "Oh, no! They're going to lose it!" But rather than falling off due to a malfunction or regaining stability, this wing kept accelerating until it hurtled away from this ship and at our orb. None of the cadatas in the Observatory moved as the wing flew the short distance to their ship, and bounced off the anti-gravitational magnetic field, which protected it from giant asteroids exceeding the relative speed of light during space travel. It flew with magnified force back at the likos ship. The likos lacked obstructions to hide behind in this vast expanse of space, or time to retreat. Their aim was precise for their temperament, but this meant the wing flew precisely back at their own side.

The impact punctured the already strained metal of the half-winged ship's outer structure. They could not hear the impact in the Observatory as no sounds are heard in the vacuum of space, but the crash resounded in magnifying echoes across the likos' ship. As the wing bounced off, it left a giant hole, which caused rapid air loss or depressurization. Junk across the ship went flying. A piece of debris set off a spark. Since likos and their world were alcohol-based, the single fire speck set the ship ablaze in an instant. It exploded in a sulfur-rich mess. In slow-motion, the metals inside the ship melted into a blood-red liquid emitting a blue flame. Once these metal pieces loosened, and the fire forced pressure to expand, this blast shot pieces out of its rickety machinery in all directions, including at the cadata ship. The chunks were small enough for them to be absorbed as a fuel source by the spaceship's core. Ally called the likos down on the planet again to apologize and to explain what had happened. A dead silence once again followed on the other end.

I was gazing outside my window on Likos between watching my show, speculating on the unravelling events, when I spotted the explosion in what appeared like three meteors shooting out of a single point in the night sky. I called Ally and she summarized they were on hold after the beforementioned incident. At the time, I refrained from dwelling on this, feeling a vague sense of doom in being the sole cadata the likos might express their frustration on in the aftermath. Despite this threat, I did not rise from my reclining position as… well, you can understand… With my body transitioned back to the clarity of a water-based lifeform, I am more reflective about what happened.

Our cadata spaceship's design can withstand flying into a sun, as it is firing off nuclear reactions. We had never experienced space warfare on Cadata because even if a rival space agency, or country could have launched a competing spaceship, the rival computers would have anticipated each other's strategies, and concluded it was best to stalemate, or for the non-occurrence of war. With our technologies' extreme precision, any weapon could point at a tiny target on a rival ship, and fire with the exact strength, and direction needed to obliterate this vulnerable spot. Since our systems shared multi-wavelength readers, no spaceship could cloak its location from the others. Any alien spaceship arriving from a distant planet to attack a cadata ship in orbit would have been trackable across its days, months, or years-long journey. The eventual arrival would have been hyper-predictable, and therefore defensive counter-measures would have been long pre-planned.

All advanced spacefaring civilizations must have laws prohibiting space wars. Given flawless weapons and shields on both sides, firing and blocking is a waste of fuel if a war is attempted. One side must be inferior in defense or offense for a victor to emerge. While there might be an alien civilization in the universe seeking inferior species to dominate in such combat, despite cadatas' flaws, such hostility was foreign. In retrospect, our scientists determined alcohol-based species had to be avoided on future exploratory missions, as they lacked sufficient deductive power to avoid warring against militarily far-superior foes. Even the best intentions could be misinterpreted under the fog of intoxication.

Two days later, a police team broke into my room at their own police station, breaking their own windows and doors. I was still watching the show, so this bummed me out. They chortled uncontrollably to indicate their dissatisfaction with our spaceship's unacceptably violent response. They questioned why they were failing to establish communications with Ally. I explained the solar flares would strike in a few weeks. Since the likos answered "no" to their proposal to blanket the stratosphere with suppressive chemicals to dampen the impact, our spaceship had to retreat beyond these radiating waves' range to avoid damage. They asked why I had stayed behind, and I explained I was watching the cartoon. They understood, and joined me, as they had been engulfed in its developments as well. In between all this excitement, they explained their solar panels were designed to take advantage of these flares to store up on energy, and their bodies experienced a spike in happiness as they were also energized by these innocuous blasts. If we had succeeded in blocking this radiation, it would have hurt their physiology in the long term, and caused problems in their solar grid. I asked why they refrained from informing Ally of this position. The leader expressed with a giggle they had begun composing a memo, but then ran out of energy to finish the task. I asked how they managed to engineer a spaceship given their incapacity to write a memo. They explained they were motivated by fun designs of quick-moving stuff to take them to unexplored places. A space-destined ship fell into this category, so an army of likos congregated to complete a ship on an expedited timeline. They had started construction years ago when they first received the spaceship design ideas from cadatas, so they just had to launch it when news of the stratosphere pollination arrived. The squad fell asleep beside me to the sounds of the ceaseless cartoon in the middle of this tantalizing discussion. After all this trouble, when the solar flares spike came, weeks later as expected, they caused a spike in aurora displays, and just the energy kick the squad described.

Given these observations, we cannot create a permanent settlement on Likos. These people are too volatile. Cohabitation would lead to destructive clashes. And if cadatas become likos in personality and psychology; who knows what fiendish plans they would concoct after millennia in this environment. At our latest meeting-call with the Mothership's leadership, we decided, going forward, we had to approach planets populated with intelligent species with a spaceship disguised as a natural asteroid, or another un-suspicious space object. Precaution is required until we determine the hostility-level of the dominant species. I am disappointed I failed to find a strategy to force Likos to work for cadatas. But I am also relieved the death toll from this encounter has not been as bad as Mom's outcome…

Chapter 8

Hot Diamond Rain

In 95,609 BC, Cega completed her decade of isolation on Likos, and returned to the Mothership, where she was transformed back into a cadata. While it was already decided they were not staying, around half of the crew opted to attempt a trip to the planet in their own cadata skin to double-check Cega's findings. They partied with a group of carefree locals for a couple of years, without complaining regarding the water-shortage. They had water among their ships' supplies, but learned to prefer alcohol.

Cadata scientists made a myriad of surprising discoveries during this stop. But the idea a lifeform could be alcohol-based, without disintegrating, turned blasé. Then, one day in 95,607 BC, the chief communications officer detected a signal they had hoped for since Nally deployed a light-emitting satellite in the Cadata solar system. There was a clear and precise pattern to the lights Chief Solfied detected on an instrument in the Observatory. This was a finding capable of revolutionizing cadatas' knowledge of the galaxy. It signaled a chance they were within reach of their new home. Cega and the other team leads gathered around Chief Solfied and studied the pattern as the Computer analyzed it to deduce a translation. The anticipation was ripe. Most were beaming with joy, but Cega and three other leads were apprehensive. After running an enormous number of potential linguistic meanings, the Computer derived a guess-translation from the strange fluctuations in wavelength.

Dear Aliens:

Would you like to wake up to diamonds falling from the sky? Do you enjoy hot and steamy climates, just right for an escapist vacation? Are you keen to mine inexhaustible stores of natural resources? Precious stones are one category among our many natural gifts. Come to visit or settle here on Nomuat. We welcome the full diversity of universal life. If you have caught our signal, you know our planetary coordinates. You will not be disappointed with our land of dreams, hopes and fantasies. Do not wait! Depart today! We are thrilled by the opportunity of meeting you!

Warmly Yours,

The Nomuats

The Computer reported and the programmers and Chief Solfied confirmed this signal was designed for a likos audience. It was a short set of narrow bursts, rather than a signal traveling in all directions in the universe. Thus, it was a personal invitation letter intended for the likos. This raised questions as to whether the likos had posted a beacon to invite solicitations. Since their last news from Cadata, there was no response to their cross-galactic message, so it was peculiar somebody was welcoming the likos instead. Nomuat was unlisted in the original *Catalog* because it contained an excess of extreme parameters opposite of cadatas' physiology. The surface-environment appeared hostile to all life. With a unanimous vote, the leadership decided to explore these questions by forwarding a translation to the likos in their laughter-language. The likos replied in around a month when the delegation of cadatas on the planet's surface came into their planetarium and demanded an audience. Through a holograph conference, likos astronomers explained they had received messages from nomuats, but dismissed them as interstellar junk mail. Their appetite for drunken space adventures was exhausted by their sole spaceship's two-hour explosive ricocheting against the cadatas' ship. They never responded because the astronomers assigned to execute it retired on a permanent vacation. While the likos were nonchalant about this news, the cadatas were thrilled an evolved species was grasping across the galaxy with a welcoming hand. The hostility of the likos had disturbed their sense of galactic comradery, but here was a species seeking visitors. A rejoinder was compulsory. Their travel and communication-message speeds were identical, but the journey would also allow for interaction, so cadatas committed to a physical expedition.

The first downside was Nomuat's distance of 9,655 light years away. Long-distance readings then displayed Nomuat's heaviness in atmospheric carbon, a volume corresponding with diamond rain showers. Sparkling stones lack symbolic or financial value for cadatas, but diamonds are valued in construction and transportation for their strength and durability. They also discovered Nomuat was larger and thus had a twice stronger gravitational pull than Earth. Aside from the prevalence of carbon across the planet and in its air, the atmosphere was heavy in hydrogen, and helium. There was no detectable water vapor or oxygen. The surface temperature was the hottest cadatas had seen to date on any world, 1202°F at the max. The core of this planet was under tremendous heat and pressure. This was at the top range of what diamonds could tolerate without turning into graphite or igniting; at these specific degrees carbon congealed into solid rocks in a plasma-rich or aurora-drenched sky. This heat was from its proximity to a hyperactive sun, which had a higher metallicity than Earth's. *Metallicity* refers to the presence of heavier elements than helium and hydrogen.

It was obvious from these extremes why Nomuat did not qualify for Nally's *Catalog*. On the other hand, the scientists confirmed they could create gene edits to allow a cadata to survive even a constant bombardment by diamonds, and the oxygenless atmosphere. With science on their side, they set sail for Nomuat, knowing none of them would see another planetary stop. Their children would never see real land. Their grandchildren might return to grazing a natural solid surface if the first test-cadata withstood the awesome extremes in this alien world. They gathered all the extra matter they could from Likos to tap their spaceship's storage potential, and then the Computer executed the same speedy acceleration. They remained at light speed for ten millennia before commencing slowdown in 85,951 BC on entry into the Nomuat system. They were now in the middle of the Scutum Arm, having traveled along its length for most of the journey, moving away from the galactic center.

Ga Lapovy was now the chief cadatanaut. Her Second was Or Steral. The medical team was led by Doctor Na. And the ship was captained by Al Subsenar. They had spent their young lives training for this final arrival, and the start of the extensive recognizance survey. They had decided to study the local species first, before announcing their presence, unless they were observed, and contacted. The millennia they were cocooned on their spaceship was consumed in improving their technology. Thus, the newest nanobot model was leaps ahead of its predecessor in its range of assessment and speed. Despite these added skills, the nanobots failed to locate the species who reached out across the galaxy. The limited uncovered clues included ultra-locked and inaccessible living and working quarters, which might have belonged to the aliens. Captain Al sent a warm arrival notification in response to Likos' intergalactic species summons. Despite Likos' silence, cadatas commenced planetary exploration, assuming their hosts might be hibernating or away, and would later present themselves.

They had no trouble finding a second intelligent species on Nomuat, but this group did not seem to have access to advanced communication systems. They gathered extensive samples from these primitive nomuats, which they called reabs. The nanobots also gathered a spectrum of other unexpected lifeforms. However, when Doctor Na ran a series of tests on cadata animals, in preparation for editing Ga's genes to give her the reabs' appearance, he found too many complex problems that led to these creatures' deaths. Delaying the landing for exasperating anticipation was bypassed in favor of utilizing the invitation to commence spacesuit-enclosed exploration. Because the suits were microbially-isolating, and the local lifeforms had atypical genetic code structures, both Ga and Or commuted in a single standard cadata shuttle.

They ran test flights prior to the planetary trip to examine the ship-design edits instituted during the interstellar voyage. These were mini-tested in short bursts of motion inside the Mothership because they could not pop out for a test at light speed without unacceptable time- and space-related risks. Meanwhile, the entire Mothership underwent a full 10,000-year inspection once it was in a stable, floating orbit. The mechanics found few irregularities despite this being the longest non-stop stretch the ship had endured. The nanobots were cycling out over-processed matter, when they spotted a tiny imperfect crack. The flaw was corrected with an over-abundance of dangerous chemical combinations. Captain Al noticed build-up patterns during his analysis, and programmed a check for them into the Computer's algorithm. For instance, matter backup occurred in nebulas in the dense arms of the Milky Way, a path littered with unexpected volumes of invisible-to-distant-scans space dust. Insurmountable pre-problems needed a cadata's imagination to sense the likelihood of a failure in the coming 10,000 light-years. For the rest of this stop, Al and his team toiled on concept upgrades developed during the long flight, and now tested ahead of implementation.

Instead of clouds, the sky was lit with aurora displays, lightning, and the infamous diamond-forming hot regions of heavy element condensation and dispersal. There were no forests, grasslands, or other visible plant life from Ga and Or's perspective as they began the steep descent. Prior scans had determined some of the chalky rocks on the planet consumed light. But the byproducts of these reactions were disparate from the oxygen photosynthesizing organisms produce on Earth. A fascinating living-rock category varied in color and size, from building-length to microscopic. It approximated the planet's melting point, so they liquefied at every sunrise, and solidified as darkness fell. They could be spotted by their tops' pool-surface smoothness. At midday, Ga and Or emerged from their ship, parked on a shallow pool of molten rock. The Computer refrained from triggering the alarm because the lake was too shallow for the ship to sink. But when Ga and Or stepped into the melted rock, they regretted this decision. Three steps into it, their feet sank into molasses. The squishy rocks sank below their soles, delaying the next step. Their outfits' repellant qualities saved Ga and Or from the rocks engulfing and sinking their bodies as each foot was clamping onto at least fifty pounds in added weight. Scraping solidifying hot and slushy rocks, once they returned to the cool spaceship, required the robots' assistance, as they utilized vacuums, re-heating, and singular tricks to collect the deposited matter. Chunks of rock littered the main cabin after this procedure. As the Computer organized this cleanup, it also re-parked to a rickety and dirt-covered but non-melting surface. It reported it failed to anticipate melting rocks because of a lack of an active volcano in the region. Further, the temperature was outside the range correlated with this phenomenon. Instead, the radical and regular state-change was allowed by the unique combination of chemicals in these rocks.

Their second attempt to leave the ship was more successful, since the robots left a repulsion shell around the suits, just as they would have if Ga and Or were heading into a volcano. They chose this landing site because they detected a series of tunnels with an appearance akin to an underground college campus on x-rays. Digital plans for these tunnels were inaccessible to nanobots as local computers were hyper-encrypted and isolated. They had to depend on a physical map based on spaces where nanobots could travel unobstructed by locks. Dominant features included round parking lots, and sewage and chemical transport pipes. The main network of tunnels extended to a few different depths, and led between a series of giant, elongated, rectangular rooms, with a few isolated small rooms. The round hole in the ground comprising the main entrance was lined in bright lime stone. There was no visible ladder allowing for a climb down. After searching the edges for a button or a Computer to signal a lift to approach, Ga stepped on the edge of the yellow stone. It semi-melted around his foot and gripped it in place, circumventing the suit's repellant coating. First, he thought it strange they used this sticky rock on the entrance where everybody was bound to step on it, but then he took another step and realized the stuff had a secure hold on both of his feet. While his feet were stuck to the paddles, he could shift these rollers back-and-forth. He shifted forward until he was perpendicular to the in-pointing tube. He was now defying gravity without ill effects. Or mimicked this move and they were soon both trotting at a 90° angle down into the chasm. Above-ground, no objects or structures could be observed. The sun's heat necessitated much stronger reinforcement for unnatural objects exposed to sunshine. When exterior light no longer penetrated, they realized this space was lit with scattered, fluorescent, star-like specks inside of the sticky rocks. Just as they were marveling at these displays, their front feet stumbled into an opening. It was the triangular, though rounded at the tip, roof of a horizontal tunnel. They walked around this roof, and along the walls until they were back in a non-suspended position on solid ground. They continued through this level-sculpted tunnel for a time before they heard flamboyant construction noises in the distance. The tunnel opened into an enormous round hall, with a few door-like hatches leading seeming offices. The walls were carved out of precious stones sparking with an intensity identifying them as jewels rather than as the stars they appeared to be in dimmer light. A smaller side tunnel was glowing with fire at its edge. A pool of lava seemed to be burning at the back end. Perhaps, a gravity-defying sideways volcano was holding its lava suspended at the opening despite its heat and pressure.

Ga and Or had moments to study these architectural details before they were overwhelmed by fifty reabs, who swarmed them with extreme curiosity. The reabs had six striped tails on the sides of their bodies with a dexterous grabbing ability. Ga and Or had carried compressed luggage in back compartments in their suits, and they managed to put pressure on them to pop them open, tossing their nutritional-gas, spare suits, medical instruments, and various other bits of supplies. The reabs' faces were composed of a type of lamp that was their communication and vision organ. The light signals their planet sent across the stars echoed this native method of communication with light waves. Under this bulb, they also had a thick beak. They walked on their shorter and thinner legs. These had many joints in them, so they could fold into a near-circle or generate other odd shapes. They moved with their head upside-down, using their stronger-than-legs arms with broad suction-cups in place of digits for their demanding work.

Doctor Na had learned in the Lab this species evolved on a waterless planet. What counts as a shortage leading to desiccation for species such as humans and cadatas was a starting point for development for these xerophile, halogen-based lifeforms because halogens react negatively to water. Thus, halogens disintegrate on a watery planet such as Earth. Reabs' genes were built out of hydrogen (H 1), fluorine (F 9), chlorine (Cl 17), and bromine (Br 35), with sprinkles of iodine (I 53) and heavier halogen elements. Other than a lack of water, this planet also had to have several other unique environmental and chemical components to fit these unique organisms. Since reabs contained halogens and hydrogen, when they bonded, they formed acid, making them acidic to other lifeforms, including cadatas. If a cadata touched a reabs' compact, chalky-stone-consistency, pinkish-purple skin, or was exposed to intimate interactions such as a kiss, the entire embraced limb could be damaged to the point of decomposition. Reabs' systems ran by using heat to melt portions of their chalky bodies; these channels carried nutrients across the system and to individual cells in place of water or alcohol. Their chemicals and cells could metabolize the incoming food without a cooler fluid as an intermediary. Given the extreme temperatures on Nomuat, they were hyperthermophiles; hotness was their norm, so Earth or Cadata-level temperatures would have been a lethal cold. Doctor Na was surprised when he found no bacteria-like organisms across the various samples from Nomuat. He re-tested them, adding new samples from less accessible locations, and it all pointed to a complete lack of bacterial life. He ran tests on Nomuat's halogens and bacteria, and discovered the first killed-off the latter, as they acted as an extreme form of disinfectant. Reabs had evolved from more resilient organisms than bacteria, which were simplified versions of their smallest gene parts.

With the reabs up close, it became apparent the rainbow of colors splattered over their bodies were from their hard labor of extracting the stones rather than from their skin's pigmentation. Without an introduction, the reabs placed some of Ga and Or's possessions into the bucket-like containers they had been using to collect precious stones from the exposed walls. Once, Ga interjected in a theft of an entertainment computing device she had brought with her; in response, the reabs initiated a garrulous lightshow. They screeched with blinding signals at the top of their light-bulbs' capacity—they had a right to take the newcomers' possessions in exchange for the strangers being permitted to remain among them. Ga and Or decided to let the stripping process continue. They would have lost their protective suits too, had they not been stuck to their bodies with nanobonds, essential for heat, pressure, and general environment-balancing procedures. The circulation of air and water had to continue uninterrupted within these suits without necessitating even momentary removals. A single bubble contaminated by a cadata-toxic gas could have meant an instant death. An alien was unable to remove this second-skin without adopting a cadata's genome and unlocking programming of this robot disguised as a slinky, body-hugging swimsuit. The reabs took interest in cadatas' physiology, as they flapped the newcomers' wings to test their dexterity, poked at their eyes, and scratched the silky skin of their suits. After these casual physicals, without questioning why the arrivals looked so alien to their own physiology, the reabs insisted Ga and Or pick up buckets and help them select and transport stones. Ga and Or joined out of curiosity about this work-category. They discovered it was a strenuous exercise even when their suits' internal temperature could be controlled at an ideal level for such exertion. They questioned how reabs could work in 524°F heat this deep in the tunnel system.

After toiling, Ga and Or followed their workgroup through a maze of tunnels. As they neared their destination, the tunnels gained the appearance of having been dug out with suction cups instead of the precise machinery engaged in carving the circular and straight-lined geometric shapes of the business section. A portion with the top exposed to sunshine was even covered in a heavy layer of melted rocks with a mere solid plank over them. Ga and Or had to use their x-ray vision to avoid knocking into the walls, as there were few precious stones in the walls to light the way. Following the long journey, the group opened a door to a nook. Ga and Or began following them inside, but the biggest of the reabs turned and blocked the entrance with his massive chest and constantly-moving tails. As if his physique was insufficient, he also pointed a hot, green-neon plasma stick at them.

"Where do you think you're going. Don't be trespassing on my house!" he flashed.

"We were helping you all day for free," Ga replied with a lighting translation device. "Can't we stay with you guys? We're new to this planet…"

"This planet? Oh, that's why you all are so weird!" he exclaimed. "Wait here…" He retreated into the room, closing the heavy stone flap behind him. "This is Hatay, our Big Lad. He will decide," he said. When he reemerged, he pointing to Hatay, who was a tad taller, but looked much older, with deep gaps between his stony skin.

Hatay asked what they had to trade for the stay. Ga and Or had to admit they were short on gifts. Ga proposed remaining outside the door, while Or returned to the ship to retrieve offerings for this group, with which Hatay agreed. It was a long and arduous path; Or only managed to escape the labyrinth because his computing device guided his steps, and his suit created gliding propulsion through the dark channels. Or had replicated as many of the local living and edible, semi-rock-textured creatures as he could carry in the bucket he retained from their mining employment. The reabs' acceptance allowed Or and Ga to continue residing with them. The reabs ate this gift for dinner, while Or and Ga watched, lacking a need to eat solids, and their energy levels were met with nutritional-gas circulating through their suits. Satisfied with their meal, the reabs dropped into sleep, a state wherein half of their bodies remained motionless and senseless, while the other half massaged itself and its closest neighbors, scratched, tossed, or, in a few cases, kept the entire body upright in a locked position. Half of their brains, and one of each of their doubled organs also went to sleep, while the other half kept working. This indicated an over-abundance of violent predators on the planet in the period central to this species' evolution because they could not allow their bodies rest without keeping a watchful eye out for danger. They did not have any technical or convenient luxuries such as a bed or plumbing, sleeping right on the sandy floor of their cave.

Ga could not fall asleep, so after tossing for hours, she rose and retreated into the tunnel to compose a progress memo for the Mothership. As she was making 3D signs into the computing device, her narrative was interrupted by a banging shaking the pebbled ground. Suspecting their cadata spaceship's involvement, she rushed towards the commotion to check if it was an attack. Inside the responsible rectangular space, she observed a machine hacking what appeared to be a living rock with branches spreading along the surrounding inactive rocks in search of nourishment. It was unceremoniously shattered and piled onto a material-collection truck. Cadatas' nanobots failed to detect these organisms during their inspection of the planet, suggesting they might have been the last of a near-extinct species. Ga interjected by questioning why they were killing the stone-tree, and if she could appropriate a piece for study. She was barked at to retreat from the site if she wished to avoid being blasted with the next charges.

As Ga retreated to the entryway, she was caught in a throng of reabs who were streaming down a tunnel with a downward incline. At first, she was focusing on avoiding being trampled, but then she decided to join this crowd to observe what they were rushing towards at this premature hour. It was apparent they had reached the destination by the cloud of smoke and toxic fumes erupting from within a swinging latch covered in crystallized ashes. Upon entering, Ga was astonished by the sight. This was a factory producing parts needed for the construction of the impenetrable cities where the species responsible for sending the initial invitation must have lived. Due to the hot temperatures, most reactions to transform rocks into sturdier materials were chemical. But there were also furnaces and fire pits spiking temperatures still higher to mold these into the needed shapes. The noise of these being cut, drilled, singed and flattened was a roar to Ga's ears, but since reabs communicated through light displays, the audio waves were semi-perceptible vibrations. The reabs ran these machines, as well as performing manual labor on the assembling lines. Ga observed a 16-foot tall, glowing plasma door on the other end of this open space and approached it. When a reabs opened this mega-heavy and hot block, Ga followed him inside. He climbed down a long, sticky-rock 90° vertical drop. She followed until they reached another behemoth door and stepped into a heavy-smoked workshop where plasma was being produced, shaped, and then cooled into usable objects.

Before Ga could spend too long observing what they were making, she was handed some tools and given instructions on how to join the production line. It was a difficult shift without a break during which she might have sent a communique on her endeavors to the other cadatas, though she knew they would be able to use her location and data from her computing device to learn of her activities if they became concerned. As she learned more about the curious type of plasma varieties they were molding, she began inquiring as to who owns the factory, and how businesses on Nomuat were structured. Programmers on the Mothership had failed to penetrate law archives describing the government, laws and business rules societies use to control or free the actions of their populations. Ga's instructions were to infiltrate the ruling structure on Nomuat if she could not access such information otherwise. By the end of the day, her queries attracted the attention of Assembly Line Supervisor Eyath Yald, who took her aside. He said he could neither confirm nor deny if he was the owner of the factory, or if he knew who the owner of the factory was, or how the factory might be purchased. He added, if Ga had access to sufficient resources such as foods or valuables, it was possible he might provide information in exchange, perhaps even securing the purchase she desired. Ga thanked Yald for his helpful input, promising a return on the next day with the maximum transportable quantity of valuable items. Hearing she planned on dragging this stuff into the workshop, Yald redirected this plan to a meeting at his Namuat coordinates for an isolated desert location. Ga agreed and retreated to inform Or, who was delighted to become an owner instead of a mere factory worker. Or's perception of the danger of their circumstances was heightened earlier this day when a fellow stone-digger came close to death because a strange wormish creature burrowed into his feet unnoticed. It laid eggs, which matured in minutes and began crawling out. Watching the chalky skin break, and the hot fluid below bubbling out with these parasites brought Or close to fainting. But his suit increased the nutritional-gas output, and took steps to keep him on his feet. Or began questioning if those ultra-hot worms might have penetrated his suit with sufficient exploratory attempts. So he was delighted to avoid returning to this risky heavy lifting.

Ga programmed the Computer to fabricate a carrying device with extreme lifting capacity. Then, she replicated the most value-dense items within the reabs' economy. Ga and Or brought these to the scheduled meeting place. Yald accepted these, and the levitating cart they arrived on, thanked the cadatas for their business, and retreated. Later that day, Ga and Or showed up at the workshop, and began implementing new procedures and methods, assisted by their advanced knowledge of construction and plasma's properties. The reabs followed their instructions, just as they had Yald's. Yald had continued issuing commands as well. On occasion, it seemed to Ga and Or, he was contradicting their orders. But they assumed this was a linguistic misunderstanding. Regardless of who was in power, the work was completed. Thus, both local and cadata leaders were satisfied with the outcomes. Flattery aside, Ga and Or kept asking Yald about seeing the books for the business to offer wisdom on efficiency. But Yald avoided this line of questioning with the excuse he was too busy to spend time explaining nomuat bookkeeping. Before long, every time they asked him about the books, he asked them for some additional gifts. Their parked ship became a factory taking in the surrounding dirt and turned it into the objects of Yald's desire. It was no hardship for Or and Ga to keep it running non-stop. But Captain Al began to worry what impact this enormous infusion of goods to a single individual would have on Nomuat's economy.

Business continued at the same pace without cadatas reaping benefits other than empowerment from positive alterations to the sector. One day, as Ga and Or were reviewing items fresh-off the assembly line, they were approached by a ultra-dust-covered and wild-faced reabs. He introduced himself as a lead archaeologist, Professor Dlah Laneum, and enticed them with an exciting find he stumbled upon. They were curious to learn more about Nomuat's history, so they questioned him further. Professor Laneum told them to hurry in following him to the excavation site before it might be looted for its antiquities. Not wanting to procrastinate while such a travesty transpired, Ga and Or followed Professor Laneum to his stone-based vehicle. It quaked as it stumbled over the uneven ground. The trip across the scorching desert was even more turbulent, leaving Ga and Or disoriented and shaken up by the time Professor Laneum parked and waved for them to come along. He led them on a crawl through a tube inserted through a forty-feet thick crust of shifting liquid rock. At the bottom of this tube was a cave seemingly cleared by recent flooding of the same liquid rock. Once they exited the tube and more light saturated this space, it revealed itself as a buried storefront. Professor Laneum lauded his accidental discovery of this unique treasure after he found a square stone cup on the sand just above this site. He took them on a tour through the storefront's crumbling hallways and into the residential building behind it. The space seemed to have been dug out of sand and dirt; the walls and furniture were still cloaked in these sediments. Professor Laneum told them a wonderous story about prominent inhabitants living in the luxurious sand-cleaning room, and the animal skinning workshop. He spent an elongated time in the latrine, removing each piece of garbage from the pit tossed out through this disposal unit, and explaining how it was a hidden treasure-trove for archeologists.

"What we discard says so much about who we are. Our history is written in the choices we make about what to eat, or what pottery to toss aside with the season. You can see a wealthy and influential family lived and worked here. The proof of their wealth is in what they consumed, the animals they ate, the fine grade of the cuts, and the foreign, imported origins of the foods. Here, this piece," Professor Laneum began, as he showed them a few strange bones, "belongs to the illabess. It's extinct now, but was still being eaten as a delicacy in this period."

Ga and Or were mesmerized by the rich history Professor Laneum endowed to this building. He demonstrated with the lines on a piece of native fungus, the building and its treasures must have been buried three thousand years earlier. The flood of melted rocks preserved the contents in top-condition. He was proud to show them a colored-light painting on the living room's wall depicting much-loved and prized animals and cultural traditions. While he was describing the proud and virtuous reabs who inhabited this region during the period in question, a piece of pottery fell on Ga's head.

"Oh, holly of hollies!" Professor Laneum exclaimed with joy. "This piece of pottery is a thousand years older than the others. The carvings here look like they were crafted in the court of the Ruler Orkuel of the Orkuelean dynasty… What are you guys doing?"

Laneum's attention was distracted by the data and image readings they had started taking with their computing devices. They elucidated their scientific mission. They were an alien species exploring Nomuat. Laneum avoided commenting to their declaration of alien-status. Instead, he proposed a pre-sale bid from them to purchase this entire site, or at least the amazing ancient piece of pottery, which found a resting place on Ga's head. Meanwhile, Ga and Or sent the data they gathered to the Mothership's Computer, and a few nanos collected samples and took these up as well. Professor Laneum kept lecturing them about these spectacular ruins of a long-gone civilization, all the while querying them as to how much they might want to spend on the purchase.

During this discussion, the main ship's Computer finished the analysis of the dig site. The science team reviewed the data and Captain Al reported the findings to Ga and Or. The message arrived on their communication units, incomprehensible to reabs because the signals traveled on sound waves. As these analytics poured in, Professor Laneum was still glorifying the ruins. Captain Al was distraught to report they had been led into a fabricated historical structure, or rather a replica of the original discovered ancient building. It was intended to mimic the historic site to give tourists the illusion of historic significance. It was crafted with complex three-dimensional models, and digital images imitating a myriad of articles from garbage to light-paintings. Melted rocks flooded the entrance because they had flooded the original in this pattern. The substance Professor Laneum argued had been used to date the site by the thickness or thinness of the rings it formed in the coat around a piece of pottery was manufactured to replicate these age-rings, which science explained were created by fluctuations between peak summer and low winter temperatures. This was the work of historians rather than forgers, and since Professor Laneum was the sole historian or archeologist with access to the site, it was probable he was its creator. This suspicion was confirmed with a DNA analysis of the scene by the nanos. Their expressions intensified as these details came in until even Professor Laneum could not ignore their discontent and asked what was the matter. Ga delivered a stern lecture on trust in art and business, stressing how selling forged art hurts sacred cultures and histories. Outraged, she demanded what else Laneum was lying about. Laneum mimicked the fury in the expressions on her face and her gestures, refusing to tolerate their insinuations. Before she could retort, he stormed away on the sole accessible transportation unit. Ga and Or had to be rescued by a piloted shuttle from the Mothership, which escorted them to their own shuttle and out of the dehydrating desert.

On the following day, they tried to confront Supervisor Yald about Laneum's behavior. Before they finished the explanation of their scientific analysis to reach the conclusion the cave Laneum was trying to sell was fake, Yald stopped them, and gave a long tirade about foreigners coming in and insulting local reabs culture by saying it is ingenuine. Ga and Or said they were on a quest for understanding, and hoped to find sympathy with Laneum's perspective.

Yald broke his characteristic tranquility, sparking: "If you're going to question my friend, you better not come anywhere near my workshop!"

Ga and Or objected they bought it from him. He demanded they show him documentation to prove this assertion. Ga happened to have brought their local stonework with her in case of a clash, which, after the Laneum incident, was imminent. She took out a file and started searching for the contract. Cadatas had not used stone, paper or similar writing implements for millennia, so she repeated the search thrice, in case she was misreading the etched letters. When she still could not locate the relevant stone, she emptied the file on a counter and discovered a bit of chalk of the same color as the contract used to be. She asked if Yald had composed the contract on disappearing stone-paper. Yald exploded in expletives in response and hollered for his guards to escort Ga and Or out. They did so despite Ga and Or's protestations of falsification and scandal.

Once back outside the underground jumble, Ga, Or and the others on the Mothership attempted to conference to find a suitable strategy. They had thought they were building up a positive reputation and shaping the local economy, but it was all a sham. Captain Al concluded the species inhabiting Nomuat had a profound lying disorder. Cadatas were far from addicted to truth-telling, but exaggerations were reserved for gaining advantage in business negotiations. Outright fraud was frowned upon because sales of absurdities such as Verrazano bridges contradicted the dominant Potentates' interests. As the owners of such bridges and museums, they would be the victims in a potential deception by a small business rival. In this instance, the locals seemed intent on milking alien visitors for resources and production capabilities. Without Potentates' power to chastise villains such as Yald or Laneum, cadatas were disadvantaged in this falsified climate. Their invitation was thus proven to be a spam email sent to various target planets in the galaxy as a lure for a long-con theft. Despite this distasteful criminality, after three-generations-worth of travel, the crew insisted on continuing exploration of this world as a potential resettlement location. One solution was isolating themselves from the locals in an uninhabited region and building an outpost, as they continued research into gene-adaption. The confiscation of their workshop hinted the same could happen to their settlement. Thus, they decided building an enduring structures in the open, before cadatas outnumbered the local species, was futile. To avoid uncertainty, they pursued a lawsuit. The first step was determining how to file a case in Nomuat. Ga and Or queried all who exited from the mines for two days until a troop of officers intervened with an invitation for Ga and Or to follow them to the place where a lawsuit could be filed before the Great Tretures.

They took a rickety car to a spot cadatas had observed on scans but were unable to penetrate beyond its exterior. A series of upside-down stone-carved cones were scattered amidst the desert's floor. Each was composed of circular balconies with bright-plasma outlines across their circumferences. A triangle auto-opened as their lead guide, Captain Calir Tilab, neared the door, indicating the security system had either biological or visual sensors capable of recognizing a welcome member such as Tilab, while barring unwanted entry even from tiny nanobots. Upon entry, Ga and Or were forewarned by their escort they were about to meet a court of their gods. They were to avoid speaking until spoken to. They had to refrain from hostile gestures or movements. If such disrespectful acts were observed, the penalty could be an immediate death. Ga recalled the art they saw in the forged historical site was of a different species than the reabs glorified in a divine fashion here.

Inside the cone, most walls, floors and decorations were compiled from a stable and cooler on the outside plasma. Almost anywhere Ga and Or looked, they saw artistic depictions of triumphant battles, beautiful queens and the symbols of holiness floating in semi-animated hot-plasma art pieces inside of a still plasma canvas. The three-story ceiling was extra-bright, glowing in a magnetic, florescent light show. The front of the main hall was so luminous Ga and Or failed to distinguish its details until they were standing thirty feet from it. Two central glass-shaped plasma cone fountains functioned as seats for four inimitable aliens. They resembled lizards, but with ultra-thin and elongated body parts, with parachutes puffing and then deflating at a constant rhythm. Their plasma skin was reflective and semi-transparent near the surface. On the hottest end of the temperature spectrum, substances can enter the plasma state as gas is put under increasing temperatures or pressures. Different chemicals enter this state at specific temperature ranges. Plasma crystals can even be cooled to as low as a human's average internal temperature of 98°F. Plasma is common on Earth in non-thermal products at temperatures between 3140°F and 4500 K. Thermal plasma can exist at much hotter levels. Lightning, auroras, and stars are comprised of plasma. Since Ga and Or penetrated this sanctum, their nanobots were now able to fly to the deities to collect microscopic gene samples. Later lab tests showed the turapes' plasma was organized into complex helical structures with help from neural electric energy. These structures were their unique genes, and they had the same basic elements as other types of living organisms. They could divide and replicate themselves, evolve, and control their own actions as well as interact with the strange cell-like structures around them. The extreme temperatures in these cells meant the molecules in them were moving with vigorous rapidity, and thus they possessed an incredible energy level in their minds and muscles. Doctor Na estimated turapes' average lifespan was 50,000 years. This extended life, when compared with the reabs' brief and painful existence, made the turapes appear godlike. The turapes' advanced intellect allowed them to manipulate the reabs into doing their physical labor without compensation.

The four turapes, dressed in outré-thin plasma garbs, questioned Ga and Or about their complaint. However, instead of reaching a solution, the turapes invited them to spend a night in the palace. Ga and Or accepted this gracious invitation. They were escorted into a waiting room, where they spent around an hour before they were called for dinner. The table was set with extravagant dishes of simpler plasma-based lifeforms: some melted into puddles in their hot plasma plates and cups, while others maintained complex and bright-colored animalistic structures. Three strange stick-insect plasma creatures with a hundred swaying arms were at the center of the table akin to buckets of flowers. Ga and Or had to refuse the plasma snacks, explaining via their light translators they could not digest matter this hot, and they were already consuming nutritional gases inside their suits as they spoke, so they were satisfied. After this feast, they were taken into a spacious room where the four court presiders and noble turapes were preparing for sleep by curling up their elongated bodies in plasma pools. Ga and Or just prepared to nap in comfort in a diamond-littered, solid-floored part of this communal room when the turapes woke and embarked on neatening. Their sleep cycles lasted for twenty minutes.

Without much of a break, Ga and Or were then taken back to the main chamber and the trial resumed. One of the four ruling turapes, Holiness Sess, informed them aliens who were not born on Nomuat were not allowed to own property. Thus, even if it could be proven they had made unauthorized and underhanded payments to the workshop supervisor, they would not have secured a right to this business. However, the problem was deeper, as no such reabs as Yald or Laneum could be found working either at the workshop or at the display-historical-site Ga and Or indicated.

"Frankly, these names you say the reabs had who fooled you mean 'yard' and 'lane' in our language. It appears the names were invented by an individual lacking a native grasp of our language's linguistics; they are not consistent with our local naming rules, which require the title and the end of the main name must rhyme," Holiness Sess argued.

Further still, Sess proved there was no workshop in the spot they claimed it was.

The site they alleged had been faked was profoundly authentic. Sess was offended by the insinuation their history was ingenuine. He questioned if Ga and Or were even aliens, or if they were spies from a rival city sent to steal their secrets. It was even possible, he contended, rather than being wronged by these supposed "Yald" or "Laneum", they were the ones who stole the domicile and the cart of plasma. Sess even invented evidence to support these elaborate counter-allegations against the accusers. He concluded, without facts to support these conjectures: nothing could be determined, concluded, or decided. They were dismissed with a public forbiddance against being caught mentioning them. Ga, Or and the others on the Mothership listening in were stupefied by these confusing conclusions, and opted to refrain from issuing an objection in case the trial began collapsing on the two exploring cadatanauts.

Then Holiness Sess commanded Ga and Or to take them to their shuttle, so they could examine it for evidence of wrongdoing. The cadatanauts obliged, not wanting to be disrespectful to their powerful hosts, who brought around a hundred reabs and a troop of powerful-looking turapes with them on this investigation. The ship had been parked without a disguise or any significant security measure in the middle of the desert, near the location where they had left it. It surprised them there was a group of reabs surrounding it as they approached. The locals showed no visible interest in their mode of transportation in the past. The dirt on the entryway indicated the reabs had been attempting to gain entry, but the ship refused opening for non-cadata genetics. As Ga and Or approached, the shuttle auto-opened. A few dozen turapes fit snuggly, while the others crowded just outside the entryway. Sess asked for a brief introduction to the ship's operations, a request Ga and Or honored to maintain sociable relations. As this lecture progressed, Sess compared cadatas' 3D programming language to their light-based coding language. Both Ga and Or were needed for the intricacies of this discussion, so they missed a turapes triggering a rapid accumulation of matter from the planet in turbo-mode, which Ga and Or just described was inadvisable. The rapid matter suction pushed the system to experience a power-surge and then a total shutdown to allow it to process the input. With the Computer out, the turapes commanded Ga and Or to override anti-alien security measures so they would be able to pilot the ship without the need to fool it with fake genetics. Ga and Or again obliged because the cadatas on the Mothership were amazed by this strange behavior and awaited with amusement what the turapes planned on doing once they were on the cooler Mothership without insights into its operating system. As they anticipated, Sess commanded the shuttle to fly towards and to land inside the Mothership.

Just as the parking maneuver was completed and Sess ordered the shuttle's door to be opened, Doctor Na exclaimed: "No! Stop them! They're halogens!"

This warning came too late. The moment the shuttle's door opened and exposed Nomuat's halogen-based species to the metal-rich atmosphere in the spaceship, they turned into salt statues. The Doctor's forensic analysis arrived at this conclusion based on the low temperature, and the presence of water moisture in the ship's intended-for-cadatas air. It was shocking a species attracting a myriad aliens to their planet had previously failed to learn they had to wear spacesuits while exploring alien ships. Later research explained no other species had allowed the theft of a ship, or the invasion of a mothership. The thefts and frauds were contained to what Nomuat's population could grab on the surface of their own planet. Meanwhile, cadatas consumed voluminous metal-gases through their atmosphere. In contrast, there were near-zero metals in Nomuat's composition, allowing halogen lifeforms to evolve without this unpleasant saltiness. During the boarding, Doctor Nomuat rushed to revive the "invading" party, but the reaction was irreversible. It solidified the internal flow of hot liquefied halogen compounds. The transformation was expedited by the low temperatures, which were as freezing for these species as jumping in the water on the North Pole in dead-winter is for humans on Earth. The crew endured a prolonged debate regarding the proper disposal of these salt-bodies. If they remained on the ship; they would have to be recycled by the system. If they were returned to the planet; cadatas would never be allowed to return to Nomuat, having saltified the top leaders of Nomuat's biggest city. Captain Al believed a peace agreement, or a negotiation for a piece of colonizable land for cadatas were still possible. He was alone in this assertion. The others decided: if they generated this much disquiet and disruption in less than a year on the planet; cadatas were destined to go extinct, if they dared a permanent settlement. The bodies were shipped to a somewhat isolated location, and deposited in the politest manner possible. A note was included with an apology for the chemical reaction the party suffered on the cadata ship. It ended with the salutation: "Good wishes for your planet's future success." This diplomacy aside, cadatas learned mistrust was as lethal as warfare.

Chapter 9

A World Without Sunshine

A week later, on a morning in 85,949 BC, Captain Al ordered the departure maneuver in the direction of their next planetary target: Ragile. It was still further from the galactic center on the Scutum arm, and closer to the middle between the Norma and Scutum arms. The region was 500 light years above the main disk of the galaxy. It was dominated by dust and gas clouds, and younger and lighter stars. The decision to depart had been finalized with a unanimous vote from the crew. This step Captain Al deemed necessary because the planet they were now aiming for was 8,451 light years away. Everybody had to agree to abandon Nomuat, with the prospect of a new multi-generational voyage ahead. Most were convinced before Captain Al opened the floor for the discussion. The last few stragglers were convinced by the chemist's tirade against the metal-exposure problem. He argued: if they opted to stay, and convert their genes to a halogen composition; they would also have to suffer through a complete conversion to non-metal, and non-water substances in their manufacturing. The foot-draggers were too addicted to high-speed-tolerant, space-worthy, metallic machines to alter their composition into brittle stone. This conundrum turned into a pressing problem when the "gods" turned into stone before their eyes. Even if they opted to engage in extreme, suppressing warfare to enforce the cessation of lying by the locals; they feared cadatas would eventually submit to the unalterable evil nature of Nomuat. Thus, they set sail again.

As Nomuat disappeared behind the ship, they assembled to discuss a lighter problem uncovered during this brief stop. During the mission, team members had been confused by Or's name, especially when it was used after "and", as in: "Ga and Or". The two joined words appeared to express an uncertainty regarding if Ga was with somebody, or if only Ga or Or was referred to. They had begun shortening the crew's names in the following generations for the sake of making expedient naming decisions. The next generation had to be named with single letters, such as C and G. Proposals were submitted for alternative name manipulations: perhaps using half of these letters, or voicing half of the "C" sound. So, they settled on switching to using the original crewmember's name with a number to signify the number of generations removing an individual from Cadata. The youthful Ga was renamed Wocega-4. As each generation adopted a similar primary role as their ancestors, the root names became synonymous with the positions they represented; for example, Wocega was the lead pilot assigned to explore planets. The intensity this discussion attained despite its seeming insignificance should indicate to readers the extreme degree of boredom the crew was suffering on this new three-generational voyage on the heels of another marathon. Their mood was downcast as they considered the type of world their grandchildren were going to explore next, and the heightened degree of unlikelihood this would be their terminal destination.

They arrived in the Ragile's solar system in 77,497 BC, and instigated the standard research and development stage during the slowdown. In passing, this was around the time when humans first migrated out of Africa and began spreading across the rest of Earth, also setting out on a multigenerational voyage, though on a smaller physical scale. Wocega-6 was assigned the mission to explore Ragile, despite her mature age. She had known she would make the trip since birth, starting her studies on Ragile long in advance. Once close-quarter astronomical readings commenced, she was semi-living in the Observatory and the Lab, researching strategies to materialize this improbable mission.

Ragile was a helium brown dwarf, the largest sub-category among the smallest stars in the universe. Brown dwarfs form a unique group because they cannot perform nuclear fusion in their cores due to mass insufficiency; fusion requires a temperature of at least 10 million K in the core, achievable only when a larger mass is squeezed by pressure. Instead, they fuse hydrogen into helium nuclei, a reaction generating heat as a byproduct, which shoots to a star's surface before being ejected into space as radiation. At 1,400 K, this brown dwarf had a surface temperature at the top of the range. Its solar mass was .075 of Earth's sun. While this star was fusionless or stillborn to begin with, brown dwarfs exhume so little radiation they last for trillions of years. The low surface temperature is an indicator of this star's dimness, shining with a mere .01% of the sun's luminosity. Nally chose it for the *Catalog* was because it was surrounded by a dense cloud of gas and dust adding mass to this star. A brown dwarf tends to generate radiation from heat stored during its formation, but these infusions of mass created additional heat waves. It was scheduled to begin fusion in a few million years, growing into the red dwarf category and becoming a steadily luminous star suitable for habitation. Ragile was already in the habitable zone of its star for extremophile lifeforms; as the star warmed, it would increase in hospitality. There was a risk of eventual planetary overheating, in hundreds of millions or billions of years, necessitating a new out-migration. Average dwarfs have fleeting lifespans, which end when they go black after exhausting their deuterium at the 10-million-year mark. The infusion of gas had rammed the extermination clock by a billion years, and the gas-growth was accelerating.

Because Ragile had proximate, 2-day orbit around its sun; so, the cadata spaceship appeared to be flying into this star during its slowdown phase en route towards Ragile. This sun had more contrast and definition to its surface than stars of average luminosity burning in a steady, bright whitish-yellow. It was striped with red patches of darker and brighter spotted gas. Ragile's dim sun meant life needed an alternative to sunshine for energy. Nanobot expedition data confirmed pre-flight readings—hydrogen had been mixing with iron-bearing minerals through methanogenesis, a process resembling human metabolism, in shallow methane lakes across the planet's surface. These chemical reactions represented an evolutionary leap of life on Ragile as they generated energy for microorganisms with methane as a byproduct. On planets with limited heat and sunshine, this is a trick capable of sparking life; organisms utilizing it replicate and grow just as light-dependent organisms do. The abundance of methane, hydrogen and heavier elements in these pools accelerated the replication of microorganisms. Soon, it was a survival advantage in the competition for resources to join cells into larger organisms. A human begins as a single cell before rapidly multiplying into trillions of cells. Ragile lifeforms expanded until their size allowed for complex brains capable of mechanical innovations, which triggered intelligent-life indicators on cadatas' readings.

Most living organisms on Ragile were contained in isolated methane lakes and pools. Between its weak sun and a significant amount of internal heat due to its larger-than-Earth mass, surface temperatures ranged between −310°F and −265°F. The coldest temperatures occurred on the poles seeing the least sunshine, where methane was frozen solid without any liquid pools because its melting point is −296.4°F. The lake-covered regions averaged temperatures well above this lower limit, but not so high any methane had a chance to evaporate once having reached its low boiling point of −258.7°F. These numbers are approximated for an Earth single-atmospheric pressure; they were slightly different on Ragile's lower pressures, in correspondence with its lighter atmosphere. There were two oceans on opposite sides of Ragile; temperatures dropped in these bodies at lower depths, turning methane solid. Methane was only liquid on the surface of the oceans in small patches and in the lakes amidst the higher-elevation rocky regions. Methane's hydrocarbon molecules ($CH4$) are unique because they remain liquid even at 200 K below the freezing point of water. While a water-covered planet would snowball at these freezing temperatures, the odorless methane still allows molecules to move around in it, forming the interactions needed for evolution.

With the environmental concerns sorted, Wocega-6 was put through the transformation into a ragile. This time, there was no need to keep an entire half of the ship isolated for roaming because this species was genome-programmed to live in small ponds or even pools. The pool had a Ragile-appropriate atmosphere in a bubble around it, but the whole contraption fit in a single room. This species also needed less room because of its 6-inch average size. Methane was the main fluid inside ragiles' bodies; liquid methane bodies were comfortable at low temperatures. It had two mouths at the top of its head and a rectangular set of reflective surfaces across a strip around its forehead serving as an eye. The mouths opened similarly to how human eyes blink, plummeting open in an instant to allow a sensed prey to enter, and then slamming behind it. It had an external crown of thorn-like bones for vibration and sound detection. Ragiles were super-dexterous, with six arm-leg appendages and six snakelike fingers. Their natural skin color combined blue and red in a fanciful pattern, with spots of black, but they wore skin-hugging outfits stretching in approximation of wings between their sides and the tips of their arm-legs, allowing them to swim at accelerated rates. Their heads were translucent, so the pink, multi-tubular, flower-like folds of their brains were visible. This added a challenge for Doctor Nackan-6, as he had to disguise a cadata brain in these folds to avoid having an inactive ragile brain and a translucent, visible cadata brain in the same body. Given the size difference, extreme flexibility, and speed capability of this body, Wocega-6 relished her initial days swimming in her pool in this form. She adapted to catching and eating smidgen, boneless, toothless, and unbalanced creatures called dithens; they scuttled in all directions on spiderish feet barely sufficient to propel their tiny, beady, and gooey centers.

Ragiles were somewhere between apes and humans on the intelligence scale. A unique feature to their brainpower was in each lake's isolated evolution of variant ragiles. They differed in their physical appearance, leg-arm numbers, and the texture of their skin. There was a lack of interaction between these pools in the 300 million years since they were isolated. Thus, a scientist in one pool might have solved a physical or chemical law, but this knowledge was undisseminated out of this isolated community. Some lakes developed advanced domiciles, others speedy transportation devices, others flamboyant and sophisticated fashion. Some were going through an industrial revolution, while others had long passed this stage. None of these isolated societies had discovered a vehicle of sufficient strength and endurance to push out of their lake and to migrate to a neighboring lake. The ultra-low temperature and the need to invent and modify technology in a methane liquid were a couple obstacles stalling their progress. In parallel, while a few isolated scientists had invented star observation-devices with greater exactitude than how they appeared through the blanket of methane, none had yet discovered a method for breaking from the planet's gravity and into space. Linguistics also evolved separately, forming a disjointed diversity of sign, oral and fluorescent communication approaches.

Captain Nally-6 chose the top techno-advanced and largest in surface-area lake on Ragile for their initial contact location. Given ragiles' size, designing a shuttle convertible into one of their lake-faring underwater ships proved impossible. They had launched tiny nanobots through space at light-speed without complications, but a one- to six-foot spaceship with the shabby appearance of the locals' vessels and full of near-freezing liquid methane burned in the atmosphere on numerous tests, even when reinforced with the strongest materials in the galaxy. As a workaround, Nally-6 proposed parking the shuttle just outside of the lake. They would then extract the tiny submergible vessel with Wocega-6 in it, giving it the capacity to drive over to the methane pool on a disguised mechanism. The use of a see-through, plastic-consistency and scaly surface, and the internal paddle-propulsion devices promised to convince ragiles this was a vessel of their own design.

The shuttle was pre-programmed to park Wocega-6 on the scheduled stop. Even if she wanted to take manual controls, she would have struggled due to the size differentiation between the equipment and her. Wocega-6 looked on at the dim reddish light blanketing Ragile. The line between the outer and inner atmosphere was a semi-distinguishable, reddening and brightening glow. Even with the sun out, the sky was as dark as when a full moon is out at night on Earth. The sun was enormous, but looked like a red glow lamp covered in foggy spots. Near the horizon line, the darkened sky was littered with a fog as light hit the asteroid belt, and dust and gas debris scattered between Ragile and its sun. This fog gained radiance at night, dominating the view among the brighter stars.

Most planets in the *Catalog* had at least one moon circling them, but not Ragile. This absence made the surfaces of the methane lakes motionless and tide-less whenever wind-speed was low. The sun's contribution amounted to tiny ripples due to its distance; this impact is mentionable because Ragile is closer to its sun than the other considered planet. The stillness and purity of the approaching lake mesmerized Wocega-6 into suspecting she was still in her room and diving into a lab-constructed idyllic simulation. The absence of a moon was troubling because an Earth-sized planet without it can begin to wobble, losing its spin's stability until, in some unpleasant but rare cases, it could flip over. Such flips can accompany earthquakes and tsunamis approaching the strength that destroyed Cadata's environment; even the crust can crack under this pressure. On the bright side, Ragile's spin was consistent and wobble-less, so their billion-year forecast for the odds it would flip was benign. Evidence indicates Mars has flipped because its tiny moons failed to stabilize it, contributing to the drying of its water-oceans. A moon might seem to be a reflective lamp or a mere inspiration for fairytales, but planets without them suffer in their absence.

The ship began the landing maneuver at a gentler speed than usual because of the slippery, icy surface it was about to land on, and because the touchdown had to be quiet enough not to attract attention of the ragiles in the lake. This sluggishness allowed Wocega-6 to contemplate the scene of the sparkling methane and nitrogen ice fields consuming the bulk of the view. They were lined with channels and ridges of various thicknesses, and with strange patterns to their paths and intersections. The lines in the skin of the planet suggested it had been warmer and cooler before, as surface ice melted and then froze over again, and lakes grew and then contracted. This was consistent with the hot period near the birth of its sun, and then the irregular spikes in temperature as the asteroid belt and debris collapsed to feed its energy output. If the temperatures had been constant since near its birth, the surface would have been dominated by asteroid impact craters, but there were few of these in Wocega-6's line of vision. The rest must have been covered by the migrating or churning ice sheets. The most dynamic feature was a methane river, a recent peak-summer formation, which was carrying tall, rectangular, shady-yellow nitrogen glaciers with crushing force, while small, water-ice-comprised icebergs were slammed into the margins of this narrow passage. Since methane and nitrogen lack the cohesion in their solid state to form mountains, most of the planet was flat, aside for the ridges, and a few mountainesque water-ice protrusions rising out of the planet's bedrock crust. The uniformity of the planet's surface, without breaks between land and oceans to twist and disturb the transfer of heat, and without mountains to push air up or down, made the weather extra-stable and storms exceptional.

Wocega-6 was lost in this examination of Ragile's topography, when the Computer quaked incomprehensible phrases and shut itself off. The backup system kicked in to keep the ship from plummeting, but the route the ship was supposed to take to land had been lost with the Computer. Wocega-6 could see the status of this failure on a projected backup screen monitoring the key system. The mega-low temperature failure alert was flashing because the ship encountered a heavy presence of methane and still more volatile gases in the dense atmosphere. The Mothership's design withstood absolute-zero. The shuttle could fly through outer space, but with weakened protections. The long abandonment with only internal, brief test-flights left this low-temp kink in the system; it was weened out during this first complex operation. Wocega-6 had to take manual control of the ship despite impediments. On previous world-visits, there was a pilot at the helm capable of applying his or her tentacles or wings to a visual computer screen or an input device utilizing an alternate sense. By contrast, in this instance, Wocega-6 was locked in a mini-vehicle inside the larger shuttle, restricting her from the capacity to reach out to type in a code with her leg-arms. She managed to handle this disability because she had devised alternative maneuvers for manual-operation of the fish-vessel while cocooned during the pre-landing months. She rolled across the counter she had been resting on, across the length of a wall and then over to the main computer screen. She tried making signs with her little leg-arms to issue commands, but the system did not recognize them, as her genetic code was not coming across from inside of her enclosure, and her leg-arms were too tiny to be scannable as control-issuers. To solve this difficulty, she began making signs by flying her entire vehicle in patterns she would have made with her wing in her cadata-sized body. She zigzagged and climbed steeply up and down as she drew the 3D signals of their programming language. It was a tedious, slow process, prolonged by the smallest typos. She managed to input the command for the ship to approach the surface. She judged it would have been tougher to dock with the Mothership. The maneuver was slowed because she failed to regain full landing-controls. Just as she was resigning herself to a crash-landing on a random ice sheet, she recalled the ancient method of aligning a ship's path to star charts. She had miniature star charts in her mini-vessel because they were less revealing of alien-origins than the 3D star charts from the Computer's databank. The navigation lecture had explained how recognizable stars in the sky could be sighted, aligned and charted into the Computer to help it to orient itself if its access to its own data was cut off. She executed several calculations on their relative positioning, angles, and time of the year in her head—a skill rare among cadatas, but a required course in cadatanauts' standard training. She began by entering the location for the brightest star in the sky by clicking on it in the touch-sensitive see-through wall of the ship. Then, she flew over to the others, and performed the other essential steps. Just as she was about to press the spot for the last of the dozen stars required to re-established the Computer's spatial awareness, the ship drifted by a degree, perhaps due to a gust of wind-pressure. This shift meant the star-spots she had entered were now a few degrees off, and she had to repeat the steps. Finally, the alignment was complete, and the Computer seemed to have resumed a steady descent. But the angle was too steep, straying from the intended lake Wocega-6 had been aiming for. Wocega-6 began entering an additional code to alert the Computer to these irregularities. But then she spotted the readings showed the ship was about to touch down. The ship acted as if it was slowing down to contact the ground. But since there was no ground, it kept sinking in its imagination until the system believed it was over 3,000 feet under water. Wocega-6 brought enough systems online, for the Computer to react to this depth by inflating itself with excess air to make itself buoyant, forcing it to pop into the air. Wocega-6 was stumped as it started paddling a set of fins it extracted. This switch to swimming in mid-air meant the system stopped providing the thrust slowing the ship's fall in a controlled manner. Without this restraint, it began free-falling, putting an enormous g-force on Wocega-6's tiny body. She managed to activate the gravity controls inside of her ship, but just after it was tossed into a wall. The protective gel in the crafty vessel's wall (a remnant cadata design) protected her from a serious injury. As her tiny vessel regained gravitational and spatial control, she flew over to the manual flight control board, switched the controls from the disturbed Computer to this device, and began flying her ship in patterns to correct the ship into a slowdown and to regain a sense for its actual height in relation to the planet. Her goal was to prove to the system they were still in the air and not 3,000 feet under water.

The height readings were accepted by the system around 200 feet above the surface. In response, the shuttle commenced a rapid slow-down. Wocega-6 feared the shuttle might crash into the ice if it was a few feet off, so she brought it to a near-stop 20 feet ahead of this potential collision-point. She manually calculated the angle to the target, the direction of approach, and the amount of thrust needed to reach the target without a violent collision. However, the sum of these calculations was overwhelming, so Woce-ga-6 opted to tap the "gas" forward by a foot, stop, tap it a bit more, and so on. She strained to eyeball the ice below to estimate the distance during this jumpy procedure. This was difficult for a cadatanaut semi-familiar with the appearance the foot-equivalent cadata measure the Computer utilized. It would all have been impossible if the shuttle had not been see-through (when necessary) on all sides, as obstructions to her field of vision would have been disastrous. At length, she hit the ice sheet with the lower tip of the shuttle. This was their spherical shuttle model, so it began to spin and wobble without the capacity to stabilize on its tip. Wocega-6 flew over to a manual latch and released the holding platform, also known as the self-constructing docking bay, which auto-dropped to secure and stabilize the ship's position. With the emergency landing secured, Woce-ga-6 began more extensive repairs to the Computer. A few fixes released the on-board robots, who took over the Computer repair to restore full operational capability. Woce-ga-6 left the robots to this work, as she bobbed out of the bigger ship in her little vehicle.

The ship drove across the ice and then fell over the edge into Lake Cuthog. Above the liquid methane surface, the natural landscape looked as it might have looked on Earth at dusk under a blood moon. Under it, objects yellowed in tint closer to the lake's bottom because the other colors failed to penetrate the dense methane. Wocega-6 had to depend on her x-ray and non-visible wavelengths sights here because even with ragiles' heightened sense of visible light perception, the liquid environment clouded and rippled the view, making objects further away imperceptible to the naked eye. The cold was imperceptible to Wocega-6 despite the below-freezing-for-water temperature in the methane filling her vehicle because her body was designed to withstand it. However, observing the ice outlines of the lake put her under psychological stress regarding swimming in a pool genome-opposite to the temperature ideal for cadatas.

Her vessel sank somewhat and then remained suspended and afloat just below the surface. She used this marker in the voyage to eat a few replicated dithens to garner energy post-kerfuffle. With needs sorted, Wocega-6 pressed the vessel into a slow dive toward the bottom of the lake. She entered the water in a region depopulated because of a toxic factory leak two decades earlier. The rest of the lake was crowded with life, most of which was simpler organisms, with a few strong and intelligent fresh-methane creatures swimming, crawling, or drifting around. Ragiles' vehicles and lake-plant-based clothing separated them from the throng, but they co-existed with other species. If a human spots a lion and a crocodile on a city street, he or she is unnerved, and calls animal-control. In contrast, on Ragile, the dominant species was defensively armed against predators with reinforced vehicles and weapons, but they refrained from exterminating competing species because scarcity of space in the lake truncated resettlement options.

Ragiles' houses were disguised by the vegetation, rocks, sand, and icy objects scattered across the floor. The plants shared more commonalities with animals than with Earth's photosynthesizing organisms. There was semi-sufficient light penetrating to the bottom of the lake for forms to be distinguishable; this volume of light was inadequate to feed vegetation with photosynthesis. The surviving, basic, small organisms fed through chemical reactions with the basic elements abundant in the surrounding liquids and solids. Bigger creatures fed on them. The largest among them was ten inches because the habitat forbade expansion. Ragiles' buildings were built with diverse construction

methods. The biggest one in the center of the lake was composed of sand, and had a series of compressed, curving lines spread out like the spiral arms of a galaxy, with sealable
door openings. If, after a prolonged closure, one opened, sand and pebbles, piled over it
in the interim, floated upwards. Smaller houses were so bright-colored, they were vibrant
even in Ragile's dim sunshine. They had bumpy, multi-curved shapes with narrow single
openings in the middle. Their material component was still alive, and growing slowly. It
had to be trimmed and re-shaped, but otherwise it was perpetual, the whole regenerating
even as parts of it faded with time. Still other houses were drilled out of ice, taking on
either decorative or unadorned shapes depending on if they were constructed by melting
or moving ice, or out of the permafrost of ancient and immobile cliffs. A few houses
were excavated below overgrown, finger-like, translucent, alive, and poisonous-to-the-
touch tubes. The tubes served as natural home security systems to keep out predators.
The construction methods showed knowledge of chemistry and physics to solidify and
strengthen the structures. The large sand dome was used, in part, as a science research
and development institution.

Wocega-6 was recognized as an outsider the moment she was spotted because the
small size of the lake meant an absence of strangers. She said, a strange, natural occur-
rence had catapulted her across the external ice and into a neighboring lake. Even the
idea of other lakes on their planet surprised the ragiles. They regarded her as a proph-
et from beyond, and welcomed her into their community. They were prone to share
with each other, and this meant Wocega-6 had a place to live and plenty of local food.
Wocega-6 sent regular messages back to the Mothership across the following decade to
confirm life on Ragile was idyllic and peaceful. While it was always cold, there were no
major changes in the environment, such as storms or tornadoes, to disrupt the quiet
little lake. Pathogens were unheard of, as any pathogen festering in the past had long
regressed as ragiles formed immunities to it through the generations. The strange chemical makeup and low-pressure environment allowed the local species to live 500 years or
more in excellent health. Outside the lakes, there was an enormous stretch of unsettled
land cadatas' climate-controlled buildings could have populated. Thus, as the decade
ended, Captain Nally-6 and the leads decided Ragile was habitable. They had to commence a costlier exploration by sending half of their crew to various lakes to determine if
problems might surface, which were foreign to Lake Cuthog. 250 shuttles deployed for
the various major lakes across Ragile. Among them was Ortack-6, who chose a lake near
Wocega's. Ortack-6 was too eager to instigate an adventure on the only planet he would
encounter in his lifetime, so he parked his shuttle closer to the outline of the lake than
the others dared. As he was skidding across the ice and diving into the liquid methane, a
group of ragiles touring the further reaches of the lake spotted a strange, giant, metallic
object landing and causing trepidation of the ice above. Then, they kept vigilant watch
as Ortack-6 broke the surface of the methane. He was driving a vehicle mimicking
local transportation, but their unit would have failed to withstand a skid through the
windy and colder conditions above. While other lake dwellers gave the newcomers the
benefit of belief, after seeing this entrance, the residents of Lake Kippey were adamant
Ortack-6 was a foreign being or even an alien because his ship came from the sky. They
investigated Ortack-6's vehicle, and discovered hidden, complex features they aborted
trying to operate. So, they insisted Ortack-6 had to explain the origins of this outlandish

technology.

Ortack-6 forwarded this non-coercive and urgently-pleading request to the Mothership. The kippeys pled their destitution as they lacked rewards outside of housing and food for their toil. If Ortack-6 could teach them to fly, surely, he was obligated by galactic alien morals to do so. Nally-6, Doctor Nackan-6 and the others debated for months, while Ortack-6 stayed-on as a guest in Lake Kippey despite the strained relations. The final decision was billions of cadatas' arrival and settlement on the planet would inspire them to flaunt their technology; thus, introducing at some of it during this test-phase was necessary to exclude potential technology-related afflictions or conflicts. Ortack-6 began the experiment with a lecture on the foreign flying capabilities of his ragile-in-appearance vessel. He escorted elder kippeys on tours of the shuttle's interior. He also flew them in rapid orbits around the planet to explain planetary geography to a species without prior access to this sky-perspective. They were mesmerized by it; this knowledge re-wrote their scientific and geographic histories. Once a group of kippeys were trained in this new external transportation method, they asked to visit their neighboring lakes to establish relations with other ragiles. Once again, since such communications would have been necessary if the cadatas finalized the resettlement decision, the kippeys were approved to visit their neighbors to introduce themselves and to figure out how they might work together. In the following decades, ragiles established interconnections through cadata robots' buildup of airworthy-vehicles gifted to travel-curious ragiles.

The ragiles' first concern was learning the distinct communication systems developed in isolation across the planet. They also worked to begin an exchange in goods and services to catch up with technological advances made elsewhere. Cadatas did not burden ragiles with too much of their technological knowledge because when they tried to explain it, the content proved to be too advanced, and thus bored their alien pupils into stupification.

As technological and business progress appeared to be imminent in this hodgepodge environment, under this façade, several of the biggest lakes began having heated debates with their largest neighbors about frontiers. The concept of a border never occurred to them before, but now they had access to the expanse of ice above, and their exploration of and trade routes through these wild domains raised the question of ownership. Concepts such as nationhood began to be broached. Among others, the elders of Lake Kippey and Lake Cuthog began negotiations on a line in the ice in the middle of their two lakes fair for both sides. One of the reasons an equal split was difficult was the presence of a desirable set of ridges rich in minerals and rocks near the middle. Since these ridges were raised high above water-level, the sides failed to decide if they should be halved as if they were flat, or counted in above-water-level volume instead. Additionally, since they both had to travel through the other's territory to trade with the each other or with lakes only accessible with this through-passage, they began proposing tariffs, crossing fees, and restrictions on times of day and of year when a crossing could be attempted. They even put guard vehicles in critical locations to block unapproved or unpaid-for travel.

While these border disputes were brewing, a sense of anxiety about political status quo also heightened. Larger lakes and manufacturers began expressing the desire for dominance by sprouting their goods, services and ideas to the smaller entities and individuals. The pressures on consumers to purchase the latest tools a lake was propagandizing were grander to equivalent devices elsewhere. Problematically, the biggest and mightiest lakes were deciding on the parameters of techno-advanced devices. They shaped these definitions around their existing pitches, even if they were inferior to ideas from uninfluential rivals. Problems with sales tactics went unresolved because ragiles lacked pre-integration governing structures. Their science institutions were alone in leading their societies' intellectual development. There was no need for courts of law, legislative bodies, or police forces because local crimes such as murder were unheard of. The sudden amalgamation sparked violent national feelings in a segment of the population sensitive to manipulations by those who wanted to gain power by controlling other ragiles. These self-designated rulers stressed their nationality's differences from other variants of ragiles from other lakes. Expressions of species-dominance in publications spread this message even to ragiles without prior feelings of anti-outsider "pride". Such public discussions led to encouraging foreigners to withdraw if caught visiting a lake other than their birthplace. Trade made such visits from foreigners essential, so the problem could not be solved by blocking all borders. The problem intensified when several lakes formed a "defensive" bloc because the hostile actions from the opposing bloc necessitated their joining to fight for mutual, regional interests.

In the following decades, these blocks began expressing their divergences in terms of ideology. The cuthogs believed in sharing houses and all other needed resources across their network of extended families, and they expected to be able to share resources if they were visiting another lake in their bloc. Historically, the kippeys had strict ownership laws restricting trespassing on another's house. Under globalization's onslaught, the kippeys stretched this idea to the extreme, restricting ownership of land inside the lakes, prohibiting non-owners from gathering fish or stones. This split the region into two blocs: one Shared, and the other Distinct in their philosophic beliefs. While this began as a discussion in a string of books, articles, and speeches; it soon turned violent, as outright war commenced over unwanted cohabitations, and unpaid-for excavations. They began crafting projectile weapons: inventing a new industry unimaginable to previously isolated ragile scientists. Both sides adamantly came to believe they were obligated battle for as long as the other ideology existed. The other side's philosophy became a demon to be destroyed, out of fear its spread would massacre all. Despite the warring, both made official propagandistic protestations of peaceful intentions. The caveat for peace was the other side's death, or their surrender of demonic beliefs, in favor of the rival's "superior" ideas.

From the first signs of conflict, cadatas commenced bringing the sides to negotiate a peace agreement. But despite surface concessions, each attempt to establish borders and rules merely widened chasms. For example, during the Weapons Confinement Dialogs, cadatas led a discussion on refraining from using weapons-of-mass-destruction capable of destroying thousands of ragile lives. However, cadata organizers were ignorant of the representatives' sponsorships from the major weapons manufacturers in their blocs. The companies' profits from weapons were tied to continuous warfare. On another occasion, even without cadata intervention, a couple of lake factions even attempted to hold secret negotiations to allow freer movements between them. This peace agreement was derailed when a spy from the third lake region discovered this dialogue. The conspirators were accused of deceit, and bad-faith towards the other lakes. They were attempting to form a joint bloc, instead of striving for peaceful coexistence.

A hundred years into the cadatas' stay on Ragile, in 77,394 BC, they realized the social rift between the sects and races on this planet was permanent. They could not reverse the process, such as by keeping them from moving between the lakes. They could not order ragiles to stop fighting. The warfare was growing deadlier each year. The image of their planet as a little yellowish ball from space failed to inspire planetary unity. Instead, Ragile ignited in non-stop strife. It was time for cadatas to give up on Ragile, and seek out calmer waters.

Chapter 10

Life Among the Clouds

The failures from the previous few trips at last convinced the crew they had to venture further from the galactic center. They decided to cross the Scutum Arm to reach the less populated middle-point between it and the minor Sagittarius Arm. Their destination, Nefo, was 11,021 light years away. Their scheduled arrival year was 66,373 BC. Wocega-8 had been born in 70,394 BC. She anticipated this would be her sole expedition, as she turned 4,021-years-old when they entered Nefo's system.

Nefo was unique in its class of gas giants because of its eclectic atmosphere, and relatively low temperatures. It was twice farther from its sun than Earth is to its Sun. And Nefo is 22 times larger than Earth. Its surface temperature hovered under 420°F. A year lasted 500 days. Readings indicated extra-strong, revolving winds pushing ultra-dense and molecularly heterogenous clouds in a continuous cycle away from the dark side of the planet to the one exposed to sunshine. There was a tipping point in each day's cycle, at around midday, when winds accelerated to 300mph, leading to rapid buildups of atmospheric gases, after which the winds died down to a calm 50mph, a low point when clouds dispersed. The gas giant, Elpile, which Putair visited in the Cadata solar system, was covered in a near-pure liquid metallic hydrogen ocean. In contrast, a crusty layer of heavier minerals blanketed Nefo's liquid helium ocean. Minerals saturate Nefo's atmosphere. Clouds rain minerals in fiery eruptions. Reactions occur when the planet's hot surface is hit with clumps of high-speed falling materials at the precipice of turning into gemstones, such as rubies and sapphires. These corundum minerals sink to the planet's core by moving past the bulk of the planet's gaseous structure, and congeal there into these gemstones. Some of these heavy stones later erupt as an assortment of molten lava unique to Nefo. They pile up into mountains above the mineral crust in volcanically-active regions. The stones are lit on fire as they rain down on the planet through the thick atmosphere. This was an easy planet to spot from Cadata because, like brown dwarfs, gas giants emit infrared radiation. Like other giants, Nefo formed when gases merged, forming a solid core. This gas-cloud formed in a region high in heavy-elements because it was a remnant of a supernova metal-fusion from a first-generation star. A massive star about to go supernova finishes burning its hydrogen into helium, before burning the helium, and then continues fusing chemicals into heavier metals, including carbon, oxygen, silicon and at last iron. A supernova leaves onion-peel layers of different elements on the outside, as it works its way closer to the core. Most of the universe's heavier metals formed inside these pre-supernova burns. The stars' explosions scatter a cloud of elements as far as the energy of the blast can reach. This matter is reused in the creation of complex animated and inanimate components of the universe, including galaxies and animals. These post-supernova chemicals were pulled back together inside of Nefo and its new star. The most massive giants can be 79 times as large as Jupiter's mass, or near the size of a brown dwarf. They can also be as far as 9,000 Earth-distances or AU from their sun. Like most giants, Nefo had so much mass it had attracted 56 moons and several smaller satellites. Most circling objects were destined to fall into the planet, while new satellites were added from objects as big as sucked-in neighboring planets.

As usual, when they were in the slowdown phase, Doctor Nackan-8 received data and genetic materials from Nefo and began preparing the necessary research to transform Wocega-8 into a nefo. The planet was desirable because it registered as possessing intelligent lifeforms capable of complex wireless communication methods. This was curious because it was exceptional to detect a lifeform on a gas giant. Such lifeforms were common on gas giants' moons, but the giants themselves lacked land to stand on or oceans to swim in, so finding life capable of evolving in these extreme conditions shocked cadata researchers. Thus, this was an important stop, despite its remote galactic location, for understanding the nature of life.

The actual biological structure of the nefos exceeded Captain Nally-8, Doctor Nackan-8, and the others' expectations on the scale of uniqueness. These beings took advantage of the chemical group most abundant on Nefo—noble gases. Water-based life would boil away into light gases in standard temperatures of its atmosphere. Noble gases are known as such because of their highbrow or standoffish lack of reactivity with other elements. Because noble gases have full outer shell electric configurations, they are resistant to donating an electron into a bond. While they function in intricate patterns inside nefos' bodies, most of their elemental composition is inert, and chemically stable when afloat. Despite these limitations, the hot temperature, and the thinness of the air allows some elements to bind into complex structures, mirroring Earth's bacteria, or viruses in terms of their cell-count and rapid-replication qualities. Thus, the middle-region of Nefo's atmosphere is crowded with gas lifeforms, from the microscopic to nefos, which are the size of an Earth whale.

Despite this nature, the combination of chemicals, pressure, and temperature conditions in the skies of Nefo designed a setting for nefos' evolution. Their genes were dominated by the lightest noble gases: helium (He 2), neon (Ne 10), argon (Ar 18) and krypton (Kr 36), with tiny quantities of heavier gases such as xenon (Xe 54). If they were composed purely of noble gases, these creatures would have been odorless, colorless, and barely reactive with the chemicals in the surrounding air. Un-reactivity would have hindered their ability to grow and reproduce. On Earth, these gases are also toxic to life. Radon causes cancer. Kr and Xe are hazardous radioisotopes, produced during nuclear fusion. For cadatas, this group had a lighter connotation because cadatas consume gases, and our ships are coated with the heaviest noble gas, which is hyper-strong in a solidified state, oganesson. Its dense presence on Nefo contributed to the planet's position at the top of the list in the *Catalog*.

The major confusion solved with actual planetary samples was how these noble gases managed to bind together, a mixture observed elsewhere in extreme temperature conditions or in the presence of a binding agent. Nefo's low-for-a-giant temperature worked in their favor, as measurements of over 440°F would have led to weaker helium atom bonds. Helium was the lightest, and thus the commonest ingredient in nefos' genes. Xenon best attaches to hydrogen and oxygen on Earth at −378°F. And liquid helium can form a protective barrier around other noble gases at temperatures between near absolute zero or –458.41°F and –411.43°F. This is where the special chemical mixture in the atmosphere allowed nefos to break these ultra-low ideal temperatures. Thus, they could interact on the other end of the heat spectrum. The noble gases in these beings also profited from their innate ability to effortlessly bind with other molecules common in their air. These other molecules were the glue holding the noble gas cells, making them as inseparable as the cells of a solidified lifeform. Empty space is more abundant than cells in between matter in both human and cadata cells. Our bodies appear to be a single impermeable unit, but invisible gaps bulk up all matter. Gas-life takes this airiness to extremity.

Nefos' noble gas atoms were confined in cages of noble metals or porous metal-or-ganic insoluble solids, with some carbon, hydrogen, oxygen, and other binding agents mixed in. On Earth, xenon is best-suited for binding with many of these heavy metals. The conditions in Nefo's atmosphere, at around 60 miles above the planet's liquid-hy-drogen surface, were similar to Earth's. The height allowed for reactions impossible in nature on Earth. One of the metals forming bonds with one or more noble gases on Earth in extreme temperatures is gold (Au), clumping into chemical compounds such as $ArAuCl$, $KrAuCl$, $AuXe$ and $NeAuF$. Nefo's atmosphere is rich in gold among a variety of precious metals able to form bonds with noble gases. Temperatures at a certain height-range of Nefo's atmosphere are higher than on its surface, spiking at 2000°F, but this is a mere measure of the molecules' speed, rather than how hot this region would feel to an exposed biological body. The air density is much lower at this height than at Earth's crust, and even more so near the surface of a gas giant such as Nefo. This low density would generate a false low temperature on a thermometer. The particles inside of nefos' bodies were compressed by their systems into nanoparticles to cram more gas, and this also brought their internal melting and boiling points down. This way, metals such as gold could interact with the noble gases in a similar gaseous form, in which they could form bonds and process energy as combined units. Gold's boiling point at 1-Earth-at-mosphere is 5576°F, but it drops down to around 1000°F when particles are tiny. It would lack the sting of a burn with so few free air molecules bouncing outside of the mating and work clusters the nefos organized. Nefos' biology was capable of handling temperatures in these artificial clusters. The outer cell cage is ultra-thin. The combination of gas interiors and thin exteriors generates organisms notable for their gilded color and condensed-cloud shapes. While they appear to be clouds, their bodies are hyper-bound; otherwise, they would disintegrate into rain, falling to the planet. While noble gases function in intricate patterns inside of nefos' bodies, they are inert and chemically stable while floating through the thin air. Nefos control the substances inside their nebulous bodies by auto-maintaining their internal temperature and pressure in the ranges where they perform at peak-capacity.

Heightened gravity also contributed to the success of this unique lifeform. The grav-itational pull on the surface of the planet is five times higher than on Earth, but this is diminished to 80%-strength at 60 miles up, or only four times stronger than on Earth. While humans would suffer bone-damage at these levels, the gaseous species are un-affected because of their low mass. The higher gravity allows nefos' microbes to bind with ease. It also decreases their membrane's ability to resist inflow of molecules, thus allowing for a heightened production of biomass in larger and denser cells. Molecules can enter a cell if its membrane can break them down into digestible parts. Alternatively, they can penetrate with help from a component equivalent to a protein channel, which acts as a keyhole through which only needed molecules can enter. The membrane is there to block access to large, harmful, or unnecessary molecules. Energy can exit these cells once the molecules are broken and ready to be transported to the part of the body in need of them. Nefoes are also toughened by their high atmosphere's tendency to be dry and radiant. It necessitates nefos' resistance to desiccation, exposure to harmful UV light, and other forms of radiation.

After the team experimented with the production of straightforward, miniscule gaseous organisms, they commenced on transforming Wocega-8 into a nefo. Her diary is rich in curious sensations from this procedure, and what followed. I will let her narrate the remainder of the Nefo expedition.

Wocega-8's Diary Entries

66,372 BC

I have been in a continuous state of shock across this first year on Nefo. We could not have imagined life's conditions on this nebulous planet. Prior to the trip to Nefo, we suspected it was built for extraordinary lifeforms. But as we received readings commanding me to transform into a cloud, I grasped I was out of my realm. I grew terrified of disembodiment. I had practiced gliding on my cadata wings in simulations, and these felt natural, but there is a momentous leap between sailing through the air for a few moments and becoming air. For most lifeforms, when your body disintegrates into the same consistency as the atmosphere, you are long-dead. Choosing the incorporeal was akin to volunteering to fly to a distant solar system on a civilization's first attempt to break through the atmosphere. This was the first attempt to mutate a cadata's cells on this quantity of parameters: size, pressure and temperature tolerance, composition, energy processing, and just about every other dimension of existence. I can't imagine a more daunting experiment. Mass is what gives a being heft and stability on a planet. It determines if life bounces off a surface, or struggles to crawl to carry its bulk. It is altogether different to leave all surfaces to float miles above them. Beyond leaving a luxurious chair during a move to an inhospitable environment, this transformation approached abandoning matter itself.

To gain confidence in my aptitude for this challenge, I asked Doctor Nackan-8 the questions I imagined might prove why it was not a certain suicide. What safeguards were in place against my conscience failing to assemble in this cloud form? He replied: my brain would be transported last, after the rest of me was intact as a gas. If the attempt failed, he would replicate my cadata body and populate it with my intact brain. I would be ignorant of my interim dismemberment. The procedure for the eventual transformation of my brain into a gas in Doctor Nackan's pre-transition briefing was uncommon. The standard procedure was to maintain a cadata's brain chemistry intact to avoid the potential loss of intellect or processing ability due to transitioning to a brain of a less intelligent alien species. He explained the impossibility of making a genetically-cadata brain invisible inside a near-transparent nefo body. Even if it was imperceptible on the visual spectrum, nefos saw wave vibrations rather than visible light, so they would sense a strange solid object inside me, negating our attempt at subterfuge. Opting for a solid brain equated with retaining my entire cadata body, as it would be as-easy to discover. Doctor Nackan-8 reassured me nefos possessed advanced, computerish intellects. I would retain undetectable cadata qualities, which, he theorized, should be sufficient for a fruitful exploration. Another horrifying aspect of this expedition was me being dumped out from the sky, and learning to fly in this haphazard, unsimulated atmosphere to avoid crashing to the mineral crust, or floating away into the airless stratosphere. One reason my ancestors have been chosen for these transformations is because we have repeated the pilot training earning us the rank of top flight-technician on the spaceship. However, instead of asking me to master an unimaginable high-tech submarine or lava-surfing shuttle, the leadership was asking me to learn how to pilot my own cloudy body. I did my best to talk Captain Nally-8 and the others out of using me, stressing I was too clumsy and absentminded to acquire this skill. As I strained for bids from crewmembers who dreamt of this opportunity, Nally-8 became entrenched in rejecting my summons. My terror of turning into a nefo only improved my candidacy from Nally-8's perspective. A cocky pilot, without an appreciation for the challenge, would refrain from a ruckus of complaints. I had been training for thousands of years, and I was going to complete this mission even if Nally had to shove me out of the ship without my permission.

The research conducted pre-transformation only scraped the rust off the realities of being a nefo. As the scientists worked on me inside the transformation machine, I lost sensations of pain and all nerve-carried feelings. Nefos did not need to feel pain because a sharp and hard object could penetrate through their center without damaging their body. As the nanobots swooped in and facilitated my cells' mutation, I expanded to fill the container. As the nanobots began operating on my brain, a yearning awoke to receive communications on my new wireless and voiceless wavelengths, designed to receive and send messages across an inter-tribe or an inter-national network. The complete silence in space on this frequency was like being submerged in a water tank to test sound-deprivation. I sent several incomprehensible messages, trying to catch the communication officer's attention, before regrouping my thoughts and being able to recall some basic words in Nefo. With these I composed my first message in Nefo. I repeated the send a dozen times before hitting the spaceship's standard communication frequency. It was a relief to hear the comm-officer's responses and questions regarding my state of mind. Once I mastered the commonplace messages on nefos' wireless network, I managed to explain we had to direct our nanos to explore the deeper levels of code they were sending signals on. Their messages were closer to a computer program than to the symbols for tangible and intangible lifeforms on other planets. Through exploring this planet's archives, we discovered the nefo species developed nanotechnology as a biological feature during their youthful evolution. This archive was a dense cloud of neural-network materials with elongated half-lives, allowing the cloud to remain intact for thousands of years with minimum maintenance. Future generations across the planet tapped into this data. An evolutionary advantage to change their bodies was lacking because they already possessed the right chemical and physical balance to thrive; the environment also lacked predators to push nefos into a competition. Instead of such material pursuits, they spent hundreds of millions of years developing their version of internal processing servers and communication dishes linking colonies into a unified, computing mechanism. Shockingly, I was also capable of sending a grouping of my cells outside of my body to gather resources, or to transfer my genes to an external receptacle. I could perceive what these cells were experiencing as they traveled and could direct their path. They mimicked flower pollen in function as they carried genetic materials; however, plants are unconscious of the actions and perceptions of the strands of pollen they send to float across their terrain. I was blind, but the researchers described the dense cloud of my body as having a golden glow with visible interconnections or veins stringing it together. I learned to change my shape with a few limitations on centrally-tied parts, which had to remain attached in a set order. Motion across the spaceship took an instant, as I generated propulsion by creating a mini-hurricane inside of my swayed form.

Time seemed accelerated in gaseous form; thus, I lacked a full grip on a nefo's existence before our Mothership landed on the farther side of Nefo's smallest and remotest moon, hidden from direct observation from the planet. It was an elongated and bumpy-edged double-sphere. The geology of its rocks indicated it formed when two of Nefo's moons crashed and mingled. It was a sensitive landing because the entire moon was 27 miles across and 18 miles wide. We approached at top-speed to avoid being spotted as an alien machine. We were confident nefos lacked the ability to telescopically observe the universe beyond their atmosphere, but the Computer might have missed a relevant scientific paper or a secret observatory. As we landed, I listened in to the chatter between spaceship operators and the leadership team as they discussed that, as they expected, the moon's high reflective surface meant it was covered in a thick layer of snow over a water-ice surface. The snow rose in all directions as we landed, then clung back onto the surface of the moon and our ship's exterior because it was hyper-statically attractive. Since this was my first encounter with an "actual" planet and snow, I was tempted to request temporary transformation back into a cadata to watch the beautiful sight of crispy and never-before-touched snow sparkling under the radiance of an enormous, hovering overhead disk of the sun. Then, I recognized my special mode of sensing this snow, nobody else on-board could access. I sent out a grouping of my cells on a little exploratory mission. Through them, I sensed the types of molecules they encountered in the near-empty air on this moon. Then, I received telemetry and computational analysis of the snow in a language more authentic and beautiful than data from our nanobots could have generated. While they were out there, they picked up on a growing 40-foot bulge developing on the moon's planet-facing side because its relatively tiny mass was tidally deformed by the planet. These forces tend to be visible only in tides, but because of my cells' sensitivity, I felt the ground giving way here. It is shocking this moon remained tidally unlocked; I speculated it might have experienced a recent collision with another satellite, which sent it spinning again. I called the cells back before long as I sensed they were displeased with the freezing temperatures on a rock in space without an atmosphere. It was amazing they lasted as long as they did without sheathing. Doctor Nackan-8 was ecstatic about this new-found characteristic, when I reported it.

Soon after parking on the satellite, Ortack-8 asked me to follow him into a shuttle and we took off for the planet. Ortack-8 came to a hovering stop in a sky section just far enough from the planet's biggest nefo city for me to have sufficient stamina to reach it with natural propulsion. Yet, the landing spot was distant enough to avoid being spotted as an invading foreign vehicle. I sent a few goodbye notes to Ortack-8 and the crew in case of my premature death. Then, I pushed myself through the ship's wall. It allowed my cells to pass, closing behind them to keep air from getting in or out. I was terrified I was about to start falling uncontrollably, but to my relief, nefos' bodies were designed to remain suspended at this height without effort. After this momentary delay, I pushed in the direction of the main colony. Having practiced rapid-travel inside our Mother-ship, I employed these skills to embark. I used a mountain peak geographic feature on the planet below to identify my intended destination without a Computer to plan the route. I started picking up on the buzz of communications this throng was emitting long before I otherwise perceived their presence. In part because I was unable to refrain from thinking on the frequencies they scanned, I decided to announce my arrival to avoid intruding on my future hosts without a self-introduction. My worries were eased when they proved disinterested in my origin and motive for approaching; instead, they queried if I intended to join their cluster. I concurred, and expedited my flight. They had detected a strange pattern in my communications, so I was escorted telepathically to their Empress for a meeting.

The city was enclosed in a dense cloud of living organisms, lacking self-awareness, or mindfulness. They existed in states between a bacterium, and an insect: compressed into the protective walls of the city by nefos' ingenuity. I examined the city's unique resource-collecting infrastructure, waste disposal, and the methods for provision of oth-er necessities. Upon entering this realm, I had begun sensing my self-possession being overpowered by an external force. At first, the being was pulling me towards its location with more force than the three nefos who had taken me by-the-wing. As I entered her throne room, a wave of magnanimous feelings washed over my holey cells.

Empress Rina the Enormous explained in a few lines of code: she had central com-mand over the nefos in her domain. In other words, she was orchestrating their individ-ual movements across the atmosphere. Some were sent to gather resources. Others fixed the walls, or the infrastructure across the city. And a good many were tasked to spend their nights mating with the Empress: the sole reproducing female of the group. It was strangely enjoyable to forget about my typical sense of self-importance, self-preservation, or just "self", and to surrender to Empress the Enormous' demands. Everybody was swayed by her will, as if she was wind itself, as the entire massive cloud formation, which comprised the city, migrated in patterns of her design.

I performed the first few tasks the Empress assigned with outstanding marks. One of my brightest achievements was applying my cadata knowledge to engineer improve-ments in their waste disposal pipe. This strange invention made me unique in a land short on innovations because the population was focused on following the Empress' di-rections. Thus, Empress the Enormous invited me to become one of her Ladies-in-Wait-ing. I was allowed to rest at night outside her chamber, as the army of devotees marched in to lay their seeds. This hubbub exhausted her, so she snoozed through the second half of the rituals. I kept watch for wayward behavior; for example, I chased a fop out when he began decrying his boundless love for Rina. She was actively inspiring this type of love in her followers, but it was in bad taste to voice these feelings. In the morning, it was my solemn duty to bring the Empress a cloud of energizing molecules, which she inhaled for rejuvenation before shuffling her gigantic form out of her fluffy bed. She was around a hundred times bigger than everybody else in the city, me included. On a side note, if we had made the mistake of making my body empress-sized, I would have been executed for treasonous insult against the Empress' grandeur on my first day.

After breakfast, I escorted the Empress on her daily viewing of the Dazzling Knacks Exhibition, an event to strengthen the perception of the Empress as cultured and engaged with Nefo's history. I vividly recall a sculpture of a couple of nefos, their bodies preserved after death in a hugging position. Another unforgettable jam was outlined in ruby and sapphire, and depicted a powerful nefo Empress charging her head at a crowd of nefarious, tiny, malignant animals pestering the skies of a city. She tossed them in an explosive mushroom formation. An abstract composition portrayed eight nefos merged into a single creature with connecting links between their cloudy appendages. The required-stop sculpture at the entrance to the exhibition depicted a life-sized Empress, with her glorious rolls of clouds, and gem-laden adornments, sitting victoriously on a replica of one of their ancient cities.

A parade of courtiers followed around a hundred steps behind and viewed the same pieces with every expression of interest. At the end of the procession, the Empress would retreat to the Throne Room, which displayed an empty throne to the viewing public whenever she was not occupying it to give an audience. During these meetings, nobles and wealthy nefos requested transformations or consistency from the Empress' policies. The number of these rich nefoes was limited because the Empress retained ventures' profits. This unfair distribution persisted because she was charged with setting the portions. Most of the wealth she shared with this court was with her chief procreating partners (who she spoiled with affection) and the influential noble families. Most of the city's nefos were the Empress' siblings or offspring because Empresses birthed most of their populations. The Empress' court was comprised of rival Empresses and their closest advisers from cities and towns experiencing depopulation or revolt, and thus taking shelter in the Empress' domain. These supplicants' interests focused on their own territorial expansion, regaining control over rebellious workers, and the attainment of distinguishing luxuries such as shiny abodes. The last item on these nobles' agenda was the needs of the common nefos sleeping in the open air.

After a strenuous morning among her nefos, Empress Rina always changed her dress with my assistance. The fabrics preferred at the court were airy, but somehow the royal tailors managed unusual designs, in many ways freer in their radical statements and in their shapes than fashion on matter-centric planets. One of the Empress' favorite outfits had a skin-hugging, ruby-encrusted lower half, and three circular, silver-heavy, fluffy, and protruding clouds on the top half disguising the features of her face. Another dress was a pink-dusted, half-spherical cloud covering just her lower half, with precise, round holes piercing it in near-immodest places. If the Empress was planning on a tour of higher or lower regions of the atmosphere, she had a practical, parachute-like, fluffy contraption she wore, which acted as a weight to bring her down, or as an air catching balloon to raise her higher, so she could avoid straining her internal propulsion system given her extraordinary bulk. And on days when she felt extra-peckish, but had to venture out to oversee her domain, she wore a giant molecular-supplement cloud she gradually ate as the day progressed, so it had to be just large enough to fulfill the Empress' appetite before it revealed her private parts. If the Empress was planning to stay in, she wore a cloud clinging to her skin and serving to protect it against air dust, rather than against others perceiving the details of her nude form. Nudity, in general, was unproblematic on Nefo because the nefos lacked visible, offensive, or protruding sex-organs. Both sex and birth occurred in near-see-through organisms; these were a matter of exchanging molecules sprayed out and collected like pollen falling on flowers; it would be strange for flowers to be modest. Many of the dresses were designed to reflect and catch light in manners forming halos, beams of streaming light, or case shadows over a nefo's form.

Once dressed, Empress Rina the Enormous attended to her newly-produced off-spring, who were kept in a semi-enclosed playground, around a thousand at a time, until they reached maturity and became full-time laborers. She had to supervise their progress in this facility to assure they were susceptible to her commands and learning to be self-less. She would toss some of the snacks she had been munching on in a sweeping gesture, and watch her little babies scurrying on their knees or jumping by flapping their little wing-like flaps to grab the tasty bits from as high or as low as they happened to travel. They had little shelter or clothing here, but over the generations, they had invented some games utilizing the surrounding clouds. If these bored them, they were encouraged to begin learning a trade by working on a mining expedition, or gathering hazy overladen or muscle-forming carts of useful minerals from gas formations within flying-distance of their "home". Unless they were on work-release, the children were forbidden to leave the crowded, putrid, and waste-infested "playground", to restrain them from developing independent-thinking through learning from the outside world. The Empress would watch me and the others handing off her latest batch of infants to the elder kids super-vising this establishment when a new litter was produced weekly.

The next stop on the schedule was the Amusement Station, where the Empress took her dinner, as its chefs specialized in serving strange puffs, whiffs, and twirls in shapes and smells challenging to the imagination. While she ate, she watched little monkey and squirrel-like animals perform acrobatics possible due to the lightness of their forms and the unlimited all-directional movements of their bodies. She controlled these jumps her-self, making them perform little dances and gymnastics she popularized in court balls.

From there she stepped into the Globe Beating Amphitheater for a game. The field was globular, with as much up and down potential as side-to-side. There were hyper-con-centrated cloud spheres, positioned in the middle of this space. A player had to use a shield to strike a sphere in a cluster with the greatest number to propel them towards several gold, circular lines signifying winning pockets. Each of the two opposing players had their own set of balls, and lost points if they hit an opponent's ball into a pocket. It was dizzying to play this game because you just saw the bright open sky above or the turbulent, hurricane-laden surface of the planet if you were hitting a ball downwards. I bore this constant suspension because of a lifetime in weightlessness on our spaceship. A cadata who just left our home planet's gravity would have failed to adapt with grace. The Empress played this and other games by her own rules, always leading to her victory. For example, she would declare the opponent's balls were now her own, or she could hand-carry a ball into a pocket, or she could send a pet-monkey to nudge a ball into a golden enclosure with its face.

Then came the part of the day those unfond-of-work dreaded, when the Empress retreated to her study to meditate and assess the progress on the tasks she had delegated to the workers. She had to assign new tasks to some, while pushing others to work faster, and yet others had to be placed in retraining programs. Some had to be individually directed, while most were controlled by regional managers, who sent status updates to the Empress. She had to send regular guiding wireless communications to these managers, as well as to other wealthy business operators. Other than business matters, Empress Rina was also obligated to compose sorrowful letters of distress or condolence to the relations of wealthy supporter who died or suffered a crippling illness.

On one such splendid occasion, I helped to edit and transcribe her note to her cousin, Bably, who had been granted the roll of Empress for a rival city. In this note, Rina informed her relative to avoid seduction by their fancy clothing designs, or their flatteries and footsie massages. Instead of being too timorous, betraying her immaturity, Bably was to be decisive in her orders to keep her Empire from spiraling into anarchy. The Empress ended this note by suggesting to Bably a new version of her Empire's flag; it had been a cloudy color and shape combination including the image of a popular food source creature in her region. Empress Rina lectured, using animal symbols was outmoded and thus showed poor taste. A design with the simple lines of squares or circles would be easier to identify in an inter-city conflict. Frankly, Empress Rina was a bit jealous of Empress Bably's luxurious, new-construction cloud-castle, which she had considered replicating in her own city before judging this luxury-level as beyond her means. Empress Bably did not perceive this rivalry, and followed Empress Rina's advice, altering the flag to a series of gold balls on a fabric of wavy ruby lines.

The schedule was tossed into disarray ahead of Empress Rina's 210th anniversary of accession to the throne. The tailor had to be advised on the most suitable gilded style for the Empress' robe, sash, and waist-circling crown. An army of sculptors were ordered to create a new statue of the Empress highlighting her powerful features and showing her in a conquering pose, with thousands of tiny followers clamoring around her feet. The arriving empresses and foreign dignitaries had to be housed in their individual cloud-suits in the castle. A new set of recipes had to be invented for the supper to display Empress Rina's extraordinary, innovative power. And the directors of the night's entertainments were putting their crew through repeating dress rehearsals to avoid all possible mishaps on the stage.

The event began with a procession of dignitaries from their upper-level restful rooms down a light whirlwind and into the spacious main hall. The size of structures had little bearing on the cost because architects could dilute humongous structures with airy construction materials to the point of inperceivability. Each of the empresses making a grand and whirling entrance was wearing a costume defying gravity and the power of their designer's fancy. The empresses sat on flowering golden chairs of various heights, symbolizing measurements of each ruler's relative governed population size and wealth. Empress Rina was last to arrive, and took the tallest petalled throne. Instead of chatting with Empress Bably, who sat on her right, Empress Rina stared straight ahead with her perception pincers, as if busy directing the servers busying with serving delicacies. This long stare served to push Empress Bably out of her comfort, and to make her look still more timid and respectful towards Empress Rina than usual. A troop of muscular generals was also standing behind Empress Rina's seat to further intensify her intimidation factor for everybody in attendance. The entertainment began with a firework display achievable solely in this bizarre, gaseous environment. The fiery, multi-colored explosions shot out of a cannon and transformed as they burned into shapes resembling animals, famous nefos, and finally the Empress herself. Acrobatic dance and theatrical numbers followed. Just as the guests finished their desserts, the Empress sent a buzzing wireless signal announcing she was ready to speak. She then gave a lengthy speech on the past hundred-year's ambitious and never-before-seen achievements. She squeezed in every construction, sewage, renovation, and cleanup project she oversaw. There was one for each day of the reign. Thus, the speech stretched into five hours: a harsh timespan to endure when even telepathic sounds and movements in seats were forbidden as offensive. When she proceeded to the second half of the elapsed year, we heard a whizzing sound and turned our antennas upwards in its direction. A fire ball larger than those fired in the fireworks display was shooting out of the stratosphere and heading straight for the castle. It broke through the clouds composing the roof, whizzed three feet away from Empress Rina's flowing gown, and then shot through the grayish floor. It caused a miniature, smoky explosion upon collision with the crust, before sinking and melting into the watery hydrogen layer. I was afraid the Empress would be outraged at such an intrusion, but she started beaming with delight, and sent an order to her people to engage the standard meteor shower procedures. There were several meteor showers occurring on the same day every year, and then there were some of these spontaneous showers that took place because a comet had lost some of its shell from the sun's heat. Another meteor shot through the hall, penetrated a servant, and continued toward the planet. While most of the retainer's molecules repositioned, the fire burned through essential bits, damage to which caused him to drop the gas containers he was carrying. He was disabled from further court-service. Meanwhile, a crowd of Empress Rina's miners stormed the hall and set up trap concoctions. Their rough appearance distressed the wealthy guests. Facing the threat of a meteoric impact was not the party they had envisioned. So, they dispersed from the castle to their respective jurisdictions, assuming Empress Rina ended the celebration when she invited this riffraff in. Empress Rina acted as if she was venturing on a spectacular hunt, as her workers commenced fevered running towards the approaching meteors to catch as many of their molecules as their airy nets could hold. Across the planet, these regular meteor showers served as a replenishable source of resources; otherwise, the thin air lacked the minerals needed to build the grand structures Empress Rina imagined. These meteors sparked molecules, burning in the atmosphere. These were the initial ingredients for life's formation on the nefos' planet at that height. Catching chunks of matter with metallic strings increased in sophistication in rhythm with its assistance in growing empires' wealth. These meteors originated in an asteroid belt circling Nefo, which fed the planet's hundred-rock meteor storms. The excellent hunt generated enough precious stones for Empress Rina to reconsider the delayed expansion of her castle. Thus, she was in a heightened mood, as she retreated to her rooms, with me just in front to dust the lingering meteor-smoke out of her way.

66,321 BC

Back on the Mothership, the 52 years I spent on Nefo weighed heavier on my restored cadata psyche, than how they felt to an obedient nefo. It is difficult to accept how our Nefo experiment ended. I struggle to identify the mistakes that veered us into disaster. For the sake of my self-esteem, I hope our fate turned because of Hurricane Hunay of 66,363 BC, just as my first decade on Nefo closed without major genetic or pathogenic damage to myself or the nefos. When it struck, Captain Nally-8 was preparing half of our ship for genetic editing, aiming for them to inhabit a series of cities for the multi-cultural stage of our exploration. If they had commenced, the death-toll for cadatas would have been horrific.

While water takes up 70 percent of Earth's surface, the chemicals making life possible on Nefo are spread across its atmosphere. All organisms on Nefo have biological processes for the dispensation of noble gases instead of liquids, so rather than excreting sweat and swallowing water to moisten their cells, they seek out the particles in the air their bodies can turn into energy. It goes unnoticed by nefos in Empress Rina's city, but the whole city is in perpetual motion across the sky as the midday wind currents push it at 300mph towards the other side of the planet, before dying down to 50 mph, only to pick up again on the following day. Similar wind velocities were common on Cadata across most of its history, and they allowed cadatas to glide between otherwise isolated islands. The fabric of both cadatas' and nefos' bodies are adapted to high winds. For the latter, strong wind can pass through their airy skins, or push them along without causing harm. This is because nefos lack a sensory system to detect an excessive force pushing them across the sky. These wind-tolerant features have an extreme limit stretching to the highest wind speeds nefos experienced across their pre-history. An extreme wind event can harm their infrastructure and bodies if it breaks records. Untested wind extremes were outside nefos' natural evolutionary adaptations.

We began monitoring messages from nefos living on the other side of the planet a couple of days prior to landfall. The communications were from those who were just outside the strongest center of the storm. They were still seeing devastating winds ripping apart their otherwise wind-resistant constructions. Hurricane Hunay began so far from us Empress Rina's meteorological team was certain it would break up in the soup of gas separating us. It avoided rolling over major cities, merely hitting small towns during this trip. The royal scientists were startled when Hunay picked up strength and made a rapid spiral turn, bringing it in half-a-day to the edge of our city. Pink liquid rubies rain started falling in the evening of first day. Empress Rina refused the evacuation, as she anticipated her nefos would be essential for minimizing the losses and cleaning once the storm subsided. The first time I was in a storm on Nefo, it was startling because the billows of condensation arrived in thick shadows over our heads, but also went right through the city, and danced far below our feet. Unlike their standard heavy winds, storm-gales undid loose design elements from buildings. On average-sized rocky planets, the big hurricanes fail to climb above 50,000 feet (or 10 miles), but on Nefo, because it is a gas giant, they are mega-thick, reaching up to 70-miles above the mineral crust, or well above the region where most of Nefo's cities are stationed. Hunay was the most aggressive storm we witnessed since the original Nally Subsenar began the study of this planet tens of thousands of years earlier. If we detected a similar storm prior to landing, we have skipped stopping on Nefo.

Next midday, the winds began accelerating, and we felt Hunay's power. I even began questioning if it was a force of nature, or if Hunay was a conscious being. If there were so many animals of varied intellectual capacity composed of air here, perhaps a giant series of clouds also had an intellectual awareness. It seemed to have marched across the planet to hit us with malicious intent. Empress Rina laughed when asked about the clouds' scheming powers. These types of speculations were fodder for children's tales, and I was to seize wasting time in fantasies. It was difficult to avoid straying into my imaginings as Hunay began lifting feeble-tied or open-field goods and animals by the time the sun moved to a 45° angle. The full impact hit us by sunset, with violent winds of up to 429mph. Even some of our wireless transmissions were difficult to hear at this intensity because of the interference. It was as if Hunay was tearing the fabric of the atmosphere open. We hid in a saferoom in the central layer of the castle with Empress Rina and top nobles and officials, as the metal-lined columns of the castle began to give way under the pressure. Within a couple of hours, most of the castle was demolished, except for most of the room we were in, and the foundation. But this was not the end of it, as winds peaked at midnight. We drifted into sleep, as winds waned across the night, returning to normal speeds just after daybreak. Empress Rina was disheveled as we rose to study the aftermath in the breaking light. Despite her ruffled attire, she called a meeting of the surviving powerful and moneyed in the same room, since we lacked certainty in the structural integrity of other potential locales. A dozen runners sped across the city to collect death and injury statistics, submitting these telepathically to Treasurer Tectup, who drummed them out to the rest of us in a verbose report with annotations and historic comparisons. Hunay had won a 50,000-year record, with a death toll of 3,463 nefos in our region. The homeless left exposed outdoors failed to find a reinforced structure to hide in and died. There were even some casualties inside of the castle, as some of the courtiers refused to leave their rooms for the bunker. As we were trying to comprehend these staggering losses, good news entered in the form of a royal, who had been counted as "missing, presumed dead": Prince Rimmol. He recounted how he was caught inside of a small room lifted by a gust and carried at top-speed hundreds of miles from the city. He was unhurt by this violent relocation, and managed to return as soon as he could overcome the strength of the wind. Rimmol was invigorated by this misadventure and offered his assistance with the reconstruction. Empress Rina was less enthused by the prospect of an immediate commencement of the rebuilding effort. She already deployed the surviving laborers to pick up litter, reorganize the remains of the buildings, and commence reconstruction of utility structures, without which life in the city would have dimmed within days. But she hesitated to order the full rebuilding of the city's houses, businesses, and inessential adjuncts. Through mind-control she convinced them they deserved only the most essential reconstructive projects for keeping them alive; once nutrition and weatherization were assured, the bulk of the rebuilding effort was dedicated to the construction of a new castle for the Empress and her court. This project stretched on across the following decades, while her nefos experienced heightened starvation and illness rates. As nefos were required to perform hard labor, the Empress' troops dawdled in the same busy-work without refreshing their military training. They also failed to utilize this time to rebuild the broken walls meant to protect the city, thus weakening their attack-vulnerable position.

Once the immediate impact of Hurricane Hunay was assessed, the half of our spaceship scheduled to undergo gene-editing proceeded to have this procedure done. They then settled across most major cities of Nefo, with a couple (including Ortack-8) joining my own city. I missed that little fellow. We had exchanged messages in side-folders to my main cadata progress reports across my first experimental decade. I expected the others would love Nefo as much as I did, but within hours of being released and summoned to join their respective Empresses, our cadatas began reporting dissatisfaction. Most were assigned common laborer duties, which kept them engaged in all light hours with cleaning, building or otherwise enriching a city, where they were not even assigned a bed to sleep in. Perhaps because the tasks were strenuous and sonorous, a majority felt offended by surrendering their will to their respective Empress. I lacked such concerns, and described a freeing sensation of surrendering my desires for the greater good. In retrospect, I should have pushed against these feelings of tranquility, and explained the social hierarchy with more clarity in my official reports. Surprised by negative feedback, Nally-8 froze while deciding if on whether to treat this as a major political and philosophical problem. The cadata leadership team back on the spaceship spent countless meetings discussing if free will is irrelevant when cadatas' survival is at stake. The question was tabled until they could gather sufficient data to understand the impact long-term mind-control had on a large sample of cadatas.

These test decades dragged for most of us. Despite spending the previous thousands of years locked in a single spaceship, by 66,325 BC, those remaining on the Mothership and those exploring the planet experienced nerve-strain. Cadatas were split in two. Half was asking Nally-8 to call the experiment a success and to send a transmission to Cadata confirming Nefo was suitable for a full cadata resettlement. The other half deemed it a failure, and called for a new multi-generational relocation. To help answer this puzzle, cadatas on the Mothership decided to pitch in by engaging in a competition, the winner of which would invent a method to survive in Nefo's atmosphere unedited. Their body-hugging spacesuits were allowed. The task was to invent machinery to suspend them in the air 60-miles up, while collecting the needed nutritional gases, and providing for cadatas' needs in these harsh conditions. They began running experiments in an uninhabited section of the sky. They began working on this problem in the Lab on the side of the ship still acclimated to the bodies of the cadatas turned into nefos. Scientific and mathematical questions had to be overcome. They had to test elements such as their standard spacesuit's reaction to the strange gas interactions promoted by this atmosphere. They discovered the nanos and molecules in this suit had to be mutated not solely to tolerate high heat and pressure, but also to stop them from disintegrating because of spontaneous reactions with the chemicals in Nefo's air. Once the chemistry was sorted, they began building a suspension-in-the-air system requiring minimum energy consumption, while keeping a cadata steady, in line with the regular air waves, to mimic a nefo's seamless motion with it. Building a propulsion system imitating nefos' ability to move without turbulence at extreme speeds in any direction proved difficult because these speeds created high-g's. These had to be counter-balanced in liquid, or solid creatures, while they were imperceptible in a nefo without a nervous system.

Once Lab tests met minimum safety requirements, this group of rivals began experimenting inside of the actual atmosphere. The first such test ended in a cadata tumbling towards the planet, before he was caught by a teammate cadata who had designed a powerful jetpack device allowing him to maneuver with sufficient control. The second set of tests ended when a cadata was scratched in a hubbub by a rival. The friction created enough stress in his suit to commence rapid decomposition, halted once he returned to the shuttle jump-off point. The next jumps were less catastrophic, as each team worked to perfect their design. This creative bustle might have birthed solutions for the daunting challenges of a relocation of unedited cadatas. However, in their excitement, this group failed to appreciate the intricate intellect of the nefos. An occasional lone cadata dropping from the sky to join a colony failed to raise alarm. Then, they were hyper-alerted by an occupation of a dozen battling, communicating, and helter-skelter-moving cadatas imitating a series of bombs disturbing the roof of Nefo's atmosphere. I sensed a strange buzzing without understanding its implications. Empress Rina, Empress Bably, together with most empire-leaders understood these signals as intrusions by an alien force. The signatures these intruders were making were chemically, biologically, and technologically foreign to the realm they dedicated their lives to telepathically studying.

Upon discovering these restless frequency spikes, the empresses launched a joint attack against the mysterious enemy disturbing their peace. The nature and hostility of the invaders was inessential. The greatest obstacle was their advanced minds blocking all telepathic attempts. If the empresses failed to show their strength against these invaders, their ruling positions would have been weakened. Empress Rina was terrified of doubts in her leadership because her city was still in shambles after Hunay.

When the empresses' combined armies struck, a single unfortunate cadata was outside the shuttle, doing flips in a turbo-charged spacesuit. Computer Programmer Nepow was hopelessly swept up by the gust of wind the nefo-army generated; then, his suit was impaired by an explosion of chemical gases a foot from his position. They failed to anticipate this abundance of reacting and burning chemicals, so despite advancements, the suit failed and Nepow perished. His death fired up the others, who watched with hatred as the nefos attempted to disturb their shuttle with similar gusts and explosions. The entirety of the cadatas' spaceship fleet was impervious to the strongest winds on any known planet; they had been designed to fly through lava and to skim the surface of suns, so explosives were ineffective. The continued close-proximity bombardments eventually snapped the experimenters' patience. They fired back at these troops with triple gust-force. Then, they tossed retained chemicals from their ship's engine coating. When the wind and smoke cleared, most of the nefos' army had perished. Overcoming speechlessness, these cadatas queried Captain Nally-8 on the best maneuver to undertake after winning a battle commenced without her permission. She discussed the options with her War Council. Some proposed playing possum, taking the shuttle back into space, and hoping the empresses would refrain from investigating the matter further. Others argued they had to send an apology and an explanation to the empresses for these hostile actions, stressing this was the work of a few rogue cadatas, and perhaps offering them up for punishment. The winning argument was for utilizing the victory to gain control over these tyrannical rulers. They had to place strict sanctions on the empresses under the threat of annihilation to demonstrate cadatas' might. The empresses' shock at losing a battle to an outwardly inferior, solid lifeform induced them to allow cadatas to direct their political decisions.

Across the following four years, nefos' extreme form of totalitarian government was toppled. Under the leadership of engrained-among-nefoes cadatas such as myself, we coaxed, or ordered the empresses to stop controlling the minds of their subjects. Restrictions were placed on mind-control. When an empress was caught using it, she would have a device placed on her head to block controlling communications. Instead, the empresses began *asking* their subjects to perform the labors needed to keep the cities running. While this freedom promised a new democratic society, we learned (absent the strings of control) these laborers desired freedom, and wildness in the open sky. A castle or walls were inessential in keeping them safe. They found adequate nourishment in the open air. As a result, the cities were abandoned, and turned into disintegrating wastelands. While these wanderings were harmless, these free individuals also turned hostile upon crossing paths with other lifeforms, fighting to eat, or to dominate. We tried to remedy this state by giving mind-control powers back to the empresses. But the lack of telepathic contact in the interim years had degraded their abilities. And the distances between them and their herd also made a reintroduction of tyranny impossible. By 66,321 BC, we were overseeing a gas giant, with a once controlled, and intelligent lifeform hovering over it, which was now infertile and mindless.

The loss of Empress Rina's sway over my mind exceeded my emotional limits. If Doctor Nackan-8 had refrained from reassembling my body back into a cadata a year into this isolation, I would have gone wild myself. It is hopeless to imagine settling this dangerous wilderness. We must sail forward.

Chapter 11

A Planet Sailing Through Deep Space

The decision was made to search for friendlier skies. The team picked the nearest and most promising world, Bhasab. They began the 5,375-year journey, which took them to the other side of the Sagittarius Arm, in the direction away from the galactic bulge. Following a changeover in generations, they began deacceleration in 60,947 BC. Wocega-8 had died in 65,398 BC. She was succeeded by Wocega-9, who was now advanced in age, nearing half-a-millennium from her lifespan limit. Her age prompted a minority of the crew to campaign against her undergoing gene editing. The majority overruled, arguing it was better to send a wise cadata, than a newborn.

Bhasab is a free-floating terrestrial planet. A rocky Super Earth hurtling at a third of the speed of light in orbit around Milky Way's center. No sun, black hole, or any gravitational body directly holds it, unlike typical planets traveling in tight loops. The universe contains equal portions of loose, and bound planets. When a sun explodes, its planets are either vaporized, or tossed out of the system. There are numerous violent events capable of dislodging a planet: a collision between neighboring planets or with a moon, or a disruption in solar balance due to a sudden loss of one of the stars in a binary system. What is rarer is for a planet to maintain a habitable zone-distance for billions of years, as Earth has done. Thus, Bhasab's sun-lessness was unattractive to our scientists, unlike the detection of life and rudimentary technologies in its environment. Space-aviating planets can approach absolute zero if their internal activity stops. Excess heat can kill life on planets; thus, gas giants and brown dwarfs maintain uninhabitable temperatures even without a parent star. The life signs Bhasab exhibited had to have helped it to maintain a thick atmosphere even in the void of space. These biological processes were assisted by Bhasab's internal radiation from the molten core's geothermal radioisotope decay and gravity-induced friction, and its large but not humongous size at 1.4 Earth radii, with a surface gravity of 2.1g. This means a human's weight there would be twice heavier. Bhasab is a high-density planet with a large, iron-dominant core, high in metallicity, or in chemicals heavier than hydrogen and helium, such as silicon. It is split into similar portions of rocks to metals with Earth. Heat bounces back into the planet because the dense atmosphere, rich in oxygen and hydrogen, is putting pressure on and thickening far-infrared radiation. A deficiency in sun-originating radiation is a drawback. This absence is also advantageous, as a floating planet avoids impacts from a sun's ultraviolet light, which would have degraded or blown away from the atmosphere lighter elements such as hydrogen. The greenhouse effect would have decreased despite the gravitational forces exerted by a Super Earth-sized planet. While a free planet's capacity to rotate without a sun is counter intuitive, this category rotates faster because they are unrestrained by tidal friction. All objects with angular momentum in the universe rotate. Bhasab makes a full rotation in around four hours; this spin generates a geodynamo-created magneto-sphere, shielding it from head-on asteroid impacts, and further reinforcing atmospheric retention. The planet is also enlivened by the heightened volcanic activity across its sea-floors and ground, and by the resulting rapid and powerful movements of its thin plates as they are pushed aside and together by the stress of new eruptions. The hydrothermal vents in the sea, which are a byproduct of water seeping down into cracks in the crust, were prime real estate for the sea-life survivalists of the first million years in interstellar space, before they developed more advanced adaptations. Another astronomical ingredient distinguishing Bhasab is its three large moons, which further heat the planet by pulling it in different directions with their gravity, thus generating pressure resulting in heightened geological tidal forces, among other beneficial reactions.

Bhasab formed during the gas phase of its solar system, out of the protoplanetary disk. Life evolved at its natural rhythm across the following 7 billion years. All this time, Bhasab's sun was moving closer to a black hole; when it reached a gravity-sinkhole point, the hole swallowed it. The violent spin the sun suffered as it circled the black hole while falling into it created a hurtling motion on the further-out planets, which dislodged and were tossed out, either into the hole, or out into interstellar space. At speeds above a third of the speed of light, Bhasab and the other catapulted planets are among the fastest mass-holding objects in the universe. Mysterious massless entities can exceed light's speed limit; while light itself travels at none-other but its own light's speed. The jolt of expulsion from their spot in their solar system, the unstable fluctuations in temperature and untold environmental catastrophes meant a mere fraction of lifeforms survived until the tipping point when Bhasab reached a new equilibrium. The species on the planet, when cadatas reached it, had evolved in these open-space conditions across a 250-million-year journey. Moving at this steady rate and without major "intelligent"-life-induced environmental changes, a planet such as Bhasab can retain a constant surface environment, so cadatas estimated they could settle here for at least three-quarters-of-a-billion years, while they contemplated more permanent alternatives.

The nanobots and robots flew ahead of the ship to the planet as usual. However, their computers began sending back strange, contradictory evidence requiring a cadata scientist's analysis. Back on Cadata, Nally had detected intelligence-designed, biological, or technological signals emitted from Bhasab. Despite this promise, the 3D videos the scientists reviewed lacked animals or any moving, talking, seeing or otherwise active and perceptive lifeforms capable of developing intelligence. It was a beautiful, lush planet with a variety of vegetation and similar geological features to Cadata. What could this absence of moving beings be implying? The scientists proposed conspiratorial theories. One proposed, beyond detecting cadatas from a light-year away, bhasabs' advances allowed them to cloak this espionage. The top computer programmer argued they might have boarded spaceships to hide in a masking segment of the moon once they were alerted to their approach. The evidence against the latter was the continuation of signals arriving from across the planet. The nanobots failed to identify signs of life upon zooming into targets. Doctor Nackan-9 became convinced the bhasabs were invisible to their eye, but rapid and random samples of air around the life signatures were taken without spiking any complex genetic readings. Then, after a long night of scrutinizing every shot, a lab technician sped up a video to arrive at a segment curious because of its depiction of wind acceleration. As the rapid images began pouring in, the tech was mesmerized, staring with awe at an otherworldly scene; yet, when paused, it was the same time and place. The plants in the shot were crawling, climbing, stretching, searching, catching, feeding, drinking, and making sounds, which registered as languages by the Computer as it commenced deciphering their meanings. It was a vibrant ecosystem rich in diverse and strange life. This lifeform moved slower than an animal-type species. Evolutionary science demonstrates how, unlike these creatures, animals compete in violence to survive, necessitating increases in speed and strength to triumph. The term "plant" is troublesome in this context because these plants lack a sun to trigger photosynthesis, and display strategic behavioral intelligence unlike this branch of life in its broader galactic definition. Before the tech concluded the report, the Computer dissected the patterns in this species' communications, now detectable at the proper speed. The Computer also ran a pre-programmed subroutine, which built a unique translation tool to foster inter-species communications. As the cadata crew gathered to listen to the first translations, the contents of the plants' discussions appeared trivial; despite this, the scientists were delighted to have solved the conundrum of the missing intelligent species. They were sending wave-messages across great distances, but the nanobots failed to locate significant scientific or technological knowledge. They possessed the skills required for the springs of intellectual evolution: the ability to use tools, to work in groups, and to perform rudimentary math. This was not the cloaked spaceship revelation cadatas were hoping for, but it justified the journey to this idyllic planet.

The scientific team's study of life on Bhasab proved enriching, as they benefited from a category of organisms generating their own food from their environment. The capacity to spawn nutritional resources from the dead has been handy across the rest of the Mothership's journey. The hot underwater volcanic regions were not the sole sites where primary life survived. The other major springing point was in the crevices of rocks and in dirty puddles. Small microbes, such as bacteria-like organisms, had always lived here, and without much competition, they expanded into multi-cellular, complex, moss-like beings. The mosses started excreting greenhouse gases, which sparked planetary warming, until the intelligent and diverse sea plants emerged out of the water and found plenty to consume among these mosses. Without radiation from the sun to spark feedback between water and air, these species relied on chemical reactions to find energy sources in bits of simple elements or other lifeforms. Unlike mosses, most of these plants consumed and utilized water. They also mimicked plants because most lacked a nerve sensory system, akin to plants on Earth. There were plenty of battles for survival on Bhasab between species, but the ancestors of these species had few such rivals, and so had not evolved this sense to avoid danger. Instead, these plants generate chemicals responsive to proximate objects of interest, such as toxins to poison enemies or aphrodisiacs to attract pollinators. They send wave transmissions to neighbors whenever their chemical sensors detect a threat or a friend. Their type of sensory system also includes mechano-sensing of pressure signals and smells. Substances exhuming a strong smell, or touches or presses into their skin generate hormones inside the affected plant. If the smell or touch registered is associated with danger, the plant produces repulsion chemicals or violenter responses necessary to meet the threat.

Their variety and complexity were surprising given the time since they had emerged from the oceans. This rapid evolution was a symptom of a hyper-nutritionally-rich environment with more resources than their exploding population could have consumed in the next billion years. It would boggle the imagination to attempt to explain the range of forms the Bhasab plants took. Their shapes are extraneous to the leaves, spikes, flowers, branches, petals, and other units of flora found on Earth. The limits of the English language mean there are no words for undefined and unrelated categories to Earth's Tree of Life. Thus, I must employ these narrow Earth terms across the rest of this chapter, including in my translation of Wocega-9's diary. Meanwhile, on the chemical level, the presence of these creatures meant Bhasab's atmosphere was rich in oxygen (their waste product), a necessary gas for both humans and cadatas, and thus another point bringing Bhasab to the top of the habitable-world list. For a species accustomed to life on a solar-bound planet, the constant night on Bhasab would have been disturbing. Because cadatas were used to seeing artificial light inside their spaceship across their journey, the blackness was less familiar than a bright glow disappearing in cycles. The species on Bhasab adapted to this lack of light by evolving vision on the other ends of the wavelength spectrum. Thus, they could detect heat radiation, vibrations, and trickier changes in their environment. Since they were all blind, by Earth-standards, their skins lacked the types of bright-color displays Earth's plants and animals use to communicate with those they wish to repel or attract. When cadata scientists shined lights on them, their skin had some color variations because of its chemical composition. But these colors reflected veins, muscles, or other compositional variations, instead of serving as decorations, or threats. Scientific research into the tissues of these plants uncovered their ability to raise and drop their internal temperature by using their food to power this change if the region around them grew too cold or too hot for their taste. Unlike Earth's plants, these creatures' tissues include a unique muscular and bone structure to support them against the heavier gravity on Bhasab, which would otherwise cause them to drop posture-less to the ground. They possessed a skin layer with advanced mobility and anti-moisture-loss protections. Their digestive system contained a myriad of components performing specialized functions such as carrying, breaking down, separating, and burning the new water and nutrients. Most of Bhasab's species had advanced intelligence and biological systems. Like on Bhasab is impacted by its backwards temperatures. The main heat source is the radiation inside the core of the planet, rather than sunshine shooting down at it. So, plants tend to compete for space just below the ground, more so than trying to become the tallest tree in a crowded jungle. Still, since there is barely enough room for everybody at the ground-hugging level, plants spread their tops to varied heights to stretch their muscles. The lack of a light source above them also means the creatures on Bhasab are less straight in the skyward direction. They are dominated by branches, roots, trunks, flowers, and other parts, moving in wild shapes towards warmth, food, and water. On average, Bhasab's plants are higher than Earth's because its soil is richer in nutrients, allowing for rapid and consistent growth. The tallest plants on Bhasab are those that find regions without competing plants, so they can consume all nutrients, allowing unstoppable growth potential. These loners tend to develop on isolated, distant islands, unreachable by the pollen dispersal method. Another reason for their longevity (some living to 3,000 years) is the consistency of the weather. Without a sun, Bhasab has a single season, which echoes Earth's late spring. Without a winter, plants do not semi-die every year. Thus, their growth is unstinted by these types of retractions. Their leaves do not fall off. Thus, they can keep developing, and expanding, instead of regrowing.

Once the general biological rules dominating Bhasab were understood, they decided on one of the species for Wocega-9 to model. These bhasabs, as cadatas decided to call them, were the most powerful, aggressive, and strategic plants on the planet. Unlike most species satisfied to live in isolation, or in gentle competition, the bhasabs waged continuous warfare with rival forests over boundary lines. They also fought with members of their own sector over a type of taxation. They had the largest relative brain size, and were capable of more rapid exploratory movements. Given the strangeness of a transition from animal to plant, I will allow Wocega-9's diary to describe what ensued.

Wocega-9's Diary Entries

60,941 BC

Even before ascertaining bhasabs' properties, I anticipated undergoing an unprecedented transformation. My ancestors turned into gases and sea creatures. So I assumed the bhasabs' physiology would not surprise me. I feared low odds of success on this planet, after nine generations of failure. Why are we discounting planets with similar genetic elements to our own after merely two plagues? We might break this rule on Bhasab, which matches our origin-world in its air's oxygen composition, and similar, lighter chemicals in its dominant species' genes. This breach is troubling because if our science team opted to risk crew or Bhasab's life-extinction, they should have chosen a world attached to a solar system. But no, here we are testing a planet so restless, it fled into the void. Then, I was given an illustrated and annotated report on the bhasabs, with a biological breakdown of how life would be different once I was a plant. Becoming an immobile plant is a recurring nightmare for pilots. And traveling at an inch-per-minute speed would madden even a housewife on a shopping trip. The first virtual simulation of becoming a plant was traumatizing. After emerging from it, I begged Ortack-9 to take the assignment. Doctor Nackan-9 attempted calming me with the argument that once I became a bhasab, the slow pace would feel natural due to changes in my internal-clock. Animals yearn for movement. Even inside our bodies, our organs beat, lurch, pump, and churn. Animals are driven by instinct to continue to the next step in a hunt. Nackan-9 elucidated, the slower pulse of water trickling across my trunk would set the tone for the speed-of-movement innate on Bhasab.

For the transformation pod, Doctor Nackan designed a plant pot, and filled it on the lower end with moist, molecule-rich dirt and rocks from Bhasab's soil. Once I squeezed into this unit, the procedure began. Because of shrinkage, by the transformation's end, my body disappeared in the dirt. As the change happened, I felt my body shrinking and changing into a fibrous, dense consistency. Darkness fell as I sank into the ground; I lost my sense of smell, sight, and awareness of the temperature and touch against my skin. Then, I realized there was a pocket of nourishing nutrients my-length away, and before I reached a conscious decision, I felt a little root shoot out of my lower end and stream through the soil toward this treat. Just as it was near the object, I felt communication vibrations. Our linguist trained me in bhasabs' communication wavelengths. As

I strained to translate the meaning, they sounded like high-pitched, rapid squeaking. I kept reaching for my dinner, as communication seemed to be failing. This went on for a while before the speed changed, and the linguist announced he forgot to turn on the slowing translation feature. I sent a weak little signal back to confirm the message was received. He spent the next few hours working on improving my messaging abilities until we could carry on a comprehensible discussion. I asked why I was submerged in the ground, and he reminded me of the section in my transformation manual I had skipped, which indicated I would be "born" as a seed, so my sudden arrival in the bhasabs' sanctum would appear as a natural migration of a wavered babe. At least, I had retained my cadata brain in a condensed size and with deliberately slowed synapses. In the following days, my body developed in accordance with bhasabs' natural rate of growth. I initiated sprouting roots to pull my weight through the soil. When the tip of my top root broke the cover, I was overwhelmed with the "sight" of the Lab, the technicians and Doctor Nackan-9 at work, the smells of the chemicals they were testing, and the other readings my body was interpreting. I announced my arrival to the communication specialist in the Observatory, and this message made it to the science team, who crowded around to watch what the baby (me) would do. My negligence in reading the fine print for this experiment also meant I missed how I would fly down to the planet from the moon where the Mothership parked in this tiny and semi-maneuverable form. I refrained from reading the mission report to silence the temptation to jump out into space. I drowned this urge in an excess of slothmoths.

With so much free time in the soil, I began questioning if I would be capable of flying a shuttle from the moon to the target site in this debilitating condition. Our top engineer entered, at this moment, to demonstrate his proposed shuttle design. He placed a pod twice larger than my length on the ground an inch away from me. It was a circular, bumpy seed with a miniature parachute at the top. Doctor Nackan-9 explained this was the shell bhasab babies were born in. Instead of a mere disposable piece of light bark used to float between regions for dispersal, my shell was a miniature version of the shuttle. Seeing this little nut, and imagining the prospect of flying it through an explosive atmospheric entry terrified me. This was beyond past experiences guiding the Mothership by a black hole in my tenure as the chief pilot. I asked how its size could be securing, even with the strongest coating in the universe, against the extreme speed of this voyage, which burns to-cinders hundreds-of-feet-wide asteroids prior to reaching land. The engineer went off on a lecture about aerodynamics, how a tiny size creates less friction, how the parachute at the top was a propulsion module running on hyper-powerful fuel capable of bringing the unit to a rapid-stop, while maintaining safe g's inside… He finished the lecture, and instead of waiting for it to play at my speed, he left the room to attend a pre-flight meeting. I strained to maintain focus as I listened to this drivel alone. Then, we executed two simulated test-flights across the moon. The programmer entered the trip trajectory, speeds, and other elements about these trips into the mini system. To spare me from having to take manual control. We did test it inside for how I might be able to direct the manuals with my tiny baby sprouts, but it was tricky. Making 3D shorthand signs, with my tiny tentacles, to change the terminal destination, in case of a spontaneous storm-development, or another disaster, took me at least ten minutes. A period equivalent to the entire flight from an average moon to a planet's surface. One solution the programmer and engineer agreed on was slowing the descent to an hour to allow time for me to redirect the seed-shuttle if the on-board Computer and the Mothership's remote link crashed. They also ran planet-specific scenario subroutines to anticipate a range of piloting disasters. When natural training to expedite my sprouts failed to amass sufficient results, we considered granting me super-speed, or a regular cadata's speed of movement. We decided against this because perception of the slow life on Bhasab at cadata-speed would have maddened me with boredom.

Since I had reached the weight-limit for a seedling to fly its little parachute through the air, the team packaged and released me to attempt a flight. I was blind to the exterior beyond the shuttle because my perception of objects was based on chemical and vibrational senses, and empty space is lacking in these. Intellectually, I imagined the blackness. Then, I pondered how we were not slowing down to zero on Bhasab as on other planets because of its motion at a third of light's speed. In fact, if it had been moving in the same direction as our approach, it would be impossible to catch it without accelerating beyond light's speed. Then, I contemplated what the lack of days and years on Bhasab meant. If these species lacked perception of visible light, the time of day, year and season were inconsequential. Just the same cool, moist, springy weather across the planet, without poles or equators in the extremes. I had studied a spatial map of the planet's surface the lab techs created, with multi-chemical features perceivable to my sensors. I knew I was about to land on the largest continent on Bhasab. It was covered in forested hills, outside "walking" distance to the nearest oceanic beach.

These reflections flashed by, and I was starting to feel as if this voyage would be humdrum, when the Computer sent an abbreviated alert message to me—we were heading into an unforeseen ring of debris. The Computer was programmed to avoid space debris harmful to a standard cadata-sized shuttle; mid-incident, it recalculated and realized the tiny shell we were traveling in was in danger of being squashed by a swarm of tiny micrometeorites. The Computer narrated—two, insignificant for a standard-shuttle, asteroids collided, forming a cloud of debris blocking our path. While we were in the thick of this problem, I saw the evasive maneuver the Computer executed before it finished reporting on the cloud problem. I tried to move my sprout to correct a close near-impact, but the Computer ignored this motion because I was minutes behind the necessary instantaneous reactions. It had flown this close to a chunk of dust because there were two bigger micro-asteroids hidden behind it only avoidable at this trajectory. Then, the Computer asked if I wanted to collect this debris as fuel. The seed-shuttle was programmed for auto-consumption of impacting micrometeorites as matter for energy-generation. But overconsumption of matter overloaded the storage system. Once gobbled, the Computer realized an inability to condense it all, causing an expulsion of the excess. However, this toss-out would destabilize the tiny seed-shuttle, sending it into an uncontrollable spin, or into an exponential acceleration in the wrong direction to a speed where the g-force was uncontainable with anti-gravity mechanisms. The shorthand for these rules was a few cadata symbols, but even these arrived long after the Computer completed calculating the cloud's density and the safe matter intake. Based on these numbers, the Computer answered its own question. There would be excess matter on the planet, so refueling pre-landing was inefficient. It avoided retaining matter from space and evaded collisions even with the smallest speck of dust. The ideal movement-pattern the Computer calculated was unnatural for a cadata or a bhasab, as it reeled in all directions, made spiraling sweeps, then moved at fraction-precision degrees to an astronomical-point direction, before jumping up and down as if it was on hot coals, then moving backwards after all the work of going forward. It behaved in a manner reasonable solely from a pure-logic and pure-math study of the data, rather than a cadata's sense of what constituted beautiful, elegant, graceful or sophisticated flight maneuvers. Regardless of how I felt about the Computer's lack of poise, I was relieved when this noxious cloud was behind us, and we flew into the tip of the atmosphere. The Computer remained alert, and had to create a strong anti-turbulence force as we moved through the curly and thick cloud-cover because droplets of moisture in these matched my size.

When the Computer described the forest we were landing on, my recent fright made me fearful bhasabs would be hostile to my arrival. Research proved they allowed migrant seeds to settle in regions under their control. But would they be nurturing, or would I be isolated from their mainstream society? My panic on this occasion reminded me of my first social-mixers as a child. These strained emotions matched my new age group. Genetic rejuvenation made me feel as small as I looked. Just as I felt a little teardrop of water form around the tip of my trunk, I felt the ship beginning to hover, in imitation of a bhasab seedling floating across the forest air. This gliding motion commenced as soon as the Computer calculated we were within the nearest plant's sensory range. We hoped they would not question how I flew so high into the sky. Our fears were alleviated when plants appeared oblivious to the nosedive of my seed-shuttle at the edge of their soft-ground, grassy meadow, a foot away from the nearest cluster of bhasabs.

The Computer refrained from auto-ejecting me out of sensitivity for my emotional hesitations. I was nervous the lab-techs missed an environmental component about to boil me alive or poison my skin. The air in my mini-spaceship approximated Bhasab's. However, our sensor-probes might have missed unidentified essential air components. My hesitation was also tied to a case of germaphobia. I was accustomed to immaculate, robot-cleaned spaceship floors. Instead, the seed-shuttle's floor was disguised in a thick dirt layer; I had to crawl across it, as seeds do, to emerge out of its side in the seedy manner, without arousing suspicion from the overlooking bhasabs. Our engineers designed a core to this spaceship that behaved as if it had the consistency of bark around a seed. After I made it over to the pre-determined exit location and touched the wall to trigger the transformation, it split open with a loud crack. I had set the air inside to a temperature a bit closer to cadata-norms than to bhasab-averages, so it was warmer than the air outside, and this difference generated a mystical and dramatic fog in which I made my entry with my little sprout into this new world. I planted my sprout into the natural soil. It was crowded with life, one- and multi-cell organisms, decaying plant matter, bouncing molecules; all this activity overwhelmed my sensitive environment-reading organs. Our nanobots failed to gather sufficient soil samples to recreate every organism in the bhasab-soil on the Mothership. Genetically birthing this swarm of foreign organisms would have overwhelmed our scientists. I adjusted to this constant flow of input as I began a slow migration from the seed-shuttle, which I had to abandon in its cracked state, to the nearest bhasab tree. My fear of rejection returned across this week-long sojourn. In parallel, my width grew, at the natural bhasab rate. Nearing the goal, I was proud of my self-education regarding ideal nutrient and water extraction out of the soil I was crawling through.

The bhasabs ignored me until my sprout contacted one of their roots. I was conscious of the impending impact, hoping brushing against it would elicit a response, so I could begin integrating into the bhasabs' society. The response from them on the vibration frequency was to watch out for roots and to respect private soil property. I apologized, and explained I was new to the area. I needed a review of the local rules of land ownership, and unwritten rules of behavior and manners unknowable for seedlings. Dooh (as this bhasab said I could call him) was surprised a young thing like myself could communicate with such a wide vocabulary, albeit with a strange accent he could not place. I thanked him, without explaining the origin of this strange wisdom. My vocab range was influenced by the nanobots intertwined with my brain, which retained computerized data in addition to biological memory-storage. This function fit the bhasabs' biology because their memories are passed down to seeds by a parent tree in neural cells, duplicated to allow the young to access knowledge from earlier generations. This memory inheritance contributed to bhasabs dominance among the planet's species. They could share locations of the richest soils, and develop sophisticating communication and tool usage strategies across several generations.

Dooh indoctrinated me into their system of governance, which prioritized dominating habitats against rival species by truncating their nutritional supplies or moisture, crowding them out with the bhasabs' long-reaching root systems, or otherwise thorny-ing their survival-strides in bhasab-scent marked territories. I championed this fight in part because I needed an abundance of resources to grow from a seed into a tree; this goal might have failed if I avoided aggressive competition. I lacked guilt upon starving a blade of grass or a fern because sympathizing with trees as my evolutionary-equals remained difficult after thousands of years as an animal. Our Mothership used plants to filter our air and for beautification, but even if we could have communicated with our plants in their rudimentary communication methods, such discussions would have been unproductive. When I was hungry and searching for nourishment in the cold and moldy soil, I started finding parallels with the struggles of these rival species, but I resisted these feelings as some species block-out the rights of animals if they consume these members of their own evolutionary branch.

As I grew, my trunk became thick, and bone-like. I gained enough power in the muscles of my roots to raise most out of the ground, and to use them like stilts, or legs to expedite my movements across the terrain. This raised posture kept my trunk away from harmful parasites in the ground. I bonded with other youths as we went on joint campaigns, or played together, wrestling to find out whose muscles were strongest. The games were either fights, or tussles. Sometimes we bet a prized food morsel on who could reach a spot in a region otherwise inaccessible because of its steepness, rockiness, nutrient-voidness, or other dangers. One of my closest pals, Rheta, once dared me to dip my roots into a lake at the bottom of soaring cliffs on three sides and a steep-banked river on the fourth. The walls were composed of the strongest and densest rocks in the region, making it arduous to dig through these layers to access the bottom, where the water was. All previous attempts had been hard-won burrows through these rocks. So, Rheta expected I would elongate a burrow left by a previous competitor in this challenge. I decided on a novice approach. I climbed to the edge of the cliff and dropped a couple of my roots over it, so they were suspended in the open air. It was a 200-foot drop to the bottom; my roots were two feet. Rheta was callous in joking at my expense: my roots or the lake drying was likelier than the success of my plan. She went off to play with the others in richer soils, but I doggedly remained. I kept the rest of my roots in moist and nutritional soil, and moved these energy sources into my suspended roots. I kept at it for two months, despite Rheta's eventual pleas to surrender on this puerile bet and rejoin their games. I took it as a point of pride to stick with this science project. I still felt like an outsider, and I needed to find a way to prove my worth to the community. On the final day, the entire neighborhood of bhasabs crowded around me as my elongated roots touched the surface of the water. I gulped to fill my dehydrated muscles. I had created a hook-line to catch this precious water, and several grownups mimicked this tactic going forward.

My length now towered above low-lying species. I sampled the air with tongue-like cushions on my leafy hands. I had explored most curious spots, and creatures in our region. So, I decided I had to explore beyond this enclosure to gain a broader comprehension of life on Bhasab for the benefit of other cadatas, who might fear living in a restrictive land plot. I told Rheta I was about to head out. Because she believed I would fail to know a friend from a foe in the wider world, she said she was compelled to accompany me.

Rheta's movements were indelicate, as she plowed forward, grabbing unfamiliar trunks with her pawy roots. She taught me a trick to explore new ground faster. She used a specialized, flexible branch I had neglected out of ignorance. This branch possessed enhanced growth-speed, and thus could maneuver forward to explore if the rest of the body should move in a given direction at its slower and high-energy-consumption pace. She taught me how to extend the edge of this branch with exercise into a little needle-sword. Because this appendage could move faster than most lifeforms on Bhasab, it could be used as an offensive or defensive weapon to gain territory or avoid dangers on our path. It first came in handy when a couple of bhasabs stopped us and demanded a nutritional toll payment to continue. A few well-aimed jabs from Rheta's sword and they retreated, surrendering nutrients they had appropriated from another traveler. In the greater wilderness, I gained an appreciation for Bhasab's continuous rainfall. Precipitation was our nourishment, but it also chilled my bones. It was easier to tolerate this cool sensation during perpetual motion. However, this was when the ground was at its softest. In spots, it was so muddy and travel-raked, the heavy winds bent my trunk sideways until my roots made the soft soil bulge under the pressure of each gust. This happened to most trees going down the same natural highway we were on. So, in these conditions, it looked as if this road was breathing. In a few instances, trees were upturned or flopped over by the pressure, forcing them to spend days on regaining their footing. The gusts were most problematic for less mobile plants such as the flowery cipits with their long but useless teeth inside their lipless mouths, which evolved for permanent adhesion to rocks; they fed on the rocks' organism-colonies. While stationery, they sank their teeth into a bacteria-infested rock, and this habitat allowed them to reach their full potential. However, the bhasabs had kicked many of them out of these cushy habitats, leaving them in constant migration along this highway to new uncontested rocks.

I imagined I was invincible with a comrade as aggressive as Rheta by my side. I felt like a member of the dominant species, a group adamant about protecting their own. Bhasabs fed on any living system capable of nourishing its organisms. They had evolved legs to stand up and "run" after prey in conditions where few other species had caught up by evolving defensive mechanisms. Without a violent rival, bhasabs' habitat expanded for a couple thousand years. I learned the hard way, some of these "other" species, dismissed by cadata-researchers, also possessed extraordinary offensive methods. I expected Rheta would warn or defend me against dangers, but she was also ill-educated regarding these treacherous lamigs. We were crawling along a little side passage, following a delicious fungus population, when we sensed conflagrations in the air ahead. It had not rained for a couple of days, so the ground and the grass around us were dry. I realized, the forest-burning just ahead would soon reach us because there was no natural boundary, such as a river or a stone wall, to block its onslaught. Rheta proposed docking underground, or "running" in the opposite direction. The latter was laughable, though bhasabs do not have an equivalent expression of merriment to laughter, as they talk in frequency rather than with reflex-prone vocal cords. I instructed Rheta on my alternative plan. We used our fast-moving exploratory branches to shovel a pile of leaves and moist, mushy bits from the forest's floor into circles around our bodies. Then, we sucked in the surrounding accessible water with our roots. We sprayed some of this water at the earth under us to turn it into a wet sponge. The rest, we used to swell our bodies into water-balloons. Rheta whined about appearing ridiculous as she was about to die. I assured her this was our best option as we braced for the wall of fire splintering and scorching towards us. I felt Rheta disconnecting communication with me temporarily, a gesture equivalent to a cadata closing her eyes to avoid staring at an approaching threat. I kept alert in case I might have had time to spray some more water at the problem. The smell of smoke became overwhelming, so I tried to generate air molecules through the filter of the soil. The wall came right up to our protruding, spongy moats, and then circled. I kept my vulnerable bits in this cover, hunching my trunk close to the ground. A few sparks nearly hit us. I splashed water out of my roots at little fires scorching a couple of my leaves. I also squirted a few drops at Rheta's leaves, as she was too frozen to respond herself. I heard the panicked messages from the surrounding plants as they failed to find escape routes. I started to feel like the lone fireman in a firestorm. Alas, it was too late to help. Once proximate fuel sources were consumed, the fire died. Rheta was distressed, and argued for retreating to the main road. I insisted on learning who was the guilty party who sparked this fire. I was afraid a cadata was responsible. This cadata might have descended to the planet to perform an unsanctioned chemical experiment, which terminated in conflagrations. I questioned if we might have missed an intelligent species capable of fire-building. A lack of visible lightning ahead of the event struck out the odds of a natural fire-starting cause. Concerned about appearing afraid, Rheta followed me without protestations. We moved forward over the embers. Three hills and five turns later, we reached the clearing housing the culprits. They were a group of bloated water-tank trees.

I had a nanobot on me record their behavior as we approached. I later determined they rubbed branches together to create sparks, which erupted into forest-killing fires. They set these off whenever their habitat became crowded with competing species. Competitors were exterminated in these conflagrations, leaving the water and nutrients for these recoiling, and flame-resistant lamigs. If they did catch fire themselves, they could shed their thick bark, discarding this external skin to protect the moist inner skin hugging their water jugs. This bark peeled, and regrew in cycles, even absent a consuming-fire.

They spotted us, but their movements were much slower than ours, so we could have backed away or attempted some mad charge at them before they could shift a few branches in our direction. I asked Rheta what she thought we should do. She went off on a long tirade on how we had an obligation to rid the forest of these demented destroyers; if we didn't take them out, they might spread in territory, invading bhasabs' grounds. She just could not think of a way to hurt them if even the heat of a fire failed to leave a harmful mark. As she was screaming in bitter vibrations, I began doodling in a drying mud patch at the side of the road. I started by scraping out the trunk of a now black and lifeless tree with its branches spread in a fountain over the ground. Then, I added a little still-burning ember, which looked like a patch of mushroom-like fungus was growing in its place prior to the inferno. I sketched dramatic clouds of smoke hovering over the remains. My branch ran into bits of gray soot as I expanded my "canvas". So, I inserted a layer of this substance with the appearance of a layer of melting dirty snow. In the middle of all this, I depicted the bourgeoning tank-trunks and the blooming top-leaves of the murderous lamigs. My senses detected a river a dozen hills ahead. The fire was stopped by this water source. Fresh grass and trees survived on the farther side of this water body. Thus, I doodled these into my composition as a water-safety-line with greenery along a soot-sprayed cliff. I felt dying sparks sending puffs of smoke in a scattered pattern across the hills ahead of this river, and penciled them in. Without a sun to light this high-contrast scene, to eyesight it was a black, smoky night, but my bhasab senses interpreted the chemicals in the smoke as recently-deceased organisms. The lines these ghosts formed over the landscape impacted my psyche beyond the visual realm of art. I had to express this deep angst, and this drawing, while inferior to the circumstances, was my sole outlet. I spent many centuries working on my sculpting and painting while we were on the voyage to reach Bhasab. I even created a textured model of this planet's surface with color variances for the different chemicals' dominance in different regions. It eased my strained nerves to do it by hand rather than just having the Computer print a 3D model. Textual-sensing the outlines of this burned land and its tragic story in the mud helped me avoid the violent anger Rheta erupted in.

Unlike a scene of a battle won by a malevolent foe, the sketch in the mud was a story of survival of the strong. If these creatures were animals like cadatas, it would be easier to find them guilty of repugnant war crimes. However, was this fire, by definition, "natural", when it was sparked by a lifeform evolved with this destructive burning offensive strategy against encroachments onto its territory? What good would it do for me to join Rheta, and her army of bhasabs mustered to destroy this band of fire-starters? If we had come across a planet where a species like this had wiped out all other lifeforms, would it be right for us to interfere by killing the aggressors, and restoring to life those they had destroyed? What about an individual bird killing an individual mouse? Would it be moral to snatch the mouse out of the claws of death? After a lifetime on a spaceship where war was a textbook theory, I lacked the physical capacity to engage in it. Even if I judged lamigs were intelligent enough to overcome "natural" violent instincts, and thus responsible for their crimes, I lacked a triggering motive to participate in the destruction of these lifeforms. I explained the latter part of this avalanche of contemplations to Rheta, and to my surprise, instead of arguing the matter, or dropping it, she insisted this was an uncrossable line between our ethical philosophies. She dug her roots in, determined to return to the bhasab settlement to gather troops for a hot war against these arsonists. If I was uninclined to assist the campaign, I could continue without her, or return to our forest to await the outcome. It would have been horrifying to wait in fear of their extermination. Wrath would also fall on me for surviving if I used my anti-war sentiments to avoid conscription. Still, I opted to continue a peaceful journey alone, as my primary goal was understanding the rules of life on Bhasab. I hoped, while I was away, the war would transform into a death-defying skirmish.

The role my drawing project played in deflecting me onto a solitary journey, prompted me to further explore this craft. I was questing after beauty. Rheta missed my depiction of this dreadful scene. I pictured it in my mind's eye based on evidence from the gaps in the mud gathered with my chemical and sonic sensors. Without art forms inherent to bhasabs, she failed to imagine my clumsy movements mimicked the fleeting sight… This art genre was futile because the mud clumped and concealed my lines moments after I etched them. They were destined for erasure by the next rain, or the tread of the next traveler.

The pursuit was resurrected by my accompanying nanobots as they captured and transmitted to the Mothership for archiving images of these drawings. Being alone made it easier for me to send and receive messages with the spaceship. I avoided tuning into regular communications on this frequency while I was among bhasabs. I knew these were detectable because I had identified cadatas' signals with my bhasab senses. Away from bhasabs, it was now safe for me to communicate with the ship. I recorded a portion of this extensive diary entry during this period. The cadata research team was enthusiastic about the environmental tests they were receiving through nanobots. They discounted the forest fire as irrelevant in their settlement decision. Without such moral quandaries, the vote was unanimous for settling Bhasab.

I followed the river across hills to keep an eye on the surviving forest on the other side. This body of water acted to calm the turbulence of breaking out on my own, away from my communal "sister" Rheta. My memories had jumped back to our band's adventures, so I missed the widening of the river until it opened into a lake. I had studied emerged plants at the edges of the lake-puddles in the wood where I lived with the bhasabs, but this lake was just large and shallow enough for it to be overpopulated with floating and submerged plant species. Having dipped my roots into a lake before, without hesitation, I approached the edge and dipped my roots. My biology prevented me from full-submersion due to the threat of water overflowing my system. It was safe enough to keep half of my root-feet in this murky fresh bouillabaisse. The biodiversity of zinc, copper and ammonia-rich microbes in this mix was superior to elsewhere on Bhasab. The most curious of these was the ivecha, a species floating on the surface to feed, and hovering above it to relocate. It created buoyancy through prolonged oxygen retention after inhalation. It held its breath to build this waterproof, air-filled cellulous sac, which quadrupled its size when inflated. I had started recoiling to the edge of the lake to record this creature's attributes, when I spotted another unique critter. It was a riopho, a fish-like lake-moss, which navigated by maneuvering its fin-like hair in waves over their oval surfaces. I picked up on it gobbling microbes and small plants with every dive toward the muddy bottom of the lake. One of these dives shifted my attention to a submerged plant, the trury—a single, elongated leaf—but unlike Cadata's leaves, it was conscious; it brushed against a riopho's back with intent to eat the barnacles nestled there with its suction-cup hairs. Also on the floor, I discovered a minep, a hollow-on-the-inside, cylindrical plant with a dense, pink sap of nutrition-rich "blood" inside the thick, bony outer structure. This blood accumulated from its continuous fishing for large organisms by creating suction to pull them inside the hollow opening before breaking them down into the pink minerals. Another oddity floated by. Its composition reminded me of the alien species turning into salt once exposed to metals in our air. In this case, it was a species, called the owduls, with bones composed of near-pure copper, a property giving it a jewelry-consistency; it was covered with thousands of sensing-hairs on the exterior. Given the size of an average planet, our nanobots observe a fraction of species out of a diverse ecosphere, so these were useful findings for our geneticists, as they worked to understand how life functioned on Bhasab. Satiated on these memories, I emerged on somewhat dry land, and created a sketch of these plants in action.

After a few days at the lake, I noticed a bhasab making his way to the edge of the water on the other side. It was collecting a bucket of duckweed-like plants from the water's surface, and then retreating into the forest. Exploring a new bhasab colony was high on my priorities list, so I circled the lake and headed in the direction from whence the bhasab had come. Eventually, I emerged in a field stretching to the horizon, covered in rows of agricultural crops. This was a bizarre concept, as it equated with cadatas keeping fellow animals in rows for the slaughter... In retrospect, it appeared outlandish because cadatas consume air rather than brethren animals... This begs the question: what portion of the universe's lifeforms are consuming species in their own Tree of Life branch? I spotted a young bhasab worker in this field, Voby, who might have been the same bhasab from the lake, though he had been too far for me to memorize his exact scent. I approached, offering my greetings, and inquiring as to the nature of his enterprise. He explained it had been their practice for millennia to grow their consumption-preferred plants near their homes to avoid migrating to gather them. I asked him to walk me through the species he was nurturing, and he obliged, happy to explain his trade.

Voby was thrilled to show me the hoobkash, a plant composed of an edible, nourishing material, the folds of which hid its self-fertilized, red in sweetness, moist, squishy, and liplike spores. This was a special time for this row of hoobkash because they were all pregnant for the first time. This meant they were near the end of their 100-year lifespan. The hoobkash was small and semi-conscious or in hibernation across this period, but now it speed-attracted the energy needed for this eruption in mass. The rapid accumulation and the upcoming expulsion of these enormous seeds would exhaust these hoobkash to death; none would survive the birthing process. When Voby specified one of their top producers was about to shed 10 million tiny hoobkash, I began to understand how this was no ordinary reproductive burden. Having 10 million of anything pop out from an organism would scare even the most stoic. For now, the hoobkash were keeping their round bud leaves tight-bound around this precious load. I asked how they kept this volume of new arrivals from taking over other crops and overrunning the region. He agreed it could become a problem, but only if he failed to spot a forthcoming hoobkash eruption, and her babes managed to bounce away and hide in the forest or in dense-cover sections of the farm. If they failed to collect most of the yield, the following season could see the neighboring forest and most other species on the farm die off to make way for the hoobkash youths. In the miniscule chance of spillage, they had erudite weeding methods to locate and extract runaways. The reward was in the succulent taste of these fruits, a semi-match for bhasabs' physiological needs.

I needed to learn more about bhasabs' agricultural practices. I asked Voby if I could find employment on their farm, and he led me over to his bosses, who laid their roots in a circular forested area in the center of the fields. They contemplated and made the big decisions from this post, while the younger bhasabs rummaged through the fields to keep the crop healthy and to collect or seed it at the appropriate times. Since this work paid in food and the allowance to plant my roots in their settlement, they welcomed me to join their service. I was placed on shumru-duty. These were plants with a foul smell and appearance akin to a decomposed corpse. Thus, I perceived why they were short on help in this department. These shumrus were also semi-intelligent, and made sporadic joint-runs for the property line. I had to run after them and re-direct their course with zinc bars pressed into the soil; they could not tolerate zinc, so they could not pass these, and had to wait until one of them was knocked out by the wind, or some other accident, to attempt a new bolt. My conscience allowed me to keep them prisoner with the argument, if they lacked the intelligence to shift those zinc blocks with another plant's leaf, they deserved consumption. When we had a watery, sweet, tangy fruit for dinner in the central forest a couple of weeks into my tenure, I was surprised to be informed, this was what grew inside these grotesque-looking shumrus. I would have abstained if they told me in advance, but as it happened, I ate a lot of them afterwards.

I was sleeping in a little huddle with the other bhasabs, my roots covered in earth and entangled with theirs, as we tried to develop a close-knit, protective wall with our trunks. My trunk now approached full-adult-size. As we started to stir, I sensed our bodies had collected the necessary nutrients and had discarded the toxic waste products. Then, I acknowledged a foreign growth on my bark. I studied the sensation and discovered a warew had burrowed up my roots through my trunk and sprouted out of one of my branches. I felt weakness in the muscles it had penetrated, as if it was sucking out the carbohydrates or equivalent nutrients as it gnawed at me. I showed the little toothy bud to Voby, as he started stretching his roots out of the ground. He went over to a little herbal patch near the resting circle and dug up a special herb. He squeezed the juice out of it onto the warew until it shrank and died. I considered activating my nanobots to solve this problem. Instead, I opted for a lesson in re-growing this tissue without cadata technologies. I believed I was helping to stem a horrific parasite when I later applied this herb to plants in our fields suffering from the warews.

Time flew in these first five years on Bhasab. I hope to have more time for reflection, and to upload data, and analysis, in the following five years of my experimental decade. In a blink, my cadata crewmates might join me in sensing their roots enmeshed in soil.

60,412 BC

I just heard my bones groaning, and cracking within my bloated trunk. My fellow bhasabs are bewildered by my advanced-aging indicators at my mere 534 years. I have been comforted by their care since returning to the forest, where I bloomed out of a seed. I returned to live a quiet life among friends, as cadatas continue an intensive investigation of this planet's habitability. My adopted brethren are ignorant of my cadata background, or of my proximity to the edge of a cadata-lifespan. These elements have degenerated me before my "natural" bhasab lifespan. I am petrified that after centuries of living in the shadows, cadatas will announce their intrusion with a glowing parade. We are about to push our will, our technology, and our philosophy of "progress" on the bhasabs, even though cadatas are responsible for a heavy share of evil and destruction across the universe. Perhaps, the approach of my own demise is widening my pessimism, but I am tempted to climb into the Mothership to travel 534 years back in time to redirect this mission and myself to the next planetary stop.

While I was away, Rheta and her troop engaged the lamigs. Rheta shared the firefighting tricks I taught her, and they survived the firestorm the lamigs tossed at them. Then, they overwhelmed the lamigs' tight formation and cut them down. It was a brief battle. Rheta refrained from seeking other lamigs for similar trimmings. I am relieved to have stayed away, as the destruction of perhaps the last members of a species, even a murderous one, would have echoed with bitterness on my conscience.

Ten years into my stay, half of our traveling cadatas underwent a transformation into bhasabs. A few turned into other intelligent species holding strong positions here. One of them, Ortack, joined my wood. He is also climbing towards an age-limit. So, I am glad we had a chance to share some good times in our old age. This second wave of settlement lasted for 50-years, when theoretical experiments into un-gene-edited cadata survival were executed. Amazingly, the scientific team determined Bhasab was safe for cadatas even without gene-editing. Unedited settlement required immunization, reacclimation and depressurization. To be on the safe side, unedited cadatas began descending and living in shuttles. This is where the current problem entered.

The time to describe how it happened, or why it had to be avoided is short. My sap has reddened. It is streaming out of my bark, congealing into black clots, or streaming down into muddy pools turning a bright red. My skin would have regenerated to seal these wounds, but the degree of cellular degeneration was too great. Another branch just cracked, and fell to the ground. I just carry on, as if I lost a hat I never intended to wear. My skin is further weakened by overgrowth of parasitic species of mold, fungus, stranglers, and trunk burrowers. Too many invade to repel all with herbs. They form most of the mass covering my degraded muscles, and bones. I have not moved from the same spot in a couple of years now. So, I have drunken up most of the water and nutrients here, leaving the ground around me desert-dry. My roots are sticking out of this desolate ground, waiting for the next rain, which seems to filter through this dirt, without sticking to it. This drought has also crumpled my leaves from a springy, powerful overgrowth into dry, retracted dropping clumps, leaving my branches bare, as if they are victims of a snowless winter. My consciousness will depart. Proximate living organisms will migrate. My meaty bits will rot. A white, twisted skeleton, an echo of my form, will remain. Since bhasabs live for two thousand years, few such dead remnants survive intact beyond double the length of their lifetimes. Those who persevered, are still suspended in the posture of their last breath.

Before my mind goes, I have attempted interjecting in the debate shooting over bhasabs' planet-wide communication networks. Without betraying my alien identity, I have failed to find the words to explain why the path the majority of bhasab leaders have chosen is sealing a dire future for the planet. Am I repeating myself without explaining? Hm… It is my fault. I proved Bhasab habitable. I left my seedling spaceship exposed in our meadow. I was wrong to assume the indetection of the oddities of my non-degradable shuttle across the centuries would keep it innocuous. My fault? But did all those cadatas have to start coming down in undisguised shuttles, just roaming around in their animalistic forms, showing off their technological marvels to these impressionable creatures? When I was younger, I thought this would be a joyous time. Our cultures would merge. We would share our lives. We would turn this rich oxygen air back into carbon dioxide, so Bhasab's atmosphere would be replenished in a cycle of exchange adding hundreds of millions of years to its habitability range. As I peer into the future, beyond my time, I foresee this garden will fall from our encroachment. I'm going in circles again. Did I mention, seeing cadatas' shuttles, the bhasabs noted their chemical signatures and hunted for these to pinpoint hidden encampments? This was how they found the cadata signature from my seed-shuttle, accessible because it had cracked open. So, my neighbors have snuck their sprouts in there, and have been studying its composition and technology. They already started trying to find chemicals to imitate its composition. They realized their vision would be enhanced there if they were smaller, so they sent in a seed, and this little guy found a display I left exposed. If they learned my genes unlock the controls; they might force me to turn it on. I checked with the cadata team on the Mothership, and they confirmed that it is still operational, and the on-board robots are still active. Though those robots have been too tiny to help me even during my serious emergencies. The cadatas are now chatting on the communication system about giving the bhasabs transportation vehicles and letting them have a greater speed capability for developing their planet. The argument is: if we are to co-exist; we must share our technological advantages to create an equitable dependency… I hear this pulsing beat drumming through my capillaries, the ones unblocked from my cracks of congealed sap clots… I can't seem to recall what I was going to say… If that noise would just stop… maybe I could remember…

Wocega-10's Diary Entries

60,412 BC

My mother's death was anticipated since my birth. I was birthed from her embryo when she began showing signs of rapid decline. The awareness of her looming demise is my first memory. I lived with it in the five years on this spaceship. Minutes earlier, I was notified in a memo of her passing. She was unresponsive to status-check calls, so nanobots investigated and discovered she was beyond resuscitation. I am trying to grapple with the miniscule connection between me and my biological mother. As a cadata, I must detach from family and friendship entanglements, but my mother's mission across the last half-millennium has been to commune with bhasabs, to connect and grow into their social network. Why were her feelings for them stronger in her final hours than her sympathy for my needs as her daughter? I asked Captain Nally-10 if I could travel to Bhasab to be with her, but she argued it was unsafe for both of us to be there. Nally-10 is new like me. I think she's full of it. She was scared if I ran into trouble, she would solution-deficient. Even if transforming into a bhasab was a medical no-no for me, Mommy or Chief Pilot Wocega-9 should have allowed herself to be returned to our Mothership, for doctors to apply our best technology to extend her life. The one communication Mother sent to me is a letter in vibrational-Bhasab, which our translation device transcribed into Cadata. She insisted on my mindfulness towards the rights of foreign species. She commanded me to protect Bhasab from a technological overthrow. This mumbo-jumbo is the result of the senility afflicting her. With this letter, she also sent an eerie, bristled, dry-flower wreath, referring to it in the letter as an example of the preservation-worthy, natural and native beauty of Bhasab life. I'm looking at it now. I've suspended it at the top of my room. This prominent position has meant I stare at it often while struggling to fall asleep. It's visible against the glare of half-hibernating techno-gadget. I'm never going to wear it. It's creepy, dead bits of once-living plants. I have studies Mother's diary to understand her better, and it also portrays a character marked by distance, wrapped up with her work, unavailable to any distraction from her research (including me).

Then again, I'm unrepresentative of an affectionate character I expected to find in a mother. I think it's the cadata-style education system creating this disconnect from others. I've enjoyed boarding the Mothership and spending my free time on catching up with the lessons I'll need as a space pilot and explorer. So much so I'm worried: when I migrate to Bhasab (in edited form) I'll have to rough-play, as mother did. Wrestling roots with those wild bhasabs sounds frightful. The other youth on our Mothership is Ortack, who is two years older. It's just been the two of us in most entry-level classes. I can't imagine Ortack's reaction, if I propose we wrestle, or run after each other, without the intention to improve our mobility, in an organized class exercise.

I don't want to think about it. I want to look forward to the scientific research bit of the looming departure for Bhasab. As we were discussing in class today, it takes Bhasab around 90 years to travel between neighboring stars. We've been parked on Bhasab's moon for over 500 years at this rate. We passed five suns and ten times as many planets. Usually, we would be moving at light speed, so there would be no room for stops, but because the planet is going at a third of light speed, we have popped out of this traveling planet's mini-system into the solar structures we are passing. Every new stop helps us develop ideas on genetics, chemical interactions, geological change, and terraforming.

Our class on Bhasab's environment was a horror show. I fail to grasp why we are finalizing plans for permanent settlement here when reasonable creatures on Bhasab should be scheming for an escape from this heedless, flying object. Will it crash into some gravitational body, like a black hole? Will it lose its atmosphere to a solar wind blast from a massive, close-passing star? In all directions of this speck, there is an expanding, near-zero cold void of space. We have been cautious on the moon, without an atmosphere for a blanket, unfiltered solar radiation and wind, galactic radiation, and a relative lack of gravity. We refrained from activating artificial gravity in the Mothership on the moon to conserve energy in a period when the ship collecting less matter than in flight. These discomforts motivated even un-gene-edited cadatas to insist on moving to the planet. And think of the chemical wasteland we are flying through. Unlike inside a solar system, where we could hop over to a neighboring planet to restock, the trip to reach the nearest planet while on a free-floating planet takes up more energy than it could generate. If we run short on fuel, we would hope for an impact with a comet, or for intersecting a warming star in our drift through space.

I am recalling our last stop now because I am anxious about memorizing necessities for the trip to this techno-undeveloped planet. How would knowledge about the galaxy help there? Perhaps, it will keep me sane, avoiding following my mother into just being a tree helpless to assist with the scientific studies. So, the experiments they ran on the black hole are now cycling in my head. I should be focusing on geology and herbology. But I can't. So, I'll think through it here. For a couple of months, the team kept sending packages on little recyclable shuttles into the black hole to record readings as they were eaten up. They knew since arrival of Bhasab's expulsion from its system because of a tug-of-war with a black hole. Thus, upon passing a system with this astronomical phenomenon, they had to test the physics of how this expulsion happened, if it might have impacted Bhasab's composition, and if it bent its space-time characteristics. Shockingly, in a million years of space travel, we have failed to comprehend the origins of the universe. How could galaxies have been compressed into the space of an atom at the start, and if they were, why would this be the beginning? Other than gravity, what could cause matter to keep compressing beyond atoms' force of repulsion? The molecules did not reproduce themselves, creating matter where there was none… It's all so confusing. Maybe I just don't understand some of the upper-level theories yet. A theory I like argues the universe ends by merging into a single, infinite-mass black hole. The Milky Way already contains a couple dozen swarming black holes in its center. Each of them lost speed as it tugged on stars in its path, and this expended energy made it lose speed, and this plummeted it down into the galaxy's center, just like the other trapped black holes collecting there. So, the theory goes, more and more black holes will plummet down there until their combined strength begins to pull separated galaxies towards each other. All galaxies will be clumped into a single giant ball. When no external matter is left for consumption, this universe-sized hole would undergo a transformation to push its matter and energy back out, re-creating the universe, which then expands to repeat the cycle. Then again, infinitely compressed matter? What science fiction! If the fabric of space-time is forced to generate invisible space into which all this stuff is falling, it is likelier the scientists are composing fantastical reasons their theories work despite evidence to the contrary. It is difficult to imagine this is the only universe. Probability is still slimmer universes can exist on parallel, or simultaneous planes. Two universes cannot occupy the same space and time. The closest universe must be just far enough that light from it cannot reach this universe. Life looking outwards from the Milky Way can see as far as 14 billion light years away. Thus, in theory, this is the point when our universe was created. Is this even true? Maybe the start of the universe is 30 billion light years away, or even infinite, and light from there will just never reach us. Our science teacher tells us the silliest theories. Imagine, a duplicate universe parallel to our own. A duplicate Wocega-10 lives there, but she is enamored with extreme sports instead of science. This is *maxime absurdum*. An infinite set of molecular combinations during a universal-creation means two isolated creations cannot generate two similar species. To achieve this improbable statistic, there would have to be a near-infinite number of universes in the same plane as our own—not a dozen. An infinite number of molecules in the same millimeter of space where a single molecule is in our universe? The weight of these infinite molecules would generate a black hole in each of the smallest units of space-time. The probability of two potential universes having two similar planets in the same portion of their predominantly-empty space is miniscule. Relying on our clones' potential choices is escapist. No parallel universe holds mirror-creatures to ourselves. At most, physical laws would remain constant in two birthed in isolation universes. Why do I understand this, and my teacher does not? Then, he tells us, during the Big Bang, the universe expanded faster than the speed of light because empty space can beat the galactic speed limit. Another trippy theory imagines a super White Hole as the flipside of a black hole. This anti-hole is supposed to extrude light, which escapes in infinite flight from the singularity in the infinite-matter black hole, whose function it is to prevent light's escape. A hole shooting out light at an infinite scale has never been observed. You would recognize it if you see a sky with an all-directional light-source penetrating across the universe. Then again, a White Hole is what I imagine if the universe's matter collapsed into a single event-horizon, before shooting back out. The outflow is a type of White Hole. Though I envision it happening after the inflow ends, or fills the Black Hole to the max, rather than both happening together: with a universe being created on one end, and shrinking down to near-emptiness on the other. And what's this about a mere ten-billionth of the original matter in the universe surviving the Big Bang? If the White and Black Hole theory is possible, it is nonsensical to suppose the universe shrinks to a ten-billionth of its size with each new rebirth. Why would this density of matter fit into the tiny speck out of which the universe formed? I am determined to solve these enigmas in my lifetime. Meanwhile, I have been amused by watching test shuttles disappearing into a black circle. These experiments explained Bhasab's interstellar movement, and the residual post-black-hole impact on its gravitational pull.

59,904 BC

I've just re-read my last entry. How naïve. The realities of space research, and exploration have proven to be slated for failure. It's as if there is a vortex of bad luck in the middle of our spaceship. The worst possible events always occur. We have spent 1,042 years on Bhasab. In this span, it has traveled 340 light years around the Milky Way's galactic center. Other than this continued circular motion, we have achieved scarce progress towards our goal of a permanent settlement of a planet in the galaxy less damaged than our dying Cadata.

After my ambitious composition, I underwent a gene-edit into a bhasab. Children my age should refrain from turning into plants. Slow movement opposes the nature of young creatures; just as the old must abstain from flying in cartwheels. I believe this was the first time they attempted an edit on a cadata under a few hundred years old. In theory, they could have edited my genes while I was an embryo, but then again, they could have just adopted a bhasab baby and taught it cadata science. Gene-editing is performed

on a mature cadata for him or her to possess the full intellectual capacity of our ancestors, while adopting a local-species disguise. They crammed an avalanche of schooling
into my first five years. Still, the remaining gaps in my knowledge proved detrimental
in instances too fleeing to search for the answer in an encyclopedia. Meanwhile, they
sent me to the same forest where Mother died. Her corpse became a statue haunting my
mission. I tried breaking out of there, and going on a little "adventure" like she did. But I
lacked interest in studying one plant, or another. Maybe I was desensitized to the miracle
of planethood because I was born on Bhasab's moon, with its side visible past the moon's
dark outline. As we embark on millennia in outer space, I am disturbed with grief over
never again touching a natural patch of ground.

When I touched down in a seed-shuttle on Bhasab, our scientists were developing
plans for permanent buildings and cities in drier regions. Researchers were buzzing in
anticipation of the announcement to be sent to Cadata welcoming our home world's
population to join us. This light-speed transmission needed 25,000 light years to reach
its target. Our hopes were dashed before this dispatch could departed. Everybody but
Mother had interpreted the bhasabs' intellectual outgrowth as progress. They located my
shuttle soon after I landed. We had not changed the design, so it looked like Mother's.
They found both and connected them. They dissected the vehicles, placing bright seeds
inside allowing them to comprehend their composition and functions. My bosses on the
Mothership insisted on silence on this matter, hoping the bhasabs would fail to grasp
our superior technology. The bhasabs failed to link my arrival with my seed-shuttle. So,
I avoided an interrogation. They just learned they could combine parts into a protective
shell seemingly intended for travel. This triggered their scientific imaginations. Within
a century, they invented a little vehicle running on burning trees; these conflagrations
generated heat and steam, which caused propulsion. Bhasabs could soon purchase these
wooden boxes dragged wastefully over the ground, without tires or hovering ability to
minimize this extreme drag force. Ownership of these vehicles allowed bhasabs speedy
cross-planetary travel. Those who remained in the pre-vehicular ages continued crawling
at plant-speed. Drivers gained an extreme survival advantage. Crawlers invested their
nutritional reserves into acquiring vehicles to travel fast enough to access nutrients in
isolated regions. Rival bhasabs started gathering around these previously under-utilized
nutrient-rich spots, generating enmity, which erupted into warfare between groups for
dominance. Soon, fires from warfare and over-gathering to feed vehicles deforested these
hot-spots. Meanwhile, the agricultural ventures generating a steady yield for millenni-
ums were inundated with arduous production tasks such as wood-burning, picking,
spreading, grinding and machine-repair. Machine cost was so prohibitive farms across
Bhasab merged into eight, giant farming enterprises. These farm-factories worked the
land until most of it became infertile. They kept working it, even if they were forced to
discard most of their produced crops to out-produce their rivals.

Fifty years ago, I was returning from an excursion to my favorite brook, where I
spent years recording my astronomical ideas. A disaster was imprinted on the fabric of
this idyllic spot. Only rare, scattered blades of grass and bony bushes remained. The
flora was extinguished—used up as fuel or for vehicles components. It was also killed by
the increase of carbon dioxide in the air, which heated the planet to intolerable levels.
Reaching our bhasab settlement, I discovered the bone-corpses of half of the friends I

left three years earlier. The other half was riding around in earth-laden vehicles, searching for added fuel to burn to feed these mechanical beasts. I walked to the spot with bad reception for bhasabs' communications, and called the Mothership with this report. Captain Nally-10 confirmed my trepidations. An atmospheric switch had been brewing for centuries since the bhasabs started burning wood for fuel. The last three years saw a rapid acceleration of this process. Much of the precious oxygen needed for cadatas to live without masks on Bhasab had been consumed in the flames. The death rate for all species was astronomical and growing worse. Bhasab was slated to become uninhabitable if trends continued. Our science team devised a proposal promising to reverse this trend. They sent down a fleet of renewable energy, ground-transport, cadata-designed vehicles. Cadatas on Bhasab were tasked with distributing these for free to bhasabs utilizing plant-gobbling, fire-run vehicles. Our planners assumed, since they were going bankrupt from spending their resources to acquire inefficient cars, they would be overjoyed to win free cars operating on solar and air fuels, which would be free to operate forever. To our shock, the bhasabs rejected these gifts because they were convinced their own models were "cooler". Their advertising culture programmed them to believe superiority over neighbors came through displays of the biggest clouds of machine-generated smoke. We even attempted designing fake smoke into our cadata vehicle models, but they rejected these too because our smoke smelled too clean on the molecular level. This gift of free-floating bliss is ablaze beyond the point where terraforming could have saved the sizzling remains.

Exasperated, we departed in our over-developed moon base, accelerating to light speed in the direction of our next ephemeral stop. As we speed away, I am recalling the final moments on Bhasab, prior to taking off in the oversized shuttle designed for my full-grown bhasab-form. I was straining to recollect the computer code for the looming trip. My spaceship was hovering just off the ground. My imagination roamed from the dreaded departure, to the surrounding landscape, illuminated on the visual wavelength under the glare of my shuttle's lights. I saw it because the nanobots had begun rebuilding the organ responsible for vision in a cadata. This was my first visual perception of Bhasab, since studying pictures when I was five, before my gene-editing transformation. I spotted the brook I enjoyed writing by, and even glimpsed a large lake on the horizon. This body of water was coming alive in puffy clouds because global warming had reactivated the steaming air vent springs.

I was saying goodbye to this awe-inspiring world, when a giant, new-model, plant-burning, mosquito-like flying mechanism sunk its predatory gadgets—which I perceived as claws and a beak—into the side of my shuttle. I doubt he intended to hit me. He must have lost control during a maneuver because of lax flight-licensing standards. He lacked the skills to dodge me, when I appeared in the clearing ahead of him. My shuttle's core needed a moment to categorize his machine as an intrusive flying object, just as it would an asteroid, or a micrometeorite in space. Then, the core reacted by breaking into molecules this inventive flying machine, and the bhasab directing it, consuming both as fuel. The shuttle was low on matter, and needed it for the planned flight to the moon. Beyond this senseless added loss of life for the conscience of the driver, this low-altitude reaction was hazardous for my shuttle. In the confusion, my shuttle might have misinterpreted the entire planet as belonging to this intrusive flying object. It could have gobbled up a chunk out of its crust. And the starved shuttle had eaten too much matter in this process, and thus shot some of it as violent projectiles at the dry dirt below.

Was all this my fault? It is cadatas' fault. As Mom said, we are a plague poisoning the worlds we invade. I will spend the coming millennia on describing the philosophy behind our mistakes thus far, and on suggesting strategies for avoiding these going forward. Short of black holes, are cadatas more gluttonous than anything alien natures have invented?

Chapter 12

Rival Species on Neighboring Worlds

The spaceship departed from Bhasab in 59,904 BC. They had a near-ten-millennia voyage ahead. None would set foot on a planet again. Nally-10 did her best to find diversions for the crew by assigning innovative entertainment projects. 3D games fail to approach the multi-dimensional space-time and multi-sensory entertainment systems they invented. However, once the first among those raised on Bhasab started dying, their capacity for self-distraction seized. They bent under the knowledge they too would be recycled into energy or matter. Life is designed to exist in proximity with planets. The rhythms of birth, life, and death correspond to the cycles of suns, planets, and moons. These cycles cannot be duplicated by even the most advanced lighting or setting replicators. The gloom enveloping the crew did not stem in a nutrient deficiency: fixable with supplements. It was not a symptom of zero-gravity strains on unused muscles: preventable with exercise. Investigators of this depressed state weighed the ingredients involved in natural seasonal change, responsible for regulating mood, by impacting the eye's cornea. Despite mimicking intricate details, such as wind movement, and molecular fluctuations in the air generated by the bloom and death of plants, the culprit of this infectious clinical depression remained undetectable.

In 55,555 BC, Wocega-10 began the morning with a diary entry, which has been studied by cadata psychiatrists in countless books since. The following transcript is presented in its entirety to avoid contaminating the evidence with emissions.

"Doctor Nackan-10 has been pestering me with his prescriptions of happy pills and light therapy. He complains, I withdraw from the crew into an emotional black hole. I never emerge, and anybody wanting to commune with me must join me. The black hole metaphor has been swirling in my mind for a few days now. I've decided to review the data from the black hole drop-in experiments I assisted in the Bhasab system. I hope to understand the concept better to refute the Doctor's hypothesis. As you can see in these files, the raw numbers hold significant meanings."

Wocega-10 attached the data files to the recording.

"The test with the longest total-time offers the most extensive set of data points for analysis. Here, the mock-shuttle accelerates. From its perspective, it is emitting a signal at equal time intervals. To us, watching the fall from a safe distance inside a gravity-resistant science-station-shuttle, it is slowing down. We observe slowing because the mock-shuttle is traveling so fast, the distance the signal must travel back to us is elongating. The space-time around it is slower because it is traveling through a black hole, by definition, a mass-generated indentation in the fabric of space-time. Seemingly, even the speed of light slows inside, dimming the transmitted light. Then, the shuttle outwardly drops its speed to zero, as if suspended at the event horizon. No further signals from its beacon can reach us. The telemetry from inside the shuttle indicates it is being stretched radially and compressed perpendicularly. These counter pressures increase the g-force. Our reinforced test shuttle withstands the hole's force at the five-hundred-kilometer mark, a depth where all normal matter would be shredded. The remainder of the fall to cross the event horizon lasts 2 milliseconds. Despite our efforts, we never received data from the last fraction of a millisecond. The final images are of pure blackness ahead and blindingly-bright whiteness behind the shuttle. Theory dictates, stars and black holes alike freeze in time at this event horizon. Fictitious dribble! Scientists have profited from a gap in data to fill it with fantasies. Available statistics prove such suspension is impossible. Can millions of stars coexist in a single, microscopic compressed space? The shuttle might have begun hyper-accelerating at the horizon, discharging out of a white hole on the other end of the universe. Millions of years before we arrived, how was Bhasab influenced by a hole's catapulting? Bhasab's hole was a stellar-mass black hole, the smallest variety. It formed from the core-collapse of a supergiant neutron star in a binary system. It overate mass from its companion until it collapsed, under the pressure of its weight, into a singularity. It then commenced tossing the remaining neutron star and the planets circling them. Bhasab dislodged at a third of light-speed…

"Contemplating the nature of black holes is maddening. I shouldn't have descended into this black hole vortex… Life on Bhasab survived even this extreme proximity to this rarely-overcome monster… It survived hundreds of millions of years in the coldness of space. Beyond surviving, it even evolved into an idyllic forest with limited intellectual intervention. And within a few hundred years, a blink in its history, we arrived and even without the ability to stop time or space via a black hole-scale event, we stopped life on Bhasab. We can't even say: we succeeded in an intentional destruction of life out of a malevolent psychosis… We just did it with sheer stupidity. And as if Bhasab was an insignificant bug on the galactic scale, we are looking back at the data we collected from its solar system to improve what might not be a concrete science, but rather a fictitious interpretation of the physics of the universe. The black hole resulted in a rebirth of life on Bhasab, so the numbers from our tests of black hole physics are irrelevant to the question we should be asking before infesting another world with death. Where are the numbers explaining what about cadatas is innately destructive to Life? The reports of our stops across this colonization voyage mimic an infectious disease outbreak; catadas are the invading cause of the outbreak; we kill even when we refrain from spreading actual ailments. But why am I shifting blame onto all cadatas when the seed for the catastrophe on Bhasab was mine alone? If I did not abandon my exposed shuttle, they would not have discovered the science capable of burning their forests for fuel. Instead of diving into my astronomical fantasies, I should have realized the impact agricultural mass production was having. But even when I raised the alarm, and was willing to go to extraordinary measures to reverse the escalating environmental crisis, there was insufficient resources in minds or assets being invested by the cadata higherups to affect a reversal. As I was lobbying for them to stop the bhasabs from using the plant-burning vehicles, they were saying the bhasabs had free will to use any vehicle design they chose. When our grandchildren step on Pallos and its neighboring Byddwr, they will face similar choices. As they review the history of our actions on Bhasab, their takeaway will be, we were politically in the right. We were right to experiment with co-habitation on a foreign planet. We were right to expose the locals to technological ideas they lacked the philosophical capacity to predict the long-term impact of. We were right to then employ a policy of non-intervention, minimum oversight, and the deployment of educational suggestions as to how they might solve the problems we inspired. How will they manage to destroy two worlds in unison, a water world and an icy one? Even the wisest plotter of world destruction would have failed to match out lethal outcomes. History will look back, and ask: why I did not give the next generation instructions on the laws, and procedures we must enact to stop ourselves, before the next mistake. Short of refraining from penetrating another planet's atmosphere with our shuttle technology, or fighting a war to force the cessation of utilization of our technology by an alien species, what rules can hope to solve a future catastrophe with undefined parameters? I am responsible for killing all Life on a planet. What punishment do we have in our lawbooks for this crime? No intergalactic, overseeing agency is levying tickets for mis-executed settlement missions that are terminal to an alien lifeform. If we alone can punish ourselves for crimes against a species, why am I alone in lobbying to institute a punishment system? Perhaps our ship and crew had to return to Cadata after the first disaster to be confined in detainment or we should have self-destructed to atone. But our mission must continue because the survival of cadatas is at stake. Our own planet approaches downfall as we search for a planet capable of surviving our settlement. I spent thousands of years in search for an element uniting our destructive missions. I conclude, we are destructive because we are bereft of a punishment organism. My predecessor pilots, all Wocegas, have been the drivers of death, and yet their lives after mass-homicide were softened in technological comforts on our spaceship amidst admiring coworkers. Knowledge of personal repercussions for negative impacts on Bhasab might have altered my choices. I must leave a legacy for my yet-unborn child, Wocega-11. Her mission must be to continue life for all, instead of joining an uncontrolled spiral towards universal death. It all fuses down to Death. Somebody must die for my lesson to become historical. I cannot imprison myself in my room as punishment, nor narrow my diet to nauseous smoke. These would be trivial disruptions unworthy of the news. No, the punishment must be Death. Killing myself would meet minimum-historicity, while refraining from causing further harm to others. My mother killed herself in a way when she refused to return to the spaceship for treatment as she aged. Her decision is a footnote in this millennia-engulfing voyage. It is my moral duty to punish our spaceship, but the destruction of our research library would equate to the destruction of a mini-world: adding a crime. I must deduce who—other than me—is the responsible party, who will magnify the intensity of my final lesson to our ancestors. The leader is this designated supervisory party for a country, business, or spaceship. Thus, Captain Nally-10 deserves the brunt of the *guilty* verdict. Her decision swayed the vote in the cabinet against preventative intervention. Everybody under her could claim to have executed sanctioned orders. However, the decision was not hers alone. She acted under the advice of the cabinet and the scientific team. The mission to Bhasab was tested and planned under the supervision of Safety Controller Elleqow Comeepout-10. He instructed me on surviving living as a plant, and crafting an upgraded design for the seed shuttle to outstrip the version my mother took to Bhasab. What other than safety measures could have prevented setting Bhasab aflame? Then again, perhaps our failure to convey the message of how destructive burning resources would be was a linguistic or a medical failure, or the realm of Chief Doctor Photuate Nackan-10. If he had meticulously researched bhasabs' language and the psychology of the triggers of their emotions, our messages would have swayed instead of pestering them. Medicine could have invented a hyper-growing and multiplying species of plants to repopulate Bhasab faster than the onslaught of deforestation. Nackan-10's inaction is culpable. But medicine is a science in need of an individualistic application. Whereas, a computer program can deploy an army of robots and nanobots to execute a correction on a massive scale. Chief Programmer Guagray Raimingly-10 was at the post; he was responsible for designing a solution at the first signs of trouble, but he re-worked old codes to disguise his lazy inaction. Yet, Raimingly-10, if I questioned him, might shift the blame onto Chief Maintainer Vomere Tenievan, arguing keeping my seed-ship hidden was a maintenance task, a matter of cleaning our imprint. Planetary maintenance falls to the Chief Maintainer; however, a world would fail to fit a job description manual thick enough to encompass the tasks involved in maintaining a livable environment on an entire planet. And what about my closest friend and co-pilot, Ortack Steral? I cannot be blind to his blameworthiness because we are companions. The way he acted as we were on our final departure from Bhasab is characteristic of his attitude across our stay: he was playing games and resting; ignoring our impact on this incinerated lifeform. His mood was unaffected by our misdeeds. He is cheerfuler than before his first step on Bhasab. He talks about his test flights to prepare the team for landing on Pallos and Byddwr. He spins tales of gliding through the thick, hot air on one or a snowstorm on the other. The crew is anticipating our next experimental stop. Failing to assimilate lessons from past disasters in favor of happiness is criminal. Our smiles are shielding a graveyard. It would be criminal of me to execute a fellow crewmember in a fit of anger, but given my full control of my emotions and reasoning, I believe I can reach a just decision in this case. I condemn myself. The execution need not be concealed. An alibi for my whereabouts at the time of the act is irrelevant. No method needs to be devised to shield the ship's safety recording equipment. I am even leaving this recording to detail my motives and to confess my guilt. And there is no escape with my life intact, after exacting punishment on another. Not that I would do so, if I could. I've said I am guiltier than the rest. If I tried to escape in a shuttle; it would break into atoms, as its relative speed dropped from the speed of light to near-zero… Though maybe it would maintain the same velocity as the ship while retaining a gravitational pull to it, maintaining a constant motion. Our experiments with flinging a garbage dump mid-flight have detected atom-sized bits flying away at extreme speeds a micro-instant after the expulsion. It might be scientifically valuable if I put myself through this expulsion to test what would happen, but I would be wasting valuable matter if I took a shuttle with me. The execution ahead of me is a historical abnormality. It is unheard of for a cadata to have suffered a similar, violent defiance. None on this voyage have felt sufficient guilt for this irrevocable ending. If I raise these concerns with the council or with Nally-10, they would interpret my discontent as a personal ailment. There is no external court capable of adopting the role of assassin…"

Wocega-10 stopped this discussion to conduct a search through the digital archive for methods employed in the few instances on Cadata when a murder took place outside of warfare.

"Cadatas are arduous kills, as you all know. Body parts grow back. Nanos in internal fluids seal wounds. Past murders have all been lengthy and tactical procedures. I don't think we've had an accidental death on Cadata since the nano-integration. One complicating factor is the lack of weapons strong enough for the job inside this spaceship, and the Computer will not replicate a weapon on-command due to safety protocols. Weapon fire threatening to a cadata would also blow a hole in the ship. At the speed of light, a particle breaking through our shield because of this hole would explode the spaceship. Short of an explosive or shredding weapon, a chair or a lamp could poke a hole or create a dent in a cadata's head without causing lasting damage. A single office contains fatal materials—the science and medical lab. Because I cannot judge the relative guilt among those I have described as guilt-worthy, the party among them who will be in this death-nourishing place, Doctor Nackan-10, must be chosen as the receiver of the verdict."

Wocega-10 ran a search for the poisons available in the lab's collection, and decided on one used to dissolve a solid into a liquid even when the surrounding room remained at a temperature unconducive to this transition. She did not stop to add a farewell or a concluding argument, but proceeded to the Lab, armed with a judgment and a formula for its implementation. The Computer's system recorded her progress down the hallways, entry into the Lab, and the following events. After this incident critics debated giving the Computer the ability to intervene if another cadata recorded intentions to carry out a murder, but they decided this was too intrusive on cadatas' free will. Voters feared, given such preventative powers, the Computer might have started refusing to serve cadatas intoxicants or stopped them from accessing rooms of their sexual partners if they had already reached the normal quantity of encounters per a set period. The Computer registered the danger, but since it estimated it applied to a single lifeform, rather than the integrity of the ship, it did not issue warnings when Wocega-10 greeted Nackan-10, who was busy studying under a microscope a tiny synthesized lifeform they were expecting in the atmosphere observed from afar on the water world they were approaching. Wocega-10 floated (since they were in zero-g) over to the digital spherical-tube closet holding the premade active chemicals relevant to the ongoing set of experiments in the Lab. She made the signs for the formula she needed and a container full of this mercury-like liquid appeared in the dispenser. She placed it inside of a special spraying device, made a leap, by bouncing off a wall, to Nackan-10's station. From two feet away, she carried out a consistent spray to cover most of Nackan-10's body on the side facing her. The reaction was instantaneous as the cells throughout liquefied and became droplets spraying out due to the released pressure. When their potential energy was exhausted, they remained suspended in the air. Nackan-10 did not feel any pain through this process as the nerves would not have had time to register the immediate state-change. The nanobots strove to repair these droplets, but the chemicals kept them liquefied. Essential data and biological remains for recompositing were lost.

Nackan-10's decomposition triggered a biological spill warning in the Computer's safety system, so the maintenance team, led by Chief Maintainer Vomere Tenievan, arrived at the scene a minute later. Wocega-10 had floated away from the spreading droplets, lost in unvoiced thoughts. She was still holding the lethal spray can. The Computer announced to the maintenance team: "Nackan-10's matter has liquefied and is forming a dangerous spill in the Lab, which can cause a disease outbreak if it is not contained. How would you like me to proceed?" The Computer's system was designed to eject dangerous matter from the spaceship, but it had a protocol to refrain from doing so with body parts belonging to a cadata without first checking with a commanding officer.

"Why has Nackan-10 liquefied?" Tenievan-10 asked the Computer.

"Wocega-10 sprayed him with a liquifying agent," the Computer reported back.

This exchange was played over the speakers to the entire ship because the emergency signal triggered monitoring alarms. Hearing the last comment brought in the security team, Nally-10, the council and as many other crew members as could fit inside the Lab.

"Please keep away from the remains," the Computer requested, as the crowd came within inches of some of the droplets to examine them.

The group migrated to the main auditorium. Wocega-10 followed the crowd unprompted. Nally-10 ordered the Computer to collect and test the remains from across the Lab once it was empty of cadatas. Once in the auditorium, they played the videos of Wocega-10's comments and actions. Several minutes of silence followed as each tried to process what had happened. This was the first murder on the flight, so the crew did not have a procedure or a suitable emotional response prepared. Nally-10 broke the silence by observing they should revive one of Nackan-10's embryos to secure a new lead doctor upon its maturation. Wocega-10 became flustered with anger upon hearing this digression. She repeated the ideas she expressed in her recorded confession, arguing they had to acknowledge their criminal culpability in the deaths befalling the planets they abandoned. This was a strange proposal for the others.

Given their lengthy lifespan, it would have been an extreme hardship for the others if any one of them seized useful service across thousands of years. They were already imprisoned on a small spaceship without any place to escape to for stretches of five to ten thousand years between stops. The punishment most common on Earth, imprisonment, was not considered feasible on the spaceship. They were already locked in single rooms to perform individualized work or to rest. They only formed groups for the mandated two hours of exercise, for the communal meals, and for meetings. An identical plan, but without work-duties to benefit others would have been a burden for their society; it would be a conviction to leisure rather than a grueling moral lesson for the perpetrator. Capital punishment was also absurd because after living for a few thousand years, they were unanimously looking forward to their eventual death. Thus, if it came a few centuries sooner than at the 5,000-year mark; it would have been a relief for the perp. Meanwhile, the doctors would have had to breed and teach this cadata's offspring, investing time and energy into his or her development. Monetary penalties were also impossible because goods and services were shared in their commune. The sole exception to their equal-sized rooms was if a member mutated into a mega-species, which demanded a matching enormous living space. Everybody partook in the same array of gas-nutrients, and had access to 3D printers to fabricate the types of furniture and goods they needed. The limit to infinite consumption were physical matter-weight standards, according to which cadatas were limited to a maximum number of pounds in their rooms to minimize spaceship propulsion-energy expenditure. Whether this matter appeared luxurious or cheap depended on the owner's decorative imagination. Confiscating matter allowance from one cadata to offer it to another as a reward failed because each cadatanaut still needed a bed, a shower, and the other components essential for survival. These consumed similar volumes regardless of their designs. In the first generation in space, the crew concluded ground-based laws became nonsensical in space. Think about the crimes punishable on Earth; now, imagine any of these taking place on a confined cadata spaceship. Violent tendencies could emerge in personal crimes such as assault, battery, or homicide. These are hyper-discoverable with a ship's monitoring system. Viciousness was in part screened out of the genome for space-faring cadatas. A tendency towards homicide would have led to the deaths of the entire crew. Wocega-10 came to a logical, rather than instinctive, decision to kill because irrational rage was counter to her natural encoding. Violence had a suppressed role in the space-cadatas genes as they were obligated to engage in military conflict if they faced a violent rejection on an explored planet. Despite this reserve of aggressive reflexes, crimes are counter to the design of a spaceship. For example, taking another crewmember into your room and stopping them from leaving would be more of a social call rather than kidnapping if neither of you can leave the ship. Because cadatas are sexless, or lack the separation in sex organs as humans do, rape among cadatas is a bizarre phenomenon, parallel to a human woman raping another woman. It is possible, but rare. The Computer reports all non-consensual violent acts, so a rape is auto-publicized. It is awkward to hear homosexual rumors in a small town. If this town was inescapable, the shameful "rapist" stamp would sour the remainder of an extended lifetime. The fear of disgrace has prevented such misbehavior. Since property was shared, property crimes, such as theft or burglary, were an absurd undertaking. This would be akin to stealing paper towels out of a dispenser in a dormitory: since they belong to the community, it is legal for everybody to partake. Taking a treasured possession out of somebody's room would have been pointless because the Computer, and its nanobots would locate the missing item, and auto-return it to the correct owner. Arranging a fire on the spaceship was impossible because the Computer auto-extinguished conflagrations threatening the safety of the ship. While financial crimes were as common as breathing on Cadata, they were difficult to conceptualize in a spaceship's communal economy. Granted an assigned role from birth, there was no need to blackmail a supervisor to move ahead. They lacked access to purchase luxuries with dirty money. Fabricable luxuries were offered by the ship for free. Thus, money laundering was nonsensical. They were allowed to forge any art or sculpture masterpiece the Computer had in its database from Cadata or a visited planet. These forgeries lacked significant value. Such value is translatable into a monetary reward motivating such mimicry on Earth. The forgeries' sole value was in the volume of space they required be deducted from a cadata's volume allowance. Selling sex was absurd in a communal space without poor cadatas willing to utilize their sexual organs in exchange for food or shelter. Cadatas might have carried out massaging or cooking services for others in exchange for sexual stimulation, but these practices are labeled as "romance" on Earth. Lifetime spacefarers never utilized external vehicles, which might have endangered them if they drove under-the-influence. Having open containers or public intoxication on deck or in the observatory was a social requirement to unify the team, so this was an unbannable entertainment. Since cadatas matured from infants to full-grown cadatas within a couple of years, laws against alcohol consumption and other limits on minors on Earth were unnecessary on the ship. By the time a cadata considered such activities, he or she was bio-developed to partake without adverse physiological effects.

They discussed the implications of the first homicide. They lacked a criminal code, a punishment system, a prison system, or a government branch with overseeing responsibility. Overwhelmed, they settled on inaction. They entreated Wocega-10 to abstain from murder. Excessive slaughter could have emptied the spaceship, ending the expedition. Wocega-10 begged for capital punishment. Nobody sided with this solution, dismissing Wocega-10 as a stir-crazy pilot restrained from flight mid-stop. While her replacement, Wocega-11, was growing, one of them was obligated to surrender scientific or intellectual duties to practice digital, simulated flight, a practice now linked to potential insanity. The meeting dispersed, leaving Wocega-10 alone in the spacious auditorium. No sanctions or movement restrictions were placed on her, though the Computer was monitoring her actions. Minutes passed as Wocega-10 floated aimlessly, before arriving at a revelation. She ordered the Computer to disable the protocol preventing a cadata who has hit an exterior spaceship wall from passing through it, during flight. Then, she stepped through the closest wall. She emerged in open space. First, her body continued traveling at the same speed as the spaceship because it was attracted to the ship's artificial gravity field. Wocega-10 had a concluding thought for the recorders still tracking her activities, but the lack of air outside prevented vocalization. Just then, her nanobots began failing to keep her body warm despite the near-absolute-zero temperature in the vacuum of space further from the shell of the spaceship. The drop was instantaneous. Systems collapsed. Wocega-10 disintegrated into tiny ice crystals, which were recycled by the ship's matter collecting field.

***

An entire generation was born and died between the Wocega-10 incident, and their arrival in the Pallos system in 50,068 BC. That year they began an eight-year deacceleration cycle. This was a slower landing schedule because they previously detected signals suggesting life on one of these planets could have been intelligent enough to develop space-viewing equipment. This technology could detect the Mothership, if it approached any closer than the edge of the solar system at its standard light-speed. While the signals suggesting intelligence were inconclusive, this solar system was also of scientific interest because it had generated at least two planets with intelligent lifeforms. Unnatural signals requiring intelligent design were confirmed by preliminary tests. There might have been other life-supporting planets in this cluster, a hypothesis requiring for testing the deployment of shuttles to the fifteen large planetary contenders. The missions strove to discover why both worlds formed intelligent life. Were their ingredients fostering advanced-life-support in this solar system. They first entered the orbit of the planet farthest from the system's sun. It was so rich in graphite rock, it was black, or unreflective. Then, they explored a planet orbiting in the opposite direction from its star. This strange orbit indicated a gas giant far-from-the-sun migrated inwards. As its orbit shrunk, it collided with a heavier version of this planet, which survived with a fraction of mass, changing its direction of spin. The planet neighboring the two of primary interest to the expedition was an Earth-sized world with ten moons. The distance between these celestial bodies was just wide enough to avoid impact. Two moons had collided head-on, forming a mini-asteroid-belt. The techs concluded the rest of these moons will collide in the following two million years. Each added moon increases the odds a dangerous object will spin inwards; these massive collisions tend to end in the transformation of the host-planet into floating debris. The quantity of potential disrupters in this solar system was worrying. However, the gas giant responsible for dislodging the counter-spinning planet seemed to have settled in a new equilibrium position. And the little planet and its moons were bigger threats to each other than to neighboring planets. Thus, the system was more suitable for prolonged habitability than other systems in the *Catalog* they could have visited in this galactic neighborhood.

The focal two planets displayed opposite geological and environmental readings, so the scientific team anticipated they would house divergent species. However, life might have evolved on one of them and migrated to the other. The latter would become apparent if the genes of the two intelligent species proved to have common ancestors. Both were around the same size and distance from their sun as Earth, and were classified as rocky planets, despite both having little rock on their surfaces. Byddwr is a water world, but the water is a mere blanket around the rocky Earth-like interior. Most water worlds are dominated in mass by this liquid, spreading it in layers of super-liquid states. The second planet, Pallos, the one further from the sun, is covered in ice and snow. Pallos has three rather large moons with a 2:1 resonance ratio between them, so the innermost moon orbits twice for every time the middle-moon makes a full cycle; with three of these cycles affecting the planet, there should be several daily high and low tides, but these are not as apparent as they would be on Earth because of the solidified, cold surface. The distance between the Pallos and Byddwr is only .013 AU, or five times greater than the .0026 AU distance between Earth and its Moon.

The nanobots and robots deployed at the start of the slowdown had difficulty spotting the intelligent lifeforms they needed gene samples from on Pallos. After a month of full deployment of the search agents, they found a solitary hunter and collected the microscopic samples and visual images to guide the genetic-editing process. This hunter somehow managed to drop out of the nanobots' investigative range prior to reaching a shelter in the ice, so they lacked images of this species' hidden living quarters.

Armed with the genetic tests, Doctor Nackan-12's team faced a decision on if they would make slight or major adjustments to avoid a new calamitous end. They commenced deliberations without the new pre-destined pilot. Wocega-12 was born in 50,592 BC. She learned of her grandmother's cognitive decline from books and videos. Wocega-12 was scholastic, interested in planetary weather patterns and political systems, instead of favoring high-speed piloting like her ancestors. A minority of doctors and professors on the committee supported a hiring process to fill Wocega's position with a cadata predisposed toward adventurous travel. They cautioned Wocega's gene-line was corrupted with extremist and explosive predilections. Another sect theorized the two-planet system should be explored in unison, necessitating two pilots. If Wocega-12 had crazed genes, her nerves would be soothed on Byddwr, as it had the characteristics of an idyllic retreat. At least one visited water world was more turbulent up-close than it seemed from space. They could pull Wocega-12 out if the social situation deteriorated. On the other hand, if she turned homicidal, leaving her on this foreign planet would be best for keeping cadatas safe. Meanwhile, Wocega's co-pilot, Ortack Steral-12, could travel to Pallos, the world classified as hazardous. This was the sole practical plan because Wocega-12's stoicism was unmatched, and a call for alternative volunteers for this mission returned zero qualified applicants.

Running two genetic transformations at the same time required two separate sections of the ship transforming into environments matching Byddwr and Pallos. Wocega-12's compartment needed more room because of the 30-foot wingspan dimensions of the Byddwr species. It also split its time between ocean swimming, and air flying. Thus, to allow Wocega-12 to practice these skills, sufficient empty air-space, and water depth had to be reserved. The byddwrs' distinguishing feature was an ironing board growth on the back retractable or deployable to frighten rivals. The species also had four wings, with a couple of these stretching to the fin-leg-like structures in the front. Their bodies' unique surface texture (unlike feathers or scales) was suitable for both flying and swimming. They also had a lizard-tail, and a long feeding device at the end of an elongated neck. While the byddwrs were more voluminous, they were nigh-hollow on the inside, a characteristic making them hyper-aerodynamic. This lightness was achieved by their lack of bones or other weighty internal supporting structures. Even their digestive system processed food at an accelerated speed to avoid it weighing them down in flight. They were also reedy in the connecting torso between their wings and feet; this center stretched into a gliding apparatus in flight. Their bodies' composition was looser than an average species, with more empty space between cells. Each cell was also comprised of lighter chemicals. Byddwrs maintained prolonged flight and swam long-distance due to the power concentrated in their wing, leg, and back muscles. Hollowness is commoner in insects than in birds. Like insects, they expanded by consuming their atmosphere's over-abundant oxygen. They breathed through openings in their skin, instead of needing lungs to process air.

Wocega-12 struggled in physical therapy after this transformation. Robots had recorded video showing byddwrs controlling their lank bodies just with the narrow muscle-groups in their appendages. Wocega ascertained their youth was consumed in learning this skill, just as a ballet dancer learns to dance with grace and precision on pointe. Without an experienced trainer, Wocega-12 mimicked the harvest of this dance. To safeguard against serious injury, the air-filled practice dome was outfitted with heavier wall padding to minimize the pain of impacts as she bounced against them. It was also mind-straining to switching between fin-driven swimming and hopping out of the water into flight. They needed the long eight-year stretch for this training to reach a point where Wocega-12 was unhandicapped in this outwardly textbook-designed for multi-environmental movement form. Learning how to manipulate tools and engage in sign-communication with her wing-hands was also a matter of repetitive practice. Wocega-12 lacked time for reverie. She fell asleep within minutes of completing the day's training. A month before arrival, she discovered she was enjoying rapid flight, favoring trick-twists at the top of the dome. She preferred grueling practice to wild sporting action.

While the byddwrs were taller than pallos, Doctor Nackan-12 was twitchier about engineering a pallos in his lab than a byddwr. When questioned, he explained a hyper-aggressive species was needed to survive as a carnivore on an ice planet. His hypothesis about their aggression was confirmed by the extreme length of this creature's claw-like devices and heightened aggression-linked chemical levels in its brain. In fact, Nackan-12 was so distraught by the compilation of these characteristics, he petitioned to send a single exploratory cadata to Byddwr. The leadership unanimously overruled this bid. After ten millennia in space, it was criminal to dismiss even an implausible target. In fact, one reason to travel to this solar system was this duality. Other options had a single bet within thousands of light years. Nally-12 held a secret meeting to discuss demoting Nackan-12 and giving his leadership role to one of his supporting techs because his mind was enfeebled after Nackan-10 was murdered by Wocega-10. Nackan-12 had spent a few hundred years prior to this pre-landing research phase on an extended autopsy of Wocega-10's digitized remains. Wocega-10's original bones had disintegrated, but the Computer retained a molecular-level 4D diagram of Wocega's interior from the moment before death. Nackan-12 concluded the gene-edit to turn Wocega-10 back into a cadata after a prolonged slowed-time existence as a tree had driven Wocega-10 into a mental state conducive to murder. He replayed Wocega-10's final message until he devised a scheme to institute loftier safety precautions. The committee granted him oversight over health and safety protocols. The task of anticipating all potential problems, given the randomness of past medical, scientific, political, social and psychological catastrophes, forced him into a state of perpetual petrification. The Computer edited his List of Probable Doom Scenarios. When Doctor Nackan-12 queried the Computer on the protocol to prevent all derived scenarios, the Computer's summary-answer was to stop exploring. This finding depressed Nackan-12, sapping him of motivation; despondent, he even refrained from updating the gene-editing and pre-landing procedures.

Ortack-12's genetic transformation was shocking for the team involved in this process. Nackan-12 was the least surprised among them with the vicious hostility Ortack-12 began displaying even as he was emerging from the gene-editing machine. He roared as he tossed a table covered with lab equipment out of the way, nearly hitting two medical assistants. Pallos were a hirsute species, with their features disguised by thick fur. Before the crew saw Ortack-12, they smelled his approaching stench. The odor intensified in temperatures above zero °C, outside the optimal range for his dense fur and thick skin. Physical therapy was straining for the trainers because Ortack-12 pushed, shoved, roared, threatened, or violated those who dared proximity. He even attacked the robots cleaning after him; the tidying was needed because he tossed remnants of his food and utilized possessions in haphazard piles along the parameter of his spacious ice-cave. As this dwelling's odor intensified, cadata technicians and therapists began wearing masks despite the air being bio-compatible for cadatas. Upon investigation, the source of the stench was discovered to be a seafood Doctor Nackan-12 replicated to mimic a pallos' diet. This fishy creature was rank in a liquid. Its toxic, prey-stinging, digestive sacs exploded once exposed to pressure lower than underwater. Captain Nally-12 banned production of this fish for the safety of the crew. Its toxins were lethal to cadatas in high concentrations when inhaled or ingested.

They carefully kept the ice-cave just below the average temperatures on Pallos. If it rose above this point, Ortack-12 began sweating at a rate extreme in other species, with water streaming out of his skin in a manner akin to a shower opening. This extraordinary sweat-rate was designed by nature for rapid-cooling of a pallos' body to the low temperature at which it functioned at maximum-capability. When external temperature was a few degrees above zero, pallos' bodies converted water into air through evaporation, allowing heat to escape through this transfer. This was easier in dry conditions, where air was thirsty for moisture. Snow storms indicated pallos had to stay indoors. They were indifferent to the precipitation. Problematically, the high air moisture level responsible for snow-formation also caused pallos to overheat during vigorous exercises, such as running after prey. This property also explained the pallos' dislike for clothing: their faces and sensitive body parts had to be exposed to air for this exchange.

Ortack-12 and Wocega-12 were unfit mentally and physically to commence the settlement experiments. Despite their state, the ship's completion of the asteroid-mimicking deacceleration stage in 50,060 BC was the deadline ending their training. As a test of her underlying sanity, Wocega-12 was asked to pilot the Mothership in its landing on the far-side of the closest moon to Pallos. Wocega-12 adjusted the landing protocol to mimic an asteroid collision. She adjusted the speed and direction during the landing to correct Computer-generated misalignments. The spaceship kept approaching the moon until it created a shockwave in front of it to push sufficient dirt and rocks up from the moon's surface to make its appearance consistent with an asteroid impact. In the cover this debris cloud generated, the spaceship express-deaccelerated to near-zero just before it soft-landed on the polished-by-the-blast ground. Internal gravity stabilizers minimized the force of the resulting g-force.

Wocega-12 and the exploratory team suited and play-hopped across the moon to stretch after a lifetime in space. This moon's mass was sufficient to house an iron-rich core. This core was clothed in a molten silicate mantle. A thin silicate crust draped its surface. Geological tests indicated the chemicals in its layers were mixed, revealing a history of repeated melting. Despite the surface cooling in the last billion years, lava overflowed from geyser-volcanoes, forming temporary, semi-liquid lakes. The vents spewed yellowish-orange sulfur and sulfur dioxide rains, which rush-metamorphosed from liquid to steam and hissed out of cracks with ferocious velocity. Even in shielding suits, the team avoided these vents because the steam could catapult them to dangerous heights in the low-gravity environment. The atmosphere's thinness generated forceful dust storms because of the inadequate pressure to suppress rapid shifts in air temperature and wind direction. The team experienced this first-hand when they air-glided into a dust storm. In the first three hours, they marveled at the beatific sulfur rain and the bright-yellow outline of the moon. Then, the crashing sulfur and lava clumped to their spacesuits, lowering visibility and audibility, so they retreated. Given the reports they brought back, few cadatas opted to go outdoors here for regular exercise, but some braved these elements just to touch natural ground. By the time the moon made its two-day orbit around Pallos, Wocega-12 and Ortack-12 were prepped and packed for the missions and were shuttled out. Ortack-12 struggled to maintain sufficient composure to write computer code. So, his shuttle was pre-programmed for the landing.

***

Wocega-12's flight to Byddwr required her active participation because of the nature of the primary mode of assisted air transportation on this planet: ballooning. Byddwrs' wings allowed them to fly at a low altitude. But if they wanted to fly a great distance at a faster speed, or at a higher altitude; they had to depend on the technological advantages of light air's buoyancy in heavier atmospheric air-density. This atmosphere reached to 30 miles into the sky. This is equivalent in height to the tip of the stratosphere on Earth. Byddwr had a thicker atmosphere than on Earth, which gave at least 10 extra miles of flyable space, up to the 40-mile mark. Byddwr's thin atmosphere required airplanes with larger wingspans than on Earth. They also demanded greater thrust to maintain flight. Earth-style, huge, metal lumps failed to remain airborne without wings a dozen times greater than the length of the plane. In contrast, a balloon's light material (rather than metallic) weight, engrained buoyancy and wind-propulsion made them ideal for Byddwr's atmosphere.

Wocega-12 took off from the spaceship in one of their standard shuttles and flew over from Pallos' moon to Byddwr. Once she arrived at the edge of the atmosphere, she began a gradual descent. The engineers disguised a petite shuttle model in reflective cloud-shading to minimize the radiation it emitted during energy-processing. Wocega-12 landed in a spot where a mushroom of clouds had flattened and condensed against the ozone barrier for the privacy offered by this cloud-cover. Concealment was necessary if byddwrs possessed advanced air vehicle detection technologies cadatas might have failed to discover during their observation phase. At the 40-mile point, Wocega-12 triggered the transformation procedure and the shuttle mutated its outward appearance. The hardware was squeezed into a flat stand below Wocega-12's feet-fins. The top expanded pre-condensed helium into a long, rectangular balloon, a common shape on Byddwr because it is stretchable, unlike round or angular shapes. They were decorated with bright pictures of admired byddwrs in flight. Their platforms were rail-less because byddwr's flight-capacity made a potential fall non-life-threatening. Wocega-12 refrained from testing a jump out of a balloon because of the risk of freezing or rapid pressure fluctuations.

99% of Byddwr's atmosphere is below the 40-mile line, necessitating an artificial air supply when traveling at the tip of this pyramid. If Wocega-12 transformed the shuttle in the thermosphere, or at 50 miles, the surrounding pressure would have been an ultra-low .01 hPa. Even if she could survive exposure to the crispy negative temperature, the low pressure would mean too few molecules to operate the byddwr air-condensing device. Isolated molecules in the thermosphere are energetic because they absorb an excess of energetic rays, keeping this kinetic energy longer because of their lower collision rate. This retention gives these molecules a high reading on human temperature-measuring devices. Despite this deceptive "hot" temperature, the low air density means a near-absence of heat per cubic foot. Exposure to this frozen-in-immobility atmosphere is dangerous to life with a required minimum internal temperature.

Dense air creates weather. Therefore, without air, wind and storms cannot form. Going above wind currents deprives a balloon of a natural propulsion force. Another obstacle to balloon flight above the troposphere is the insufficiency of breathable air, necessitating gas masks and then pressure-suits. The balloon was outfitted with standard byddwr air-condensing devices packaging the thin air at high altitudes into the proportions necessary for byddwrs' survival; this was preferable to carrying a fresh air reserve from the surface as an added weight. In the moments after the transformation, Wocega-12 felt warm because she was inside an ozone-layer-equivalent layer, which retained much of the sun's incoming ultraviolet light. As the balloon lost altitude, she began shivering. To prevent hypothermia, she dressed in a blanketing, cushioned dress.

Wind was strongest just below the height where it seized. It lurched Wocega-12 in its direction. Wind, in general, is created by horizontal pressure variations. If the pressure at the top of the Rocky Mountains is 600 hPa is lower than the 1030 hPa at sea level; this contrast forces air to flow from the region with the higher pressure to the lower one. Byddwr's surface was free of colossal mountains. In place of altitudinal variations, the planet's semi-elongated shape, and contrasting hot and cold-water currents (due to underwater ridges and valleys) created sufficient air pressure differentiation to generate the wind-speeds compulsory for flight. The fastest-flowing air currents were at the top of the troposphere because of a lack of obstacles such as waves or land masses in the wind's path. Without such resistance, it maintained top 100-knot speeds in lengthy, one-directional jet streams, circling the planet at fixed altitude ranges. Byddwrs used these streams for rapid global transportation.

She adjusted a fin to align the travel-direction with the distant byddwr city. She checked the gas-heating device; its fluctuations lowered (during cooling) and raised (during heating) the balloon. If this heater broke, the helium in the balloon could condense into a cloud, causing it to plummet. Once the condensation was critical, even losing weight-bags would only generate slight altitude adjustments.

Studying the reddish hue of the landscape below, Wocega-12 contemplated the moist sensation the air produced on her skin. Even at midday, elements of this scene were reddened and yellowed. In contrast, this water would have been blue, and these clouds white on Earth. The reddening was generated by the thicker atmosphere, which only higher wavelengths could penetrate. The planet appeared to be in a perpetual sunset. Through this diffused, glassy mirror, Wocega-12 spotted unusual species living at that atmospheric height, close to the vacuum of space, with sparse air or organisms to feed on. One of these lifeforms is ephrams—tiny, elongated-balloon creatures with compositions dominated by air, enveloped in a thin blanket of skin for cohesion and side wings for directionality. They approached Wocega-12's balloon, anticipating a feeding. The flock grew. Two dropped uncouth loads onto her platform, inspiring repulsion in Wocega-12 despite their outward cuteness. The ephrams' bodies auto-regulated their internal air temperature, a quality inspiring byddwrs to invent balloon-travel millennia earlier. Lower down, Wocega-12 encountered species of insects feeding on microbes and seeds sprayed into the wind by ocean-surface plants.

The research robots secured Byddwr flight maps with unseen features to ease manual navigation. Wocega-12 refrained from delegating navigation to the hidden cadata Computer. It could have steered the balloon with millimeter-precision with a miniature propulsion engine requiring a drop of matter to circle the planet. Instead, she practiced native-byddwr navigation, anticipating she would need this skill if she had to drive a byddwr passenger in the future. She studied the specialized maps in advance. Now, hovering miles away from a turbulent ocean, she realized the significance of the regions with pockets of rising warm air called thermals. The heating device had a limited power supply, so if the balloon started losing altitude, she could push it back up by changing her intended direction towards the nearest thermal, and letting its energy contribute to her upswing. These warm air streams were invisible to the naked eye, but they remained in the same regions because of underwater features affecting the air temperature. Woce-ga-12 retained gene-edited cadata vision allowing her to spot heat-differences, and thus to utilize these streams to perform tricks only an advanced byddwr pilot might have dared. A byddwr would spot the edges of these formations because this is where clouds tend to form.

As the balloon departed from the heated ozone layer, the temperature dropped to below-zero, sending a chill across Wocega-12's skin. She turned up the heater to avoid balloon air-freezing. She refrained from maximizing the heat to benefit from the temperature drop to decrease height speedier across this cold zone to reach the warmer air region near the ocean. During this altitude-loss, she saw ice crystals forming virga streams in the air surrounding her. This virga was blown by upper-level winds, creating veil-like cirrus clouds. They were picturesque, so Wocega-12 headed for them, but as she penetrated this veil the larger-than-Earth's crystals stung her skin, inspiring her to avoid these sublime but chilly temptresses going forward.

The lack of land features such as continents below focused Wocega-12's vision on the strange patterns in the behavior of clouds and deviant atmospheric elements. The denser atmosphere and heavier moisture content formed and dissipated cloud pockets, which built into 6-mile towers in minutes, creating the illusion a magician was attempting to build a castle in the sky, but was dissatisfied with each new iteration of the columns, and was smashing these as they neared completion.

When Wocega-12 spotted balloons in the sky and free-gliding byddwrs, she checked the measurements and confirmed, as her skin indicated, it was much warmer at this low altitude. It was always pleasant in the air within two miles of the ocean, without major temperature fluctuations between day and night.

There was a light breeze as Wocega-12's balloon patted-down and floated on the water's surface. Wocega-12 deflated the balloon to keep it from being pushed too far by the wind while she continued her survey underwater. It was a gullible culture, without need for keys or security locks to avoid balloon theft. And if a balloon floated too far from the initial parking spot, byddwrs would go out of their way to reunite the owner and the vehicle. As Wocega-12 dipped her fin-foot into the water, she recalled an ocean heats five times slower than soil. It gave her a chill until she dived in, which cooled her skin to better match this lower temperature.

The balloon flight was tranquil. The attitudes of the byddwrs, who Wocega-12 en-countered as she swam towards the underwater city destination, were laidback and wel-coming. Unlike on other planets, nobody asked Wocega-12 where she was from, or what business she had in their neighborhood. When she asked for employment, she was ques-tioned on her skills. When she explained she had few of these, she was taken to a weav-ing line, where the skins used in byddwr balloons were being crafted, before they were sent up to the surface to be dried and painted. As her skills grew, she was promoted to the respected painter role. When she was on the geologically lower tasks, she slept with fellow workers in mud-baths on the shallow ocean floor, curling into shells filled with a moist, marshy substance purifying to the skin. By the time Wocega-12 was promoted to painter, she joined the "upper" dwelling flocks, who slept upside down by grabbing onto the top masts of their balloons. Since she already had a balloon when she arrived, she could have opted for this sleep-location from the start, but she preferred to socialize with her neighboring coworkers, and the slimy underwater bedding was more comfortable to a byddwr's boneless frame than the awkward upside-down maneuver. Wocega-12 was amazed fellow workers refrained from displaying jealousy towards her balloon. They also refrained from interrogating why she was granted this luxury, awarded to painters after their first year of excellent service. They also failed to see oddity in her ignorance on topics residents of this planet knew since childhood. For example, she had to research their methods for sporadic mid-flight naps. They napped on-instinct during prolonged flights since childhood, so they struggled to explain the rules behind this process to Wocega-12 in the technical language she preferred. After arduous study, Wocega-12 managed to snooze for a few minutes on material-collection flights to neighboring cities; she considered these instances as peaks of her cultural and physical acclamation.

As a painter, she was invited to join a few government meetings. She even witnessed planetary Assembly meetings, which were supervised by the two Guides who shared the top leadership position on Byddwr. She made a few helpful suggestions on more efficient resource-use, and this kept her on the leadership's list of prized advisors. Overall, while Wocega-12 worked from dusk-till-dawn, this lifestyle appeared laid-back. Continuous worked failed to register as stressful or strenuous. The work-environment was further enhanced by mutual-support of co-workers, in place of struggling to outdo the competition. However, stress entered Wocega-12's life as she was promoted into leadership roles. She learned the quantity of goods they were producing outstripped what their planet could ever consume. Because it was her job to understand the society she was researching, near the end of her experimental ten-year stay, she started pushing leaders to explain where this extra stuff was going, or why it was being produced. The answer turned an idyllic relaxation retreat, into another cadata-nudged-disaster.

***

Ortack-12 departed after Wocega-12 reported a safe-landing on Byddwr. He had been standing-by in case she needed a rescue pilot. The odds of a catastrophic crash on a first exploration were high. Because Ortack-12 was already gene-edited into a species from this solar system, it was less invasive to send him, than a team with a cadata appearance. They had not detected inter-species mingling between the two systems, but the inaccessible parts of Pallos hinted at a higher level of technological development than the surface suggested. Nackan-12 administered a few adjustments to Ortack-12's genes and chemical composition prior to sending him on the flight to Pallos because their tests indicated Ortack-12 might have lacked the patience needed for planetary exploration if he was an exacter biological replica of a pallos. He risked standing-out among the hyper-aggressive pallos if his character was softened. If he instead joined the natives' revelries, he would be neglecting cadatas' tight-scheduled scientific tests to determine if Pallos was settlement-compatible. As it stood, Ortack-12 made regular recorded reports across his flight to Pallos, which was shorter than Wocega-12's as he just had to drop down from one of its moons.

The planet looked like a foggy, carved, ferny snow-marble from the moon, and as Ortack-12 neared it, the snow-covered mountains, valleys, and ice lakes became somewhat visible through the dense, dark clouds enveloping most of its surface. Geological tests indicated Pallos was undergoing an ice age after a short warming, when intelligent beings developed in pleasanter conditions. Now the average temperature was hovering at just below water-freezing level. In general, a rocky world cannot be less than 1.8 times of Earth's size without losing hydrogen and helium from its atmosphere; if these chemicals are absent, heavier chemicals fail to cling to the air until it turns into an airless planet. Pallos was a bit smaller than Earth, and a bit further from its star of similar brightness, so it retained a semi-dense atmosphere. However, the snow and clouds enveloping it during its latest ice age, were 89% reflective, a high albedo percentage, so light bounced off instead of staying near the planet, keeping its average temperature low. Its ice was less reflective, at around 55%. Though most small ice lakes were also covered with snow. Thus, open ice comprised a small percentage of the landscape.

While Pallos' atmosphere was intact, the air had less mass and pressure to it; low air pressure turned its weather volatile, leading to greater cyclonic weather systems, which were dominated by snow storms, hail, and high winds at its lower temperatures. The shuttle was disguised from the outside to look like a snowed-over hillock. The bulk of the flight was pre-programmed because the team anticipated his pallos-attitude would inspire Ortack-12 to destroy his control console, or to skydive prior to reaching a safe altitude. However, Ortack-12 retained final control functions. Upon penetrating through a thick layer of dirty clouds, he saw enormous snowflakes hitting the ship; these mesmerized him, so he switched to manual to fly into a region where these were swirling. Ortack-12 flew pirouettes, figure eights, barrel rolls, rolling scissors and spirals, approaching the ground before near-miss recoveries, which turned into new spiral patterns. He also executed maneuvers inconceivable with human flying machines, so they lack equivalents in Earth's languages. Regardless of their names, these countered cadata protocols designed to disguise space flight towards and near a planet under exploration. They had spent eight years soft-landing on Pallos' moon in an asteroid-mimicking manner to minimize detection odds by the local species. Ortack-12 was tossing these efforts away in minutes of manic flight prone to trigger regional detectors, even if these were as rudimentary as lookouts scanning the skies without enhancing equipment. He tossed caution aside to glare at giant snowflakes. His pallos instincts matched a child's, despite his brain remaining in its cadata form.

Data about the size of these snowflakes had been in the reports he should have read prior to departure, but their actual appearance and the strange ways they seemed to play in the air despite their dimensions were far more mesmerizing. These reports explain what is of more interest, in retrospect, to scientists. Pallos' snowflakes were enlarged because of a few unusual characteristics of this planet's atmosphere. For comparison, cloud water droplets on Earth are 10 μm, while raindrops are 1 mm, a growth of millionfold in volume; so, to make giant snowflakes even more astronomical volume magnification is necessary. A driving force of this growth was Pallos' over-abundance of aerosols, or small solid particles entering their air before the onset of the Ice Age. Snowflakes form when water molecules attach to aerosols. Aerosols around Pallos were bits of dust, smoke from fires, pollution from factories, and salt and other seawater spray components. These readings suggested rampant pollution prior to the downturn in temperatures. Some of this happened in the ten millennia cadatas needed to reach the system; its readings were purer when cadatas departed for it. Despite alternations, they decided a wasteland was preferred over commencing a new journey without an inspection. If Pallos was capable of a downward dive in a few years, it might be pliable to swift terraforming back to its fruitful condition. Pallos' aerosols were so large, they must have caused respiratory problems for living organisms on the surface. Minor species developed evolutionary adaptations to withstand breathing these into their lungs. The dominant species retreated under the snow or into artificially or naturally air-filtered environments. The bigger the aerosol clump, the more massive the snowflakes forming on it, as it has more surface area for water to freeze on. Pallos' poles were colder than its equator, so this temperature contrast migrated snow clouds. These clouds formed when water managed to melt at a few degrees above 0°C, and then evaporated; this water vapor crystalized in the colder air high above the planet's surface. The snowflakes' growth was enhanced by the high humidity and temperatures hovering not far below freezing, as this kept the snow soft and sticky enough to overgrow without gaining excessive weight. Because Pallos has weaker gravity than Earth, it has a greater updraft, an important ingredient separating the formation of snow from hail; larger icy planets experience chunkier ice and vaster hail storms, whereas smaller planets allow for this soft, sticky, and fluffy snow-consistency, which can turn into giant-fluffy-snowflakes at this narrow border between extreme below-zero and well-above-zero planets. The volume of the snowflakes is further enhanced by the rapidly dropping temperature at the tip of the snow-forming clouds on Pallos. Because Pallos' atmosphere is thinner, there is less molecular motion at the top of this part of its atmosphere, so temperatures plummet to −50°C or lower; these conditions allow for the formation of supercooled liquids, which donate their moisture to grow a three-dimensional, porcupine, giant snowflakes. The supercooled liquids facilitate the process of smaller snowflakes sticking together into larger crystals through aggregation. Pallos' lower gravity also enhances these ice crystals' fragmentation resistance, avoiding a premature fall before it can reach the larger volume.

From the ground, now exposed to the elements without a pallos-style full-body-covering outfit, Ortack-12 saw these snowflakes otherwise. Each was the size of a human hand. Being hit with one was unlike a hit from a snowball on Earth; it was a heavy blow rather than a snowflake tinkle. When it hit skin or snowed-over ground, it fragmented into an explosion of ice splinters. Snowflakes on Earth fall into the following categories: thin plates, needles, hollow columns, and dendrites with complex branches. The giant snowflakes on Pallos are primarily the latter, but with knottier designs than Earth's atmosphere allows. Given the larger volume potential, the supercooled liquids can create complex and random variations, including non-symmetrical shapes. Flakes on Earth are said to be hyper-symmetrical. H_2O mixes with several other chemicals as well as with much more abundant aerosols on Pallos, and this unpredictable mixture creates inconsistent snow structures.

Ortack-12 might have continued to play in this hyper-snow forever, but at the tip of a spectacular upward spiral, his system stalled and the shuttle began plummeting. The Computer deployed nanobots and robots to the affected area, and they fixed the iced energy-processing mechanism. Human fiction repeats a trope wherein a biological hero fixes a mechanical problem of this variety mid-space-flight. However, anybody who attempts in practice to untangle even a minor loop in a parachute during a skydive will testify fixing a basic car engine in these seconds is impossible. Cadatas survived traveling as far in the galaxy as we have due to instant-assistance from our robotic army, deployed by the Computer in pre-programmed emergency scenarios. It would have controlled the ice-level through automatic, localized exterior-heating, but this planet had a sudden change in temperature in its atmosphere, so the Computer's predictions failed to match this reality. The supercooled liquid froze on contact with the body of the shuttle, and began forming some of those giant snowflakes like barnacles on a whale but with incredible rapidity. If an expert cadata mechanic attempted to step outside to fix this problem in a top-grade spacesuit, the suit's exterior would have solidified in an instant, necessitating nanobots to unfreeze it. While warming the location, the Computer also gathered data on these causes for the malfunction, and created a subroutine for avoiding it on future flights through this section of the atmosphere. At around the same time as the Computer finished this automated programming process, the ice was steamed off and the shuttle's functionality was returned. Then, the Computer auto-set it to hover 232-feet above-ground. If the system was a mite slower, or if the freeze occurred at a tad lower altitude, Ortack-12 would have died from crash-landing into the ocean.

Ortack-12 heeded with impatience Computer's oral incident summary report, also broadcast to the Mothership's leadership, who were tuned into these tribulations. Then, Ortack-12 ordered the Computer to finish landing at the pre-determined spot on the planet, as he reclined for a nap. It was as light out as the climate on Pallos allowed when Ortack-12 awoke. He gathered the compressed bag of items he needed for the expedition, all outward mimicries of local goods. It was a larger reserve than on prior planetary explorations because preliminary tests failed to affix a location of an entryway into a settlement with greater than a single resident pallos. At least one door-like location was buried under several feet of snow, but no attempt seemed to have been made to dig it out. They did not know if Ortack-12 would need an hour or years to find the nearest village, but they knew Ortack-12 had to make this trip on his pallos-feet to avoid spooking local inhabitants. They placed traps to detect and block mechanical and living creatures from uncovering their domiciles. Ortack-12 grumpily abandoned the shuttle, which executed a concealment maneuver to blend it into the landscape minutes afterwards. Ortack-12's body was designed for this climate, so he walked until sunset without tiring. He had a large, condensed-meat dinner, watching the last rays of the sun. Then, he dug a hole in the snow mimicking a site the computerized explorations had located. He had some cadata luxuries shielding him from weather extremes inside this hole. They maintained a low temperature optimum for a pallos' physiology. He spent the following three weeks in a continuous walking, eating, and sleeping pattern without discoveries, except for spotting wild game. However, since use of cadata weapons was forbidden, having failed to catch the game with his bare body, he continued eating his condensed supply. Then, one morning he spotted what looked like smoke on the foggy horizon. He sped up to a run to reach it before whoever started the fire fled; upon arriving, he found snow fluffed to conceal footprints, and no evident source of a fire in sight.

A week later, Ortack-12 ran out of condensed water, and had to drink the local snow. The snow tasted dirty, but Ortack-12 was so thirsty, this went undiscerned. This strange taste gained significance when a few hours later he felt its painful purgative effect on his bowels. He had chemical-testing equipment with him to inspect the quality and to filter out toxic elements from this frozen water before drinking it, but his pallos-nature prevented utilization of caution. After the illness struck, Ortack-12 gave up on these displays of endurance and allowed the cadata tech to purify whatever he was about to consume.

It would benefit a dramatic narrative for Ortack-12 to discover an amazing method to sniff out where the pallos were hiding at the very moment his food supply is dwindling to the point of necessitating a return to the shuttle for resupply. In actuality, the scientific team and the nanobots deserved the credit for the reversal in Ortack-12's fortune. They had continued the survey using the real-life data Ortack-12 was gathering across this lengthy hike to come up with a better model for spotting pallos across the planet. The few sightings they had made were all at around the same time of day. They noticed Ortack-12 woke unassisted and preferred commencing walks at just before the meteorological sunset. He was capable of walking non-stop because he returned for a long night's rest a couple of hours after sunset. Nally-12 and Doctor Nackan-12 were complaining about this short workday when they realized any behavioral pattern was relevant for understanding this hard-to-find species. With the question precisely-worded, the Computer determined the pallos had a biology conducive to this unique over-sleep cycle. They hibernated for most of the day and night, emerging into the cold near the meteorological sunset. This is the moment, on any planet, when the incoming solar radiation becomes less than the outgoing infrared radiation; this is when temperatures reach their maximum peak. This always happens before the actual sunset. The pallos had started to evolve more acute night-vision to see better in this waning light. Since a peak is the point at the start of a decline, temperature always begins falling at the meteorological sunset, but slowly enough the pallos can function until a couple of hours after visible sunset. Once they understood this pattern, the team organized the data on pallos-spottings, and directed Ortack-12 to the closest village. Ortack-12 made a few circles without finding inhabitants, so he initiated screeching from frustration. The scientific cadata team assumed he had doomed the effort as the pallos would never reveal themselves if there was a hollering madman approaching. They began plotting Ortack-12's retreat to the shuttle, or perhaps even the idea of sending a robot with the supplies to minimize lost travel-time. Just then, a group of pallos emerged from the snow, and pushed and pulled on Ortack-12 in a strange, bearish greeting. Then, they shoved him towards a slight, messy, somewhat snowed-over entrance and pushed him down this shoot. Apparently, no self-respecting pallos would have allowed other pallos to escape from his grasp (as they did when he spotted their fire) without raising enough noise to cause an avalanche. When he did become boisterous, they accepted him as a compatriot, and invited him into their community. He might have had a softened response to weeks of unsuccessful exploration because of the cadata features Doctor Nackan-12 added to keep Ortack-12 "civilized".

The doors into the underground tunnels were biometrically-triggered, and this was the reason cadatas' tech failed to gain access. The tiny robots and nanobots in Ortack-12's baggage and on his person, spread across this tunnel network upon his entry to craft a map of this hidden world. Meanwhile, the pallos took Ortack-12 to their feeding quarters. Nobody at this common feeding station—cluttered with chopped carcasses of local species—looked, or acted intellectually-inclined, as they groaned and grabbed food. The nanobots learned this group-living arrangement was unusual for the pallos, who preferred living alone in isolated outposts. This was a shelter for those who were too old or too weak to hunt for themselves. The complex included an engineering and a mechanical government branch, which controlled this planetary segment. This agency included a few outlier pallos, who had ideas about biosignature readers and advanced travel technology. These small groups of braniacs had pushed the planet so it registered as advanced on cadatas' galactic scale. The science labs and military offices had additional protections on their entry points, so the nanobots were unable to penetrate these. Given this limitation, the science team informed Ortack-12, his legwork was necessary for them to gain intelligence. Ortack-12 set out on this espionage mission by continuing to blend in with the locals. Because most indoor-hours were spent by the sheltered pallos on feeding and sleeping, Ortack-12 felt obliged to join them in their meals. The problem was this on-the-surface simple undertaking was Ortack-12's recollection of his sickening snow-drinking. It terrified him to consume foul-smelling and bug-infested meat dragged in by hunting parties, and deemed low-grade enough to dispense to the unemployed. Ortack-12 mumbled an order directing his nanobots to test the meat for dangerous pathogens. It turned out he was right to worry as this meat could only be digested by locals because they developed immunity to its pathogen volume and diversity. The Computer system controlling these nanobots auto-calculated the medical treatment via application of probiotic equivalents necessary for Ortack-12's digestive system to process ultra-toxic food going forward.

Ortack-12's taste-sensors appreciated this visually-repelling food. The living quarters Ortack-12 were also tested for lethal bugs and destructive organisms. The common room was toxic-smelling, and grotesque amidst the clutter of garbage and waste lining it. It was a social sin to clean waste, which had to be left wherever it was deposited. During meat consumption, bones were tossed over the shoulder or at other pallos, rather than into any kind of a waste basket. Most pallos deliberately slept on a pile of garbage because it enhanced their personal-scent. As an anti-social species, the more repelling a pallos smelled, the more he or she was succeeding in achieving their full potential. If they all smelled foul, they were interaction was improbable; if they never exchanged words, peace was possible.

Given these deterrents, the pallos were keeping to themselves in their communal sleeping and living room cave. But Ortack-12's mission was to interact with them to gather information. He resisted this task for a couple of days, but then decided remaining in the icy crevice without developments was worse than whatever might unravel if he proceeded with questioning his shelter-mates. He had a light grasp of the pallos' language, but the pallos around him also had a low vocabulary storage, so he did manage to carry out some conversations. The first few of these did not gather much intelligence aside from hunting strategies from their youth, methods for processing meat for storage, and the size of the biggest snowflake they had ever seen. Then, he tried to talk with one of the younger pallos in the shelter, Tickus, who proved to be in no mood for interrogation. Tickus had lost a territorial competition to a rival, which left him with serious wounds, from which he had been recovering at the shelter for a year. Without listening to Ortack-12's questions, Tickus challenged him to a death-match, interpreting a rival male's approach as a challenge to his capacity to his capacity to dominate a territory of his own. Ortack-12 tried to apologize and to explain he did not want to fight Tickus, but Tickus was set on it. Observers decided, by interacting with Tickus, Ortack-12 suggested they had to settle the question of who held the power position between them. When Ortack-12 was further informed the forthcoming fight would be to the death, he once again commanded his nanobots to help enhance his odds of survival. Ortack-12 was somewhat smaller than Tickus in height and girth, so the nanobots swiftly enlarged his muscles, achieving a larger body-mass than Tickus in two hours. During which Ortack-12 ate nonstop to provide the matter essential for this conversion. He tried to find a dark corner for this feast, to avoid being spotted "doping". While some of the pallos glanced at him sideways when he emerged for the fight bulging with musculature, none of them objected as it would have been against their code to complain about potential cheating. While in most human films, the hero would spare the evil challenger, in this reality, Ortack-12 killed Tickus in the ensuing brief fight. Ortack-12 relied on fighting strategies he learned from alien cultures to perform flips, kicks, pushes, and tricks unimagined by pallos. The fight concluded when Ortack-12 knocked Tickus' head into a chunk of ice. Tickus laid a few faltering punches on Ortack-12 as this barrage was unraveling. Ortack-12 did not consider stopping to ask if Tickus would consider surrendering before the final blow because leaving the enemy a moment to gather his strength and come up with a better strategy would have left him vulnerable to death. The crowd cheered, and Ortack-12 seemed to be climbing in their respect, but then he acted in a manner grotesque and taboo in pallos' culture. He complained to a meat distributing shelter worker about being attacked and intimidated into the fight by Tickus. Ortack-12 thought this was necessary to explain a death at his hands, but he learned killing in a duel was legal and honorable on Pallos, but complaining about the challenge was illegal and disrespectful.

Since cadatas were still trying to grasp the Wocega-10 homicide-suicide incident, they watched with interest as Ortack-12 was escorted into a courtroom for a hearing on the charge of Protesting Against Combat Trial. Cadatas followed these proceedings, hoping to be inspired with ideas for improving their own criminal code. The cadata research team was surprised to learn Ortack-12 was forbidden from defending himself. Other than being told the general charge, he was not informed of the facts on which the accusation was based, and the other side had no legal responsibility to make this clear to any of the involved parties. The main hearing unraveled minutes after Ortack-12 was accused in a public room crowded with spectators. The pallos to whom Ortack-12 made the complaint repeated Ortack-12's words with embellishments designed to make Ortack-12 appear to be hyper-critical of the pallos' culture; he stressed Ortack-12's whining and unbecoming manner of groveling to escape a righteous challenge. After this indictment, Ortack-12 was informed by the judge the verdict about to be delivered was un-appealable. Ortack-12 asked for time to read their Criminal Code to plan a defense for his actions.[4] The Judge shouted back: it was an "act of war" to threaten to read the Criminal Code while attempting to reduce guilt. Ortack-12 asked what the punishment for Protesting Against Combat was. He was told, all punishments on Pallos were capital. Protesting Against Combat was among the most serious crimes in this un-seeable Code because it was a crime against the security of the state. This category contained over

---

4   This system does not provide free lawyers to indigent defenders, in contrast with most countries on Earth.

half of its thousands of articles. The Judge asked Ortack-12 to approach the bench. He clarified the sole road for Ortack-12 to flee capital punishment in this case was through a goods confiscation. Ortack-12 offered his bag to the Judge, but the Judge hinted with signals it would take an added series of gifts to his personal coffers to resolve the matter. Ortack-12 agreed, and asked for a chance to gather some gifts for the Judge from a reserve he said he had gathered outside their underground compound. Since the shelters of the region were connected and there was no way for Ortack-12 to survive if he could not make it on his own and asked for help from the shelter earlier, he was allowed to go back out into the cold to retrieve these gifts on his own, or without a guard to make sure he complied.

Once Ortack-12 was back outside, he asked for cadatas to retrieve him in a shuttle, assuming, since his capital punishment was on the table, the experiment was concluded. To his surprise, Nally-12 and the others opted to send him back in with the promised gifts. They decided it was sufficient to place a high-tech robot on Ortack-12 to assist with fighting his way out, if the gifts were judged wanting. Ortack-12 agreed once the team reminded him how furious he was at the Judge, to the point of wanting to watch the robot massacre the goons guarding the courthouse to aid his escape. The cadatas replicated the types of gifts (meat, clothing, and weapons) mimicking valuables on Pallos. Laden with a bag full of these, Ortack-12 returned to the place where he exited a tunnel, but this opening was now untraceable, even though the cadatas had marked its GPS-equivalent on their map. Ortack-12 started hollering insults again, and a pallos emerged to recover him and push him down an entry shoot. He was escorted to a more private, smaller cave, where the Judge and a group of politicians led a chalk-melting ritual. They turned on a mechanical device, which triggered the melting of a chalk block, and then each inhaled the fumes emitting from this block. The light-blue chalk turned into a fire-lava texture and began falling into itself, releasing a cloud of smoke smelling as toxic as it was. The nanobots warned prolonged inhalation of the substances would be lethal, but a few breaths were survivable, so Ortack-12 breathed it in to avoid committing a new crime against the state. Once they were all intoxicated by these fumes, they passed Ortack-12's gifts around the circle. As they did so, they insulted the low quality of these gifts and Ortack-12 for daring to make such sparse offerings. Then, they demand more gifts. Ortack-12's stress approached a violent response, which seemed welcomed by this justice system's approval of duels. The cadata researched team warned against a new duel through the comms, insisting he accepted the insults and exited to retrieve a still larger pre-replicated gift-package. Ortack-12 obliged and after one more chalk-burning ritual and a new round of insults, the Court accepted these as a settlement for his transgression, and released him.

As the cadatas observe this system, we created our own laws, called the 50,060 BC Criminal Procedure Code. The use of capital punishment for all crimes was inspiring to some. But most of the professors and jurists on the review-board recommended gentle laws to discourage a similar incident happening to Wocega-10's. Our continued inability to find a suitable planet, and the growing psychosis among the crew necessitated theoretical repercussions. However, these have remained unutilized given the tiny population on our spaceship, a group which might be the last of our species if all who remained on Cadata have died with our old planet.

After the trial, the Judge, Eeboth, approached Ortack-12 and proposed they go on a hunting trip. Eeboth appreciated the gifts. He realized Ortack-12 strayed into trouble at the shelter because he was attempting to bond with pallos who valued their solitude. Eeboth discussed this problem with the politicians participating in the burning, and concluded Ortack-12 had to be an orphan without a mother to teach him solitary-hunter survival skills required for able-bodied pallos. Eeboth volunteered to teach Ortack-12 these skills now to keep him from returning to the Court. It was their ritual to criticize the gifts they were receiving; in fact, they were overwhelmed by the quantity and value of the gifts Ortack-12 gave in the first round, so the second set put them in his debt. It was a challenge for Ortack-12 to venture out near Eeboth. But the cold closed their sinuses. So, they could not smell each other across most of their expedition.

Eeboth directed Ortack-12 in the procedure for rapid-construction of snow-caves for night-stays during the search for good hunting grounds. Most of an adult pallos' life was spent on these solitary hunts. Eeboth expressed worry Ortack-12 would spoil the hunt by making unwanted noises, scaring away prey, a concern Ortack-12 attempted to remedy by mimicking Eeboth stealth movements. They are un-cautious by Earth-standards as a pallos' body is prone to loud digestive and perspiratory noises even at rest. Eeboth explained strategies for finding immense herds of animals grazing on peculiar, one-stalk shrubs penetrating the thick layer of snow even in near-zero temperatures. In fact, one way to follow these herds was to find pre-eaten shrubs before identifying nearby regions with fresh shrubs, where the herd might migrate next. At the peak-temperature point, these solitary buds bloomed into mini-fruits only to die in two hours from the cold, before a new rebirth on the following day. When they would happen upon a herd, they would each chase one of the beasts on-foot, catch them between their thick claws, and then tear them down with their teeth. This was the sole acceptable method to kill an organism running free in the open; using a weapon against an unarmed non-pallos was taboo; killing an unarmed pallos was acceptable because it was also taboo to let oneself be murdered without mounting a defense. The duo traveled around a mile per day following these scattered buds, camping in cramped snow compartments at night. Ortack-12 did his best to detach from inhaling Eeboth's smell in these times by scowling at the three moons in the sky as they adorned various waning or waxing forms.

One of these nights, Ortack-12 was tossing from the stench and the rough icy-snow below his back. Just as he lost consciousness, the ground they were lying on started quaking. Eeboth just had enough time to grab their bags, and to push Ortack-12 out of their snow-trap. They scrambled up a newly-forming incline amidst the tremors. Chunks of what started to feel like glass were shattering and breaking just as their feet took off for a higher position. Once they sensed they had gained solid ground again, they stopped and turned to study what had transpired. They had parked on a snow-roofed ice lake. This ice sheet had broken and formed a ravine. It was a local summer and temperatures were just a bit above zero, so the ice weakened somewhat, but it was still too cold for any liquid water to be visible, and these summers only lasted a couple of months. Eeboth acknowledged the peril they were frequently trotting on weakened ice, but major cracks such as the one they saw that night were rare enough to risk it. It would have been dishonorable for Eeboth to mention the odds to Ortack-12, as this would display criminal cowardice. In fact, since the first rays of the morning sun were breaking, Eeboth proposed returning down the ruinous, semi-shattering ravine to attempt catching "ice creatures." These were not fish by human definition. These were self-heating organisms burrowing through the ice like worms, and reaching significant snake-like dimensions in these covered ice lakes in the summer, when the region bloomed with ice seaweed and the organisms feeding on it. Ortack-12 let Eeboth descend first. Several large chunks of ice cracked under his feet, but he just hopped to another piece as they did so. Since Eeboth started glancing at Ortack-12 with glee, suggesting he knew all-along Ortack-12 was easy to frighten, Ortack-12 gave up caution and followed him down. When Eeboth mentioned fishing, Ortack-12 pictured a fishing style common to warm planets with liquid bodies teaming with edible creatures, but Eeboth demonstrated the method required on Pallos. Eeboth took out an axe-like device and began hacking with it at the ice by pulling it all the way back over his shoulder and then smashing it with full force into the ice. The task was eased by the ready-formed crack exposing a generous surface of the interior ice. Eeboth was spotting shadows of the "ice creatures" and hacking in their direction. Through slow motion ice-hacking in front of their heads, Eeboth caught up with each creature he pursued. Seeing the ice reached its breaking potential, Ortack-12 followed Eeboth's example, and also hacked pieces of the rime, in pursuit of the "ice creatures". He was too cautious at this task, so he reached one in the time it took Eeboth to catch a dozen. In fact, Eeboth would have abandoned the task once he had plenty to eat for a few days. He continued because to be respectful, he had to wait until Ortack-12 caught a game. They were dragging these carcasses around with them across the following few days, spiking the party's combined stench to the max. Just as the supply of the "ice creatures" was running low, Ortack-12 spotted an animal trail for the first time, and showed it with pride to Eeboth. He failed to find the appreciation he expected. Instead, Eeboth explained this trail belonged to the biggest and most aggressive predatory species on Pallos. It had multiple lethal weapons engrained in its natural biology, including poisoned spears, strangling tail-ropes, and metallic rather than calcium-based claws. Eeboth began explaining the various anticipated dangers in an attempted slaughter of an ouks, but Ortack-12 diverted the conversation to the sky, asking about the oddly-dark and menacing clouds. Eeboth mused at the clouds, gaging their characteristics before concluding a major snowstorm was forthcoming. Snowflakes were at their largest in the summer on Pallos, and being hit with a lot of these could be as dangerous as facing an ouk. Noticing Ortack-12 had become speechless and stiff, Eeboth dug out a shelter. They settled in it for the night. Just as they did this, the massive-snowflake bombardment began. It caved in a portion of their dugout, crashing on top of Ortack-12's feet, but he dug himself out without complications. Eeboth insisted on following the dangerous ouk trail in the morning. To Ortack-12's relief, the trail showed signs the ouks had migrated out of the region, sensing the coming summer storms. Other game also appeared to have fled in the night as they seized observing herds, gleaming mere isolated, feeble beasts. Eeboth had to admit their hunting season was over. They tracked back to the compound. Eeboth gained a heavy judgment-load for a season when even functioning hunters were forced to join the shelter because game was sparse in the wild.

***

By 50,051 BC, the Mothership had remained stationary on the yellow moon of Pallos for so many years, its exterior had to undergo a special cleaning to free it from the lava and ash dropping on it in steady millennia of accumulation. It was an exciting time for this home-base because they were receiving volumes of data on Pallos and Byddwr. The analysis of this data could assist with acclimating their technology and physiology to these conditions. One day, at the volcanic activity peak, the hissing gases and the dropping lava's sound matched magnified crickets on Earth. These signs became white noise, or noise occurring at a constant, flat frequency spectrum. It was always raucous, but this constant bombardment was so consistent, sound-receptors screened it all out as irrelevant.

The third-ranking pilot among the spacefaring cadatas, Nurry Loody-12, was inhaling a new fragrance of a nutritional gas mix the laboratory just released, while reclining in a position semi-touched the soft seat below because of the moon's weak gravity. He was glancing at the latest findings from tests of the microbiome of some of the big-game species on Pallos, but his attention was also split into a study of the moonscape outside the transparent walls of the Observatory. On one of these glances at a lava-flow a couple of miles away, Nurry saw a shadow glide across the side of the hill the spaceship was topping. In the first instance, this apparition was discarded from his consideration, but then he questioned it. He had never seen a shadow this size gliding in the air on the moon unless it was a flying device he had deployed for testing. He knew he was not running a test flight, and it would have been strange if somebody else sent out a probe without submitting the idea and the test parameters for his review. Loody-12 considered it might have been a piece of metal cracking off the surface of their ship and floating away because of low-g conditions. Then, Loody-12 realized their ship was composed of a material capable of withstanding light-speed. Why would it crack because of a light breeze on its parked surface? A micrometeorite might have hit the spaceship and then burned up from the impact, but its flight would have been more rapid. The shadow had moved around a part of the ship lacking in transparency due to its containment of private quarters. He contemplated letting the appearance slide as if it was a bubble of gas or a cloud; though, clouds could not form in this moon's thin atmosphere. Then, he decided the team needed a scientific mystery to contemplate. So, he made a ship-wide announcement he was turning all walls transparent; those engaging in private acts should adjust their behavior to prepare for oversight. A few minutes later, he followed up and turned out 360° transparency. Doctor Nackan-12, Nally-12 and the lab techs joined Loody-12 on a full round to search for strange external movements or objects. The group was chatting garrulously behind Loody-12's shoulder, without examining the outdoors, when Loody-12 spotted an Unidentified Flying Object. It had a slow spin rate, but it was a mechanical device rather than a natural lifeform. It blended with surroundings because it was camouflaged in shades of yellow to match the colors of the surrounding hills. Loody-12 froze to study it, and the rest of the team also stopped and fell silent, as they tried to gather what it meant. After watching for a minute or so, they spotted a distinct flash, and these flash signals continued at equal intervals. Loody-12 sent out a group of nanobots to catch up to this device and run some tests. The group huddled around the incoming findings. The object was maintaining a trajectory circling their spaceship. On average, it remained too far from the ship to be spotted with the naked eye, but came too close on the one spin when Loody-12 observed it. The Computer outlined its complex structures and matched its technology to the engineering styles they had observed on Pallos. It shocked the team they failed to spot this potential counter-espionage in a decade there. It now seemed they had been discovered by the pallos, who had hidden space-flight capability. This was the first time cadatas encountered counter-intelligence, so they reacted with panic symptoms; many proposed shooting this mysterious surveyor down.

Cooler cadatas prevailed, and Loody-12 was assigned to perform a spacewalk to apprehend, or contact the UFO. They could have sent a robot instead, but even with artificial intelligence it would have failed to respond with appropriate instinctive politeness or moral fortitude if the UFO had an unanticipated reaction. The decision was further eased by Loody-12's intense interest in volunteering for this mission. He was eager to see it for himself. And he was convinced it was a great honor to be the first cadata to interact with a space-faring device from an intelligent alien species. He also thought it was necessary to show respect towards these space-rivals with a personal rather than a mechanical touch. Moments later, he was suited up and hopping along the rocky moon, an easier proposition than if they had to make this journey while hovering in space. The device's outer layer consisted of the moon's yellow stones. Loody-12 first moved in the opposite direction and made a few haphazard turns to avoid spooking it before taking a few short glides on his cadata-wings towards the object and jumping on top of it. The object's propeller turned rapidly, pushing Loody-12 into the air with it before his weight overpowered its capabilities and it descended to the bumpy ground, with him on top. With it secured below his body, Loody-12 studied its sides for potential explosive, or otherwise offensive components. None were apparent, so Loody-12 struggled with it as he guided it towards the spaceship. After coming through the wall, he forcefully guided it into the Lab, where the techs secured it with restraints. They feared the chance of a self-destruct explosion from internal components engineered in foreign patterns. Thus, most team members remained on the opposite side of the ship, as the robots and nanobots ran a series of tests, unscrewing and un-welding the components of this device. These probes determined, this device and atypical hardware on Pallos used a form of wireless communication cadatas had not witnessed before, which kept this advanced technology invisible to cadatas' surveys. The team attempted to communicate, in the pallos language, into this device to establish direct discussion with the pallos. But this did not inspire a response. The dissection of the device determined it was capable of incoming and outgoing communication. So, the silence was political, rather than a technological malfunction. The engineers isolated the device in a sound and vision-resistant closet to assure the pallos were prevented from further espionage, now from inside their ship. With it hidden away, they decided to request for Ortack-12 to engage in a spy-campaign leading to comprehension of the pallos' motivations; they appeared unwilling to share these with cadatas without such subterfuge. Alternatively, they could have sent a mission in a shuttle brandishing its alien nature to Pallos to negotiate for settlement, but they did not know if the spy-device originated from an isolated tech businessman, a radicalized country, or a group of pallos misaligned from the planet's common knowledge and view of the universe. A majority of pallos were in the human-equivalent of a Stone Age. For these masses, seeing a spaceship landing might have produced a violent, instead of an intellectual response.

***

Ortack-12's dominant activity in his years of solitude was hunting to develop a qualified pallos reputation. The rest of the time was consumed in assisting Judge Eeboth in his trials, as a kind of volunteer-clerk. Eeboth needed Ortack-12's help in the hunting off-season because he saw more clients when the shelters became overcrowded. Under stress, residents complained about each other and their accommodations, breaking their unique amoral code.

When Ortack-12 received the orders from cadatas to investigate who on Pallos might be capable of space travel, he was surprised, and proceeded with great curiosity. He deliberately showed Eeboth he was mechanically-inclined by breaking his court clock with three mini counts inside of it measuring Pallos' three moon cycles. Then, he volunteered to fix it, setting the component he disconnected back in place. In the following days, Ortack-12 integrated his technical knowledge into his chats with Eeboth. When the seed was planted, Ortack-12 mentioned he wished to contribute to Pallos with his knowledge. Eeboth nodded, and on the following day returned to Ortack-12 with an idea. Perhaps, Pallos might benefit from joining the Army as a mechanic, a skillset they were always short on. The Army had a large unit stationed at the same underground complex with the Court, so Eeboth was able to secure the job for Ortack-12 via his contacts there.

Ortack-12's training in advanced sciences across hundreds of years as a cadatanaut made fixes on pallos' equipment seamless, so he was offered greater responsibilities. When he designed a mini flying device matching the types of propellers used on Pallos, he was advanced into a secretive wing of the Army. The group possessed one of the only constructions with walls composed from a material other than ice. He was warned to refrain from sharing with everybody including Eeboth what he learned in the special division. The need for secrecy became apparent when he found a set of computers in the room. This was the first time he saw an object resembling technology such as a television screen, or audio/visual broadcasting devices on Pallos. The staff in this office had an appearance contrasting with the other pallos' grooming style. Their hair was uniformly straight-trimmed, with a few areas around the eyes shaved bare to maximize computer-screen viewing. They also had to cut their claws to allow for easier typing. A dozen even wore clothing to hide the private parts exposed due to the shortened haircuts.

He made himself indispensable by suggesting an improvement in these computer's chips, so he was allowed to browse through the data on them under the cover of trying to discover better data processing methods. A folder labeled "Space Travel" was easy to spot and accessible with his credentials. Three days of review explained the questions mystifying cadatas. The pallos were using an advanced form of radiation, noise, and signal blockers to keep their spaceships from appearing on cadatas' radar. Regular technology across the planet was allowed to transmit signals, and these attracted cadatas' attention from thousands of light years away, but advanced technology related to computing and space travel was undetectable without knowing its exact frequencies and dampening techniques. A subfolder called "Aliens" detailed the data the pallos had collected on cadatas' moon-parked spaceship. They had discovered it five years earlier, but had only sent three temporary probes to circle it for a few hours while the spaceship's internal lights were off. The cadatas' Computer disregarded the signal-blocked probes as irrelevant chunks of rock or as bits of flying lava. The cadatas' recent capture of their last probe disturbed their space program developers because it indicated the aliens could stealing their sole observation probe. They had not captured images of the spaceship's interior, nor caught cadatas outside the spaceship, in the few hours their space-flying devices could capture on their limited energy-stores. The unknowns allowed them to speculate the aliens were horrifying, odorless, and toothless monsters. A couple of horrific sketches approximating these aliens' appearance were included. Cadatas were right to send in a spy because the pallos were considering sending a preemptive strike to keep the aliens from invading and colonizing their terrain. This strike and the discovery of an alien presence were kept from the general pallos public. It was broadcast to a tiny class of intellectuals, and business owners, who inbred among themselves, and maintained a hold on power. Since the other pallos were illiterate, they could not have comprehended a broadcast, if one had been issued. Ortack-12 was treated with a bit of suspicion, at first, because he was not from a wealthy intelligentsia family. So, he would not have received the relevant schooling. And yet he was more capable with technology than the best in this circle. But since a leading characteristic among the pallos was accepting statements as facts for the sake of not dwelling on the unknown, they chose to ignore this incongruity. Another folder labeled "Byddwr" informed Ortack-12, the primary peace-time function of this space division was traveling to Byddwr and collecting the main commodity enriching the pallos' upper-class, and sponsoring their extraordinary technological advances. The information Ortack-12 was allowed to access was too limited, so he was encouraged by the cadata leadership to volunteer to join the space traveling units to gather first-hand intelligence on the pallos' operations.

Ortack-12 was taken through grueling physical training and some classes on shuttle piloting before being assigned to a shuttle and sent to perform flight exercises in space with a second pilot to correct him if he forgot a procedure. These shuttles were tiny, so there was just enough room for the two pilots' bodies, the equipment, a little box for sleeping, and another box for bodily functions. The safety tests they ran assessed Ortack-12's capacity for stoicism. In one disaster scenario, a shuttle tossed an explosive device at Ortack-12's shuttle, and he had to manually avoid its path. On one occasion, the internal Computer stopped functioning. Ortack-12 assumed it was another test, so he was tranquil as he fixed the hardware component sparking out of alignment by replacing it with a supplemental reactor. Just as he finished the fix, his copilot, Komure, explained it was no test; the component had failed and they could have lost the ship's pressure and atmosphere. Every flight into space was a test of Ortack-12's capacity for short-term starvation because they lacked space to carry-on giant slaughtered animals composing a standard pallos meal plan. Ortack-12 enjoyed being weightless again after over seventeen years under gravity's pressure. He played in their tight quarters, leaving equipment in the air prior to landing and watching it drop to the back of the room as he started to feel the g-force on his own body from Pallos' pull.

After six months of training, Ortack-12 was assigned to an active unit and sent with Komure on a mission to Byddwr; he was to avoid speaking with the byddwrs, and to let Komure lead the interaction. Ortack-12's chief function was operating the flight, as he had demonstrated advanced essential for survival skills in this department. They had a gentle liftoff and a few days of smooth sailing in space. On the morning before the scheduled landing, Ortack-12 was startled out of a nap by the sound of a metal chunk breaking off the side of the shuttle and slamming into what must have been the engine as it reacted with a shaking mini explosion. This set them on an uncontrolled spin, threatening to once again collide with the loose metal chunk also hurdling through space by their side. Komure was overseeing the monitor while Ortack-12 was away. However, Komure lacked a hypothesis on the culprit; he did not see an asteroid or another space object capable of striking them. Komure speculated, they might have taken fire from the moon-aliens. Ortack-12 was sure they would not fire at his ship. He still checked readings for mechanical flying objects in the region. He found none. With this concern alleviated, Ortack-12 focused on evaluating the damage the fracture and impact had caused. Window and Computer observations indicated: one of their three main engines had failed. Ortack-12 fought nausea and disorientation from the extreme g-force of the continuing rotation to activate a reserve thruster to stop the ship's violent spinning and to push it sideways, away from the chopping metal blade. With the ship stabilized, Ortack-12 shifted his attention to fixing an impacted isolated depressurization in their sleeping cabin. He used up most of their fuel reserve to regain the speed and direction indicated in their flight-plan; Komure assured him this was acceptable because they could refuel at Byddwr, which was pallos' sole fuel producer and supplier. Ortack-12 sealed the room, preventing them from utilizing it for the remainder of the flight. It continued leaking air and pressure through a micro-hole. A heavy pungent smell of burning plastic, dye and nylon flooded their cabin. Smoke even clouded their window, impairing their visibility during the landing. Ortack-12 touched down safely on an above-water floating platform reserved for these shuttles on Byddwr. He put on a sealed bag over his head full of air from Pallos, which had a different composition than the air on Byddwr; thus, with his pallos physiology, Ortack-12 needed this oversized head balloon to breathe. Fighting to balance on the bobbing platform and the heavy air balloon, Ortack-12 inserted sealant into the hole in the engine as a temporary solution for the trip back; advanced equipment for a full repair was inaccessible on Byddwr. The walls in this sleeping compartment were charred, discolored, and warped by the heat breaking through the opening on the inside. The damage extra-striking on the singed and dented exterior. Ortack-12's inspection determined a piece of the ship fractured earlier, hanging on because it was glued on by the previous mechanic. Ortack-12 wrote up a scathing report stressing the need for the ship's surface to be devoid of such imperfections to prevent repeated chirpings and failures. He was careful to avoid accusing the other mechanic to prevent being re-charged for complaining.

With these tests sorted, Ortack-12 joined Komure on the sole-mission-objective business trip. They took a balloon moored at the side of the platform to the nearest city, where Komure announced their arrival. A delegation surfaced from the ocean to greet them. Ortack-12 spotted Wocega-12 in this delegation. They listened to Komure and the byddwr leader discussing the resupply they were preparing to load. When the group settled in for a feast to converse in a casual setting, Ortack-12 gulped giant pieces of meat before departing with Wocega-12 for a balloon ride under the excuse he wanted to explore this new planet as a tourist. During this trip, they put together the pieces of these planets' story, as they were indistinct without the other half. Ortack-12 knew they would be collecting supplies on Byddwr. He was oblivious to this being a forced tax collection system the byddwrs complied with because the pallos subjugated them through military raids. The onset of the Ice Age forced intellectual pallos to invest in the invention of space-faring shuttles in the hope of resettling to their neighbor-planet. They were so proximate; it was an enormous ball in the sky with distinguishable sea features. This proximity inspired many generations of pallos with the fantasy of leaping far enough into the sky to reach it before it planted the seeds for the mechanics of how this could be achieved. However, when they arrived, they discovered the air was incompatible with their chemical needs. They could not bring a large enough air balloon to spend more than a couple of days per visit. They also learned their bodies were dysfunctional at higher heat levels without uninvested medical interventions. So, they were prevented from relocating to this pleasant planet. The pallos settled on muscling the byddwrs to supply their desperate needs, using the raw materials and manufacturing capabilities to advance their space age. Ortack-12 expressed his confusion regarding pallos' investment of excessive brain power, aggravation, stress, and warring hours merely to collect taxes utilized to feed this overworked machine. Meanwhile, Wocega-12 was ignorant she was assisting with slave labor imprinted into their interplanetary peace treaty. She had been busy singing, dancing, and creating engaging and meditative craft projects. On average, the byddwrs were living longer and healthier lives, while most pallos died in violent confrontations, such as attempting to bite a wild beast to death for their supper.

During these discussions, Ortack-12 and Wocega-12 had drifted back to the feast location. They had planned to partake in refreshments. But as they parked the balloon on the water, they discovered they were facing a furious Komure, who held weapons in his arms pointing at both. He chained them and tossed them into crammed sleeping quarters, tightening a byddwr air bag over Wocega-12's head as the air in the ship was set to pallos' needs. While it was rare for him to be assigned as the chief pilot, he still managed to navigate the shuttle to Pallos without complications; even the patched-up hole retained its liquid; this would have been catastrophic to the two guests relying on the air it would have drained. Upon landing, the two disguised cadatas were taken into a holding cave.

While they were languishing in prison, the cadatas on the yellow moon watched with amazement as five massive missiles were deployed from Pallos at their parked spaceship. Their shield could withstand flying through a sun full of these types of nuclear explosions, so they were not concerned about the impacts damaging the ship. They started worrying the moon might crack into pieces as the crater surrounding their ship deepened with each hit, and lava started bursting in gigantic eruptions from under the thin crust. The cadata crew knew their ship would survive being flooded with lava, and the moon cracking into pieces below them. The gravity dampeners would keep them from feeling the g-force of collision impacts. And their shield could detach clamping lava-growths. It was worrisome they had to abandon the yellow moon after this attack because its air was irreparably damaged by the radiation. But they could just take an orbit around Pallos, as now subterfuge was unnecessary. The pallos stopped firing after the fifth shot because as the massive dust storm triggered by the explosions settled, the pallos observed the ship's polished surface remained unscathed.

Failing to impact the aliens with this attack, the pallos gathered in a private courtroom, and called on Wocega-12 and Ortack-12 for statements. The justices explained the motivation for their espionage on the cadata ship and the explosive attack. The pallos had witnessed the cadata spaceship extracting rocks and dust from their yellow moon. The pallos had forced the byddwrs to surrender their resources and labor, so they were threatened by an alien species conquering their moon and mimicking this colonial behavior as they stole their resources. This appropriation appeared to be an act of war from an invading alien battleship. The pallos mistrusted Ortack-12, so Komure planted a listening device on him. He heard the discussion between him and Wocega-12 during their balloon ride. They had chatted about their life in space, their genetic transformations into local species, and their espionage agendas. Given the presence of an actual alien ship at the location they mentioned was interpreted as factual rather than as the ramblings of delusionals. Now knowing the Mothership was impenetrable, the pallos lacked an alternative recourse to punishing the two cadatas they apprehended. Wocega-12 and Ortack-12 confirmed they were cadatas and answered questions about their intentions in the solar system. The argued they were a peaceful species searching for a home, and willing to exchange their abilities and resources for the right to cohabitate on either Pallos or Byddwr.

After being given a chance to ask a question in their turn, Ortack-12 asked why the pallos perceived the mining of dirt on the yellow moon as a threat. The justices explained this dust was rich in a metal that was a key ingredient in the exterior of their space shuttles. While small quantities of it were also present on Pallos and Byddwr, on the first it had to be mined in specks by digging through enormous ice mountains, and on the other it was in granules under a heavy layer of sediment at the bottom of an ocean. The pallos had sent the probe that discovered the cadatas' spaceship to the yellow moon to collect this metal for a new shuttle construction project. Ortack-12 explained this was an unnecessary metal for cadatas. Thus, having learned its value, they would refrain from extracting it. He argued for a peaceful agreement on the type of dirt lacking in value for the pallos, or on their departure for unreachable-to-pallos planets to collect resources for cadatas' development in the pallos planetary region. The justices replied, there was no useless dirt to be had on its planets. They had plans for every drop. Instead of being willing to give a grain away, they needed the cadatas to pay for their past stay with their spaceship. Ortack-12 tried to object pallos lacked the capacity to fly their spaceship because of its advanced systems. He was certain cadatas would never surrender their ship to any planet because it was their interstellar home. He also stressed cadatas would not negotiate for his life, or Wocega-12's because they had retained their genes, and thus could grow the next generation of the Wocega and Ortack birth-line, even if both died.

When pressed what, if not their ship, cadatas were willing to offer in exchange for staying, Ortack-12 proposed to terraform Pallos, to resolve its harsh, icy environment. Cadatas could deploy an army of robots to plant hundreds of thousands of trees daily in regions with the fertility-fostering nutrients in the ice. They would also heat large sections of the ice to create an extraordinary spike in water vapor in the atmosphere. These droplets of water would produce a heating effect, increasing the average planetary temperature by around 30°C. A justice reminded Ortack-12: pallos' physiology cannot withstand extreme heat. They would sweat-to-death, like a shower with a limited water supply gradually losing its moisture reserve. Ortack-12 acknowledged the flaw in this plan, and proposed another: CO_2 eating bacteria could terraform Byddwr to increase its oxygen level, allowing the pallos to breathe without air balloons. The byddwrs breathed in a chemical other than oxygen and carbon dioxide. So, they would survive this mutation. Heightened levels of carbon dioxide would lead to warming, but only by a few degrees. The average temperature would remain near 0°C: a level safe for the pallos.

Ortack-12 was so engulfed in this explanation, after all those years without intellectual engagement, he had not noticed the justices had paled, and were glaring at him with psychotic hysteria. When he glanced up at them and spotted this dissatisfaction, he stopped the lecture. Without the preceding silence while listening, the justices began yelling over each other. Ortack-12's statement had proven cadatas were plotting to destroy pallos by heating them, or poisoning their air. The wisdom of cadatas' science was more intimidating than if they had shot back thousands of missiles at Pallos. One justice even threatened to fire their remaining missiles at their own planet, and at Byddwr until both were uninhabitable to life; if cadatas failed to vamoose away from their system, taking with them their malicious science. Since these justices already near-destroyed one moon with the metal needed for space exploration, the cadata team was certain pallos would follow-through on the execution of this suicidal threat. Nally-12 had been determined to discover a method for bribing, or muscling the pallos. But now, she recalled the smoldering mess left behind on Bhasab. She realized she could not authorize pushing these erratic pallos to a similar brink.

Nally-12 ordered a robot to fly to the surface to collect the two "prisoners".

The robot flew through the tunnels. It collapsed the door to the courtroom. Then, it created a metallic bubble around the two cadatas, freed them from their restraints, and flew them in this little ship to the moon. Without further delay, the Mothership departed an instant after their arrival. It began a rapid acceleration in the direction of their next scheduled planetary stop, thousands of light years away.

Chapter 13

Are Two Suns Better than One?

The crew entered the next solar system in 39,365 BC. While Eekalah was the only potentially-habitable planet in this region, it spun around two curious suns. Systems with more than one sun are at least twice likelier to undergo a violent star-death. But the atmosphere of the planet under investigation in this binary appeared near-identical to Cadata's. The crust composition also included the essential chemicals to cadatas' manufacturing methods. What was unapparent from Cadata's long-distance probes, but was revealed once the team sent nanobots to Eekalah on their approach, was this planet's overpopulation by a single species responsible for expunging the rest of the evolutionary tree. With every new element cadatas learned about Eekalah, a few team members switched to voting against even exploring it. By traveling to Eekalah, they had reached the farthest point from the galactic center cadatas attained on this elongated journey through the Milky Way. When they looked through their transparent walls at the sky, during their approach to Eekalah, it was dark, sprayed with distant stars, and galaxies. Those who still supported the expedition argued: even if the settlement was futile, they had a responsibility to cadatas' astronomical research to understand how stars, planets and life functioned this far from the galactic core. Identified abnormalities might have helped future cadatas comparing habitable planets close to the populous core, as opposed to these dark zones, with hundreds, or even thousands of light years between stars. They argued, the trip could have been avoided with a re-examination of this system before departing from Pallos. Instead, Nally-12 rushed with the decision, and directed them to the next planet in the *Catalog*. She failed to conduct a scientific review of how this system had changed in the interim since the first Nally in her hereditary-line selected it for the top of the *Catalog*. The crux of the problem was the two stars started approaching each other, and exchanging mass in a pattern signaling they lacked a promising future. Wocega-14 was 1,300 years old when the cadatas began their 1-year deacceleration on the approach to Eekalah. She sided with the sect favoring exploration, even if it proved lethal to the crew because the next stop would have fallen outside her lifetime.

Their parking maneuver had been shortened to a single year because a uniquely weak technology-signal was detected on Eekalah. They slowed from near-light-speed to near-zero in a geostationary satellite orbit. They disguised their exterior to have the appearance of an asteroid, in case anybody on the planet was looking up. Unlike on prior landings, they experienced a technical problem in the slow-down year, which necessitated parking farther from the planet. To run maintenance, they opted for a stationary orbit. It was stagnant in relation to a specified location on the planet. They were still moving at the same angular speed as the planet, but this speed did not need much added propulsion without an atmosphere to slow them down through friction from a constant speed. This ease of transport is achieved at a distance where the gravitational pull from the planet cancels the centrifugal force. If the ship was any closer to the planet and not propelling itself to regain height, it would have been in a continuous freefall. On Earth, at the equator, the geostationary orbit is at 22,300 miles up; in contrast, Earth's radius is 4,000 miles. Eekalah was somewhat larger than Earth. So, they had to orbit somewhat higher from the surface. On most other planets, cadatas descended closer to the planet. Eventually, on Earth they would have been in a polar-orbit at 500 miles from the crust. At this lower level, and within the scope of air resistance from an atmosphere, a space-ship needs occasional engine bursts to remain "still", while the planet rotates below. This position is ideal for observation on a planet without space-viewing capabilities, as it allows cadatas to take images of a new strip of the planet with each circle it makes around the globe.

The problem necessitating a low-energy-consuming orbit involved the spaceship's matter gathering mechanism. The error was spotted by the Computer when the ship exited light-speed. In response, the Computer under the scientific team's supervision deployed robots to investigate the cause. Their tests determined it should have been functioning normally, but the system repeatedly failed to consume the micro debris floating within the ship's range. Cadata engineers and administrators alike also could not spot an abnormality on the images the robots sent of the equipment's external and internal components. They had enough matter in reserve to deaccelerate and park on Eekalah, but they had to find a solution to this crisis if they hoped to leave this planet at some point. Wocega-14 was in the final stages of her genetic transformation into an eekalah, so Ortack-14 was assigned to a manual spacewalk with hope his touch of the affected areas might intuitively explain what machines failed to interpret. They considered creating a three-dimensional replica of the affected area for this purpose, but Ortack-14 was eager to participate in at least one historical mission in his piloting career.

The problems with matter-accumulation were related to the spaceship's external gravitational field responsible for restraining a spacewalker from floating away by attracting him or her back if floating out of a specified range. The potential for the automatic attraction system to fail meant they had to secure Ortack-14 with the added measure of a tear-resistant rope utilizable to climb back, if he lost contact with the ship. This was the first spacewalk a live-cadata performed since departure from Cadata, so they asked Ortack-14 to report on his sensations and encounters during this excursion. His first observation was of a strange smell, as if somebody was welding metal or burning ozone nearby. This was unusual because they were well above the planet's ozone layer. Instead, the team attributed it to a malfunctioning component or perhaps to the smell of outer space, which Ortack-14 might have been exaggerating after all those years indoors.

Ortack-14 wore a propulsion unit on his back. He propelled with it through the ship's self-sealing wall and flew to a section of the external matter-gathering shell with the main system vein. They believed this spot caused the congestion. Tiny bursts propelled him by great distances in these near-zero-g conditions as the lack of an atmosphere meant no drag on Ortack-14's suit. He reported having no sensation of movement as he zoomed along the side of the ship. They had just spent a year deaccelerating at a g-force equivalent to the one they would find on the planet. Gravity reacts with components in the human ear to tell the brain if it is moving in relation to other objects. In outer space, these detectors fail to pick up such movements, and this disagreement between the brain thinking an astronaut is motionless and the visual movement in relation to other objects is what causes space-sickness. Ortack-14 was not sickened by this phenomenon because he had spent his life until the last year in zero-g, so these conditions were more inherent for his body than even slight degrees of gravitational pull. As Ortack-14 neared the presumed congestion point, he realized a component seemingly irrelevant on the images was the sole suspect feature across the flat surface of the ship. The spaceship had liquefying-on-command walls. There were no conventional "doors" for entry or egress. One exception was a drop-door with a latch, added by two engineers as a joke during a slothmoth-laden party. The engineers convinced Nally post-insertion this latch was necessary in case the walls' liquifying mechanism malfunctioned. Ortack-14 now fathomed the scans had indicated matter flow was impeded by the abnormal materials composing this outmoded latch. The engineers had used recycled metal scraps rather than the ultra-resistant chemical mix dominating the rest of the ship's skin. It was fortuitous this component had not allowed a meteorite through at light-speed, an error capable of causing the total loss of the spaceship. The Computer had not identified this material as problematic because its external gravitational shield repelled, slowed or processed every atom, making an accidental collision with the surface impossible. The team now realized the bombardment of nuclear impacts on Pallos' yellow moon had corroded this weak metal until it created the blockage. Nuclear energy and matter from the blasts was absorbed and caused a type of artery-clogging because the system was not designed to take on five nukes in rapid succession. Seeing the corrosion startled Ortack-14 with the insight he had volunteered for exposure to toxic nuclear waste. In response to this voiced complaint, the Computer assured him his spacesuit was sufficient to thwart hazardous materials. He asked if the earlier ozone-like smell was a sign the corrosive materials initiated a leak in the suit, but the Computer insisted his perception of odor was psychosomatic. The Computer judged Ortack-14 had based his description on space air smell narratives from books. The suit's system failed to register even microscopic holes in the armor, or toxic substances entering the air supply system.

Since Ortack-14 was displaying paranoid symptoms, the team proposed substituting him with robots to replace this latch. But Ortack-14 was reassured enough by the Computer's argument to believe he could perform at least a few steps of the maintenance process without robotic assistance. The scientific team agreed it was impossible for Ortack-14 to achieve this because nobody had performed such manual maintenance since their departure from Cadata. However, this improbability turned the attempt into an inquisitive experiment, so Ortack-14's proposal was approved. Ortack-14 was handed a cutting tool through a wall and used it to chisel an outline around the door because the nuclear reactions had welded it in place. Since the spaceship had to be protected from the inside, they could not push the door out into space. Ortack-14 had to gently extract it to minimize damage to the surrounding nanobot-rich material. The door weighed the equivalent of around 1,400 pounds on Earth, but was weightless in microgravity. The problem for Ortack-14 was not its burdensome heaviness. In weightless conditions without other complications, he could have lifted it dozens of times out of its sockets. The snag was its bulk, as he struggled with wrapping his little cadata hand-wings around it. He extended them at 45° angles to create some leverage and flexibility to do a chest pass to wiggle out and swing the door into space. Anything with mass has inertia, even at zero-g. Ortack-14 pivoted the door around the center of its mass in the middle of its length. Ortack-14 acted tenderly in this maneuver, shifting it by a few millimeters on one side, and then the other. Five hours of this, his frustration at the slow progress boiled as he pulled with excessive aggression on the door until it slipped out of its coating before Ortack-14 could shuffle around its side or to escape its path. Thus, the door served as a massive projectile, which yanked violently on Ortack-14's tethering rope, breaking it. The door and Ortack-14 flew away from the ship at the velocity this jerk generated. Ortack-14 panicked and triggered his jetpack to thrust him into the door and to propel both back at the ship. He had sufficient time to realize rapid flight towards the spaceship could prompt the engines to consume him, and the door. The engines consumed all matter, such as an asteroid, approaching within range of an impact. He just managed to fly around the door in a few seconds to push it in the opposite direction, causing it to slow down. Then, he gave it a tad more thrust and pushed this stubborn door into space, where it would burn on descent through Eekalah's atmosphere. The last stage of the procedure required the robots. They had to fill the gap the door left with a distinctive layer matching the surrounding core mechanism. Ortack-14 drifted back inside the ship during this process because the heat and the chemical and physical reactions it generated endangered proximate, exposed lifeforms. The rest of the procedure was executed without incident. The ship's surface was smoothed. A test of matter-absorption showed no abnormalities. The team re-designed the shield's program, so the system would auto-filter similar access of nuclear materials in future attacks, even if a new arterial block formed in an unforeseen, vulnerable section.

With this emergency sorted, the cadatas began analyzing the trouble brewing in the binary suns dominating the sky. Back on Cadata, when Eekalah was entered into the *Catalog*, the two stars were maintaining a .22 AU distance between them, at a stable 41-day orbit around a common center of mass. Eekalah was circling both at a shorter distance than the one between Earth and its sun, making a round in 229 days. Now, over 60 thousand years later, the suns' proximity to each other gave them the appearance of a single elongated object. This sudden change in the stars' closeness proved any system with more than a single sun introduced an unacceptable risk of chaotic interactions between the stars: including collisions, collapses, dislodgings, and explosions. The two stars might have maintained a stable orbit for millions of years, but logic dictated the main objective for the stop was sending shuttles close to this binary to understand how these systems operated at the late stage of their lifecycle. This was the first binary system explored by a cadata, so there was substantial potential for scientific discoveries.

While Wocega-14 undertook her settlement tryout on Eekalah, a dozen shuttles moved within a hundred miles of the stars' loops to deploy nanobot probes to collect readings at maximum resolution. The cadatanauts remained far enough to escape if one of the stars commenced going nova. Probes indicated this binary system began with two distant stars with an orbital period of 6 years. They began approaching each other because Star 1 evolved quicker, attracting more gas from the region and thus gaining more mass than the main-sequence Star 2, until it turned into a red giant. Counterintuitively, the smaller star began capturing mass from Star 1 at this juncture, until the smaller star's Roche lobe shrank. The two stars were now close enough for them to share an extended gaseous atmosphere, making them into a contact binary with a connecting stream column of hot gas. This process continued until, in the stretch of cadatas' journey across the Milky Way, their orbital period shrunk to only a .2-day interval, with only the diameter equal to the larger star between them. This was lethal proximity, which gravitationally deformed into teardrop-shaped pulsating tidal bulges both of their outer layers. They had also begun to lock with their faces towards each other, a process slowing their rhythmic pulsations. This stage was accompanied by dramatic spills of gas from one star to the other, intimidating to watch from the proximity of the shuttles' orbits. Each spill was felt as heat waves down on Eekalah. Cadatas' extraordinary shields kept the shuttles from suffering these extreme light wave strikes.

Wocega-14 persevered in her transformation and training to fit into the eekalah culture despite overhearing these frightful results about the melding binary. Her body gained height and muscle. She now had six powerful but short legs at the bottom of an elongated upright crocodile-like torso, and six mini snake-like fingers in place of each of the arms at the top. Eekalah had a rocky and rugged terrain, so species on it evolved with powerful legs for balance and stabilization. Her face had teeth-like hard surfaces in unexpected places, with some serving decorative functions. Because she had six lower feet, she discovered she could lock them in place. This kept her back straight without exerting added energy. She could sleep with locked feet in the standing position. It was more uncomfortable for her to lie down to sleep because the feet and hands were protruding in all directions. She squashed at least a couple of them in any chosen lying position. One of the positives of this species' biology was its air requirements matched those on Cadata. Thus, the team did not have to isolate Wocega-14 on a side of the ship all to herself. She often ran across the main hallways and around the large auditorium because her body required substantially more exercise than a cadata's. Her rapid approaches and the pouncing sound she made with her multiple feet while cracking them over the metallic floors made every cadata jump out of her way, even after several passes. Her constant heavy salivation also troubled cadatas. Spittle formed as she breathed with her mouth open on these runs, as if spotting each of them made her hungry. About a month into testing this new body, Wocega-14 developed a massive hump, a giant forehead, a beard, and male genitalia. These hormonal changes were accompanied by Wocega-14 beginning to aggressively pursue members of both sexes among the cadatas. If she saw a couple together, she would shove the more dominant cadata partner with her snake fingers, growling and challenging him to a duel over the female. These characteristics troubled most of the crew, so Doctor Nackan-14 ran a study on these mutations. It explained this was a normal transformation among the eekalahs whenever there was a shortage of dominant males. While this science was curious, the team was only satisfied they had a resolution when they arrived in orbit around Eekalah and Wocega-14 took a shuttle to the planet.

The shuttle was disguised on the exterior as a rocky formation to blend with the surroundings at the scheduled landing site. The pre-programmed flight path had turns and altitude fluctuations because there was a dense blanket of asteroids orbiting Eekalah. The recent turmoil between the two suns had dislodged the closest planet to them and as it drifted away, it struck the next planet over creating an asteroid belt between Eekalah and its suns. Some of these rocks were flying out of this belt and hitting Eekalah, which was grabbing them and pulling them into its own mini belt. After Wocega-14 cleared this debris, if she kept the lights off in her shuttle, she could see these zodiacal lights as a faint glow reflecting sunlight. Knowledge of this new asteroid belt would have added a reason to avoid this system because planets in the path of an asteroid belt are unlikely to develop life, as evolution is interrupted by life-destroying massive impacts. One of the astronomical exceptions to this rule is a moon near a gas giant because big asteroids are likelier to hit the giant planet's mass than a moon or another satellite circling it.

Looking down, Wocega-14 saw much smaller oceans than the ones on Earth, with much darker continents. Most of the landmass was covered in dark, almost-black rocks with soil between them, which was previously rich with plants. But the eekalahs consumed most of these greener objects, leaving just the dark blanket. This raised the temperature across the planet as water and rocks a max albedo of 10%, retaining most light hitting these surfaces. Regardless of the substance of the objects on the surface of a planet, darker colors absorb and retain light, and light penetrates down into oceans and stays there. A black planet or a "black body" (if artificially painted or composed of black rocks) absorbs radiation at maximum efficiency. It is much hotter than it would be at the same proximity to its sun than if it was instead covered with ice or another non-absorbent color. A black body can reach a surface-temperature approaching a sun's, or 6,000 K. When Nally added this planet to the *Catalog*, it was green on the outside, with colder icy regions on the poles offsetting this effect, but the eekalahs had consumed most plants, and drank a disproportionate quantity of the water. So, it was much hotter now.

One of the difficulties cadatas had with landing on Eekalah was finding a spot distant enough from all eekalahs, to avoid being spotted flying in from the sky. The eekalahs multiplied until there was an eekalah for every 100-feet. They settled on landing on a dump-site, in the middle of two garbage hills, even creating artificial garbage debris and attaching it to the ship's exterior to camouflage it into this landscape. There were still plenty of eekalahs gathering trash elsewhere on this dump village, but this region had a toxic smell, so few ventured there. During Wocega-14 time on the planet, the spot increased desirability. When she next saw the shuttle, she had to command the Computer to pull it out of a mountain of garbage to access it.

The cause of this strife was the eekalahs' adaptability. They were shrewd with powerful builds. At the onset of their development, they were efficient at trapping prey. Being superior to the other species on its planet at first helped eekalahs gain control over food sources and to expand into hostile territories. However, this dominance accelerated until the eekalahs began eating into extinction the other species of animals and plants. During their evolution they had begun producing around 10,000 new egg babies in each season. When this number was first reached, most babies would have been killed by competing predators. But at a later stage in their development, there were no predators left for eekalahs to compete with. Then, near a hundred percent of these eggs matured. Each generation multiplied the population of eekalahs exponentially. These hungry youths eventually devoured the other species across Eekalah. Whenever all food sources ran out in a region, the illiterate youths (abandoned by parents out of fear they might be eaten) began eating each other, as unborn sharks do in their mothers' stomachs. Through these types of massacres, the eekalahs survived the other lifeforms on the planet, except for micro-organisms too small for effortless-consumption. Eekalah scientists were working to capture and bottle microbes to make them available for consumption with some successful products already on the market. With a single species, the planet's atmosphere mutated, as the chemicals eekalahs breathed were sucked out quicker than they could be restored, and their toxic exhaled chemicals overloaded the filtering lungs of the planet. Air thinned to dangerous levels and further aggravating global warming.

Wocega-14 parked her shuttle, watched it camouflage into the garbage piles, and took a short stroll through the dump when she spotted three eekalahs. She had studied the local language and was prepared to engage in a conversation with them, but without a word, they started running towards her. After a lifetime on a communal cadata spaceship, she assumed they approached with eagerness to greet her into their fold. But as she kept watching, she spotted their faces were covered in a blue liquid, which was an internal fluid in eekalahs. When she detected the open-meat eekalah scent they reeked with, she acknowledged they saw her as a meal. Glad she practiced running back on the spaceship, Wocega-14 took off in the opposite direction. But while she had practiced running in a controlled, sterile environment, they had been hunting each other from birth, so they soon caught up with her. Hollywood films would have you believe a strong and disciplined pilot could fight three brute assailants with bare hands and legs, but Wocega-14 judged, by their speed and muscular composition, trickery was needed to escape this encounter. She gave up on protecting cadatas' scientific secrets by turning into a weapon a robot she was carrying in her bag for emergencies. Once it was charged, she stopped, faced them, and shot them with precision befitting hundreds of years of simulated shooting practice. She considered burying these bodies to avoid an investigation uncovering the strange nature of the wounds, but she saw six smaller eekalahs peaking from around a garbage mound. The building appetite with which they were probing the bodies made Wocega-14 realize these vultures were preparing to clean the crime scene. Wocega-14 retreated from this bone-cleaning and kept walking through the garbage field, which was composed of unnourishing items such as the indigestible parts of bones and the eekalahs' vehicles, as all edible matter was consumed, regardless of decomposition. It might have been prudent to turn back after the first chase, but Wocega-14 interpreted it as a success since she managed to survive it. The eekalah genes were clouding her judgment, so he was eager for new confrontations, and had a tinge of cannibalistic hunger in the hourly death-defying fights she endured to explore the region. Her indoctrinated cadata morality prevented her from eating these eekalahs. Instead, she had her robot prepare food for her. However, the robot might have gathered the ingredients in this food from decomposing eekalahs because there was scarce edible matter aside from the eekalahs on this landmass. She was concerned when she fell asleep on Eekalah for the first time, as she imagined she might be defenseless if an eekalah snuck up on her while she was unconscious. But she was attacked in the first half-hour of this attempted standing-rest, learning eekalahs' had a fight-instinct even in their sleep. She began punching and kicking when she sensed an eekalah's breath on her skin, before it made contact. She managed to win the fight. It was difficult for Wocega-14 to enjoy "nature" on Eekalah as a walk by an ocean or across a mountain side at sunset just turned into a homicidal, rageful fight to the death.

Wocega-14 lingered in the border region between mountains and valleys, or mountains and water bodies because one of her research objectives was determining the environmental impact of species uniformity, and effects of this problem were most apparent in these regions. One of these sides of a mountain faced an ocean, so as Wocega-14 stood at the bottom of its steep incline, she felt strong gales from incoming trade winds. Soil readings indicated this spot previously had an excess of moisture feeding an abundance of greenery to carpet this mountainside. The other side of the same mountain had always been desiccated because desert conditions dominated the region. While there was still a near-constant cloud-cover at the top of the mountain, the rain falling from these clouds was dropping on infertile soil, which was not regenerating the pre-existing jungle. Despite these desolate conditions, it was still a beautiful scenic location. There were rapid rain and dryness cycles because little of the water was being absorbed by the soil, so most of it was evaporating back into the sky. With renewing rainfall, the sky was erupting in rainbows, one from each cloud. These rainbows disappeared within moments of rain stopping; there were plenty of other rains and sprouting rainbows in the surrounding mountainsides in these instances. Being attacked by a gang of cannibals with dozens of rainbows as a backdrop was disquieting even for Wocega-14.

Despite these extreme threats to her safety, Wocega-14 acclimated to this environment and stayed on Eekalah without appearing to have developed PTSD or other signs of a nervous disorder. The cadatas on the Mothership were astonished when, after a stretch in this warzone, she proposed cadatas eradicate the entire eekalah species, treating it as an invasive weed. She recommended re-planting the planet with the native, extinct, peaceful species of plants and variant organisms. Nally concluded Wocega-14 had cracked under the pressure, becoming overwhelmed by the paranoid belief everybody on Eekalah was out to kill her. Believing they were all evil opened the door to the conviction they deserved to die. They tried to encourage her to consume some anti-depressants to help her see positive elements in the eekalahs. Even while on the strongest mega-anti-depressants in cadatas' arsenal, Wocega-14 tried to prove her mind was irrelevant to the debate. She insisted exterminating the eekalahs was for the good of the galaxy. To show they were listening, the cadata scientific team ran a study on the impact this plan would have on Eekalah. They discovered the millennia with a single-species ecosystem had destroyed so much of the soil and air needed for a habitable planet, terraforming Eekalah would take more effort, time and energy than terraforming a lifeless planet, such as Mars.

Wocega-14 began developing a solution to make Eekalah habitable for cadatas, but just then cadatas who were parked in shuttles next to Eekalah's suns withdrew rapidly from this hot spot and rejoined their Mothership. Wocega-14 was ordered to return to her shuttle and fly back to the spaceship as well. A lot of garbage was tossed upwards for her to make this getaway. Just as the last of the latecomers parked, the Mothership began an extremely rapid acceleration process, which would have crushed the cadatas aboard if not for the anti-gravitational internal field keeping them at 1g.

As they sped towards light speed in the direction of their next stop, they saw Eekalah's stars brightening despite their increasing distance from them. The team gathered in the auditorium to listen to the surveyors' explanation as to what had transpired. In the last few days, the suns had ejected their envelope, which separated and became a surrounding planetary nebula. This loss of matter to feed on cooled Star 1 down to a white dwarf. It started eating mass from Star 2, while they seemed to be spinning further apart, a sign that could have signaled longevity, so the researchers did not panic. Because they were intermediate-mass stars, statistically, they might have settled into being two carbon-oxygen white dwarfs in a 30-second orbit. But the worst soon happened as Star 2 overflowed its Roche lobe and dissolved into a heavy disk, which was devoured by Star 1. Overeating spiked Star 1's weight to break the Chandrasekhar limit. It could have instantly exploded as a Type Ia supernova, killing not merely the researchers, but all life on Eekalah. Instead, it started to sizzle out in a dwarf nova. Star 1's brightness increased tenfold. The shuttles hovering nearby were hit by a powerful set of ultraviolet waves. The ships' systems auto-recorded these impacts as call-backs and returned to the Mothership at top-speed. They arrived just before these ultraviolet waves would have caught up with them, a day after Star 1 began going nova. By this point, the Mothership had only seen the visible light-speed wavelength of the explosion. Thankfully, alarms on the shuttles had traveled to Eekalah at light-speed, triggering Wocega-14's rapid preparation for departure. Cadatas' ships could have withstood this first set of waves, but might not have survived continued bombardment across 15 days they discovered these lasted before a quiet interval of 100 days ensued, followed by another cycle of 15 days of powerful radiation. The robots cadatas left on Eekalah reported temperatures more than doubled with the first ultraviolet wave, so the first 15-day bombardment killed the eekalahs with heatstroke and radiation. The timing of this dwarf nova was suspicious for some cadatas, who insinuated the probing shuttles near the binary, there to run tests, might have triggered an acceleration of the death-cycle. Some even suspected the leading researchers engineered the speed-up of the mass transfer to take readings inside of an unraveling nova, a new experiment for cadatas. The researchers repudiated these claims, but some of the data capable of solving this mystery suspiciously went missing.

Just when it seemed cadatas escaped the nova, without damage to their fleet, the Computer informed them of a hazardous liquid spill. It deployed the robots to fix it. Cadatas were ordered to avoid the affected area surrounding a shuttle parked inside the Mothership, after surviving the near-sun impacts from ultraviolet waves. Cadatas were discouraged from approaching the region, even in their spacesuits with air masks because, just like the shuttle's shell, the suits would have been inefficient for blocking the dangerous substance from seeping into their air supply, or from disintegrating their suits.

The mechanics who were in the shuttle-parking area, when the spill occurred, went through a purifying procedure. They discarded their clothing. These outfits were disintegrated into molecules to assure flakes with fatal properties were neutralized. The altered molecules could be reintegrated into the spaceship's matter supply. The mechanics endured an intense version of their regular air shower, which was programmed to collect damaging chemicals into a reinforced storage container.

Matter stores in the shuttle had been irradiated, leading them to overheat into dangerous, unstable free-floating, bright-green liquids, which were seeping from tiny holes in the shuttle's shield, as it was corroding from their toxicity. The shuttle's sensitive equipment was decomposing. The liquids were starting to stick to the gravitational "floor" of the Mothership, and to penetrate it. This degradation was moving towards the Mothership's propulsion and energy processing arteries. On the surface, the damage appeared as acid stains in the floor. To avoid fuels concentrating in a single spot, and burrowing through the ship, quicker than the robots could safely collect this dangerous matter, the Computer switched the Mothership to zero-g. Now, instead of pooling under the shuttle, the droplets formed liquid balls, floating across the parking garage space, where shuttles were maintained. They formed perfect spheres because at near zero-g, a liquid's surface tension dominates, in the absence of the opposing force of gravity. These bubbles burst when they hit a wall, or any object, disintegrating some of the contacted material. Cadatas were warned not to fly through greenish clouds, or bubbles across the Mothership, in case one of these had broken loose.

The Computer struggled with calculating if sealing leaks, or collecting the spilled liquids was of primary importance. So, it deployed the robots to perform both. Cadatas oversaw this process from the Observatory, and issued to the Computer moralistic ideas on gradation of process importance. The Computer engaged its available tricks to control this branching accident. It used the air system to blow bubbles towards safe collecting bins, to contain this matter, without letting it burst into smaller fragments. These bins were designed by the engineers, and the Computer for this disaster out of a chemical that auto-attracted these toxic molecules. The Computer also deployed a frozen ammonia-equivalent chemical to cool the overheated shuttle, and these liquids into frozen solids. Solidity was these compounds' safe-to-touch, non-corrosive state. During execution, this process was proven to be more challenging than the Computer estimated. So, cadatas' leadership decided to expel the damaged shuttle. This was dangerously wasteful because the expulsion would lose the enormous matter-volume of a shuttle.

The robots gathered into a chain, and pushed the shuttle through a wall, without turning on the shuttle's propulsion. The Computer theorized that allowing the shuttle to push itself out would have sparked an explosion, with all those hot chemicals floating around. The expulsion was successful. At least the shuttle did not bounce into the side of the Mothership. They saw it drifting away, as the pilot switched the speed from deacceleration to acceleration. To maximize precaution, the Computer gathered the recovered corrosive elements, and expelled them too. This massive matter loss meant they had to stop four solar systems later to fuel on matter from a moon that matched the required elemental composition. The Computer kept deep-cleaning the spaceship for several months afterwards, until it repaired the resulting cracks and corrosion.

Chapter 14

Life Among the Metalloids

As the nova exploded, the Computer had directed the Mothership automatically to the next planet in the *Catalog*: Ahutababha. It was an especially controversial choice. Ahutababha had an extremely high degree of eccentricity. This eccentricity was pushing it to a distant part of its solar system for a segment of the year. Then, it approached too near its sun for another portion of its annual orbit. Most planets across the galaxy have irregular orbits. This majority was predominantly screened out of the *Catalog* because life is unlikelier to survive in an environment with a spiked chance of collisions between planets. An elongated trajectory sends a planet on an intersection course with the other orbits. The orbital abnormality would have been an automatic disqualifier. But the viewing equipment also detected signals of advanced-intelligence life from this distant and abnormal planet. Cadatas had been moving towards denser galactic regions in the interim since its inclusion. However, there were still few alternative contestants at that distance from the Milky Way's center. Ahutababha presented the most interesting scientific specimen among many frightful options. The landing was the longest across the Mothership's galactic journey because the data indicated they needed to arrive at the start of the warming period, when the atmosphere began to recuperate from winter. In its coldest form, this planet was as useful for cadatas' research, and settlement prospects as any icy rock lost in the universe. They began a slow 9-year asteroid parking maneuver in 29,407 BC. They arrived in orbit around Ahutababha in 29,398 BC. Wocega-16 was in her prime, at 1,312-years-old, as she began an eccentric transformation.[5] Initial probes returned with technological ideas even cadatas had not imagined. Thus, the trip became more promising with each piece of incoming data and materials.

On Earth, the seasons do not change because of Earth's proximity to the Sun. Instead, during Earth's summers, the Sun's angle hits the planet with direct sunshine across a longer day. Direct sunlight allows for greater heat absorption, while winter's low-sun days lose most incoming heat. Earth migrates closer and further from its sun, but this minor eccentricity has a smaller impact on regional temperatures than the angle of ray-impact. Ahutababha experienced extreme temperature fluctuations because the changes in the distance to the sun were enormous. Ahutababha was 65 times further away from its sun on the coldest extreme of the elongated 20-months trajectory. These changes were more apparent in Ahutababha's oceans as they turned from frozen solid at their greatest depth to hovering at just below and just above the boiling point, until the top layer of the oceans boiled out into the atmosphere. At the perihelion, or its closest point to its sun, the record-setting observed high temperatures neared 390°F, so a water-based ocean would have evaporated. Instead, hydrogen was bound to a metalloid chemical, boron, in

---

5   This transformation is described later in this chapter.

this ocean, a chemical combination with a higher boiling point resistant to evaporation. Borazine is an Earth-equivalent chemical, which combines hydrogen, nitrogen, and boron in a stable hexagonal structure; the mix creates a stable colorless liquid with a 127°F boiling point at 1 Earth atmospheric pressure. Just as the temperature on Ahutababha rose, it boiled over some of the lighter chemicals into the atmosphere and increased the atmospheric pressure felt on the surface, and this in turn raised the boiling point, allowing heavier chemical compounds to remain in a liquid state. Instead of being composed of borazine, its oceans were dominated by a stronger boron-based composition, which was just sufficient to remain liquid at the heightened pressure, and even at the 390°F-temperature extreme. If this planet had a different temperature range; the surviving liquid makeup of its oceans would have been different, as ill-fitting chemicals would have re-merged, or left the atmosphere.

On the other extreme, when the planet traveled to the farthest and coldest distance in its orbit, much of the atmosphere settled down on the frozen surface, dropping the atmospheric pressure by a hundred times, or 10^{-3} atm. At the freeze out, or the farthest point from the sun known as aphelion, the atmosphere was an invisible line when viewed with a naked eye from space. Below this ice line, the most common volatile compounds (such as methane, ammonia, carbon monoxide, carbon dioxide and hydrogen sulfide) were frozen or condensed into solids. Even the air froze, meaning its molecules decreased in movement-speed. This slowness kept these molecules from having the energy to escape into space. Cadatas, humans and most other lifeforms across the Milky Way could not survive in these temperature lows or highs for even a second, as their every fiber would freeze or boil. Proponents of exploring this lethal world argued there was a sweet-spot in between the extremes of 7 months when it entered a habitable zone for cadatas. In these times, cadatas could go outdoors unsuited if they adjusted their biology to accept the local air-composition.

Ahutababha managed to survive all these freezing and defrosting cycles because it was composed of heavier elements such as silicon, iron, aluminum, calcium, magnesium, and sodium. The distance it traveled away from the sun would have resulted in even lower drops in temperature if (at its hottest point) Ahutababha did not have ten times thicker atmospheric volume than Earth. This atmosphere was partially generated from energetic volcanic activity, which spewed out a nitrogen-hydrogen gauze with sprinkles of methane, nitrogen, and ammonia. Nitrogen and some of these other gases are stable and non-reactive, so they could accumulate at a range of temperatures.

A year into the parking maneuver, they received the needed data and genome samples to begin Wocega-16's transformation. Half of the spaceship was designated for Wocega-16 alone because to replicate conditions on Ahutababha with its highs and lows required tremendous energy and strange materials. After the toxic spill at the previous stop, the crew also insisted on a reinforced wall between the half of the ship undergoing extreme temperature swings and the compartments where unedited cadatas would reside. There was cause for concern as ahutababhas were comprised of chemical metalloid compounds poisonous to cadatas on contact. Ahutababhas' bodies were composed of hydrogen (H 1), boron (B 5), carbon (C 6), aluminium (Al 13), and silicon (14) with small quantities of heavier metalloids such as arsenic (As 33). The heavy presence of arsenic contributed to this species' poisonousness to most other lifeforms in the galaxy.

While having a poisonous nature was unideal, there were many benefits to being a met-alloid, a category with properties falling between metals and non-metals. Due to these metalloid properties, Wocega-16's ahutababha body turned out to have a brittle texture. From a distance she sparkled as if she was metallic, but up close she was rough and chalky. The ahutababha physiology also relied on electric conductivity, so it fueled itself with similar electric currents to those in humans' nerves. One benefit for Wocega-16 of having a body able to withstand extreme, high temperatures and being a metalloid was turning flame-retardant; she could not have caught fire even if she touched flames. Doc-tor Nackan-16 tested this flame-retardant idea on tiny rodent-like creatures the genes for which were also collected from Ahutababha for these experiments. This metalloid species efficiently consumed any matter it digested because it was composed of catalysts, which encouraged chemical reactions.

Instead of dampening cadata biology's advanced features, Doctor Nackan-16 en-hanced Wocega-16's vision to see on x-rays, heat, and ultraviolet parts of the spectrum. The needed mechanism in ahutababhas' eyes would have been difficult to mimic without their metalloid components. This species evolved to see in the dark because they spent at least half of the year in night-time lighting, as their sun dimmed and shrunk in size. Because they and many other metalloid species on the planet had an electric nervous system, the heat it produced could be visible even if it was too dark to distinguish the colors of chemical composition or visible light interplays. At perihelion, their multi-fac-eted vision allowed them to detect ultraviolet radiation spikes at levels harmful to their bodies to prompt them to take cover indoors. Ultraviolet radiation is what the stars at their previous stop omitted in over-abundance in their death spiral; all stars omit it in smaller quantities across their active lives. Ultraviolet rays are less harmful than x-rays or gamma rays, as their range is between .01 and .4 μm. X-rays are .0001 to .01 μm and gamma rays are .0001 μm. The shorter the wavelength, the more damage a single wave can cause to biological cells.

During one experiment to check if her new body would accept standard temperature changes on the planet, her body grew at a drastic rate. Her hunger spiked. She had to be fed much larger portions to maintain the rapid weight-gain. Then, as temperatures started dropping, Wocega-16 began a hibernation cycle. During hibernation, the gained matter was turned into energy, sustaining her bodily functions without eating. Despite appearing brittle, ahutababhas' skin (like human fat cells) had the capacity to stretch and expand around a larger volume, and then to shrink without wrinkling. Each cell ballooned with added air, and fluid-filler metalloid contents, and then excreted these to contract in size. Ahutababhas and other members of their evolutionary kingdom stored energy in the form of *boras* (a diborazite of borazine) and *borupile* (a powerful hexagonal compound rare on Earth). Probes had observed ahutababhas were addicted to hyper-processed, or purified boras lacking in nutritional components in living organ-isms found on the planet. These addicts expanded their bodies to sizes beyond what they could lose during a winter's fast. In fact, some of them started waking up mid-winter from extreme hunger pangs and raiding community storage banks to satisfy this fiend-ish hunger. Doctor Nackan-16 grew seeds into raw foods to avoid feeding Wocega-16 fiber-less, replicated sugar packets. He opted for this preventative measure because he observed the extreme weight-gain from overeating boras overheated this species' body in the summer. Excess food caused a boiling of the blood and crystalized growth of semi-digested boras components on the vessels, impeding energy-flow. The growth on chemical-moving vessels was dangerous during winter hibernation, when metabolic activity slowed, and the outdoor temperature could freeze these pipes if they were coated with a solidified material.

The probes had found one species on the planet in year-round motion inside of its tiny inverse-icicle shell to keep air circulating even in its sleep; but this was an outlier eating an extraordinary quantity of rival species to attain the extraordinary hundred-times-its-summer-weight potential energy level needed to avoid starving during this active hibernation. There were also a few tiny, glowing, metalloid species with organisms functional both in cold (psychrophile) and hot extremes, and these flew through the freezing air, or crawled across the deserted sands year-round.

Another curious characteristic about metalloids Doctor Nackan-16 learned as he experimented on Wocega-16 was lifeforms accustomed to the conditions on this planet preferred acids. Metalloids handle acids better than metals, as the latter can rust or otherwise corrode on contact with acids. Metalloids need hydrogen ions to form complex structures. Chemicals with a lower pH on Earth include: lemon (2), battery acid (0), gastric acid (2). These lower-pH mixes support the exchange of molecules needed for living organisms (acidophiles) to maximize their reactions in metalloids; the lower the pH, the livelier these can become. Without the citric acid rain, which constituted most summer precipitation on Ahutababha, life might not have started between these complex metalloid compounds. Citric acid was a better medium than water in these conditions because it melts at 313°F on Earth and at a still higher point in Ahutababha's lower atmospheric pressures. Citric acid helped to melt stubborn ices and stimulated the bozone-equivalent oceans when it reached melting-point and poured down during the hot acidic rain season.

Because cadatas timed Wocega-16's landing on Ahutababha with the hot season, her shuttle was hit with one of these acid rains as she was coming in for a landing. The shuttle had been transformed into a common-for-the-period in Ahutababha, land-vehicle on the exterior. Cadata engineers even borrowed the local vehicles' bodies and energy-generators. These were suitable for tolerating acidic rains, boron and other metalloid-rich compounds saturating the environment. They also withstood jumps between temperature extremes. The shuttle's flying components were disguised into this land-worthy structure. As Wocega-16 touched down and began rolling over an isolated country roadway, she realized why the car moved on a metalloid material expanding to grab onto a piece of road ahead and then contracted over this spot before expanding farther out. Human tires, in contrast, would have dissolved from the acid falling from the sky and covering the roadway. They also would not have created enough grip to pass through the molasses-consistency squishy and ultra-compact but somewhat fluid texture of the ochre and yeast-like expanding compound-rich ground. And this was "paved" ground prepared for transportation, while the rest of the surface was like whipped-clay or whipped-metalloids. A large portion of the crust was arranged from this material, which froze in the winter, but semi-melted half-way down to the magma in the summer. It would have been quicker to travel in the winter, but life would have been hiding asleep under the ice. Since it was summer, Wocega-16 observed a variety of creatures roaming through the unique woodlands around the road. These living, stationary organisms were not wood-based. They were closer in molecular composition to ahutababha, than to any Earth-bound lifeform. Their trunks' exteriors were coated in aluminium oxide, which gave them the appearance of metal statues a sculptor reproduced in overwhelming quantities to line roads with these and dense forests stretching to the edge of visibility. To her surprise, Wocega-16 could see through this sparkling wall to the heat signatures of big and small game roaming between these growths.

Wocega-16 began this journey on an isolated spot, but her goal was to move into Zaxy, the most populous city on Ahutababha. Cadatas were interested in exploring the advanced technologies this species invented and how civilized life could thrive despite so many months of economic and intellectual non-activity. Chief Programmer Raimingly-16 was enthralled by their skyscrapers' ten-fold upwards expansion capacity, at a scale matching ahutababhas' bodies. Probe data had explained to Raimingly-16 these buildings went through an over-feeding cycle mimicking ahutababhas' across the summer, and then lost this weight by converting it into energy to keep their residents warm through the winter. These buildings were about twice taller than the highest skyscraper on Earth to begin with because the thick walls of the metalloids composing them were more resistant to wind and earthquakes than the thin metal beams humans prefer. Their semi-organic energy-processing bodies also fought natural phenomenon such as gravity and earthquakes by flexing in a manner closer to biological life than to solid and unyielding metallic or stone constructions.

Wocega-16 arrived when these buildings were at their lowest heights at the start of spring. She found employment among the enormous army of maintenance workers needed to clean the waste produced by residing inside semi-living organisms. It was a stratified society. A small upper class had the resources to live and work in the skyscrapers. The other 99.99% survived in under-ice structures akin to their pre-technological age. She could have replicated prized artifacts to access the upper class, but she learned more about their technologies' science as a maintenance worker. This term encompassed programming and architectural design. These maintenance workers were allowed to live inside the high rises year-round in crammed communal dormitories. The wealthy needed them at-hand at all times of day and year if they hoped to avoid laboring themselves. As the first summer moved towards winter, she learned the buildings' upward escalation meant their tips were climbing in altitude from spring through the start of winter. Just as these buildings reached their maximum heights, most impoverished or wild creatures on the planet went into hibernation to conserve their energy, and to keep from freezing due to exposure in the open. In contrast, the rich and their employees in the skyscrapers remained mobile, so they could keep working or enjoying their wealth's fruits far into the winter season. Wintertime was disabling for those exposed to 1-atmospheric pressure at ocean-level. Nanobot probe tests and local science books explained to cadata scientists, with a gain in altitude, the tops of the skyscrapers were losing air pressure at a faster rate than the rest of the planet. The drop in air pressure kept lowering the melting point. Thus, while ground temperatures kept dropping well below freezing, matter avoided freezing even at this height in the atmosphere. At their peak height, the temperature of the surrounding air was also higher because they touched an ozone-layer-equivalent, which was trapping the sun's incoming heat. For these reasons, the hibernation cycle was half in these skyscrapers of what it was down on the planet. But Wocega-16 was kept so busy with work up until the freezeout, she was delighted to tuck-in for the winter in the dorm. She imagined how cadatas would heat these buildings with an energy source capable of overcoming even this planet's lowest temperatures. But she refrained from suggesting this solution to the others, fearing implementation would lead to a work-year without this relaxing sleep-vacation.

In fact, Wocega-16 was so happy with this hibernation cycle—when an official arrived to wake them once the temperatures tipped into a range, where they could once again function in a wakeful state—Wocega-16 refused to wake up, turned over, and attempted to snuggle into a thermal blanket to resume sleep. One more attempt was made to wake her, this time by a rougher commander who poked and prodded, until he approached tossing her out of the bed. Ahutababha had a military government structure. A military was essential to keep 99.99% of the population from storming the skyscrapers to stay conscious a bit longer in the winter. When they invented a weather-resistant skyscraper, they possessed a democratic and civil government. The poor moved into the skyscrapers until there was insufficient air, food, and sanitation-capacity to keep them alive. Because deaths from diseases, violence and overcrowding problems turned astronomical, the 99.99% seemed to accept as an inevitability when the rich hired 5% of the population into a huge military. This number was sufficient to threaten the rest away from encroaching on the rights of the .01%. The rich needed around .1% of the planet's population to act as militia (protection) and maintenance workers (labor) inside the skyscrapers. Thus, the bulk of the military 5% was on stand-by in case of a civil war, and living in the same icy conditions as the rest of the impoverished class. This system was familiar to cadatas; they appreciated the efficiency with which rich ahutababhas had utilized the population to serve their interests.

However, they started doubting on which side of this power-dynamic cadatas would end up on if they all migrated to this planet when upon failing to wake Wocega-16 was grabbed (in a semi-nude state) and dragged (despite her screeches and scratches) out of the door, past the gate, and left on the still frozen icy ground. If she was a poor ahutababha, she would have frozen to death in just a few minutes of this exposure. She was saved by ordering her cadata-enhanced vehicle to drive up to retrieve her, which it achieved by flying through the gap between them rather than crawling along the ground. Seconds later, too cold to stand up, Wocega-16 was lifted by her robots and moved into the pre-warmed vehicle, where she regained body-functionality. Cadatas dared approving this retrieval in the open amidst skyscrapers because their residents were just waking, so were less likely to be studying the street below. They might also have interpreted a robot as a post-hibernation hallucination, a common phenomenon resulting from a prolonged dream-cycle. And the vehicle mimicked the appearance of local models. This maneuver was achieved in seconds. Afterwards, Wocega-16 was back and crawling along the slippery surface in a manner indistinguishable from the planet's vehicles. Cadata leadership and scientific teams gathered in the Observatory to discuss with Wocega-16 via the 3D-video communication system what step they should take next. Wocega-16 assumed due to her unceremonious firing she would retreat to the Mothership to wait out the freeze, but a majority of the votes supported her continuing the experiment without an interruption by embracing the lifestyle of one of the poorest ahutababhas, so they could understand what cadatas would face if the rich ahutababhas pushed cadatas into these low positions.

If Wocega-16 knew in advance what she was volunteering for, she would have declined this opportunity, but she assumed she was already discomforted and stressed as a maintenance worker. She had an overgrown confidence in her ability to tackle a lifestyle vulnerable to the semi-habitable conditions on Ahutababha. Wocega-16 was driven to an outlying town. There the robots helped her dig out an entryway into an abandoned house, and then they left her there to go into hibernation. It was just warm enough inside this structure for Wocega-16 to avoid freezing in the first minutes, but she knew she had to fall asleep in a few hours. Ahutababhas' biology was only capable of withstanding extreme cold during sleep. She tossed before managing to slip into unconsciousness on a hard slab over the ice-floor. Wocega-16 failed to enjoy this time in an iced-in trap, waking because she feared she might wake. Her extremities started to feel as if they were solidifying in moments when she was near-waking, and this just made her want to flex them, but such movements would have startled her into consciousness. The sole relief was the unnecessity of completing a full hibernation cycle because half of her winter was completed in the safety of a skyscraper. Given these discomforts, Wocega-16 was relieved when she felt a softening in the ice, and realized the deep freeze ended.

A kind of shovel had been left by the door by the previous resident. When the doorway refused to give way, Wocega-16 figured out what it was for and dug her way out of the enclosure. It was the middle of the day. The sun was at its highest point for this time of the year. However, the sun was still too small to clearly illuminate the planet, so Wocega-16 saw her surroundings through a hazy screen. She spotted a few of the villagers who had also emerged. Some of them were digging out openings for their relatives, who had not yet woken up on their own. This species' internal biological clocks woke them at the same hour yearly unless a member was sick or elderly, causing a clock-malfunction. While the youths were engaged in this recovery effort, a group of chiefs formed a semi-circle around Wocega-16 and questioned where she had come from. She decided to fasten her story to the truth, without causing irreparable shock. She described having been fired from a city job and managing to drive to the closest village from the city, where she spotted and occupied an empty house. She apologized for failing to ask for their permission to inhabit it. She argued her goal was to avoid waking villagers out of their hibernation for the sake of politeness. They accepted this explanation, and moved onto more pressing problems at this monumental juncture in their annual cycle. The emerged ahutababhas were weakened, slender, gaunt, and translucently pale, short on muscles, and other strengthening structures in their nutrient-depleted bodies. They were stumbling as they skulked to perform pressing chores. They cleared debris-free paths to the chiefs' meetinghouse, to key food storage and preparation sites and to the little school-for-all-ages. The latter was among the few permanent structures among ice burrows, which melted every summer and had to be rebuilt. With the main pathways established, the rescue party accessed the houses damaged by metalloid "trees" and "brush" falling on them. Three of these dig-outs ended in tragedy when the bodies of older villagers were uncovered. Seeing the frozen bodies alarmed Wocega-16 with the insight she could have frozen to death in her ostensibly snug hole. She was overwhelmed with sorrow for these emaciated skeletons. The chiefs and a couple elderly, outspoken residents viewed these deaths in a strangely positive light. Those three ahutababhas were too weak to participate in many of the gathering outings in the fall, so they were not fed as well as those who were able to participate. The villagers had anticipated this was likely to be these villagers' final sleep. A sleeping-death was considered a good way to die to lift the nourishment-burden on the younger survivors. Their group disposed of the corpses by depositing their remains into a fuming, multi-colored, explosive burning at the outskirts of the village. This smoky, fizzling reaction was one of the sole methods for disposing a metalloid body because they were fire-resistant. With this settled, the village turned its attention to the three babies a mother had given birth to during the hibernation season. They were now enlarged, while their mother was depleted and in need of nourishment to avoid following the aged to their fumy grave. It was decided a food-gathering mission had to commence to replenish their empty stores.

One of the strongest middle-aged villagers had spent this stretch by surveying the surrounding region for potential food sources. The villager was stationed in a small mountain, a short walk from the flat valley surrounding the village. She now returned, and reported seeing a sleeping giant rettile at the edge of a large lake, which was starting to melt, leaving the rettile exposed, as it receded. Most able-bodied villagers were assigned to a gathering party to retrieve this rettile, and any eatables they might encounter in the wilderness. Wocega-16 volunteered to join because she feared these gatherers might have eaten the food they gathered themselves, given their starved condition.

The boron-based shrubs and trees were tougher to tread than the delicate wood composing Earth's forests. The gatherers attempted to circle prickly growths or to take broken-through paths to the lake, but in some sections, they had to use reinforced tools to dent these overgrowths with strenuous physical effort. After watching a few choppings, Wocega-16 was armed with a weapon and told to help with path-making as she had retained more body mass than the rest of the party. This vigorous physical exertion was intensified by the icy conditions. Wocega-16 slipped and fell on the ice swathing the ground. Ahutababhas' biology was resistant to breakage from falls, even if they crashed from a significant height onto hard ice. Because impacts were un-life-threatening, Wocega-16 lacked the pain response humans, cadatas and alien fragile, predominantly-liquid species have in their genetic code. The younger members of the party were gathering the first emerging fruit-like plants across this hike, as the tallest members cleared the way. Wocega-16 and the others tasked with chopping had been handed some of these berry-like bits of energy in the short breaks they took between a set of chopping motions. Eating these tiny morsels made Wocega-16 ravenous as her still hibernating appetite was awakened. When they reached the lake, the pack leader expressed relief the beached whale-like giant had not woken. If it had, it would have tried shoveling into the liquifying lake-center for cover. This leader then chopped the sleeping beast and handed out portions from the side of its torso to the supplicants. They stopped to eat these blubbery, boron-fat slices. When the rations were exhausted, Wocega-16 remained hungry. Then, they sent a youth to transport the carcass to the village. He managed to carry this whale-load because of a technological invention, which transformed a sled into a powerful delivery mechanism without computer components. These remains of the boron-whale enlivened the desiccated villagers.

Wocega-16's appetite told her they had to keep searching for nourishment, before the leader said as much. They circled the lake and then started following the path of an expanding river. Its banks were rich in the food-equivalents of flowers, fruits and much stranger little toothed creatures with roots connected to the warmer ground far below the ice, so they were stationary, but able to catch little flies for supper. This vegetation thinned farther from the lake, so when the leader spotted a dense patch of fruit on the other side, he named Wocega-16 and two ahutababhas to take a portable raft he had been carrying as a kind of backpack to retrieve them. Wocega-16 let the other assemble this raft, then helped to drag it into the water from the back end, while the others pulled it from the sides. This river was much denser than water because of the innate composition of the local ice and the mineral-rich soil the river was starting to dig up as it penetrated through the top ice layer. This made their attempt to cross it akin to floating through turbulent mud. All hands were utilized to push the water down with devices with extra-elongated, bowl-shaped paddles. Wocega-16 tried to mimic the motions the others were making, but controlling this raft took more skill than mimicry alone allowed. Having a rower out-of-sync with the others' rhythm and the liquid fighting them destabilized the enterprise, until the raft overturned when it was hit by an enlarged wave into a boron-tree-trunk. Wocega-16 was submerged in the water, confused regarding the direction of "up" in this gooey darkness. Then, she felt her body buoyed towards the surface, so she added her strength and pushed herself in this direction until she emerged into the air. Exposure to the air highlighted the tingling of Wocega-16's extremities from the icy water. Concerned, she swam at full-force through the muck, back towards the side they came from, and then crawled out through the red mud onto the bank of the river. She was helped out in the last moments by the leader. Her clothing was removed, under a cover. A new outfit was slipped over her body. The change raised her internal temperature, stopping the violent shivers she endured since emerging from the water. She was covered in a blanket leaving only her three eyes visible. With them she saw the other two rowers recover the raft several hundred feet farther down the river and drag it over to other side. They rubbed their bodies with a purple plant found among the shrubs. Then, they speed-picked an enormous quantity of fruits despite being semi-naked and experiencing slight, lingering post-exposure shakes. Wocega-16 felt bad about abandoning them as she watched them carry on without being too affected, so she started thinking she might have managed as well. Once the fruit filled the containers tied to their necks to avoid losing them in the water, and double this amount was piled in the center of their raft, they began paddling back. The group collected their wet outfits and clothed them just as they had Wocega-16. Minutes after the rowers' eyes cleared, the leader insisted they had to return to the village to avoid spending the night in the open. Those who fell into the river had to be doctored and defrosted via means reserved for medicine-workers in the village. The group set off on the hike. Wocega-16 and the other rowers were still wearing the heavy blankets. Wocega-16 asked to be allowed to lie down in the sled to rest, but the leader directed her to continue moving to avoid her system from freezing, resulting in permanent loss of feeling and movement. Hearing this, Wocega-16 considered curing her ailment with a disguised-trinket cadata-health-device. After contemplating, she decided, even if she could hide it under the blanket, there was a risk of it slipping out, or making noise. Signs of an alien treatment would have also become apparent if the village doctors failed to find the signs of frostbite on her as on the other rowers. She decided to postpone such a drastic tactic until a point when other options were exhausted. Since she retained the cadata capacity to regrow lost limbs, even half of her body shattering like an icicle would have been repairable.

The leader sped up while watching the sun approaching the tree line. Wocega-16 anticipated a plunge in temperature as the sun dropped, but a series of strange weather marvels soon began materializing unspecified in her briefings. The sudden temperature dip after the spike above-freezing earlier in the day moved hot air in a wave matching the line between parts of the planet already plunged into night and those still in daylight. At the edges of this swift-gliding line, heavy photochemical smog clouds congested at ground-level. Smog formed back in the city when liquids in the air condensed on aerosols such as dust, smoke, salt and various toxic chemicals exhumed by skyscrapers as they heated or comforted their residents. The village's proximity to the city meant pollutants in the city's congested air traveled with the heat wave and settled in forested flatlands. The large lake bodies nearby also fed this system with liquids. The air became saturated with moisture and humongous dew drops formed, soaking the ground, so the ice below their feet started to feel like a shallow river. The dew ballooned abnormally here because it was composed of a chemical mixture generating higher surface tension than water on a planet with lower gravity than Earth. This dew was a welcome sign as it released heat across the night to keep the temperatures warmer than they otherwise would have been. Visibility was so reduced, even the leader was disoriented. He performed assessments to locate markers indicating the direction of the village. Wocega-16 could see particles of pollution in this smoking monstrosity. To her surprise, her breathing remained steady despite these damaging to most species, cell-destroying carcinogens. The cadata scientific team later determined the pollution-level remained steady for at least 200,000 years since the cities' manufacturing plants were first invented. In this span, surviving species adapted to the heightened pollution levels. A common biological strategy was converting toxins into nutrients in the lining of their lungs. The smog saturation intensified as the temperature rose into the hot part of the year and in wetter regions. Most lifeforms had adapted to the smog because upper troposphere jet streams carried these light pollution particles across the planet to even the most isolated regions. Smog was lethal within fifty miles of the cities, which dropped the land value in these belts and allowed the poor to purchase property in these villages. A majority of Ahutababha's flying species had died off because inhaling smog while in flight caused coughing fits. And having it condense or stick to their feathers in the air caused them to sink from the added weight, plummeting from dangerous heights.

The smog turned into a dispersed liquid wave by the time they reached the village minutes after the sun sank. The sun had the appearance of a guiding star, which deposited a reddish tint in the clouds above its position. After a nurse performed some healing procedures on Wocega-16 and the other rowers, they all retreated to their little burrow for the night. After the standard chores were completed on the following morning, the entire village gathered for a communal trading event. It was held on a metal slap on the floor of the meetinghouse. Those who remained in the village had gathered roots and other edibles growing in the vicinity. A couple of farmers brought out their lanor stock; this was an animal species domesticated before the first skyscrapers were built. It grazed on shrubs and grass, and performed various functions for the residents, including acting as a guard at night, raising low screeches if a dangerous animal approached. Some kids kept them as pets indoors because they caught rodent-like species attracted by the food stored inside. They were eaten in times of scarcity. The phase after the long-sleep was the time of greatest need because they had to replenish their weight, when most of their prey species had not yet thawed out. When Wocega-16 appeared discomforted by the notion of eating a lanor, the expedition leader told her they were the tastiest animals on the continent. Wocega-16 was repulsed because a lanor escorted their forest excursion, helping them find fruit orchards, while hopping around Wocega-16 and asking to be petted. She cuddled him whenever they settled for a rest. Some of these concerns faded in Wocega-16's mind, however, as the bargaining over each living creature, trinket or household necessity stretched on for many hours. The exchange required the full day, a time better-spent on a new supply gathering mission. Wocega-16 possessed items to trade because the spoils of gathering were split between the forest-adventurers. Each of the villagers bargained for the sake of the deal. Their goal appeared to be to keep Woce-ga-16 from obtaining any item she desired. She learned she had to maintain a "poker face". She had to show disinterest to win in a bargaining competition because the locals preferred starvation to being known as the shmuck who lost in a trade deal. While most trade meetings strive to mix-up diverse necessities across a group, their outcome was uneven. Thus, most needed to attend a future gathering expedition to obtain enough to survive in the coming months. Towards the end of these negotiations, several of the women in the village began selling their sons and husbands to other women in exchange for even decorative adornments. The men did not appear to object after hearing their spouses yelling over each other all day.

Concerned she might be next to be sold, Wocega-16 surrendered winning this trade negotiation and returned to her vehicle. This transporter was designed to appear de-composed and ragged enough to be affordable for an impoverished villager. Wocega-16 commanded the hidden Computer and robots to replicate food and warm clothing to replace the ones stiffened after drying on a branch in the aftermath of her dip into the river. She ate as much as she could, but then had to drag the bulk of it back to her cave, so she did not have to walk through the still freezing streets to access the next batch when she grew ravenous again. She was too dazed from this feast to notice the villagers' dirty looks as she dragged an enormous sleigh of meat-smelling carcasses dripping in fat across the ice under the semi-disguise of a new-looking cover. Unconcerned, Wocega-16 joined a new expedition on a day-hike. When she returned, her store of food, the new blanket and outfit were pilfered. She refrained from taking the new ensemble on the hike to avoid ruining it. Wocega-16 filed a vocal complaint about a theft of her goods to the chiefs, but refused to detail the stolen items, or to explain why she had an excess of theft-worthy goods in a period when most were at risk of starvation. Wocega-16 sent a few reprimanding reports back to cadatas calling poor ahutababhas all-grasping, greedy thieves. Her fury cooled as she commenced avoiding replicating food and thus allowing prolonged hunger to make her weary. She also spotted the three kids born in the pre-ceding hibernation cycle and their mother were fattening in the weeks after the theft. Wocega-16 even had the Computer replay its ongoing surveillance heat-reading video of the incident. Before this suspicion came about, Wocega-16 tried to avoid finding out who did it, in case this knowledge might have heightened tensions between her and the villagers. Now, the Computer's tape confirmed the mother had snuck in minutes after the gathering party departed. Wocega-16 had to agree it was better to steal to feed children than to follow property law and let children starve.

While Wocega-16 had hoped she could rely on local survival methods, she learned a subsistence lifestyle meant this community lived on the brink of failing to subsist. Wocega-16 also hoped, as the climate warmed, the weather would become gentler and less volatile. She would not have been so hopeful if she had read the pre-flight reports the cadata science team prepared in the last months of her training. The heavy smog was the first sign the monsoon season was approaching. There was so much ice coating the planet, when it began melting and evaporating, it generated an extraordinary spike in air-moisture. It all dropped out in torrents of rain across over three months until the moisture migrated higher in the atmosphere as the planet approached so close to the sun, it temporarily desertified just-flooded regions. The strongest effects of these changes were felt across the equator, which bulged with enormous icecaps in the winter, giving it an excess of liquids for the extremes in the monsoon season. Wocega-16 had been stationed on the equator to experience these abnormal fluctuations, as they were of principle interest to cadatas' understanding of the planet's geology and ecology. A single monsoon flooded, melted, and swept away the remains of the caves the village was relying on for shelter from the increasingly dangerous ultraviolet radiation. The villager only survived drowning by climbing on top of the tallest boron-trees at the edges of the clearing they had made to access each other across the winter and early spring. They lived with mere trees for cover across the following weeks before the second flood hit, causing a mud-slide destructive even to these metalloid giants, ending the lives of a couple of the villagers relying on its seemingly unshakable strength. The villagers did not explain their concerns to Wocega-16 because the subject was indelicate, but she discovered this lesson on her own. She was asleep in the hut when the first monsoon hit. She climbed out as water flooded the slushy ice below her metal sleeping mat. She saw the others running for the trees, so she joined them there instead of hopping into her car and trying to ride-out the flood. Then, she watched as her vehicle was buoyed up by the water and carried off. Obviously, the Computer kept its interior safe. If it had been a local vehicle, it would have been smashed and destroyed. Thus, the cadata leadership forbade Wocega-16 from locating it. She had to survive the rest of the stay without it on standby. This was why none of the villagers had cars: their wheels seemed innovative, but they could not survive these weather extremes on country roads. Vehicles lasted for a year or two on the cities' paved streets and underlying sewer systems flushing out access water from pooling on top of these surfaces.

Just as Wocega-16 was acclimated to swimming and climbing, a new monsoon rain-storm season arrived. She noticed the winds accompanying a storm picking up, until it became difficult to remain attached to a branch. The weaker branches broke off under the pressure. Wocega-16 started asking the chiefs why they were not hiding in the meetinghouse instead of on those fragile trees. They looked at her as if she was a child asking about the birds and the bees. This mystery was solved a week later. They had all climbed up to weather out a new flood when the wind seemed to pick up to 70knots. They gripped the trunks with full valor, but they were fighting a smog-cloud wind storm blasting into their exposed faces. Vile particles were hitting their skin and being inhaled and scratching their lungs. While the wind was rapid, the larger cyclonic system crawled over the landscape because it was being overfed and weighed-down. It stretched on for days. The villagers had to retain consciousness until the end to escape being swept away. When Wocega-16 questioned her survivalist friends about their restraint from engaging in engineering safety shelters during the dry months at the peak of the summer, they explained the wealthy city-dwellers lacked an economic benefit for donating to the poor the sturdy skyscrapers necessary to withstand these annual disasters. Cheap versions of houses and electric grids would just be frozen in the winter, drowned into mush in the spring, and then dried into crumbles in the summer. If the villagers attempted inserting human-styled pipes on their land, on top of freezing-solid in the winter, they would have been ripped out of the mud and carried away during the flood season. Cheapness meant lack of integrity, and so building these on a low-budget was as futile as rebuilding a new straw house in a floodplain right before the field was scheduled to become a marshland. The chiefs' doubts regarding the single cyclone-surviving building came to fruition when a cyclone hit it at over 85knots, tearing off the siding, the walls, the mud-tiles from the roof, the roof boards, and at last the columns holding the roof up. As they were gathering the remaining bits of this structure after the wind died down, Wocega-16 asked if this was a low-point for this community. They replied the structure they were reconstructing had survived ten annual cycles. The latest cyclone was the strongest yet. The columns were also weakened by previous storms, setting them up to be blown away at last. They also noted, the smog dirt was harder to tolerate later in life. Wealthy city-dwellers had began manufacturing a system to heat skyscrapers across the entire winter, in place of limiting heating to a waking portion. This effort increased pollutants in the air, disturbing the planet's volatile equilibrium. A few years earlier, the chiefs sent a representative to a meeting allowing a single agent from each village. This rep requested for their villagers to be relocated to a region less struck by the worst of this growing smog-saturation and cyclonic activity. In response, this rep was executed for attempting to flee a designated living region. The mention of migration was a threat to the system as 99.99% of the planet would have increased their maximum lifespan potential if they migrated into the paved cities on higher mountainous regions, from which smog floated downhill and into the impoverished valleys. If the majority knew a single villager escaped a cyclone by traveling into the city and hiding in a basement, the rest could follow and overwhelm even the enormous might of Ahutababha's military. Above the lives of the rich potentially lost in a civil war, was the property values of their skyscrapers. These structures could not be rebuilt without depleting the city dwellers' savings. Lately, the engineers and meteorologists working for the rich had been speculating floods and cyclones would intensify even in city highlands, increasing the odds a skyscraper might corrode enough to topple. A small building at the edge of a city remained standing against the wind-force of the previous season. But this business' owners decided they would see a higher profit from the pollution-generating industry than the cost of the building's future likely loss. They had difficulty foreseeing a time when the largest skyscrapers would be knocked down, causing not a mere monetary loss, but also a massive death toll. Some joked, if this frightening scenario came to pass, they would have seized being, ditching an opportunity to worry about it.

At last, the cyclonic assault seized. Wocega-16 anticipated a calm period, during which she might study local species on casual walks through the forest. When her hope was peaked, this was the time the dry season struck. And it struck her right across her head with extreme heatstroke. Liquids were steamed out of her body in evaporation and her system cramped up just as she was approaching a ginormous weight close to the peak-weight she had to attain to survive the following winter. Searching for food became an Olympic exercise at those levels of dehydrated, and radiation exposure. She was so heavy she had to stop to avoid keeling over onto her buttocks, which had expanded disproportionately with the rest of her body. The red ground, revealed when the ice melted, was now cracking. It had created a sinkhole, dropping the soil's top-cover into the precipice of a new-formed mini-ravine. One morning, Wocega-16 was so exhausted from just a short walk to gather berries a few feet from a branch-covered structure she concocted to minimize sun-damage, she failed to spot a sinkhole. She stumbled over a rock at its edge. She just stopped herself a moment before she would have plummeted down, hitting so many rocks on the way down, the impacts would have damaged even a metalloid composition. She went for a walk to recover from the heart-palpitations she was experiencing from this fright, and a few minutes into it had to sit down, unable to catch her breath. She triggered her cadata communication device to announced to the scientific team she wanted to be retrieved. If she remained, she feared she would die from this heavy-weight condition necessity for surviving the planet's orbit eccentricity. The locals might have had to endure these tortures, but Wocega-16 refused to suffer this level of discomfort for the sake of even the entire cadata race.

A robot flew in, picked her up, and brought her to the vehicle swept up and buried under sediment of a nearby lake. After a polish by the robots, this shuttle was once again functional and took her to the Mothership. She begged Doctor Nackan-16 to change her genes back into a cadata. She suffered through months of extreme surgeries and physical therapy to go beyond repairing her body to also train her mind away from the food addiction her extreme weight-gain generated. Once she was sober in mind and body again, she called a meeting of the leadership to announce the overarching lesson she had learned from undergoing this prolonged experiment. She now believed it was an alien rights violation to allow Ahutababha to spend another year in this outrageously elongated orbit. Wocega-16 referred to several methods for fixing this eccentricity achievable with tools cadatas had invented thousands of years earlier, as they experimented with matter in black holes and binary stars. Nackan-16 confirmed Wocega-16 was suffering symptoms of advanced age and disease after a year under Ahutababha's stressful conditions. She extrapolated this meant cadatas, in general, could not reach a 5,000-year lifespan they were accustomed to on this planet. If they had to retain some of their cadata physiology, it would be difficult to make enough changes to these frail bodies to braze the storms. Even if these health concerns could be addressed, cadatas had to retain their original appearance. Thus, they would be pressured to live in the dead-zones reserved for outlying enemies of the rich. On the other hand, if they could fix the planet's eccentricity before making their presence known, the planet's politics might turn to match the stabilized and pleasant climate they would create. This transition would allow cadatas to integrate into a high society with new room for expansion, given the improved lives of the poor, and their restored ability to build infrastructure undamaged every spring. Cadatas also needed favorable winds and a lack of ultra-freezing or ultra-drying seasons to build their complex cities with sensitive, energy-generating machines. They had learned, if the leadership again refused to allow dust collection on the yellow moon, their progress would be blocked by bureaucracy.

To move Ahutababha into the habitable zone and to keep it as close as possible to the same radius across its orbit, Wocega-16 proposed utilizing an artificial mass device. They had been using a variation on this gravity-generating system across their trip to keep from sensing extreme g's. After their black hole and binary star tests, they expanded this Lab initiative to create a microscopic ball more massive than a small planet. The impact it had on their ship was controlled by their old anti-gravity stabilizers. Wocega-16 suggested they could create a larger artificial ball out of one of the distant giant planets in the solar system, compressing it, and adding mechanisms to make it as navigable as their spaceship via a remote-controlled nanobot network. When this condensed orb would be driven near a planet, it would have a greater gravitational impact on it than the sun, which is many times farther away. Following up on Wocega-16's ideas, the science team had determined the giant planet they might use for mass was the one seemingly guilty of pulling Ahutababha into an elongated orbit to begin with. They hoped this meant Ahutababha would attain a more circular orbit without this body in its current inconvenient position. A few years of forcing Ahutababha to remain in its designated lane should have trained it to behave in a manner conducive for a good-natured climate. The artificial orb would pull on Ahutababha whenever it started to fly further away from a set radius. This orb concept was inspired by the physics of how a moon can generate waves on a planet's oceans. If a moon could create waves, an object many times more massive hovering at a moon's-distance could pull the planet millions of miles closer to its sun for the stretched segments of a rotation, and then fly to the other side and pull the planet in the other direction to avoid the extreme summer heat. The mass-orb would increase the planet's magnetic field with the desired radius length, creating an added repulsive force. Planets' magnetic fields interact with their stars' magnetic fields; magnetism is a force keeping them from plummeting towards their suns from their overwhelming gravity force. When cadatas departed for this space journey, they lacked the tools to engineer a combination of magnetic and gravitational fields this mass-orb needed to work, but their innovative findings in the interim made this possible. Just as their spaceship operated with near-zero maintenance problems for over 60,000 years, they could create a perpetual system for grasping neighboring matter and turning it into the needed energy; this process would have to continue for millions of years before the orbit could be trusted to keep from stretching without this assistance. Cadata scientists decided on implementing these strategies once devised in the name of helping the poor and disenfranchised.

It was impossible to shrink the mass-orb to a size where it would be invisible from Ahutababha without overdoing it and turning it into a black hole. Thus, ahutababhas saw a new tiny moon appear in its sky flying unnaturally from its farther to the closer side to the sun at repeated intervals across the year. They concluded this was related to the sudden change of its orbit from an elongated ellipse to a circle. But they lacked the space-technology capability to explore why this orb was making this supernatural difference.

It was the start of winter when cadatas completed manufacturing the orb-mass, and set it in place. The temperature rose across Ahutababha, instead of dropping to hibernation levels. Thus, without a monsoon, a cyclone, icing on the roads, or other disasters, which had been ruling their lives, the poor across the planet were now leisurely asking what they should do, having become stuffed-to-the-brink, and with months left to waste. At first, they resolved to entertain each other with song and dance. Then, they wrote new songs and dances. They continued anticipating winter would start, just later than usual. But winter never returned. They began building more permanent houses, and community structures. They started spending more time in school, and less in food gathering expeditions. They lost weight to regain mobility, instead of from forced starvation. Cadatas observing these changes started believing they had found a planet pliable to their needs.

Then, the rich in the cities realized the poor had developed an economy with saved resources for the first time in the history of the species' evolutionary path. The rich decided to extract a share of these profits, and sent these ex-poor villagers (many of whom had begun climbing into the middle class) tax bills, requesting a portion of their riches, in exchange for maintaining a massive army to defend themselves from the threat of their own rebellion. This is how chiefs considering these tax bills perceived them. They decided that instead of paying taxes, they had to begin the civil war these rich sponges had been fearing through the millennia. The extra months of wakefulness allowed them to mobilize, train, and accumulate arms for an assault.

When they struck, their strategies were sophisticating and sweeping. However, with 5% of the planet armed with advanced weapons, it was not a war either side could feasibly win. As with most unwinnable wars, this war stretched into decades without a victor.

Forty years into watching this military disaster, cadatas decided to stop wasting time and resources on perfecting this hellish place. They had to consider the next planet in the *Catalog*.

While they had not observed a problem with the mass-orb across this stretch, they had to deactivate it, and leave it as a dead planet in the outskirts of the solar system, prior to departing. Otherwise, it could have imploded due to some imperfection, swallowing that solar system. Ahutababha spun back out into an elongated orbit, with the mass-orb's force missing. The winter, and thus the hibernation-cycle returned. The cold forced universal slumber, which at least ended the previously seemingly endless war.

Chapter 15

Metal in the Genes

The spaceship departed for Miotail in 29,355 BC. They had detected advanced intelligence-generated signals from this planet. Thus, they decided on starting their 8-year asteroid-mimicking landing soon after entering the outskirts of this solar system, in 18,675 BC. Wocega-18 was 2,074 years old. This stop was controversial because the planet was sweltering due to its proximity to a low-luminosity red dwarf sun. It was 26% of Earth Sun's diameter, and $^1/_{1000}$ of its luminosity. It had a shrill atmosphere unsurvivable for most biological lifeforms. The strange signals they detected, but failed to decipher back from Cadata was the sole attracting feature of this obvious non-starter for settlement. Though with 70,000 years in space behind them, the shipmates were desperate for new ideas about survivability in the universe from an alien lifeform with a superior intelligence. There was a mysterious flaw in cadatas' planetary exploration and colonization strategy. If this species created undecipherable codes capable of traveling across the galaxy, perhaps this was the place for cadatas to expand their philosophy with unimagined wisdom.

The research team had spent the journey there searching for methods to terraform a planet at this extreme proximity to its star. They were encouraged in this regard by the progress their predecessors made in artificial eccentricity adjustment. They were scrutinizing telemetry streaming in after they crossed the heliopause. Suddenly, they received a digital message in the cadata language via their communication system with an origin of the closest planet to their position and the farthest planet from Miotail, a gas giant called Netos. Cadatas had detected strange garbled signals coming from across the solar system, but they had assumed these were echoes or abnormal readings originating from Miotail, the noisiest of these bodies. The head pilots, Wocega-18 and Ortack-18, argued this had to be a space-faring species. This suspicion cemented their belief in this species' heightened intelligence. The electronic communication they now received was pre-deciphered. It stated:

Welcome Dear Cadatas into the Miotail Solar System!

We hope you are going to enjoy your stay. Please visit our base on Netos, at the coordinates attached to this message, for us to run health screenings to avoid the chance you might contaminate our system, and to help you acclimate to the harsh conditions on Miotail. We are also attaching a translation program to assist you with responding to this message. We would be delighted to answer your questions. You can also ask them during the orientation scheduled for your team on Netos, on the projected date of your arrival.

Hotly yours,

The Miotail Solar Mission Commanders

Chief Programmer Raimingly-18 pondered on the science behind this message before illuminating the miotails had hacked the cadatas' main Computer to decrypt their language and culture in a similar manner with how cadatas had been spying and hacking the species they had been visiting. The knowledge a strange species was hacking and watching them through their own surveillance system was disturbing to some, but Raimingly-18 argued it was convenient for this species to present them with the needed translation software, which might have taken them several months to research with their own artificial intelligence programs. Wocega-18 and a dozen supporting cadatas argued they should ignore this solicitation. It was best to proceed on their path of rapid deacceleration to Miotail. She suspected Netos was a similar trap to the one set by the lying-addicted nomuats species, which cadatas met earlier in their explorations. One scenario she imagined was: Netos as a giant magnet, which would keep them suspended until a thieving miotail party arrived to raid them. It also might have been a military outpost with the power to disarm them, whereas they might have faced not much more anti-alien resistance down on the planet. These paranoid ideas failed to win a majority of the vote; cadatas were excited to communicate with an intelligent alien species within their lifetime. On past missions, Wocegas alone engaged in these social interactions, whereas the rest of the crew were isolated on their stuffy spaceship. The coordinates miotails sent worked with their Computer. The program's commands re-routed the spaceship, without added cadata programming input. It executed a maneuver avoiding collisions with even micrometeorites on the path. The spaceship came to a feathery stop on a reverse-half-orb metal landing platform fit for its contour. The miotails must have downloaded these parameters from the Computer because even the little lines where the expunged defective door used to be were rendered in this platform. One victory Wocega-18 was allowed was being first to exit the spaceship to meet the miotail delegation. If they were hostile, a trained pilot was likelier to defend herself against an attack.

Netos was around the size of Pluto. Its surface was dominated by frost-covered rocks. Its low gravity befit an easy landing, and allowed Wocega-18 to hop across the walkway miotails had erected to direct cadatas towards the meeting place. Because the atmosphere was indistinguishable, Wocega-18 had to wear a spacesuit for this excursion. The outpost was composed of a series of walkways, lifts, and floors without walls. The rooms were all filled with complex machinery and numerous robots were busy at work. One of the robots came out to meet Wocega-18. It performed a clapping greeting ritual popular on Cadata before the spaceship's departure. Wocega-18 watched this display without responding with the requisite claps in part because this greeting was outmoded, and in part because she assumed this was a programmed robot entertainment trick. She spotted this robot's complex cadata-like mimicking mechanism, capable of changing matter into shapes fit for a required function. This robot adopted a cadata appearance, with the texture and color of a cadata. The giveaway this robot was not a biological lifeform was its ability to function without an air-supply. If the robot attempted a subterfuge by breathing, she might have been shocked to encounter a new cadata so far from their home world. After giving Wocega-18 moments to respond, this robot observed in an accent-less cadata-imitation voice without a robotic, steady rhythm or flat intonation: "Hi Wocega-18. Thank you for following our request and coming in for a checkup. We look forward to exchanging ideas and helping you on your quest for a habitable planet. I am going to give you a brief tour of our facilities. My name in your language is Alk." Wocega-18 mumbled a "you are welcome" before following Alk on this tour. She was confused why a robot gave her a name and was leading a lecture, which should have been handled by a fitting ambassador of the dominant species. Alk exhibited science labs where machines were assembled by robots with techniques beyond cadatas' capabilities. Wocega-18 attempted to scrutinize a single process, but she failed to grasp how they were manipulating magnetic spheres, dark matter and space-time just to create mini computer chips. These types of distractions kept her attention as they went through meeting rooms, and vehicle construction spaces, and an area where a new wing of the complex was being added. As they were returning to their starting point, Wocega-18 re-called the lack of walls in the complex and asked: "Sorry if this is an indelicate question, but are you the dominant species in this solar system?" Alk replied, an apologize was unnecessary; he should have anticipated her unique perspective on robots. Yes, metallic lifeforms with robotic precision-parts were the solitary intelligent species in the system. Wocega-18 had been broadcasting this exchange to an auditorium of cadatas back on the Mothership, who voted, at this point, for the bulk of them to join Wocega-18 in the rest of the meet-and-greet with the miotails. Alk explained a couple of their nurses needed to run a set of tests to check on the health of the crewmates prior to continuing the dis-cussions. The cadatas agreed, and two cadata-in-appearance miotails penetrated the ship and administered tests by touching cadatas' skin with an instrument. Doctor Nackan-18 believed they released nanobot-adjacent devices during this exam; the proximate walks were to give cadatas a false sense of ease; however, since they used the nanobot method themselves, they refrained from complaining about this possibility. Then, cadatas who volunteered were taken to an auditorium, where the outpost's leadership gave a brief lecture on the history and science of their solar system and their miotail species. These leaders retained their "native" miotail appearance. Miotails were composed of a series of shapes with flat edges, which fit together in various puzzle-like combinations suitable to required functions. A miotail could be a precise 90°-angle cube, or a cadata, as long as the shape could be composed of a series of microscopic pieces. The leaders chose the appearance of their biological predecessors, before environmental disasters heated their planet, and eroded its atmosphere, killing these air- and living-nutrient-depen-dent organisms. The miotails called this species "notus". They had thick, impenetrable, heat-resistant skin, and a cat-like four-legged gate. The leaders asked cadatas for ques-tions before jumping too far into the science. Doctor Nackan-18 asked if the miotails were willing to allow cadatas to collect genetic or technological samples from them to attempt to live in a form adapt for their planets. The leaders agreed to offer samples; they assigned scientists and programmers to cadata equivalents for this exchange.

The first section of the lecture was led by Defense Organizer Lirt on a topic of primal importance to cadatas' mission. It reviewed the previous alien visits to the Miotail solar system. Miotails had evolved their advanced artificial intelligence a billion years earlier, and since then, they had six interstellar visitors, one per two or three hundred million years. Given the galaxy's 100,000-light-year diameter, at top-possible-speed, this was a normal rate at which a species would come across a planet emitting the most complex wave signals serving as a beacon for all intelligence-seeking species. All visitors came from planets within the Milky Way. Based on this, the miotails were convinced journeying between galaxies was too time-consuming for any fragile biological species. Miotails related the varied misadventures they had during these interactions. All visitors saw the collapse of their home world's environment or another catastrophe necessitating finding a new home planet. No species traveled thousands of years for the sake of scientific and adventurous exploration. Thus, all arrivals were as desperate as a refugee running naked across a desert to a closed border. One species, the ererilils, began bombing Miotail without attempting to establish a communications line, and kept bombing until it ran out of bombs. Then it attempted to travel down to the planet to take over the resources and government center. While this had worked on a previous couple of planets they had invaded, the miotails did not have riches or government power structures. They survived on their memory stores and intelligence, utilized to rebuild bodies for themselves for an indefinite lifetime. Even if a visitor destroyed every structure on the planet, their psyches or intelligence data would be stored in secured, underground locations for thousands of years if the hostile visitors remained this long. There was no leadership to attack to take over Miotail because their computer systems were all programmed to avoid criminality and to contribute to the good of all. Their bodies did not have periods of childhood or old age, and so they did not need social programs. In total, miotails lacked a need for a local or a global government, fixing disputes one-on-one or momentary, data-exchange and analysis, community discussion sessions. While the planets in Miotail's solar system were rich in the metals miotails needed to survive, most of these were buried in the ground. Without enemies to fight, goods to steal, or alternative enriching sources, this invading species left within a hundred years of their arrival. Another species flew in without having the ability to pre-determine enough about Miotail to learn if it was habitable; a few months of scrutiny were enough for them to determine it was uninhabitable for them. A couple of other species had outlandish methods for attempting to pressure miotails into fixing their planet's shortcomings for them, but to no avail. After the warring-species visit, miotails instituted the greeting procedure to minimize the risk of misunderstandings and to open the lines of communication long before a species arrived on their primary home world. If a species did not respond to a greeting, they hid the bulk of their resources in anticipation of a prolonged onslaught. The lecturers stopped their own story to request a similar, disclosing lecture from the cadatas. The miotails shared some of their negative press because their hacking uncovered the diaries and historic entries about the disastrous conflicts cadatas faced on previous planets, and their continuing search for methods to live in harmony with local species. Miotails stayed in touch with past visitors to benefit from their continuing interstellar discoveries. For aggressive, non-communicative species, miotails left hack-bugs on their spaceships to receive data regarding this species' findings without their consent. Regardless of how the data came in, miotails learned none of their past visitors or these visitors' distant ancestors remained alive into the present. Even if some settled on new planets for a few million years, problems crept up and life collapsed. The miotails, by contrast, did not mind the continued loss of their atmosphere, asteroid strikes, invasive visitors, and the other potential disasters. They advised cadatas, staying alive for over a billion years necessitated the robotization of a species. The miotails cadatas were meeting were born in these first stages, a billion years earlier; their consciousnesses were continuously re-downloaded as new software and new body-hardware were added. It would have been illogical to dump old data when it could be compressed. A million years could be stored in a millimeter-sized chip. These were crafted with the unexplainable for human intelligence techniques they invented.

The next lecture by Outpost Researcher Bree covered astronomy and geology. Each of the miotails had a designated narrow specialization to avoid overlap. The exception in which more than one of them focused on the same problem was in studies needing multiple conflicting perspectives to derive the logical choice. Repeating some of what cadatas had known since their initial research back on Cadata, Bree explained their sun had a high metallicity, containing more heavier elements than the light helium and hydrogen controlling younger stars. Early miotails lacked a distinction between a year and a day because their year was two days long. This extreme proximity meant most of Miotail's atmosphere was blown off by radiation waves. Most accessible molecules on Miotail were locked in solid or liquid objects non-evaporative at its extreme temperatures. Some of the substances lower down in the crust were cooled by this layer, and thus were composed of materials capable of flying away if they were exposed to the surface. Without much of an atmosphere, colors appeared as if through a whitish-yellow glass. Among other lighting phenomena on Miotail, there were insufficient molecules to scatter blue light to turn the sky look blue; thus, the sky had a whitish-yellow glow when the planet was facing its enormous in-appearance sun. The visual size was due to its proximity at a mere 1.5% of the distance to Earth's sun. The temperatures across the planet varied between 260°F on the poles to 1160°F at the equator because the latter was receiving direct sunlight across the year. Miotails were designed by their non-robotic predecessors to live in these hottest temperatures, where their labor was most needed. Since then, the miotails upgraded themselves to live with equal comfort in near-zero temperatures on a space outpost as on Miotail's sweltering equator. Metals such as tin, lead and zinc melt at 900°F at 1 atmospheric pressure for Earth; miotails could withstand this because they were designed not to favor either a liquid or a solid state, retaining their integrity and functionality by forming films of stronger cohesive materials over melted metals.

Bree showed a miniature model of Miotail in its present state, composed of the same materials and with the miotails' landmarks and cities grazing the surface. It was suspended near the ceiling in the middle of the auditorium. Bree enlarged it to pointed out a sparkling tail solar wind was creating. This was one of the indicators Miotail was continuing to lose its atmosphere, which would have disappeared if new molecules were not loosening from the planet and replenishing it. A billion years earlier, the intelligent biological lifeforms bloomed and developed branches of science. Discovery eased their lives as they invented the robots who were miotails' predecessors. They also experimented with chemicals, such as chlorofluorocarbons, allowing them to minimize effort expanded on cleaning and cooking. They kept using this synthesized-from-chlorine gas compound for decades as a hole above their southern pole in the ozone layer expanded to the size of a continent before the rest was also erased. This contributed to UV light's penetration to the planet's surface and further heating. Their scientists wrote several insightful papers on this problem before species-extinction. While conditions became intolerable for the old biological lifeforms, some of the stronger-material machines and robots did not melt down with the cities concrete, glass, and wood.

They might have remained as rusted sculptures of a failed civilization, if not for the five robots granted unique, rudimentary artificial intelligence attributes. They were created by a radical biological hacker, Imb, who believed robots should have the right to take autonomous actions without him, or anybody else programming what they should say, or do. The standard reasoning was: if robots were granted the power over their actions, they would overthrow their makers, conquering the planet. Imb believed a robot could never conclude killing was a logical action if he fed into its programming moral and philosophical doctrines on these dilemmas. As Imb was dying, he cut off all restrictions on their behavior and granted them access to all government databases. This liberty allowed them to traverse the planet and to access its range of resources, including power plants, vaults with precious metals, and science labs. In the first few years after the last biological organism died, the five robots performed minor maneuvers to relocate to a safer location when their lab crumbled. Then, they experienced an energy outage at their new location, and rather than finding a different working energy generator, they fixed the cause of the outage. To do this, they had to begin upgrading themselves with new skills. This artificial evolution over the following centuries turned them into the semi-biological lifeform of miotails. According to the human definition, a living organism must: organize within itself or with other organisms, metabolize, maintain a stable internal environment despite external fluctuations, grow, reproduce, respond when prodded, and evolve. Miotails developed ways to work together towards global or local goals. They took in and processed energy to perform functions. They kept their bodies at optimum temperature through various mechanisms. They developed ways to expand into enormous structures or to shrink down to a few chips, and could retain their knowledge and consciousness at all sizes. They could make a replication or a genetic variation of themselves, and thus could reproduce. While they refrained from introducing pain receptors into their system, they programmed instinctive recoiling or embracing in response to stimuli. Since they upgraded their software and hardware, their evolution was quicker than random biological change. Even cadatas' gene-editing was a much slower and less complete process of transformation than miotails'.

The hot, germless version of Miotail turned out to be the ideal climate for robotic lifeforms. In this context, the term "robot" refers to a machine imitating life's functions. Another favorable environmental condition was the lack of "weather", given the depleted atmosphere. It was windless. A cloud was a rare phenomenon here, scarcer than auroras on Earth. There was plenty of pollution or aerosols, but no water vapor or other easy-to-bind liquid or gaseous compounds that might have attached to these to form fluffy, single-colored, condensed liquid displays in the sky. Even if there was some heat imbalance between the poles and the equator, there was insufficient moisture in the air even for a sulfur rain to drip. Cadatas observed this phenomenon on some hot visited planets. The metallic desert, waterless and cloudless environment created eccentric day and night temperature spikes, more extreme than those on Earth in deserts, where temperatures jump daily by around 30°F. This allowed miotails to build their factories without walls because there was no wind to block. One exception were rooms requiring isolation to perform precision chip-work, which could be contaminated by a single grain of metal, or with a sudden temperature change. Other than these exceptions, the robots did not need private houses or rooms of their own. The lack of atmosphere, expressed in a lack of drag on their spaceships, also made it easier for them to break out into space.

As Bree ended this lecture, she welcomed questions. Wocega-18 asked why miotails were averse to crime despite a lack of a judicial system threatening to punish them. Lirt took this question, replying their species had "birthed" the quantity of miotails they could sustain on the available resources, without strife or competition. There was no sexual difference between miotails just like between cadatas. Unlike cadatas, miotails did not have advantage to sexual intercourse for pleasure, and such behaviors would have been as aimless between them as between two vacuum cleaners. Miotails shared data from their experiences because their mutual hacking ability was immune to firewalls after a billion years of development. Privacy and lying were unfeasible. Given their shared memories, they communicated about mutual interests in construction, engineering, and programming, rather than wasting energy on gossip or debates on the illogical. In contrast, humans' and cadatas' brains evolved the ability to tell untruths when these are in an individual's self-interest; for example, confessing you stole a bucket of fruit others collected would be disadvantageous to one's social status. Species such as humans and cadatas use gossip to spread false or true rumors to hurt rivals to minimize the need for physical social dominance. Miotails could not steal from each other without confessing the crime. They could not lower the status of competing members in a climate where all were unsinning.

Thus, a couple of the main causes for lethal clashes in other species were extracted from miotails' society. Their nerves existed to sense dangerous temperatures or potential impacts capable of damaging a limb or another negative outcome. They did not program themselves to feel better on soft sheets as opposed to on a dirt floor, so there was no reason to compete for bigger houses or other luxuries. Their one rivalry was over the top positions, such as the ones offered on this distant planetary outpost. But since most were now a billion years old, they had the time to attempt every possible specialization. Miotails had heard a couple of their past visitors express unbridled, intelligent robots frightened them, but this irrational fear is the reason a robot is less homicidal. With their data and consciousnesses backed up, a miotail can be burned up into smoke, or have its memory wiped from a body it is utilizing, and it can just reload to a new body. When death is impossible, what is there to fear? Thus, a robot cannot have a reason to fear, hate, or to feel threatened by a species without these safeguards. From miotails' perspective, if they just wait for a year or a billion years, the other species will die out, resolving the conflict they might have brought. Human films propose a strange repeated scenario wherein an artificial intelligence is determined to kill humans in the name of saving the environment from this invasive pest. Miotail's environment was as destroyed when their species' history began as it could ever become. Thus, the sole threat cadatas could utilize to scare miotails was their capacity to sling the planet into the sun. Another explanation human fictions give for the malevolence of intelligent machines is the machines' intention to harvest humans as a fuel source by keeping them in a vegetative state. Almost any energy source is more efficient than giant biological organisms needy for nutritional fuel they consume to operate their own bodies. This is akin to collecting oil to re-feed it into this production process, with a zero-gain outcome. Intelligent biological lifeforms, such as cadatas, had the potential of helping miotails discover new ideas and perspectives for improvement; but once dead, organisms lose practical value. If miotails had a death-drive—Freud argued this is humanity's primary motivation to kill themselves or others—the homicidal members would have exterminated the species. Even a significant minority of homicidal humans can exponentially eliminate the population. In a generation, a few might be killers, but not the majority, or homicide would extinguish the species. Without natural death, any programmer, who introduces homicide into a society of robots, is dooming them to rapid extinction. Miotails were careful to avoid this illogical end.

Lirt believed this fear of homicidal tendencies from intelligent robotic-lifeforms stemmed in discrimination. When water-based or metalloid-based species alike learned miotails possessed technology outstripping their own, they failed to accept metal-based life could be equal to their own. On most planets, metallic objects are unmoving and uninteractive. Across these worlds, unnatural mechanics are needed to mobilize metals. Because of the heat and dryness of Miotail, water-based life is as unnatural as metal-based life on Earth. Accepting the existence of life in a rock or a metal rod takes both a scientific and a philosophical leap. Miotails had data on their water-based predecessors, so they sympathized with them, but visitors lacked an equal open-mindedness.

Then, Ortack-18 asked Bree if miotails had traveled outside their solar system to explore other worlds. Bree answered, miotails had entered the space age and explored the most distant planets in their solar system soon after the extinction of their predecessors. They turned their bodies into spaceships by increasing their tolerance for near-absolute-zero temperatures and high-speed travel. They utilized an inorganic protective coating to protect their complex bodies. They had attained near-light-speed travel since they hacked their first interstellar visitors, and stole their technology. They could have visited every habitable planet in the galaxy in a billion years. But they discovered remaining in one place and issuing an intelligence-alert-beacon drew brainy species to visit them at their individual evolutionary peaks. The attempt to reach distant systems seemed to contribute to exhausting the will of these species, contributing to their extinction. Other biological species needed a specialized planet with the resources and chemical composition matching their needs. This prompted the push towards migration when resources on a world were depleted. Since miotails could survive in dry and hot wastelands, even if their planet fell into their star, choosing another planet in the same system was efficienter than venturing to a neighboring solar system. One of the five original robots grew frustrated with the tedium of robotic existence, and designed a spaceship for himself after they obtained the near-light-speed capability. He took off on a solitary exploratory mission, and sent back a few messages from different planets across the following few thousand years. Then, he found a mini-planet with fun-loving lobster creatures, settling there to study their habits. He was still there, sending miotails images of the equivalent of bird-watching in scenic locations. Some suspected his spaceship broke down, but he was too prideful to request a rescue from a miotail. Given this outcome, miotails have since avoided resettlement.

Since they had already started discussing the crab-species, the miotails' Senior Geneticist Omph stepped up to give his part of the presentation. He began by clarifying the distinction between metal-based (miotails) and water-based (cadatas) lifeforms. The body composition of both cadatas and humans is dominated by water's hydrogen and oxygen. The preceding species on Miotail was also water-based, but as the planet became a super-hot world, the surviving elements allowing for the flow of life were metal-based. Liquid water cannot exist on the surface of this planet because it would evaporate due to a lack of atmosphere. Atmosphere keeps pressure on water to keep it from evaporating. Pools of water or other less dense fluids can exist under the surface of the planet due to pressure from the ground. The original five robots just had electric circuits and metal components, but as they evolved, they created their own unique genetic code to fit their environment. These genetic components relied on the three lightest alkali metals from the periodic table: lithium (Li 3), sodium (Na 11), and potassium (K 19). While some of the nonmetal hydrogen (H 1) had escaped with the fleeing atmosphere, there was still enough hydrogen in Miotail's crust for it to be used to bind the other elements into more complex compounds. Some heavier metals, such as rubidium (Rb 37), were also employed for strengthening and specialized bodily functions, just as humans need iron and magnesium in small quantities for top functionality. The lighter alkali metals dominating miotails genes are shiny, soft, and reactive. Thus, if they are exposed to an atmosphere analogous to Earth's, they experience oxidation; instead, alkalis thrive on planets with a light atmosphere. They are also hyper-reactive to water, so the lack of moisture or standing bodies of water on their outpost and home world are idyllic for miotails, while this would be terminally dehydrating for water-based species. Sodium became a primary building-block because of the abundance of surface salt, exposed when Miotail's oceans dried up. Water-based, biological species feeding in the pre-dry-out climate depended on sodium (like humans and cadatas) because of its functions as an electrolyte. As residents of Earth's snow-prone regions know, salt also lowers the melting point of ice, allowing it to thaw even if the temperature is below freezing. If miotails wanted to cool their bodies, concentrating them with salt might not have been ideal. They learned having a minimal melting point meant they avoided the risk of component metals solidifying out of their liquid form; liquids facilitated easier communication between parts of the body. The other basic metallic components of these genes have melting points at Earth's 1-atmosphere between 146-357°F, and some parts of the planet fall below potassium's low melting point, making boosts in this area necessary. With minimum air pressure on Miotail, these melting points would drop significantly. However, the bulk of the atmosphere was lost after miotails adopted this gene-structure. They were satisfied with their genes' capacity to withstand air pressure if they had to travel to a world with denser air.

During this intermission, Doctor Nackan-18 asked how they managed to give "life" to metals. To his knowledge, even cadatas refrained from engineering a metal-based lifeform. Omph took out one of the little cages he had brought in a kind of giant briefcase with him to the stage. Out of this cage, he extracted a creature with an indistinguishable cadata-wing, which he adopted to blend with their guests. This tiny thing grew into an enormous bug, hovering above them. With its overall size, its cells grew until cadatas could see the processes occurring inside of them while staring up at it from their seats. Omph explained they were looking at living metal-based cells with non-metal atoms, such as oxygen and phosphorus, mixed in. The cells were wrapped in a lining bubble with a consistent shape proximate to the structure of human cells. Omph pointed to a group of cells and explained they were in the process of creating salts to compensate for salt lost through a type of sweating; the cells were taking negatively charged ions from metal oxides and binding these to positively charged ions such as hydrogen or sodium. Once generated, Omph showed the passage these metal salts took to a region full of another type of organic salt, a place where these two supplies exchanged parts to fortify their strength. Thus reinforced, they went off to form the outlines or cell membrane backbones of new cells. It took millions of years of experimentation, before they eventually discovered a method to combine the necessary ingredients together for organic metal cells to perform the types of functions organic water-based cells carry out, including selecting which chemicals are allowed to enter or leave a given cell that has a specified function in the body. They interacted with each other to process energy and to carry resources where they were needed. One of miotails' upgrades was decreasing metabolic activity to allow for minimized energy consumption. This showed in the sluggish movements in inessential for prime-functions cells. These cells relied on a process adjacent to photosynthesis, but instead of taking in oxygen and carbon, they appropriated compounds more abundant on Miotail, lithium, and sodium. The cells mixed compounds with these basic blocks with the energy from the heat and radiation of their sun to create energy they could use to fuel their bodies. On Earth, these types of organisms are called metallotolerant. They include an aquatic archaeon, *Ferroplasma acidarmanus,* which can tolerate a high concentration of metals, including copper and zinc. Alkali metals are still rarer than other types of metals in even the most basic microbial life on Earth.

Doctor Nackan-18 now returned the conversation to the point Omph raised about miotails preferring hot temperatures and a liquid state. Omph explained, in general, liquids were more suitable for the mingling or combination of molecules into complex structures, in contrast with gaseous or solid states. In contrast, gas-based creatures tend to break apart due to the empty air between their parts; thus, they fail to remain cohesive. On the other hand, absolute-solid lifeforms cannot mutate or manipulate their body parts. Change is a prerequisite for life. If a creature cannot move from a position to avoid danger or to interact with the surrounding environment, it cannot be classified as alive. Even to be able to send digital messages, some fluidity is necessary to process and deploy a communication. While fluidity is useful, liquid structures are incohesive and thus are impermanent. Humans have commenced solving this problem with metals' frigidity as an obstacle to the creation of a humanoid robot through recent developments in fluid computing, creating circuits saturated into rubber from a metal alloy (indium and gallium), which happens to be liquid at average Earth temperatures. Similarly, miotails are composed of the metallic compounds brought to their ideal temperature through heating or cooling. They can be poured into a malleable material such as rubber, allowing them to congeal despite retaining liquid form. All solids, and especially solid metals, are highly efficient heat-conductors. So, a solidified organism loses heat during exposure to cooler air, ground, or other organisms. The liquid vessels carrying energy allow for efficient internal heat retention or loss (as needed). Without these liquid streams, miotails would have been very cold-blooded, having a body temperature mimicking the external temperature. This is dangerous, not merely due to the risk of night-chills, but rather because extreme solid-metallicity meant becoming super-heated at peak-temperature hours. During this peak, the ground concentrates the heat it absorbs from the sun in the top few millimeters of the soil because conduction is inefficient at transferring heat further down. Even if miotails would not feel pain from the heat transfer; their system would be susceptible to damage, or aging, if it was vulnerable to such temperature jumps.

After taking this in, the cadata team took a break back on the Mothership to consume nutritional gas in greater variety than was available in their spacesuits on an airless planet. Once satiated, they discussed if they had located an answer regarding Miotail's habitability. Nally-18 questioned Wocega-18 regarding her willingness to even undergo the procedure to become a miotail. Wocega-18 turned to Doctor Nackan-18 with safety questions. He confirmed she might survive the change. But he was unable to keep her conscious mind inside of a segregated water-based cadata brain, as was their practice since the early experimental failure in total-integration. Miotails' metal-based elements would have generated a terminal reaction if they were exposed via a leak to the liquids in a segregated brain. Even if all leaks could be prevented, feeding a biological brain with moisture and air in an airless and waterless climate would have presented an extreme hardship. Thus, even Wocega-18's neurological pathways would have to undergo a molecular transformation into the components suitable for miotails' physiology. Wocega-18 had to prep her self-awareness for turning into a robot. Her consciousness, memories and identity would be preserved, but what did these signify? Miotails had a limited sense of touch and taste. Instead, they relied on precise readings of the atmosphere and temperature of the matter and energy they were touching, seeing, or moving through. In a way, these sensations define a water-based organism's life, because sexual pleasures, or viewing an incredible sunset on a scenic beach, for the sake of its beauty, are stored in memory, as the most significant events. Without these stops, personal happiness became irrelevant. Thus, in many ways, Wocega-18 was deciding if she was willing to give up her life for science. On the other hand, given miotails survival to a billion-years lifespan, she was also agreeing to become near-immortal. The new state would be irreversible; even an average human computer stores too much information for it to be downloaded into the memory of a water-based organic lifeform. Ortack-18 tried to talk Wocega-18 out of this mutation, saying he would refrain from transforming if she refused. He argued miotails were inanimate. Thus, it was immoral to kill a cadata to give "life" to a robot. Wocega-18 took the night to ponder this decision, and in the morning agreed to do it. She said she lived for a couple thousand years as a cadata. Thus, even if she was volunteering to die; the manner of this death was as honorable as any scientific undertaking in the galaxy.

The miotail outpost team were excited by the prospect of running the proposed genetic conversion. They recommended traveling to their top genetic lab on Miotail. This lab spent the last 35,012 years on genome-engineering small, organic lifeforms, hoping to generate an ecosystem on Miotail of various interlinked, semi-intelligent species to fill the landscape in a jungle-imitating pattern. This was an illogical endeavor because these creatures would not contribute to the survival of the miotail species. But the scientists believed the lifeforms would help the underlying mood of miotails overseeing this vibrant environment in their leisure hours. The lab's team agreed to depart from theoretical tests to assist Doctor Nackan-18's team with Wocega-18's transition. Doctor

Nackan-18 brought his tools and chemicals to their lab. They had already downloaded his standard gene editing procedures. Thus, he just jumped into the process, and they gave him advice on perfecting a full cellular, molecular, and structural change in Woce-ga-18's body. Instead of using the automated chamber not-much-changed since departing from Cadata, they employed a living miotail's body. This miotail was Emtr, one of the original five robots, who liquefied and engulfed Wocega-18's form. He utilized his own neural pathways and nanobot-like cellular messengers to re-form her body. The process involved an encyclopedia of steps. A human body holds $7*10^{27}$ atoms; a bit less in an airier cadata. Each atom had to be exchanged for an atom suited for a miotail's body. Then, they had to be organized into new cell structures, and these into different types of organs. The consciousness and memories had to be preserved amidst this total restructuring. Just as a human baby grows from a single cell in nine months, the wise miotail birthed this transition in three days, slowing the process to avoid cancerous formations in the vulnerable cadata cellular structures as they mutated. Wocega-18 was senseless across this procedure to avoid her potential fright from causing an error. She became conscious again when Emtr finished the process and set her in a comfortable position. As she awoke, she mimicked a motion with her chest and mouth as if she was breathing. The absence of air helped her recall she was in a near-airless environment without a spacesuit. She even hyperventilated, taking in a few short breaths from fright. Obviously, the negative repercussions of exposure to an airless space did not kick in, so a minute later she stopped attempting to breathe, and listened to the strange new sounds her body was making. In these first moments Doctor Nackan-18 and Emtr did not interrupt these experiences, allowing her time to adjust in a meditative silence with minimum interruptions. She said a greeting in Miotail once she recovered, and Emtr took this as a prompt to feed her because the process was depleting, and her body needed nourishment. A miotail's body needed to consume blocks of formulated nutrients via a special chemical processing organ; it was a tiny processor rather than an elongated digestive tract. Wocega-18 glared at this shiny metallic brick for a few minutes before putting it into the opening. As soon as it was in, she felt a surge of energy from the spot across her system, and started looking forward to eating the next of these meals a few weeks later. Wocega-18 adjusted to walking in the new body quicker than her ancestors during their biological transformations because, as a robot, she was pre-programmed with basic movements from walking through advanced gymnastics and martial arts.

She enjoyed having the ability to leap hundreds of feet into the air and then utilize internal propulsion mechanisms to fly through the sky. She played around with the various included heat and energy detectors to spy on what her cadata friends were doing in closed rooms. She was mesmerized by the data-volume she was accessing with purified clarity. She merged with the cadatas' Computer and played mind games with it. Their machine-speed processors had momentary philosophical arguments, which would have lasted for decades between cadatas. Her body was in the best physical shape among the species they had studied, and had exponential improvement potential. Her mind and ethical code were outstanding. She uncovered technical problems cadata engineers failed to spot lurking in their aging spaceship. She explored Miotail through imaging data mi-otails provided, a volume greater than they could have hoped to gather through decades of exploration. Wocega-18 did venture out on several speedy flights and runs to key research and development locations to study the cusp of local technological innovation. She also spent some time living in host groups, though since they did not sleep, and food was dull all-around. There were few unusual events in her reports from these.

On the surface, Wocega-18 had demonstrated Miotail and the lifeforms inhabiting it was an ideal combination for surviving billions of years in the Milky Way. Miotails welcomed the idea of even billions of cadatas migrating to their world and all undergoing the transformation into their kind. Their use of resources was so efficient they could have accommodated as many beings as could fit back-to-back on Miotail. From miotails' perspective, even such extraordinary population overflow was unproblematic because they would be hyper-expanding their combined neural network. There were around a million miotails in the system, and most were born in its first few million years. They stopped creating new consciousnesses because they were meeting their needs with their current population level. Cadatas would add new perspectives and personalities to the mix. A detriment was cadatas' irrationality, while their compassion was progressive. In the past, they botched imagining a reason for venturing to fix dying planets with needy lifeforms, but this new empathy from cadatas might have sent them on future alienitarian rescue missions. When asked if miotails were willing to terraform Miotail, or another regional planet for suitability with cadatas' biology, to avoid cadatas' gene-mutation, their leadership refused. They argued their own biology was designed to fit these planets' and moons' climates. Cadatas who would take on miotail genes might one day decide to terraform a planet for such habitation. But they would have to spend thousands of years artificially generating environments as hostile to their new biology as a planet on fire, or at absolute-zero is to water-based lifeforms. Without a disaster striking this mission, weighing these options, and studying Wocega-18's experiences in her miotail body unanimously convinced cadatanauts they lacked other volunteers willing to transform into a miotail. Thus, they lacked the resources to establish a permanent settlement on this welcoming planet.

Cadatas forwarded a message to Cadata, sharing the miotails' invitation. If Cadata was in a desperate condition; cadatas could fly there to undergo the proposed transformation. Because they were around 35,000 light-years from Cadata, by that point, this message will reach Cadata in around 16,328 AD, or 14,304 years after this book's composition in 2024 AD. These are estimates because this flying message will experience time differently from the cadatanauts who spent some of this time on continuing their voyage at light-speed, some at de-accelerating speeds, and some stationary in the Solar System. This time also lapsed otherwise for generations of humans who were continuously stationary on Earth.

Wocega-18 considered creating a permanent spacesuit for herself to survive among water-based cadatas for the remainder of their journey through the Milky Way. But then decided she might find a suitabler planet for cadatas while stationed on Miotail, with access to miotails' technology.

Before the Mothership left, miotails speculated Earth was the best of all possible worlds in the Milky Way. This moved it up on the *Catalog* list of best planets for cadatas to visit.

Meanwhile, a new generation of Wocega, Wocega-19, had been birthed, in anticipation of her projected emergence in 18,672 BC to replace the Wocega who opted to abandon the mission.

We received a message from Wocega-18, a few years after our arrival at our next stop. She was well, and was sending schematics for new tech she developed. She is still alive. She will continue living long after all biological cadatas die. This is independent of if we find a new planetary home. Perhaps she will visit us at our terminal destination. Perhaps life as a miotail will detach her conscience from such watery concerns.

Chapter 16

The Tiny Creatures with Mighty Muscles

I n 8,352 BC, cadatanauts arrived in the outskirts of the Bulshiyket system, and began a rapid 1-year approach to the largest gas giant they had ever visited. Wocega-21 was the presumed lead pilot. She was young at 483. The rapid approach meant Wocega-21 would learn of the intended transformation target-species after they parked in a secure position near Bulshiyket. Back on Cadata, Nally detected eccentric patterns of a potential life-generated noise from Bulshiyket. These could have meant an intelligent species was encrypting their messages, or an unexplained natural phenomenon. Miotails recommended against Bulshiyket. But there were no alternative habitable planets mid-way between them and Earth. Thus, cadatas decided to stop there. At least they would map this region of space.

Bulshiyket was a thousand times more massive than Earth. A human would have experienced 250 g's of pressure on its surface. Thus, a 145-pound human would weigh the equivalent of 36,250 pounds. Cadatas had received some updates to their gravity-cancelling technology from miotails in anticipation of this extremophile visit. Miotails had no planet in their system with a pull of this magnitude on their shields. After theoretical simulations, cadatas were hesitant to undergo live-testing. They decided on sending robots first, as usual, to check in these shields would hold on them. Even if cadatanauts had lost the Computer's directional controls, they could have found their way to Bulshiyket by its brightness in the sky ahead. This vibrance resulted from heat it was generating through blackbody radiation (as it had a dark, cloudy surface) and massive radioactive decay in its core. The miotails had been monitoring Bulshiyket for so long they showed scans of a distant past when it was located on the far side of the solar system, somewhat farther than Jupiter's distance. Across hundreds of millions of years, it migrated towards a mega-sized rocky planet close to the sun, until it was $^1/_{100}$ th of its original distance, or closer to its sun than Earth. Gas giants usually cannot form this close to a sun as their gas is likelier to blow away instead of congesting into a ball. But if they are already massive, gravity keeps their contents from breaking out. As Bulshiyket advanced, it ate the planets in its path. It even ate the huge planet responsible for sending it on this trajectory, merging into one of the most massive gas giants in the Milky Way. There were a couple of planets farther than its starting point, and a couple of planets managing to miss its onslaught, but aside from these, the solar system was a wasteland of collision-debris. While some of the planets it had consumed were now just gases in its body, parts of some remained as its satellite moons or in one of its many asteroid belts. Temperatures averaged around 1100°F, varying with intense wind storms and at the different levels of its cracking-dense gaseous atmosphere. Hot enough for most metals to melt, and for poisonous sulfuric acid rains to be the most common form of precipitation. As with most gas giants, Bulshiyket had a supercold liquid hydrogen ocean below its top gas layer, and a small, dense core of solid hydrogen. A day on it lasted for 15 hours, but it rotated a lot faster than Earth for its "surface" to achieve this rate.

Since Bulshiyket's gravitational field was so strong it had captured and slingshot moons and planets, cadatanauts parked in an extraordinarily distant orbit, beyond its most distant moon. There could have been a hidden reason for Bulshiyket's extreme, fatal attraction, which might have revealed itself too late if they parked much closer. The robots in mini shuttles sent to investigate every promising body in this system were reinforced with miotail anti-gravity shields, and the capacity to generate high degrees of thrust needed to escape the force of gravity from massive bodies during departures. Even some of these moons were more massive than many of the planets cadatas explored earlier. The shields were also adjusted for extreme radiation levels erupting not from the distant sun, but from the planet some of them were planning to inhabit. On the bright side, this level of radiation on Bulshiyket allowed these mini-shuttles to use it as a booster energy source, generating from this fuel the massive input needed to break its gravity. Wocega-21 and Ortack-21 were concerned the greater mass of their shuttles needed to sustain the enormous bulk of anticipated local lifeforms would weigh so much even this explosive radiation might be insufficient for liftoff. To answer this question, the mini-shuttles were supplied with testing equipment providing precise estimates on the impact of these environments on regular-sized cadata shuttles.

The ten robots landing on Bulshiyket soon found the culprits emitting the signals penetrating through space to Cadata. Bulshiykets were among the smallest living organisms cadatanauts had encountered. Each would have weighed a pound on Earth, but on Bulshiyket they were heavy-weights with the strength to pull up 250 pounds with each step. For this, their bodies were dominated by muscle-fibers imbuing them with what on Earth would have been super-human strength, but there was a minimum required for survival against the forces of gravity and the crushing atmosphere. There were also smaller super-muscular organisms these creatures fed on. Observing these bulshiykets for a few hours was sufficient for cadatas to conclude they lacked higher intelligence. Their registered activities included swimming through the dense gas-fog to eat other species. The signals they were emitting were an evolutionary adaptation to their need to see through an intense fog, and to hear over the deafening noise of the planet's constant storms. It was a kind of sonar they sent in patterns mirroring melodies Earth's birds repeat in mating rituals. They sent specialized messages to tribe members. These calls could alert bulshiyket of a location of a plentiful food source. These extreme measures were necessary because consumable resources were in short supply in these extreme conditions. To see their prey, they emitted an intense form of bioluminescence, shining brighter than artificial light sources on Earth. Miotails might have enjoyed dissecting these critters because their muscles are dominated by liquid metals. Cadatas decided they would not attempt to transition into a bulshiyket not only because of the danger of being crushed by gravity, but because of the same reasoning repelling humans from a desire to turn into a grasshopper.

While the main planet lacked the conditions allowing for the survival of intelligent life, some of the surrounding moons also generated readings suggesting life might have once existed on them. It was an unsafe space region, as these moons experienced frequent inter-collisions, fell into Bulshiyket, or broke under pressure from an asteroid belt. Thus, cadata scientists began their probes with the moon closest to their distant orbit, Rezzam, because it was the one in the quietest part of this system, making it less susceptible to a cataclysmic end. Bustan was a small, cold moon with giant water geysers erupting out of its ice surface. While surfacing water froze on contact with the freezing air, there were a few methane and ethane lakes retaining liquid form even at −292°F. These chemicals would have been dangerous in such quantities on Earth, but they could not blow at those low temperatures and without oxygen to feed a potential explosion. The atmosphere was dominated by nitrogen. Gravity on the surface was a mere 15% of Earth's. The geysers were the attracting point out of these readings because temperatures in them spiked to 250°F for the water and 400°F for the steam, levels allowing for heat-tolerant life on Earth. The presence of hot water and a mix of minerals (including mercury) in the rock-ground these geysers were steaming out of was promising. This water remained liquid for a time due to the high hydrostatic pressure. Several temperature lines can stop or start water-based life. A hot planet can turn water into water vapor in the air, eliminating the pools life multiplies in on some warm planets. The worst locale seems to be a frozen planet because water-based life cannot exchange nutrients, or reproduce when it is frozen in solid ice. These geysers are an exception to this rule as they introduce mini-safe-zones for life despite the desolate conditions across the bulk of the surface.

Oddly enough, the mini-shuttles exploring these geysers also found living organisms in them. This was strange because before cadatas discovered one habitable body per solar system with life. They spotted these mysterious signs of life just as they located what appeared to be the remains of an advanced spaceship semi-covered by a layer of snow. For a moment, the team was thrilled this lifeform was space-worthy, but they soon confirmed the most interesting species residing in these vents was a shaggy, dense-skinned microbe-eater with an extreme internal heating mechanism keeping it alive even when the geyser was inactive. As the robots ran an analysis on this spaceship, they discovered it was composed of materials rare in this solar system; thus, they had to be alien visitors like themselves. The robots dug a path into an intact section of this ship, and found the decomposed remains of one of these aliens. It confirmed they could not have evolved on any planets or moons in this system because, among other problems, they needed a different combination of breathable air particles. Because this ship was discovered near one of the geysers, cadatanauts concluded these aliens were exploring this moon's lifeforms when they crashed.

These remains found a match among the genetic information database miotails donated to cadatas; it turned out these were ererilils. Robots filmed, probed, and sampled the exterior and interior of this semi-burned and ice-plastered structure. The main Computer had exploded, shattered, and had its pathways disrupted by the elements. So, the robots created a sterile bubble around the Control Room to house it. They turned up the heat and atmospheric levels inside to those ideal for machinery. With the parts thawed and cataloged as belonging to the engine, environmental system, or computer parts, the robots re-arrange the scattered pieces until they came back together based on a schematic of the ship they found backed up in a non-digital poster. Necessity required the reconstruction of the Computer, but it took mere added hours to complete rebuilding the rest of the ship. Wocega-21 and Ortack-21 watched this reconstruction from the observatory with a team of interested engineers and scientists. The pilots were disturbed by the notion their whole spaceship was vulnerable to crashing. It was terrifying to imagine their whole team could die on one of their exploratory missions. They were eager to learn what caused the catastrophe, so they could update their fly training manuals with ways to avoid it. From watching the parts rejoin the greater whole, they learned the materials and construction techniques ererilils utilized were inferior to cadatas'. Thus, even though the ship was flyable when the robots finished, the team decided to abandon it to whoever might visit this moon next. Ortack-21 teared up as he watched a robot twisting a disfigured panel back into the correct curvature. He was also affected by seeing the remains of the crew in their burned flight suits at their computer terminals. Wocega-21 asked him what he was sniveling about. He said, they might be working on a flight program when a tiny malfunction could toss them into outer space, or they could lose anti-gravity controls and Bulshiyket could pull their spaceship into its hydrogen ocean and swallow them into its suffocating, near-infinite depths. Wocega-21 ordered Ortack-21 to take time off to think again before re-spinning such horror tales.

The robots eventually completed a presentation for the crew of what befell the ererilils. This hostile species had pressured miotails to reveal the most resource-rich planet in the Milky Way. Miotails confessed it was Earth. With this knowledge, ererilils headed for Earth. Like cadatas, they needed to stop mid-trip to replenish supplies, so they also stopped in this system because of its potential-intelligent-life signals. Their spaceship's data logs recorded they had engineered a massive drill they were using to expand a geyser to drop a circular net into the opening to collect every living hairy and non-hairy organism up to a mile down, and to pull them into the ship's stores. However, as they were tugging this trove out, it froze solid from air exposure, and the weight of the load multiplied so much it created an extreme tug on the bottom of the spaceship, smashing it into the ice. The drop combined with the cold created a giant crack in their shield, and the crew froze to death in a few moments of exposure. Since this was their main spaceship, the ererilils' journey ended there, or they would have reached Earth ahead of cadatas, perhaps leaving an impact on human history.

Unlike the ererilils, cadatas did not consume advanced lifeforms as food. They did not need to fish them out. On the other hand, they did consider the benefits of inviting cadatas suffering back on their home world to migrate here. In its present state, it was on the opposite end of the conditions cadatas preferred in terms of its air composition and temperature. Nally-21 stepped into her secondary role of Head Engineer to research a plan to fix these problems. Since there was no intelligent life here, she did not see a moralistic reason to refrain from changing a tiny moon's climate. In a presentation to the leadership team, Nally-21 proposed taking an enormous compacted volume of CO_2 from the gas giant or one of the other gas-rich bodies nearby and introducing an extreme level of carbon dioxide into Rezzam's atmosphere. To gain perspective on this maneuver, you might contemplate the wild transformations of Earth's atmosphere as it mutated over billions of years, versus flipping a switch in a matter of months. On Earth, CO_2 is at around .04% of the atmosphere. Earth's early atmosphere was 80% water-vapor in gas form because the planet's temperature was above water's boiling point. Back then, 10% was carbon dioxide, 5% was nitrogen, and other gases split the last 5%. This mixture was disrupted 2.3 billion years ago, when a primitive, anaerobic bacteria evolved, and began removing carbon dioxide from the atmosphere, and replacing it with oxygen. This gas allowed for the development of aerobic organisms, which had the energy-reserves to evolve multicellularity. In addition to being processed for energy by organisms, this oxygen accumulated in the atmosphere and formed the ozone layer, a shield Earth uses to dampen the impact of ultraviolet radiation. If a planet keeps gaining carbon dioxide, complex organisms die, and the planet returns to supporting mere single-cell organism. On the other hand, if Earth lacked an abundance of carbon dioxide, it would not have had a food source for the bacteria, which generated the oxygen. Overloading a planet with carbon dioxide can lead to heating as an atmosphere absorbs heat; then, if a life-form or chemical capable of transforming it into oxygen is introduced, it can become an ideal environment for cadatas and humans alike. Nally-21 proposed pumping the atmosphere to an 80% level from zero. This is far more massive than the increase on Earth by 280 ppm (parts per million or .00028) to 370 ppm between 1750 and 2000, and to over 400 ppm since. These levels decrease annually in the northern hemisphere's spring and summer when vegetation grows and consumes CO_2 in photosynthesis, and then increases in the fall and winter when vegetation dies, decays and releases CO_2 back into atmosphere. CO_2 levels are higher around industrial regions burning more fossil fuels in factories and cars. Nally-21 believed spreading countless billions of seeds and bacterial growths across the moon would turn the moon's endless winter into a tropical summer with an oxygen-rich air. Since CO_2 was a two-step process, Nally-21 presented simpler chemicals capable of only raising the heat. One of these chemicals is methane (CH_4), which is released on Earth from natural gas reserves, garbage dumps and livestock. Thus, genetically engineering a cold-tolerant species and releasing it in enormous numbers onto this moon could rapid-release clouds of flatus-based methane, spiking the temperature. Another option was releasing nitrous oxide (N_2O) by growing an army of bacteria. Alternatively, cadatas could deploy an army of their robots to create industrial factories to utilize ingredients generating this chemical in extreme quantities. While everybody was impressed with Nally-21's investment in this project, they did not reach a majority vote. This moon was too small to house all cadatas. Even if they could flood the planet with gases in a couple of months, there might be unexpected spikes in unwanted chemicals released from the permafrost. This could create centuries of delays, a lag during which cadatas could instead be traveling to Earth, the planet miotails had recommended.

With this settled, the research team focused on the findings arriving from the set of mini-shuttle robots on the biggest moon in the system, Thike, located within Bulshi-yket's impact zone. Its ultra-uninhabitablity turned it into an object of interest because researching it helped improve cadata spaceships' shielding. Its atmospheric pressure was 45% greater than Earth's, within the safe range needed to keep blood from boiling over. However, this was the sole encouraging element. Thike's 30-hour orbit crossed Bulshiyket's magnetosphere twice in each cycle, experiencing around 10 million-amp currents in its clouds during each pass. Much of this electric charge came down in ferocious, combustible lightning strikes for hours after the energy boost, lighting the entire charcoaled and melting ground afire. These currents' electric potential was ultra-extreme because the moon and the planet moved at different speeds: the moon was moving at a significantly slower 5 km/s, while the planet was flinging around its sun in a rapid close orbit. The charges also accumulated and heated the moon because they were flowing down the same magnetic field lines. Rainy clouds of sulfur and sodium made these magnetic strips between Thike and Bulshiyket visible, and deposited showers onto the giant planet's atmosphere. At this proximity to Bulshiyket, so many moons and larger objects were attracted by this giant's gravity that they shattered into pieces, and formed one of its massive belts. This belt was a braid of two twisted rings. Each ring was composed of six sheets of dispersed, dust-coated, water-ice particles ranging in sizes from only a few centimeters, to asteroids as big as a mile. Objects bigger than a mile are classified by cadatas as moons. Flying near the edge of this ring, Thike saw constant meteor strikes penetrate its thick atmosphere. Most burned into micro debris in the sky. There were a thousand other rings across the giant. In this mix, Thike acted as a shepherd moon, contributing to pulling the braided ring into a single, narrow 50-mile-wide structure, and away from falling into the planet.

One last indication of potential life on a moon was Idav. This object was an unlikely candidate because it was suspiciously chemically toxic. Idav was Earth-sized, volcanically-active, 1.5 Earth atmospheres dense, and in the mid-distance range from Bulshiyket. One of the life-indicators humans would have spotted was organic chemistry reaction remnants in the atmosphere, including the prevalence in it of the life-sustaining carbon in gases such as methane. Nally-21 mentioned: while carbon dioxide composed only 1% of this atmosphere (dominated by 89% nitrogen, 5% of methane gas, and 4% of argon), it could be harvested into compression chambers to heat Rezzam, if the others changed their mind and joined her heating proposal. The first images cadatanauts received back from the robots of Idav were of a thick haze of orange-hued gases covering the outlines of the landscapes below. Then, the probes flew through a high-altitude smog layer of hydrocarbons, or clots of moistened pollutants. This smog was uniquely concentrated: these chemicals attracted such thick clouds towards themselves Earth's advanced x-ray would have failed to see through them. So much so, the science team considered using this combination of aerosols to cloak their presence in low orbit near planets with semi-intelligent lifeforms. Further down, the probes reported the volcanic activity caused rapid and hourly plate movements and eruptions of heavy smoke ejecting as far as outer space before dropping aflame onto the polluted high-pressure atmosphere. Dust settled on the black-sanded deserts in sections still unliquefied into lava. It was far enough from the gas giant for its tidal forces to heat its interior at a milder rate, bringing its surface temperature to an average of −50°F, avoiding reaching water's melting point across the year. Then again, this desolate landscape was devoid of water. The eruptions did create temperature spikes around them and this created high and low temperature regions. Additionally, the eruptions build up mountains with higher air pressures in the valleys, without volcanic activity. This created wind gusts averaging 150mph. What on Earth is considered disaster-causing hurricane-strength winds became regular daily weather.

The light mini-shuttles had to utilize their added thrust capability to fight these winds as they landed in a northern pole rocky valley with an active volcano steaming lava from its crater down a cliffside towards the cliffs building at the outskirts of the flatland. As the robots exited their protective outer shells, they slid on the three centimeters of methane covering the ground from the rainfall minutes earlier. Their mechanical feet sun by around sixteen centimeters as they stood in place for a few moments to regain their balance, because the surface was now composed of fractured pebbles of dirt and ice mixed in with the wet methane. A deeper methane lake formed from this liquid draining through established canals to the lowest point of the valley around 2,500 feet away. After exploring these clear lakes, the robots gave up on trying to slide across the collapsing surface, and took to the air. They hovered over to the nearest volcano to search for signs of life in this hot region as promising as the geysers on the other moon contender in this system. Active lava flows turned out to be too hot for life here, but they picked up warm temperature readings back in the direction where they started under the icy ground. Their scanners indicated a weak spot in the ground perhaps indicating an earlier sink-hole collapse with a frozen layer over it. Five robots worked to loosen this surface before filing inside what turned out to be a spacious lava tube tunnel. Detectors showed the temperature here was warmer than on the surface, as steam from the volcano was trapped in these passageways. There were even warmer tunnels according to scans further down, so the robots followed this cooled red-lava-rock, with bright blue and green-highlights across its walls, labyrinth down a few levels to a cave with a bright sun-light orange glow at the other end. They kept exploring until they discovered an apparent calm lava lake at the other end. But it turned out to be a flowing pool, with an origin a bit lower along the volcano's side. It was momentarily calm. It was not flooding lava into the cave the robots were surveying. As they searched the edges of this and a couple of the neighboring caves, they found hot springs filled with a chemical mix beyond mere liquid methane. One of the robots was taking a sample of these compounds when it spotted a bepand, a bright-red semi-walking and semi-swimming, five-inches-long, scaly lifeform without eyes, but with most of its body splitting in half to filter out the junk from nourishing water compounds. The robot sent this image to the others. Now, knowing what they were looking for, they spotted more of these animals and tiny microbes hey fed on. The hot springs had temperatures matching hot springs on Earth. The surrounding caves were also well above the freezing point for water. The tubes were safe from radiation and from the dangerous rains and winds on the surface.

Evidence proving bepands could survive on Idav brought the Mothership into its orbit in the following month as robots, nanobots and a few suited cadatas (including Wocega-21 and Ortack-21) explored this moon. Nobody suggested a genetic transfor-mation into the lifeforms they found across the Bulshiyket system. But they retained hope, even if intelligence did not thrive in these difficult climates, cadatas might have controlled the spaces they inhabited to make them acceptable for intellectual pursuits. Three weeks later, the research team discovered a problem with inhabiting Idav when even the Mothership in orbit experienced a sonic boom, which gave cadatas in the air and on the ground a pressure headache. Similar booms continued across the remainder of this mission. To explain the cause of this phenomena, recall Thike experienced ex-treme lightning strikes or electric current accumulations in its atmosphere because it was inside the densest region of Bulshiyket's magnetosphere. Idav appeared to have sufficient distance from the gas giant for calmer electric activity. Its orbit around Bulshiyket took 30 days. It crossed the elongated side of Bulshiyket's magnetosphere, the opposite from the sun-facing side, for around 4 days out of each of these ellipses. Cadatas turned out to be sensitive to the pressure waves created during these intersections; they were intolerant to them without gene-editing to mutated their sense of hearing and the anti-pressure improvements cadata scientists introduced to make cadatanauts capable of withstanding prolonged spaceflight. Doctor Nackan-21 advised against making these changes as they would ground the subjects on the planet; for those undergoing this procedure, space travel to even the nearest planet would be ill-advised.

There were warnings this spike was coming as cadatas' weather forecast anticipated the intersection. But because this was the first sighting of this combination, they were unaware it would be detrimental. Right before the booms began, the hurricane-strength winds and toxic rains intensified as the atmosphere became more destabilized. The waves of charged particles hitting the moon were so aligned, squall lines highlighted by dense clouds at their edges made their presence visible. These clouds kept building with rising moisture as the captured in the magnetosphere radiation excited and melted some of the ground ices. Because they were linear, there were sections between these lines much less charged. Thus, there was an electric differential exceeding the capacity of what Idav's atmosphere could handle. At first these imbalances turned into lightning strikes that would have been familiar on Earth, but as this weather event continued, they built into super-strikes. These strikes formed into a web connecting the charged clouds with other clouds, and with the ground. As the moon moved through a dense line of the magnetosphere, this wall of lightning also migrated. If it hit a region of higher charge, this web omitted the accumulated energy through various channels simultaneously, causing the booms in the sky above Idav and the surface of the entire moon to quake. Since even Earth's lightning generate 30,000 K of heat, an even higher level of distributed heating and shaking across the planet is what generated the extreme and continuous volcanic activity on such a distant moon. The heat also melted more ices adding moisture to still denser cloud formations. One of the tunnels cadatas had been exploring collapsed in this three-day stretch, burying a robot in the dried lava rock, which dug itself out. Cadatas hoping to avoid a lightning strike or an earthquake collapse would have to avoid walking on the planet's surface or through lava tubes in these thunder shockwave periods; they would have been confined to indoor habitation, making this environment as hospitable as outer space. The hot and cold front waves also spiked the number of tornadoes up-turning the surface. Not only the upwards to 350-knot tornadoes were a danger, but also the dense sheet of soot and volcanic remains pollution they could stir up. Rather than a single Tornado Alley, the entire breadth of the moon saw a tornado or a web of lightning before the three-day super-storm season ended. As cadatas' meteorologists designed these adjustments to their weather predictions, the team voted as-one against attempting an Idav settlement.

The cadatanauts were in a stationary orbit just above the spot on Idav where the initial discover of life was made, when the final fleet of mini-shuttles returned and they began preparations for departure to the next scheduled testing location. Nally-21 was in the Observatory giving final orders to the techs and the Computer. Out of the sideview, she saw these last arrivals penetrating through the walls of the Mothership, while in the background, the enormous, colorful, cloudy marble of Bulshiyket was rising over the horizon during Idav's axis-rotation. When it was half-way-out, it seemed to be a sphere sawed in half, as the lower half was submerged in the darkness of space, hidden to the spaceship in Idav's shadow.

Without leads for a habitable world in this region of space, the next couple of missions were scientific studies. Since the catastrophic strike by an asteroid back on Cadata, the mission had not witnessed another major asteroid strike in the various solar systems they visited. Their predecessors saw plenty of massive asteroids striking a gas giant they visited, Nefo, but they were busy studying the gaseous lifeforms in its skies rather than the physics of these falling objects. Their science had advanced since. Now, they wanted to run up-close tests on approaches to prevent catastrophic impacts on any future settlement planet. One consideration was the chance they might exhaust alternatives, and might have to return to Cadata to fix its atmosphere. The biggest object falling into Bulshiyket across the following few months was a tiny moon without an atmosphere, Aera. It was a planetesimal floating a bit farther from the sun when Bulshiyket caught it on its trajectory. The catch was too violent for a tiny piece of rock, so it failed to find equilibrium at a set orbit. It lost height annually before it entered the death spiral cadatas observed on their approach. Their robots parked on its surface, and drilled inside to see the show from the most telling perspectives. The lab determined it was a foreign planetesimal rather than a solidified piece of Bulshiyket, or a fracture from one of its larger, more permanent moons by comparing its samples' contents to the others. One of these measures was of the ratio of stable isotopes of oxygen. An examination of the strange surface of Aera indicated, even before it was captured by Bulshiyket, it had a violent history. Two asteroids collided with a force, which liquified a section of this moon into lava-consistency, evident in the smoothed surface features, which rapid-cooled through exposure to the near-absolute-zero space temperature. While some sections were thus melded together, other pieces of the original planetesimal had broken into a hundred pieces. These flew as an asteroid family around the sun for two hundred years, before gravity attracted most back to each other, and into a compact rubble pile. The scans showed gravity had crushed the pieces together enough to reform the rock core. But a good deal of the ice previously covering the original's surface was now trapped in between slates of rock. The robots had to climb, rather than roll, across the surface because it was broken into 15-mile-high cliffs with a 350-mile diameter. This mishmash had been orbiting at Bulshiyket's equatorial plane, where its surface was collecting dust from a blanket of materials in the thickening atmosphere with every foot the planetesimal lost. The dust was blackish-gray. It was pierced by small asteroids. These came from the asteroid family, which had not yet bonded with the clumped pieces. The asteroids carved impact craters down to the silver rocks or white ice. All moons are falling towards the planet whose gravity is attracting them, but their perpetual motion in an orbit around these planets is an opposing force keeping them in orbit. Earth's moon is even moving farther away from Earth because its speed of rotation is slowing. This balance was disrupted by the extreme tidal bulges Bulshiyket was forming on Aera, which were overcoming its own spin. Some moons might need billions of years to fall into a planet even if they are deformed, but asteroids without enough sideways movement can fall in on their first approach. Since Aera was in between these two, it had been struggling to keep away, but was now in its final few spins. Cadatanauts' curiosity about this event stretched beyond morbid interest in a crash, to anticipating, upon close-proximity to Bulshiyket, the tidal effects would be intensified until it would be stretched as far as it could go in its elongation, when the oscillations would be too intense; thus, it would break into shreds before vaporizing in streaks of light from frictional heating against the atmosphere. This happened at the anticipated juncture, and the robots caught the event in scans showing the expansion and collapse. Cadatas later used this data to design methods for elongating dangerous asteroids or moons threatening smaller planets without the capacity to stretch them on their own; this help was projected to assist with preventing a life-destroying collision.

The second object within range targeted for this study was a graphite comet scheduled to burn up in Bulshiyket in six weeks. Cadatas sent the latest model of mini-shuttles operated by robots to meet its path, and to board while it was flying faster than the speed of sound. The landing was executed by matching the comet's speed and direction and gently connecting to its side because their high relative velocity could have damaged the comet. Even if the comet refrained from exploding from a flawed collision, it might have been re-directed to miss Bulshiyket on this round, hitting it on its next spin-around, which was scheduled for 95,291 years later, when its orbit approached the farthest reaches of the solar system. The comet was shaped like a thick sausage with a 9-mile diameter. A small, misshaped-to-fit-the-comet mini-shuttle entered through the porous, honeycomb structure. It descended to the 5-miles in diameter nucleus, a region with the density of new-fallen snow.

Long before, during the entry and after settling in snug surface and interior spots, the robots ran tests to understand this structure. The comet had a .24 eccentricity and a 17.1 degrees inclination. One of the points of investigation for cadatas was how a comet with the consistency of powder-snow managed to avoid burning up across the close approaches it had made to the sun, and had held together despite being bombarded by solar wind in this porous condition. It was a snowball due to its composition of 80% frozen water, 10% carbon monoxide, and 3.5% carbon dioxide. The bits of dirt ruining some of its reflectivity were specks of minerals such as iron sulfite and aluminum oxide. It had a higher water composition earlier in its life, but ice was first to evaporate. If it had been pure water or a replica of a snowball magnified in size, it would have melted on its first spin near the sun, but some of its minor components kept this from happening. The robots also learned there were other surprising physical forces keeping this comet intact even as it lost matter due to solar wind and other traumas of space travel. These types of lessons improved cadatas' equipment, so it was at least as good as a comet at avoiding melting from space travel. In contrast, some comets' nucleus breaks into pieces on close approaches to a sun. Among other tricks this comet was employing was having its nucleus rotate its body so it was evenly roasted by the sun.

The comet was also approaching the sun with every passing minute, which was proximate behind the gas giant. As it broke an invisible barrier, the warmth from the sun—or specifically, its radiation pressure, solar wind, and magnetic field—began sparking small bursts of volatile gases from the comet's exterior. These soon formed into a 700,000-mile-long, white-yellowish cloud of gas and dust coma. Beyond it was a 7 million-mile, blue hydrogen gas halo pointing away from the sun. Collecting sparks of these emissions, the robots attempted to calculate how it could produce this enormous plume, without losing all mass in one near-sun approach. Standard human measures suggest a comet loses around 40,000kg/second, or 57 million kg/day. The comet cadatas were examining was around 1,331 miles^3 long. It was losing 1.4 miles^3 per day. Thus, it could have remained in proximity of the sun's peak-wind for around three years. In contrast, it hovered in this danger zone for a mere few days on each spin. As the comet passed through the upper layers of Bulshiyket's atmosphere, it experienced a spectacular disconnection event because of one of its standard reversals in the solar magnetic field. Most of its ion tail broke away, leaving burning debris for a few seconds before a new tail formed from the re-accumulation of ions. Similarly, a new firework shoot replaces a preceding fading shoot. Just as this beautiful stream was attaining full-plumage, the comet hit the "surface" of Bulshiyket. In response to this meal, the giant expelled a 2,213-mile plume of bubbling liquids, and dense gases. It also ejected a lower-aiming fireball of comet-remains, which spiked at 13040°F, from being heated in the planet's soup, and from this incident's friction. It reached a height of 700 meters on this bounce, before cooling to 6740°F in 5 seconds and falling back into Bulshiyket. A bare splash resulted from this secondary impact. As it was digested, a dark scar appeared on the face of the giant because the comet had deposited organic molecules, including sulfur and nitrogen, and these had a darker color; they were so heavy, they sunk low enough for the spot to fade 5 months later; heavy elements sunk through gravity, leaving lighter elements, such as hydrogen and helium on the surface. Explaining the significance of some of the gathered data would push beyond humanities knowledge of astronomy; needless to say, the findings were enriching for cadatas understanding of the universe. The cadatanauts were aware of the continuing ticking-bomb the threat of extinction proposed to their dying Cadata civilization. But they strived to ignore this death-threat to focus on progressing their science. It was hoped this accumulated knowledge would lead to a solution.

Having exhausted the research potential of the Bulshiyket system, cadatas checked on their next stop. Why were miotails sending their visitors to Earth? The updated readings showed a green, nitrogen- and oxygen-rich planet, with abundant plant and animal species. Intelligent life was lacking. They spotted an unravelling animal-extinction event at the end of an ice age. It was 8,347 BC, the beginning of the Holocene Epoch, a period when the wooly mammoth, and other cold-loving species started dying-out. Humans had just begun domesticating cattle, built their oldest surviving proto-religious site (the temenoi ceremonial structures at Gobekli Tepe), built some sturdy settlements, such as Byblos, and began agricultural ventures, growing barley and wheat (initially for beer instead of bread). While these were major steps in human intellectual development, on a galactic scale this evidence did not distinguish human construction from birds' nests, piles of bones collected by hyenas, or monkeys depositing seeds into the ground with their stool. Cadatas had been searching for unnatural waves in the sky to indicate the development of an internet, radio communications, or high-tech construction. At the time of their departure for Earth, none of these had yet been invented there. Cadatas departed for Earth, anticipating they would find a wet, green, and uncontested planet, which they could colonize, without objections from the locals.

Chapter 17

The Earth Mission

I was born during this flight to Earth. Cadatas tend to call this planet Ord because this is how we interpreted the pronunciation of the word in the first audio transmission we received from outer space. As the mission's Historian, I have been researching all narratives of cadatanauts' explorations, including our current investigation of your Solar System. Wocega-23 is younger than me. She took my history class as a youth. My Ortack ancestors held a similar role to my record keeping, educational, and history-writing duties. There are few mentions across this history of the unique contributions from these invisible historians.

Wocega-23 was born in 1124 AD. For comparison, in 1125 AD, Henry V of Germany and the Holy Roman Empire died. In 1163 AD, work began on Notre Dame. And in 1168 AD, Oxford was founded. Wocega-23 and I have been in transit to Earth across half of humanity's recorded history. The existence of "intelligent" life is one of the measures our Computer considers when setting the preference-value for a potentially-habitable planet. However, its programmed definition of "intelligence" classifies most animals on Earth as possessing a similar level with humans. Readings of Earth were malfunctioning at light-speed, preventing us from noticing technological developments humans made across the past few millennia. We detected these technologies when we began our deacceleration maneuver in 2008. Our Computer had made calculations for a rapid deacceleration to Earth. However, in the first minutes of exiting light-speed, the Computer detected digital audio-visual signals from Earth. These indicated that if the Mothership's rapid appearance had not yet been noticed by human telescopes, we might soon be spotted. To address this concern, the Mothership slowed down to the maximum speed of a natural comet speed. And the Mothership's exterior was cloaked with a comet-appearance. Then, our scientific team began brainstorming the best strategy for avoiding detection given the strange space-faring satellites and sporadic debris we detected littering the Solar System.

Our Computer downloaded the entirety of humanity's digital imprint across the previous years, including libraries of books written since the first histories and novels just a few hundred years earlier. It was remarkable to see such rapid development on a planet, so we suspected another alien species miotails directed to Earth had nudged this rate of change, but we failed to identify evidence of such visitations. Unlike in other cultures we had studied, there was a lack of cohesion or consistency among humans. It was as if they were experimenting with every type of government structure nature introduced among their lower species: they built cities in giant corporations like ants, but then some of them defended territories like lone leopards. Humans were committing the horrific offenses we had seen bring several earlier alien species to global disasters.

They were exploding star-level nuclear weapons on their own planet. They were clouding their skies with pollutants. And unlike species we interacted with who were clueless to these dangers, humans were writing countless news reports and history books on these developments. Despite understanding their own flaws, they kept escalating these faults. Our scientists were mesmerized by the Great Pacific garbage patch, an island of garbage lengthier than the state of California it paralleled. We had seen pollution on most planets as all species (even trees and bacteria) make some waste. But this was the first time we saw garbage accumulating into a semi-continent. A species was writing books about this development, but it was not taking steps to resolve this problem. It was also horrifying for us to listen to the Computer's compilations of popular music and films, which, when stitched together for patterns, could be divided into two categories sexual and violent threats and proposals. Songs shared a rhythm and films a plotline. The Computer explained these repetitions in its annotations. Death and sex were the driving forces among humans. They were driven by the consumption of poisonous, fattening foods and drugs. Civilization bloomed hundreds of years earlier, birthing multifaceted innovations, but then humans discovered methods for generating money from inactivity and devolved into pre-literate states. A few outliers were inventing new machines, science, and arts. But most of their ideas were being stolen by the mob, and were appropriated without compensation. Worse still, humans across the planet were xenophobic and had been fighting ethnic or nationalist wars for thousands of years, and they had pre-judged aliens to be enemies in their futuristic films, threatening to shoot the lot of us out of the sky. All this meant we had to disguise our arrival, mimicking the ethnic groups of the countries we inhabited. Failure to assimilate was forecast to trigger a holocaust of our kind.

We pushed dark speculations aside, and proceeded exploring your Solar System with scientific detachment. One of the reasons Nally chose Earth is its Sun was bigger than 95% of stars. This high energy output can power a cadata spaceship without the need of other sources during flights across the system. The strength of the star also allowed Earth to remain warm and habitable in a distant orbit. As we had observed in other systems, a dim star with a planet in a narrow orbit tends to shoot out too much solar wind and radiation at it, poisoning living organisms, or turning them into extremophiles focusing on survival, rather than on intellectual or moral development. If Earth did not have an atmosphere, based on its proximity to its Sun, and the Sun's radiation output; its equilibrium temperature would have been −18°C, or below the freezing point of its main life-supporting liquid: water. If unfrozen water is the benchmark, humans spotting Earth with current telescopes would conclude it is outside of the habitable zone. Cadatas judged Earth's composition and atmosphere with more precision by assuming its water vapor could absorb much of the incoming radiation. In contrast, the atmosphere's gases alone cause a slighter difference in the overall temperature. Thus, as Nally guessed, the actual temperature average on Earth is 15°C.

Also of benefit was the Sun's status as a third-generation star. It is thus composed of the dust and gas remnants of prior stars. This accumulation deposited in this Solar System heavier elements, instead of the mere hydrogen and helium that dominate first-generation stars. Nally also noted Earth's Sun oscillates or varies by only .1% across an 11-year cycle; if its oscillation was more intense, it might have turned into a pulsating star with an unsteady energy output. Weather extremes turned out to be catastrophic on several of the visited worlds. It was a relief humanity had the restraint to refrain from sending nukes into their Sun to turn on a more extreme oscillation during the psychotic hyper-development phase they were going through. Some of the drawbacks that moved Earth closer to the end of Nally's *Catalog* included its Sun being scheduled to exhaust its hydrogen fuel in around 5 billion years. It will become a red giant, first melting Earth's mountains, before engulfing the planet within its growing radius. This eventual disaster was too far into the future for some because the Milky Way galaxy was projected to collide with its neighboring galaxy, Andromeda, before this, in four billion years. Still, even the most optimistic estimates showed Earth's climate would remain habitable for a mere billion more years, not too long considering the millennia cadatas needed to begin exploring the Milky Way.

The Mothership had entered the neighboring Alpha Centauri star system in 1998, and zoomed through it for a decade before exiting light speed just outside of the Oort Cloud, .79 light years away from the Sun. It was safe enough to travel at light speed up to this distant Cloud border because even cadatas would have struggled to detect a spaceship traveling at light-speed from this distance. A telescope had to be focused on this narrow region of space and searching for extreme-speed objects. Earth's telescopes have detected over 100,000 objects within the 50-mile diameters in the Kuiper Belt, but cannot spot smaller flying objects in this distant region of space. Since humans only found Pluto's giant moon, Charon (which is a whopping half its size), in 1978, they failed to distinguish the momentary smudge our tiny spaceship left in the night sky as it deaccelerated from the visual equivalent of a ray of light. Meanwhile, if we were still going at light speed, we would have jumped from the Kuiper Belt to the Sun in a day, but we slowed to an asteroid's top-speed by Pluto's orbit, and then flew the remainder of the trip at a constant speed mimicking a lifeless rock. We had zero-gravity on the spaceship during this slow glide.

We edited our standard landing procedure for Earth's Solar System, after the meet-and-greet we had with miotails. We had gained a new understanding of the reasons for our ship's discovery on the yellow moon of Pallos. Thus, in 2008 we began a deacceleration maneuver slowing us to comet-speed. Then, we ping-ponged between planetary stops rather than heading straight for Earth. To avoid unearthing as an undecisive and erratic space object by humans, we cloaked the ship in a manner we learned was undetectable to human telescopes: we added a black, and thus non-reflective to light, coat to the ship—generating near-zero albedo and turning on an invisibility illusion. We decided, even the strongest space-stationed telescopes could not detect us with this simple trick even as close to Earth as Jupiter's orbit, but beyond this point we planned on a still more elusive tactic. We caught up at the Kuiper Belt with Eris and followed it for a bit as it moved closer to Pluto's orbit, sending a few shuttles from there to explore some other distant dwarf planets beyond Neptune's orbit. We were curious to learn why humans declassified Pluto to a dwarf planet; instead, they could have called the other gravitationally-significant bodies in this region, including Haumea, Makemake, and Eris, planets as well. Eris had just been discovered by humans in 2005. Unlike our ship, Eris is one of the most reflective objects in the Solar System because its surface and atmosphere are frozen ices. This atmosphere is scheduled to defrost into gases in a couple hundred years: thus, studying it allowed us to imagine the impacts of artificial atmosphere freezing. It reminded our scientists of the lethal eccentricity on Ahutababha. The risks on a habit-able planet were higher if its eccentricity is adjusted. However, our scientists imagined a series of experiments to run on Eris to enhance its habitability. They proposed shrinking its orbit from its current 68 AU at the farthest point. Eris is near identical with Pluto in size. It is 28% more massive. Thus, making it habitable might have created a small, inaccessible to humans, world, within the Solar System, where cadatas could have es-tablished a clandestine outpost. However, a complete terraforming necessitated shifting its orbit to near-Earth's. And humans would have noticed this extreme diversion. Even if we did not proceed with this adjustment, we understood human astronomers bet-ter through these explorations. As we understood, human scientists were so terrified of learning the name of a tenth planet, they decided to name "dwarfs" all objects of Eris and Pluto's size and distance. Given their size similarity, Eris cannot be discounted, while keeping Pluto. We also spotted Planet Nine during this analysis. However, Nally-23 ordered me not to reveal its location to you, my readers, because it would be easier for cadatas to set a future base on a planet that humans are ignorant of, as opposed to one, such as Eris, which is on your radar.

Given the hype about Pluto, we had to stop there. It was closer to Earth or to its 29.7 AU perihelion back then. Perihelion happens once in its 248.5-year orbit. This was a proximity humans utilized to send probes to photograph it. Well, we were a bit con-cerned when we learned NASA had sent New Horizons to Pluto back in 2006. Then, we realized it would not reach this destination until 2015. We were safe to explore at-our-ease in 2009. I was in the Observatory as we approached Pluto, and saw its dark side. Eris, and still farther objects that we studied lacked atmospheres. Pluto was a black ball with a blue atmospheric haze radiating from it in contrast with the sparkling sunlight coming from behind it. I squinted and distinguished outlines of ridges and mountains at the edges of this beautiful planet. It was an emotional moment for me after a lifetime in space. I imagined breathing in that strange open air, before returning to consciousness of the distance to this planetary object, and the Mothership's protective shell. Pluto was approaching Neptune's path, so we were traveling in the right direction, as we parked on Pluto to run our studies. The dwarf planet binary system between Pluto and Charon (some call Charon as Pluto's "moon") was also of interest. At $1/_8^{th}$ of Pluto's mass, it is proportionally the most massive moon in the system. We explored how these two keep from colliding. If physics was not maintaining a distance between them; they would have merged given their mass, the proximity between them, and the significant distance from other gravitational forces.

While we were taking samples and studying the geology and physics of these dwarfs, our nanobots returned from their direct flight to Earth with samples of human DNA and materials we needed to mimic humanity during our stay. Doctor Nackan-23 put Wocega-23 through the conversion process, keeping cadatas' shape-shifting ability intact as well as the structure of the cadata brain. Given humanity's preference for lying in the fictions and falsehoods permeating their lives, and the tendency to engage in violence, we anticipated shape-shifting was necessary to escape a war. The nanobots produced the needed data on how the human eye reacted to light, resonance of vocal cords, muscle reflexes while asleep or awake, and how different organs reacted to food or water intake. The big processes of sight and hearing had been observed by our predecessors on other planets, but we were surprised by this body's inefficiencies. It was strange how inadequate a human's sense of smell was, even when compared to less intelligent species on Earth, like dogs. The air across the spaceship was changed to the composition on Earth, as cadatas can live on air rich in nitrogen and oxygen with a substitute intake of a blend of nutritional gases. The air pressure was heavier than on Earth because we had maintained the higher pressure natural to Cadata across our space journey.

Wocega-23 had standard Earth pressures in her room, but could tolerate Cadata-levels when she walked through common spaces. She used a mini-depressurizing suit feeding pure oxygen. She shifted the psi when switching between the two to avoid decompression sickness. We learned Wocega-23 had a somewhat sever case of agoraphobia, or fear of open spaces, when she took a spacewalk on the next planet we visited, Neptune, the ice giant. She took five steps out of the ship, struggling with the heavy gravitational pull suppressed by the anti-gravity propulsion in the spacesuit. It was an icy, slippery surface, but the weight of Wocega-23's body would have prevented her from falling even if it was polished ice. The view ahead of Wocega-23 was a massive crystal field with hills and etched, littered features. After those first few steps, she shifted her focus from the struggle of moving her feet and arms to studying this vast expanse. The limitless potential to run in any direction for years without finding her own footsteps again overwhelmed her. She turned and shuffled back to the spaceship. While this would have been disqualifying for a land explorer or a sky pilot on Earth, the cadatas on our team were now in the twenty-third generation of spacefarers. The confinement of our spaceship was comforting and natural, and the few stops we had made for a few decades here and there were abnormalities, so we were all a bit agoraphobic. Doctor Nackan-23 came up with some orientation and concentration tricks Wocega-23 could use to regain composure if she endured another panic-event.

In parallel with Wocega-23's gene editing, the engineers were working on the shuttles we would use once we were ready to land on Earth. The Computer's search through Earth's databases determined their fastest plane, the Lockheed SR-71 Blackbird, was developed decades earlier, and had broken the top speed limit in 1976 by going 2,193mph. The schematics for this plane showed it used an absurd fuel type requiring refueling soon after takeoff, and its parts were so rare, it was retired back in 1999, without a competing speedster introduced. The engineers replaced the engines with cadata reusable, non-refueling technology, which collected matter for fuel from the atmosphere akin to the mothership. Two developers argued we should design a gas-guzzler in imitation of the original, but since only the military could refuel it mid-flight, these dissenters surrendered on these strict authenticities. Others objected to this plane altogether, arguing each of the surviving twenty planes out of the initial thirty-two built was logged with a specific parking location; and these parking destinations were museums. This was a sound argument, but Earth's circumference is 24,901 miles. Wocega-23 believed going slower than 2,193mph on a planet this size, or failing to circle the entire planet in under 12 hours, would spike a condition Doctor Nackan-23 diagnosed. Rather than a fear of wide-open spaces, it was a fear of places and situations foreign to space-life. Practically, these shuttles needed enhanced aerodynamics and speed to survive both space and atmospheric flight. The head engineer also judged its black surface would further assist our efforts to avoid adding glare to our space flyers, and thus keeping them from detection.

Looking ahead a bit, this turned out to be a troubling choice later in our exploratory mission to Earth. Our planes turned off GPS-tracking and antenna-generated broadcast-auditory messages when stealth was necessary. We flew low to hide inside of mountains and storm cloud terrain features. At other times, we had to fly at over 100 miles up to stay above the normal range for air traffic. Air traffic control could still catch us on radar as an unregistered object moving at an extreme speed once we dropped below 30 miles from the ground. If we had used the original Blackbird design, we would have had to land on a commercial air strip, and this would have created a national emergency as an unidentified plane zoomed into a busy airport at 2,193mph and then skidded to a stop at some random gate without checking in to clear a landing or a pilot's license. Even if we made the plane invisible to radar, a black plane shooting down to earth would attract attention from eyeballs on the ground. Our research pre-determined against such brutish behavior, so instead we landed these plane-shuttles in direct vertical drops. This was done somewhat like a hovering helicopter, but adding propulsion to speed downwards, and then slowing by pushing in the last few hundred feet against the air below, rather than landing by slowing sideways propellers. On occasions such as when we had to land near a major metropolis, we generated fake papers and landed using standard commercial-strip human procedures. We parked there in off-hours; however, on two occasions, a tarmac mechanic and a pilot-in-training were familiar with this famous model, and questioned why we were allowed to extract one of these out of a museum. I think on both occasions they accepted flying these monsters once as a sufficient bribe not to ask further questions. Most of our other landings and takeoffs were executed on isolated landing-strips on plots we purchased pre-arrival with the amazing human invention called "credit". We hacked into a credit card's main database, generated near-infinite credit for ourselves, and then bought hundreds of miles of land with existing air strips across the world. We had not encountered an exchange system this susceptible to alien manipulation on other planets.

Since industry had been using UAV's since 2006. We designed our robots to appear as little pipeline-inspection drones. This allowed us to bring these larger spy gadgets into areas we needed to run tests on. They could shapeshift to fit a specific function once they were concealed at an intended destination.

While the voluminous quantity of fuel unmodified Blackbirds required was shocking, the use of decomposed oil from ancient plants and animals in human transportation was more so. On Bhasab, cadatas learned these types of hyper-burns can escalate to engulf a planet's resources. We hoped to help humans transition away from this addiction. While this was a hope of ours, as usual, our leadership refused to act preventatively, waiting instead until we faced a total-extinction-level-event. Wocega-23 needed therapy to gain Zen regarding sitting in a vehicle run by rapid explosion repetitions in its engine. As a compromise, the engineers edited the structure of her Earth driving-car to mimic the noise of combustion, while it ran on reusable and non-explosive methods borrowed from our own vehicles. Wocega-23 was also outraged the car had rubber wheels under it, a material known to degrade and need replacement every few years. It was even prone to blowouts. She believed riding a balloon would have been safer. Since using metal or an advanced road-gripping non-deflating design would have stood out as alien on a roadway, the wheels were kept as they were, but a better semi-soft material than rubber was engaged to improve safety. Wocega-23 had locked herself in her room in terror for a week, after the Computer recited to her the Earth's car accident and fatality statistics. She struggled with accepting she was obligated to regularly operate this planet's tenth leading cause of death. Humans had created a death-trap machine. It had semi-effective safety measures. Their belts could prevent impacts, but they could also themselves cause fractures. And their exploding bags of air could cause lung-damage. The scientific team negotiated with Wocega-23 by inserting a cadata-designed softening-gel across the interior of her car to prevent potential injuries. Prior to this resolution, one scientist objected that the application of this cadata gel was a point-of-vulnerability, imagining a scenario where a human investigator could have studied this gel under a microscope, discovering its alien properties. Wocega-23 also struggled with adjusting to steering the car with a wheel held between her hands. Then, the engineers tried giving her a "self-driving" car. However, this turned out to be a human marketing gimmick, as it required the user to sporadically intervene upon prompting to take over navigation. Wocega-23 was too confused by this occasional incompetence of the car's programming. Wocega-23 was also mystified by the rule that she had to keep the wheels within the lines drawn at the sides of roads. If the goal was to keep a car within precise linear boundaries; the logical solution was surely to allow a program to calculate the GPS parameters of this space, as human and cadata vision and spatial orientation could not achieve consistent adherence, leaving a chance of accidental veering leading to a crash. Wocega-23 was also irritated by the two-paddle system that separated gasing from breaking, or speeding up and slowing down. She kept asking the science team why humans used two of these instead of one. Humans had figured out they only needed a single speedometer. If they had been unaware of their two-paddle flaw, they might have designed two separate speed measurements: one for the degree of speed increase, and another for the degree of speed decrease. The scientific team experimented with a single-paddle design, However, the rudimentary human-adjacent design allowed for sudden deacceleration from top-speed to zero, which generated potential concussion-force. Wocega-23 had to settle for practicing with the two paddles, until she finally stopped rapidly switching between them to violent jumps. As her driving tactics normalized, she increased the duration of her diatribes against human technology, growing firmly convinced it was the worst in the Galaxy.

I (Ortack-23) joined Wocega-23 in the exercise of learning human languages. Linguistic acquisition was one of my standard functions as the Mothership's Historian. Past historians have created dictionaries, grammatical and theoretic textbooks for each new language we encountered. The Computer compiles data on these languages, and then historians process this data by designing programs that maximize translation precision. On Earth, the Computer identified around 6,500 languages, used by as few as a dozen, to as many as over a billion people. Some languages are near-identical accents, while others are based on unique components, such as clicking or whistling. Alphabet lengths vary between 12 letters in Rotokas to over 50,000 Chinese characters. Linguistic diversity is a commonality among the planets cadatas have visited. This consistency of linguistic divergence is demonstrated by songbirds of the same species that make slight variations in their tune and pitch in distant regions. Only mechanically reproduced sounds can be identical. And just as cross-region migration still means given species of songbirds have a similar sound across the globe, human migrations have meant there are root commonalities among the big families of human languages. The presence of several distinct language systems on Earth suggests humans' tendency towards isolationism, or nationalistic societies afraid of "others". In some extremes, a unique language serves as an artificial border that prevents migration or interactions outside this language-boundary. Earth's complex geology—including the separation of continents, and impassable mountain ranges and oceans—caused a few of the major divergences.

With transportation and wireless communication methods easing humans' idea-sharing capacity, Wocega-23 was confused as to why they refrained from switching to a common language in commerce and cultural exchanges. I explained to Wocega-23 that colonialists had been laboring for centuries to shrink Earth's languages to the handful of languages favored by these colonizers. The British Empire's monopolization of the press enforced English's current standing as the world's most popularly spoken language, followed by the Chinese Empire's Mandarin, then India's Hindi. Britain managed this artificial conquest by manipulating legal systems, so that 39% of the world's countries enforce English as their official language. English has no logical superiority to its rival languages, as it is derived from the broken-Latin Old German. There are three times more unique words in Swedish and Korean than in English, thus a translation between these loses two-thirds of the intended meanings.

Cadatas had come across several territorial species in the past, including our own history of wars, and Nefo's monarchical government structure. Uniquely, on Earth, the tools of nation-building have been used as weapons of profiteers. Tyrannical human leaders have been marketing their countries' resources by designing national flags, artificially designed "antique" costumes, dances, and "ethnic" foods. For example, the now famous Chinese food dish called General Tso's chicken was designed by a chef in a New York City restaurant, and it is now even served in restaurants in China that claim to be culturally-authentic, so that most guests still eat with chop-sticks. Walter Scott invented most "traditions" he attributed to Scottish clans, such as the kilt that he claimed was common there partly to suggest Scot rebels were still wearing simple barbaric loincloth. This propaganda was designed to convince Scotland to remain in the *United* Kingdom. More simply, earthly countries have benefited from tourism. Thus, owners of tourist-frequented establishments have had a motive to underhandedly pay artists and writers to invent reasons for tourists to visit their puffed "historic" resorts, museums, buildings, pubs, and restaurants. Especially creative advertising-copywriters have introduced ideas such as alien crop circles, and faked treasure maps just to redirect wealthy tourists into isolated rural locations where profits (due to lower construction costs) can be maximized. The concept of nationalistic uniqueness is also used by human governments as a motive to funnel tax-money into billionaire business-owners' coffers in exchange for their manufacturing of weapons to kill and maim the "others". These "others" are created by developing a fictional regional or nationalistic motive for hatred against the constructed inhumanity of these enemies.

Wocega-23 and I had extensive debates about human theology. Human-centric religions classify terrestrial animals and advanced alien species as equivalently inferior beasts without souls. This logical flaw stems from accepting a fictional system of belief. Theology is an extreme in entirely unfounded trust in falsehoods. It appears to have been artificially invented to create a litmus-test to only promote those who are loyal liars, while refusing to employ or educate those who cannot claim they believe in something that has not been proven. If regional myths were authentic fictions created by isolated cultures; these might have at least been valuable as cultural artifacts. However, most religions have been plagiarized from uncredited or demonized predecessors. Muslims, Christians, and Jews war with each other over religion despite their shared use of the same Old Testament. And the Torah's authors borrowed concepts and anecdotes from older Indian religions. Back in 1869, Louis Jacolliot's *La Bible Dans L'Inde* explained that *Jezeus* is a Sancscrit expression meaning "pure divine essence", from which was derived the names of Egypt's *Isis*, the Hebrew *Josuah*, and the Christians' *Jesus*. While most of the world thus shares the delusional belief in a fictional love-based all-powerful character, humans have managed to fight wars over minor disagreements regarding this character's attributes or backstory. Religions that plagiarize predecessors tend to hold a grudge that does not merely insist believers in the original narrative will be tortured after death, but also find a motive to bomb and torture them in wars while they are alive. Theology insists that true believers must imagine the hallucination of the presence of God; and yet those who report audio-visual hallucinations are placed on mind-numbing drugs to stop seeing these aberrations. And there is a stigma on calling older religions as a financial scam, while all new religions must ethically be classified as scams by the media.

I created a program for Wocega-23 that created GPS-coordinate based digital encyclopedias that transferred only relevant information on the economics, language(s), and culture(s) of the specific jurisdiction she was in. In Australia a person exhibiting signs of a disturbance can be placed on an involuntary imprisonment or hold in a mental institution and forced to take tramadol or codeine for attention-deficit. However, if this same person takes these prescription drugs with them to Nigeria, Singapore, or Greece, they will be arrested for carrying illegal drugs.

When the Mothership first came into full-resolution proximity of Earth, our scientific team was disturbed to find smog congesting the skies above all major cities. This was before we picked up media-signals, which later explained that Earth's scientists had explained the cause of this deadly pollution, and yet businesses continued releasing these chemicals into the air. In one horrifying report, the New York State Department of Environmental Conservation issued a health warning that wildfires in Canada had created a dense blanket of health-injuring fine particulate matter. The only "advice" was to "avoid exertion outdoors". The Department was not taking actions to prevent wildfires, or to help citizens survive if their jobs required strenuous outdoors exertions. The US is becoming progressively more incompetent in dealing with wildfires, as there were twice more people killed in wildfires in 2023 at 183 than in the second deadliest year (104 in 2018). Wildfire smoke kills 16,000 people in the US annually. This pollution can cause dementia symptoms, so perhaps the governing administrators who should be taking steps to prevent this are afflicted. Five of the Solomon Islands have already entirely sunk into the sea by 2011, and a decade earlier two Kiribati islands sunk. If islands on Cadata began disappearing due to our industrial activities; this could have been the cause for cadatas to have been forced into our current cross-galaxy homeless migration. Learning of these environmental disasters on Earth, and humanity's plan to foster this death-spiral has prompted some among us to conclude we should have redirected to the next potential planet without our current exploratory stop. Wocega-23 was on the side of attempting settlement, and she proposed following the humans' strategy of escapism into interactive and static entertainment methods to forget about looming disasters.

One of the available escapes from Earth's problems for cadatanauts were via other planets in the system. This was one of the reasons we scheduled stops on most of them in the years running up to our sneaky landing. After the shaky start of her cadatanaut explorations on Neptune, Wocega-23 relaxed as she took a few walks across Uranus. This planet was similar in its icy exterior, though 10 AU closer to the Sun. I thought its external atmosphere of ammonia, sulfur and methane had a similar appearance to Earth's because of its bluish-green coloring. It has fewer natural resources per pound of its mass because only 25% of it is rock, while the majority, or 68%, is simple ices, and 11% is hydrogen and helium gas. Our science team failed to locate interesting elements on this solitary body, so we continued to Saturn after a two-day stop.

We were now close enough to Earth to distinguish it in the night sky from among the other bodies in this Solar System, including the approaching giant ball of Saturn. Earth appeared more yellowish-white, whereas distant stars in its portion of the sky had a reddish tint. In kindred with the majority opinion, I was hoping that, despite Earth's challenges, Wocega-23 would avoid an interplanetary crisis. This is our only chance of visiting a single habitable planet in our lifetime. The view of this planet in the distance renewed out hopefulness. Upon closer approach to Saturn, the sky grew bright from the myriad dazzling asteroids, moons, and gas-cloud components of the Saturn system. Here, Earth looked like just one of the stars being flooded out by the brightness of Saturn's farthest, thin, rainbow-colored rings. These objects had gravitated towards Saturn's enormous mass ($95.159 \, M_{\oplus}$). Its composition reflected the earlier stages of this Solar System, when it approached the makeup of the Sun, with an upper atmosphere low in helium (3%) and rich in the universe's lightest element, hydrogen (96%). Wocega-23 and I (Ortack-23) took a shuttle to explore the liquid metallic hydrogen ocean: a phenomenon we had seen before. Our shuttle recorded a rising temperature as we descended through the atmosphere. It was an uneventful trip, memorable only by the shifts in cloud coloration from brown, to white, to red.

We had anticipated there would be no scientific breakthroughs from our visit to Saturn, and yet we spent a relatively long time on this expedition. This and other extended scientifically-unmerited quests were added to our schedule largely because we were forced to prolong Mothership's descent to Earth by four years. It would have been just as safe for bodily-integrity to deaccelerate from light-speed four years earlier. Instead, the motive for this delay was the avoidance of being identified as an alien spacecraft. Thus, in December of 2013, we hid our Mothership on the back surface of comet P/2016 BA14. We opted for this strategy knowing this comet was headed towards Earth, and it had not yet been detected by Earth's scientists. We predicted it would make a closer approach to Earth than any comet since 1770. P/2016 BA14 is a near-Earth object and periodic comet of the Jupiter family. Jupiter is as far as it travels and this was the neighborhood where we met up with it. This comet was also chosen because it reflects only 3% of the light hitting it, a very low albedo; this was one of the reasons it remained unspotted by humans during its previous 2011 pass, until it near-hit them in 2016. Thus, we had an easier time hiding our ship on its giant, unobserved surface. Humans failed to guess its size or category, assuming it was a 125-meter asteroid, before measuring its nucleus as belonging to a 1-killemeter comet. The size was measured as an indirect inference from the size of its tail when it entered solar wind's range. To be frank, cadatanauts might have contributed to the appearance of that tail, as we did some mining while we were parked on it across those years. We needed these resources to build up more robots for our looming close exploration of Earth. While we were based on P/2016 BA14, we explored its other potential scientific uses. We learned it was large enough to generate some heat from its 37-hour rotation along its axis cycle. Its orbital period around the Sun was 5.25 years. It came as close as Earth is to the Sun, and at its farthest point reached five times this length, flying within striking distance of Jupiter. The high number of asteroids flying in the belt between Jupiter and Mars has contributed to human scientists' failure to spot the significance of our asteroid-mimicking comet. There were many other asteroids to choose from, but the rest did not depart from those farther regions, or arrived within our intended landing schedule. The team considered parking the spaceship just above Earth-calculated atmospheric edge of 62.1 nautical miles. They might have disguised the spaceship to either blend with the sky, or to appear as an asteroid caught as Earth's satellite. However, in this System, asteroids orbit the Sun, instead of small planets, such as Earth. For example, the two-lobed asteroid 2014 JO25 mimics Earth's orbit around the Sun, instead of spinning around Earth. We could have opted to stick to its rotation path, but humans study objects capable of striking them.

With the Mothership parked in a shadowy region between a few rocky hills on P/2016 BA14, shuttles were free to explore the potentially scientifically-rewarding stops over the following two and a half years. Jupiter matches Saturn in overall composition and structure. Jupiter was marked as a required stop because it is the largest planet in the Solar System. One technical question was why it has a stretched shape. Its equatorial radius is a sixth larger than its polar radius. It was also beneficial to the scientists on our team who were interested in gathering heavier elements after reserve-depleting years in space. Jupiter's dense core was rich in thick, crushed rocks, including magnesium and silicon.

The flight down offered a dramatic view of bright-colored ammonia cloud tops, followed by a layer of ammonium hydrosulfide, and then a water layer. Wocega-23 and I (Ortack-23) explored the three-centuries-old Great Red Spot together. As an anticyclonic storm, two Earths in diameter, it is propelled to circulate because it is between two atmospheric rivers flowing in the opposite directions. We ran some air pressure and wind intensity tests on this Spot. We also studied the effects of Jupiter's spectacular 19,000 times greater than Earth's magnetic field. We also explored Jupiter's moon, Io, as a potential habitable outpost, in case Earth later proved to be unsuitable. Io has a spectacular yellowish-orange surface, littered with nine active volcanoes aligned for simultaneous eruptions. While we failed to identify a single living microbe on Io, it is a gentler environment for habitability than most gas giant moons we have explored.

Our team diverted a unique set of resources and time for a full investigation of Mars. Wocega-23 volunteered for the first visit because she needed to train performing habitual tasks on a solid planet, after a millennium with only a simulated gravitational environment on the Mothership. Wocega-23 was eager to adjust to her new human body in Mars' 0.38g. A human can jump three times higher on Mars than his or her top leaping-limit on Earth. Either gravitational pulls are entirely unlike jumping in a spaceship in zero-g, or during a high-g de-acceleration.

One of the reasons I remained for an extended study on Mars is for its relatively pleasant environment. The equator can reach a bit higher than the comfortable-for-humans 68 °F in the summer. All seasons are twice longer on Mars due to its 1.88-year orbit. This is convenient for scientific analysis because similar conditions repeat for twice as many days in a row. At the poles, temperatures drop to a low of –243 °F, which is more sever on cadatas natural heat-loving biology. In equatorial summers, it is warm enough for humans to survive, but cadatas' biology would still suffer from sever hypothermia. Neither the human nor cadata-biology members of the team attempted unsuited walkabouts because of Mars' low air pressure, and a relative uniformity of air particles. Mars' air is filled with 95% carbon dioxide and 2.7% molecular nitrogen.

When Wocega-23 departed for the next expedition, I remained on Mars, committing to at least a few years of life in a modified-shuttle outpost. We refrained from parking the Mothership on Mars because humans continuously threaten that they are about to take their first voyage to Mars. Humans began publishing Mars settlement and terraforming proposals in the 1950s. An article in *Popular Science* in 1965 advertised the unmanned Mariner IV would need 8 months to travel to Mars, and as long if it could return to Earth; this article also imagined a shortly forthcoming manned voyage that would offer the crew perpetual vertigo with a constantly spinning compartment for "artificial gravity". Escapism into space-travelogue science-fiction seems to have stopping humanity from designing and building the necessary science for these missions. Instead, humans are investing in the profitable space films and video-games, and in puffing the imagined superiority of space programs.

Mars' ground is rich in molecules that can be transformed into a habitable environment. Mars' deserts are reddish because they are rich in iron, which oxidizes, or rusts when exposed to atmosphere. Its rocky mountains, and massive canyons provide minerally-diverse soils capable of supporting plants. For planting to be feasible, prerequisites would include the terraforming of the atmospheric gases to the ideal mixture for cadatas, and a change to the climate's temperature, air pressure, and moisture. As I mentioned in the introduction, Mars is ravaged by planet-wide, weeks-long, and Mt. Everest-high dust storms. Though these are not as lethal as on other planets we encountered because the low atmospheric pressure (1% of Earth's) minimizes the impact of their 100mph wind-gusts down to what feels like a 10mph breeze. With spacesuits to shield us from the lapping sand, we could hop around with ease during the worst sandstorm. Before Wocega-23 left, we regularly exercised our bodies and vocal cords outside in quieter weather, testing how low our voices sounded in these low-pressure conditions.

Between the first credited blurry telescopic observation of Mars by Galileo Galilei in 1610 and the first clear photograph was taken of its desolate sandy surface by the unmanned Mariner 4 spacecraft in 1965, humans imagined there were monstrous Martians residing on their closest planetary neighbor. Three hundred years of astronomic mythology is a powerful symbol. Thus, one of my objectives on the Mars mission is to search for possible microscopic sign of life. Finding a single-cell Martian can serve as an introductory offering to pacify humans towards alien life from elsewhere. I designed and dispatched across all accessible crevices uniquely-suited robots and nanobots.

Since I first learned I would be stationed on Mars, I have been lobbying to aggressively terraform it with a cadata-friendly atmosphere. This scheme was again impeded by the smallest chance of humans suddenly managing to visit Mars.

Meanwhile, under my outpost, I have been excavating the frozen crust down to the permafrost. This hard layer was created when water absorbed carbon dioxide and locked it in carbonate rocks. This excavation accessed a layer of water below: one necessary stage I had to complete to prove Mars deserves a full habitability study. Another stage involved exploring the canali, which were left over from a time when Mars had a denser atmosphere. Due to its current air-depleted condition, exposed liquid water has evaporated from the surface. Apparently, it can merely be stored in the permafrost solidified forms under the surface. This pattern suggests that instead of staying on the warmer equator, prolonged missions on Mars would benefit from being stationed on Mars' polar ice caps because they would be a constant reserve of buried water, with some also useful dry carbon dioxide mixed in. One puzzle I researched was how this dry cap ice sublimated every summer, leaving a tiny residual water cap, and then managed to gather enough matter from the semi-thawed, pressurized summer atmosphere to recreate a large winter ice cap. During these winters, water left in the atmosphere froze into ice-crystal clouds. On most planets with melting ice caps and a depleted atmosphere, the ice particles escape from the atmosphere, because of solar-wind exposure, until the planet is left without an evaporation-capable surface. Mars' shrinking atmosphere and water-reserves are the prime reasons Mars is naturally declining in its habitability potential. Cadatanauts had detected Mars' minimal water-presence long before our first step on Mars by performing electromagnetic scans from space. From a distance, it was easiest to spot the countless canali lines across Mars' surface that are proof of past water erosion. Another piece of evidence even humans have spotted is an impact-crater in the southern hemisphere with darker-colored sediment deposits, indicating a dried-out, and now dune-covered lakebed. Without disclosing secret alien-science, I can mention that the color of this sediment can merely indicate a water-unrelated elemental composition. And even if the canali had been formed by a liquid traveling along the surface, this substance might not have been water. This liquid could have been a compound that is more tolerant of low pressures and temperatures. My tests of the canali showed they were indeed once filled with water, which remained liquid at relatively colder temperatures on Mars than on Earth because of Mars' lower air pressure. My current research is suspended on the problem that Mars' atmosphere is so thin, it would require a massive infusion of gas to normalize it. Such atmospheric expansions typically necessitate turning a planet's solid minerals into gases. Mars' sand is primarily iron, which requires heating to $5182°F$ to turn it into a gas. One positive is that this sand also contains some solid oxygen, which is frozen at the caps, and can become gaseous by a relatively slight warming of the temperature.

While it is somewhat feasible to maximize Mars' atmosphere, it is far more difficult to fix the more problematic weakness of its magnetic field. This field might be strengthened by heating Mars' core until a significantly larger percentage of it returns to the liquid state. Without a powerful magnetic field for deflection a fatal dose of radiation would bombard inhabitants. To increase Mars' average-temperatures through intensifying the atmospheric-shield greenhouse effect, I proposed re-activating Mars' volcanic activity to allow its massive volcanoes, including Olympus Mons, to shoot out gases and particles into the air to fill in the atmosphere. Olympus is the largest volcano in the Solar System.

It burst out of a weak spot in the crust through hot-spot volcanism and kept growing because there were no moving crust plates to spread its erupting materials. It is now 2.5 times taller than Mt. Everest. Crust faulting from this volcanic activity also generated the Valles Marineris: an over 180-mile string of canyons near the equator. If Mars was still active, or could be reactivated as I propose, it would become a promising geological resource. Mars has been silent, without tectonic activity, for millions of years because its core has cooled. After sending probes into Mars' core and into potential volcanic hot spots, I concluded that reactivating a cooled core would require more energy than changing the orbit of a planet. All planets form out of superheated materials, wherein the heavy metals are compressed in the center into the active core. All cores naturally cool over time unless they are fed by new heat-producing collisions between planetary objects. Massive planets, such as Earth, are simply cooling so slowly that their suns will die before they run out of fuel. Re-igniting Mars might require creating the degree of friction that generates a gravitational collapse. This collapse might be intuitive for an overly-massive cloud of gas, but is near-impossible in the middle of a solidified planet. Collapse might be achieved by increasing the temperature or the mass at the core. An increase in temperature is the goal of this reaction, so it would be counter-productive if it is input into this equation. Mass might be infused by transferring it from the crust down into the core, but again the core would have to be melted to compress more mass into it. I am conducting further research, in case cadatas decide to later pursue this extreme emergency measure.

Despite seemingly reaching the preceding dead-end, I relentlessly proposed yet another scheme to artificially induce a habitable environment. Calculations showed another reason Mars was inhospitable was its lack of a massive moon of a similar size to Earth's to stabilize its spin. Mars has two tiny moons: Phobos (17-miles), and Deimos (10 miles). They are the size of oversized asteroids because they were asteroids before Mars caught them. Without a shepherding mass in the sky, or satellite, to keep Mars stable, its spin-axis wobbles between 0° to 60° across just a few million years. This eccentricity contributes to temperature variations, causing the crucial ice caps to disappear at the highest spin axis of 60°, and to freeze at near 0°. At 25°, Mars' current tilt is uncommonly-affable towards life's survival. It is at a rare Goldilocks-zone tilt that echoes Earth's tilt of 23.5°. If we remain on Mars for 60,000 years, its tilt will drop down to 15°. At 0°, a planet has no change-of-season, which generates extreme divergence between polar and equator temperatures, and in turn generates extreme weather. To freeze Mars' axis at an Earth-matching 23.5°, I proposed using an artificial moon. Cadatas have developed a technique of using artificial gravity to attract enormous asteroids. However, readers might recall that this process led to the asteroidal impact that drove us to abandon Cadata at the onset of this journey. Our science has evolved since this failed attempt. However, even the most precise execution would involve cadatas manipulating an enormous planetesimal to be pulled towards Mars, and then unnaturally stabilizing into a mathematically identical to Earth's Moon's orbit. We cannot anticipate what humans' reactions might be to seeing this supernatural phenomenon. An even more disturbing sight would be if we blew a moon-sized part of Mars off its surface and turned it into Mars' moon. There is obviously no explainable cause for a piece of Mars to explode outwards without theorizing some human nation caused this destruction. Alternatively, bringing a planetesimal to collide with Mars to attempt to create a break-away moon, would be an extremely dangerous astrologic event for the entire Solar System. This collision would generate an enormous quantity of debris, which would randomly fall on neighboring planets, including Earth. The size of this debris would be large enough to be life-ending impacts.

The central conundrum for our cadata expedition has been that life is precisely adapted to its environment. If alien life wants to adapt a foreign environment to its specifications; it must inevitably extinguish the conditions that fostered life for preceding settlers. The moral decision is thus for the alien to alter themselves to fit with the conditions on the new world. Though cadatas might have learned that they can never make sufficient adaptations for survival. One of the few approaches we have avoided before is migrating to an entirely lifeless world, thus allowing for its radical rebuilding to fit cadatas, without a risk of destruction for an "other".

My frustration over the unchangeable nature of Mars built until I began proposing extremely unfeasible strategies as a comic-relief. One idea was taking Mars backwards in time. Accelerating to a speed significantly over the speed of light in an orbit around Mars might have generated reverse-time motion. No organisms move faster than light in the universe, thus this speed would position the traveler ahead of the existence those observing standard light passing by. However, it would take too long to accelerate from zero to over-light-speed. This approach would only be practical if the Mothership was already moving at light-speed and only needed to speed up somewhat to exceed it. Additionally, going in a tight orbit around a small planet at light speed would introduce an extreme chance of collision with this planet. It would again be practical to instead be moving in a straight line through the Galaxy in the Mothership to experiment with time-travel. Obviously, moving at over-light-speed in a straight line would take cadatas far out of the Solar System, requiring a return journey that would erase backwards time-travel gains previously achieved. It would be far more rational to travel to the closest planet that matches the environment Mars had at its ideal point in the past, than to time-travel.

Then, I swung in the direction of very simplistic solutions. I downloaded findings related to humans' hotly-debated investigation regarding the presence of molecular life on ALH84001: the oldest meteorite ever discovered on Earth. ALH84001 formed on Mars 4.5 billion years ago. A violent collision ejected this chunk of rock into space 16 million years ago. It flew around the Solar System for a few orbits, before appearing in Earth's sky 13,000 years ago as a bright, blue-white light, which split into two glowing balls with an ear-piercing sonic boom. These balls rained down a meteor shower across a 24-mile area. The archived fragment was discovered by human explorers in an Antarctic ice sheet in 1984. Suspiciously, the claimed discovery of bacterial life in this meteorite was delayed until 1996. These signs of life were in the form of carbonate grains, which were interpreted by some human scientists as fossilized Martian microbes. By 1998, human scientists found a consensus that the pre-genetic-testing delay contaminated the sample with Earth's bacteria. The most likely scenario is that it was contaminated in the preceding 13,000 years it spent amidst life on Earth. The specific bacterial strand was Actinomycetes, which is common in Antarctica rocks, especially since the contaminated section was near this meteorite's surface. Despite this disappointment, I remain hopeful I will find evidence that life in this Solar System originated on Mars, before it was brought to Earth with an asteroid strike. If there had still been life on Mars merely

16 million years ago; then, Mars' environment would merely need to be reset to this point, instead of billions of years earlier. Theoretically, it would be impossible to prove, based on an Earth-landed meteorite, if life began on Mars because it would have landed around 4 billion years ago, and would have been the first life to appear on Earth. Thus, this organism would be identical to the earliest earth-based lifeform because it would be that same organism. Only by finding 4-billion-year-old evidence of matching life, which has remained on Mars in the interim, would scientists manage to prove earthly life originated on Mars. And if bacteria, or any lifeform could survive an extraterrestrial flight from Mars; then, most Mars-rocks would have had traces of this super-microbe, just as every rock from Earth's exposed surface has some bacteria on it. Generally, rocks lying on the surface of glaciers in Antarctica are extraterrestrial meteors because there are no mountains in that region capable of depositing earthly rocks there. If Martian "bacteria" was found on one of these rocks; then, it should have been found on many, if not all.

Earth's search for life outside this planet appears to have been impeded by a misunderstanding of the term "organic molecule". A *molecule* is a string of bound atoms. *Organic* is a compound that includes at least one carbon atom. 75% of asteroids are carbon-based, and still more have some trace of carbon. Thus, most objects found in space will have some organic molecules and carbon. Neither of these are relevant to establishing the existence of extraterrestrial life. The relevant measure is the percentage in an object of heavier or higher-on-the-periodic-table elements. The greater the elemental diversity and complexity, the more likely it is that interactions necessary to spark life will eventually occur. A gas cloud or nebula cannot collapse into a star unless it has at least .08 of solar-mass. Similarly, biogenesis cannot spark on a planet without a minimum degree of elemental prerequisites.

Meteorites hitting Earth are among the worst candidates on which to find life in the universe. At their journey's start, these objects fragmented away from an asteroid, or a moon. Meteorites are divided into categories by their components. Iron ones are rare: 3% of all meteorites striking Earth. Iron meteorites come from of an asteroid's metallic core; thus, they are composed of metals, which are dominated by up to 20% of nickel. 96% of Earth-strikers are stony-iron meteorites originating from an asteroid's core-rock; they have stony inclusions in a matrix of iron-nickel, together with larger stones such as chrondrites and achrondrites. While some molecules can survive the freeze of outer space, almost no "living" organism can survive the fireball that forms around a meteorite as it crashes through Earth's dense atmosphere. Cadatas have encountered some hot-loving species across our travels, but none favored being engulfed in flames under high atmospheric pressure and while flying at extreme g's.

I continued performing small-scale experiments on Mars, while most of our cadatanauts continued the flight on P/2016 BA14 on its approach towards Earth. Months later, I tuned into the video broadcast of the perspective from the Mothership's Observatory during the approach to Earth. In the previous years, Earth was initially merely a dot of light, which an untrained eye might have interpreted as the shine of a star. From Mars, it had the appearance of a tiny dim moon. As the Mothership drew closer, it gained the properties of an illuminated bluish-white dot. Then, a second faint tiny dot appeared as if emerging out of the body of the first. Days later, these expanded into the blue marble of the Earth, and the grayish-yellow little marble of its Moon.

On March 15, 2016, P/2016 BA14 reached its most-proximate point on its pass by Earth. This was the ideal time for Wocega-23 to take a shuttle down to the Earth-Moon system to begin her mission there. Wocega-23 had begun the standard decade-long isolated transition from cadata to human biology. Going on the solo mission to the Moon counted as part of the pre-approval delay for the eventual Earth-landing. To minimize departures from Earth, Wocega-23 began her analysis on the Moon's "dark side", or the side hidden in its locked position from Earth's observational perspective. There was no chance of detection on that side in 2016 because humanity's first landing there was the 2019 Chinese Chang'e-4 spacecraft. Though Chang'e-4 was only actively taking and sending to Earth via relay satellite pictures in its first couple of years of roving, and now, in 2024, it seems to have gone dormant. This again gives access to cadatas to use the dark-side as our base for Earth-Moon operations. Cadata shuttles are not detected by Earth's telescopes because telescopes are designed either to detect objects that are over 300-feet, or that have extreme brightness (such as a sparkling granite slab), or heat (an asteroid that is on fire from friction with an atmosphere, or proximity to a sun). If we use air-conditioning, dark exterior paint, and an under-300-feet-long design; we are invisible "dark matter". The Mothership is over this size-limit, and thus it remained attached to P/2016 BA14 during its pass by. The Mothership would have also left a noticeable imprint on the Moon's surface if it landed there. The antique 1959 photos of the farside were too unspecific to notice the change. But the 2019 digital photos would be scrutinized, if later photos introduced a sudden imprint in the regolith, or the spiky-sand.

As previously noted, Wocega-23's shuttle had been disguised as a Blackbird. This disguise was logically unsuitable for the Moon-landing stage because if humans had spotted it, they would have panicked regarding which country among them had developed space-capable Blackbird planes. It was less wasteful to risk this scientific mystery, over designing some human-adjacent space shuttle exterior, especially since there are no functioning manned shuttles to clone. Wocega-23 flew this Blackbird undetected, as it separated from P/2016 BA14, in its blackened camouflage. Its Computer executed a precise intended landing maneuver at a speed that exceeded the top-speed of a Blackbird, since its extraterrestrial position was already exposing its alien nature.

Wocega-23 reported a uniquely smooth and gentle landing because there was almost no atmosphere and no magnetic field, and thus no air resistance or magnetic interference. The exploratory probes she sent tested humanity's conclusions regarding Moon's formation. Human scientists speculate that the Moon formed when a planetesimal collided with Earth 700 million years after its formation, causing molten rock to flow out from inside its crust across the surface, creating the smooth and circular maria, or "seas", which are composed of basalts, or iron and magnesium. The seas' glassy appearance suggests their rapid cooling. The Moon was too small to have had an atmosphere, or water even in iced-form. Though it was big enough for heat in its core to pour out into the volcanic rock now covering its asteroid-cratered face.

Wocega-23 had volunteered the addition of figure skating to her human bodily acclamation practice routine. While on the Moon, she realized that she could practice her jump rotations, and height without a trampoline in the Moon's low-gravity. When there was no gravity on the Mothership, she could extend her aerial spins indefinitely. Because it has some gravity, the Moon served as mid-point for Wocega-23 to acclimate to what spins would be like closer to Earth's g-force. Body positioning and muscle-control is essential for precise execution in figure skating. Wocega-23 rightly believed that if she could achieve this control in nuanced aerial maneuvers, she would be able to execute humanoid walking and running. Humans are very judgmental regarding slight deviations in gait, such as a drag of the toe, lack of balance, limping from injury, stiffness, and crouching. An extremely straight ballerina posture can also be problematic, as it attracts attention by its alien structure. The robotic walk is classified by its rapid, straight-line motions and sudden stops. Wocega-23 needed the extra acrobatic training on the moon to develop her "natural" gait.

I remained on a regular link with Wocega-23 as she collected audio-visual direct-observation data from the Moon for our studies by flying in a flight-equipped spacesuit. Cataloging the composition and geological features of the Moon's highlands, craters and valleys took a few weeks. One study required a full night of data in direct contact with the Moon's crust. Instead of merely leaving the equipment outdoors, Wocega-23 spent that night lying in her spacesuit on the prickly reddish rocks. The view of the sky from the Moon is closer to how it looks from a spaceship than from Earth. There are no clouds or atmosphere to obstruct the crisp details of starlight. We had been chatting as we studied star formations, and then we fell silent, as I began to process the incoming night-time data. Suddenly, the video from Wocega-23's perspective trembled and she hopped up. The ground was shaking. Perhaps because we had been studying humanoid nightmares and horror-cinematography, my first suspicion was that humans noticed Wocega-23's shuttle, and had sent a bomb at it as a prophylactic measure against us aliens. Meanwhile, our instruments performed an automatic seismographic testing, and reported to us that there was a minor quake. These Moon quakes are generated by the tidal strain from Earth's gravity, and from some of the bigger meteorite collisions. Wocega-23 was incognizant of this shaking previously because her shuttle automatically cancelled out these trepidations with its stabilizers.

The Mothership remained parked on P/2016 BA14 between the near-Earth flyby in 2016 and a second more distant flyby in 2021. A telescope in Hawaii spotted it for a few months during this 2020-1 passage, and again did not detect the Mothership as an abnormal feature on its surface. The next perihelion will be on September 21, 2026. By this date, our research team is laboring to reach greater certainty regarding if remaining on Earth, or attempting to reach the next potential planet is best for our survival. Back in 2016, P/2016 BA14 continued following its trajectory, until reached a point that was more proximate to the Sun than the distance to the Sun from Earth. In this wedge between Earth and Venus, another shuttles, camouflaged as an asteroid, separated from the Mothership, and established a stable orbit around the Sun. This was a necessary scientific maneuver to establish a stable vantage point within this Solar System, to maximize the accuracy of astronomic event readings across the Galaxy. This data-collection process would have been regularly interrupted, or physically blocked by Earth's body, if this observatory was instead orbiting Earth. Alternatively, if it had approached closer to the Sun, or "landed" on the Sun; the light pollution coming from the Sun's overheated body would have flooded with statistic the relevant data it was attempting to gather from the sky. This observation post was necessary because we have exhausted the top choices from our potential habitable world list in the *Catalog*. Data that was collected on our last two stops indicated significant interstellar mutations since Nally's original readings from Cadata 100,000 years ago. If Earth proved to be an unfeasible domicile, cadatas needed a new set of best-candidates for habitability in the Milky Way, or perhaps in a neighboring galaxy.

One of the shuttles departing from the Mothership on its short loop, closer to the Sun than to Earth, was operated by Nurry Loody-23, the third-ranking pilot in the fleet. He was assigned to explore the potential for inhabiting Venus. Venus has features that are more proximate to Cadata than Earth. But, in most respects, it is inhospitable to living organisms. As Loody-23 began a rapid descent through Venus' dense atmosphere, it created a reddening lens effect, which turned the midday landscape into a sunset-like haze. The effect was generated by thick clouds. At 78% albedo, dense clouds bounce most incoming light, while the same gases retain the light achieving penetration in a runaway greenhouse. The clouds were concentrating sulfuric acid before dripping acid rain and excess lightning strikes onto the surface below. This thick cover keeps the ground temperatures hot and stable at a lead-melting 872°F. It also semi-blocks radiation from the proximate Sun because Venus lacks a magnetic field to enhance this blockage. The atmosphere is much denser at 90 atm: equivalent to conditions seen when diving 800 meters into Earth's ocean. A human scuba diver can survive a maximum record of 332 meters in a suit. It is similar in composition to Mars, with 96.5% carbon dioxide and 3.5% nitrogen. Earth matched this ratio at an early point in its evolution as well, before cyanobacteria or blue-green algae changed the balance to 21% oxygen and a mere .04% carbon dioxide. Oxygen is hyper-reactive with most elements on the periodic table, so its abundance promotes complex reactions needed for advanced lifeforms on many of the planets we explored. Human scientists have argued, high carbon dioxide levels impact cognition. We have observed nonmetal-element-based lifeforms are likelier to thrive in atmospheres with a smaller percentage of atmospheric carbon dioxide. Plenty of lifeforms tolerate a myriad of extremes. And for many planets, Earth's oxygen levels are extreme. Cadatas prefer a similar atmospheric formula to Earth's, so both Venus and Mars are outside of our preferred range. Loody-23 landed in a valley below the 5-mile-high Maat Mons volcano near the equator, which had experienced its last major hundreds-of-miles-long lava flows millions of years earlier. Humans had recorded spikes in Venus' surface temperature back in 2014. These were interpreted as renewed volcanism on Maat Mons, but they caused minor lava output. The activity sufficed to stress Loody-23 regarding his parking location because even a tiny lava flow on Mons-proportions could have flooded the region with fire in moments. Overflow of lava covered the entire surface of Venus 500 million years earlier, melting rocks and flattening surface features. After gathering surface tension readings on Mons' temporary dormancy, Loody-23 settled into the outpost and deployed robots and nanobots to test the composition of Venus' surface, crust, core, and atmosphere to evaluate its current and past states. These confirmed human findings: Venus was covered in water like young Earth, but then the Sun spiked in luminosity. The planetesimals circling the Sun hit Venus disproportionately more than the farther Earth. These violent collisions, and the heat wave created a runaway greenhouse effect, spiking its surface temperature to 2780°F, and raising the pressure to 300 atm, thrice more than its current level, and high enough to evaporate Venus' exposed water, and to melt some of its surface rocks. Life started on Earth inside moist rocks and hot water pools. These changes made it impossible for similar developments on Venus. Loody-23 began researching what impact taking a significant quantity of the atmosphere out of this equation would have on Venus. One idea was helping solar wind to blow the atmosphere out into space, and another was turning air molecules into solid rocks or liquids. The Computer had some warnings about this plan from the first few simulations. So Loody-23 invested some of his creativity to adjust the formula into a workable solution. The odds are low, but perhaps Loody-23 will find a miraculous equilibrium Venus needs to thrive. If history is an indicator, Loody-23 might intensify the existing runaway greenhouse or blow away the remaining atmosphere, pushing Venus past Earth's temperatures and towards Mars' levels.

Back in Earth's atmosphere, Wocega-23 made an orbit around Earth to take up-close scans of the planet to gain a perspective of life on this planet after a lifetime in space. Turning her entire ship's walls see-through allowed Wocega-23 to take in the details of the landscapes below and the sky above without obstructions. As she turned her head to catch elements of the changing scenes, she was almost blinded by the brightness of the Sun. She was now blindingly proximate to the center of the Solar System. Wocega-23 had never seen natural sunshine at this proximity. Her two human eyes in place of a single cadata eye tired her concentration as they tried to show her a combination of dual perspectives. She fought this disorientation during this orbit, and kept her eyes on the Sun, as it seemed to fall behind Earth, and then to rise above it at a rapid rate during a half-an-hour-long cycle. As she began her descent through the atmosphere, the color of the Sun behind her changed from a pure white to yellow. As the intensity of the Sun's rays faded, Wocega-23 spotted the white outlines of a rising full Moon on the other side of the sky. Wocega-23 had researched the best places to live on Earth. Before attempting to interact with a mass of humans in a city, she wanted to test basic-habitability odds on Earth on a small island reminiscent of Cadatas' isolated island formations. Nally-23 and the others did not object to her desire to have a restful introduction to Earth. She was allowed to make final decisions on her itinerary. She became even more convinced in this choice as she had watched videos of life on islands back on Cadata, and on Earth's popular island resorts.

She ran a search for an island with a runway to purchase, but located few feasible options. She excluded islands across most of South America and southern Africa because she preferred avoiding the South Atlantic Anomaly (SAA), an area where Earth's magnetic field is weaker, effecting satellites and spacefaring vehicles by exposing them to radiation otherwise blocked by the magnetic field. Wocege-23 anticipated she might need to fly into space on some occasions, and a greater cover meant less stress from impacts by high-energy particles (ions, electrons, and protons) on the shuttle's modified to appear humanoid engines, and on its ability to send and receive communications. The shuttle could have observed such radiation as fuel, but Wocega-23 wanted to be on the safe side.

The top choice across the rest of the globe was offered for $460 million and included 115 acres in the United Arab Emirates. On the downside, it had a set of existing waterfront homes, hotels, and retail stores. It was impossible for Wocega-23 to land her Blackbird from space at rocket-speed unobserved by the residents. She could not envision how she could get rid of all employees and guests in her one-months pre-landing timeframe. The second most-expensive island cost $55 million. It was in French Polynesia, a territory in the middle of the Pacific Ocean, which is governed democratically and autonomously, but under the umbrella of the French Republic. The advertisement stated the island was staffed with 80 employees residing on this island to maintain its infra-structure (access to water, sewage and electricity added to the spot's desirability). Woce-ga-23 decided, as the owner, she could order them to stay indoors for a meeting during her landings and covert events. The probability was low of intrusions from unaccounted boats floating in this region's waters because the closest island, Hao, was 62 miles away, and it was a coral atoll with a mere thousand residents. The closest major city-island was Tahiti, 481-miles away. This island had a 40-mile lagoon, breeding-ground for Tahitian black pearls and a variety of marine life. After making the purchase, Wocega-23 had some positive and negative surprises. She learned the island had been abandoned when its previous owner went bankrupt. There was nobody left on the island, so if she needed help, it would be up to her to hire workers from Tahiti or Hao, importing them. She employed them unseen, via a remote, temporary employment agency in Tahiti. The job was advertised as including accommodations, since the ad for the island stated housing for 80 workers was already constructed. She arrived first and used a tarp to cover her Blackbird to avoid questions or a theft of this rare machine. Other than the tarp, she had to park it on some dirty-white sand at the side of the paved runway. She walked over to the closest hut and inspected it. She was shocked by the rust blanketing metal components, such as sinks. The walls were bulging from excess-moisture and exposure to the salty spray and storms. Since it was March 16, it was hot and humid. The sand magnified the temperature of the ground Wocega-23 was walking on. She would have been comfortable in her cadata-skin, but her human biology was sweating out her water reserves and making her dizzy. The next hut she visited was missing a door and half of its window, so there were a couple of pools of standing water in some damaged parts of the flooring. She was hopeful about the main water-front houses, so she inspected them next. They also appeared significantly neater in pictures versus reality. The biggest house had two massive holes in its roof, which appeared to be two black shadows in the sky-view photos. Wocega-23 opted to sleep in her Blackbird. She doubled the number of workers to perform a full reconstruction of this island. She warned the supply-boat due to bring the first twenty workers on the following day of the living conditions. She had them purchase top-end tents. They loaded their boat with water and food sufficient for weeks on the island without re-supplies. She had to hire architects, designers, and expensive contractors to fix this island at a sufficiently expedited rate to fit her schedule. She considered just taking it easy by buying a boat for herself and living between her plane and the boat, but the decrepit conditions were offensive to her sense of order and style. She spent $100 million to flatten the shacks, and replace them with a livable man-sion, and live-in employee quarters. When this construction was finished, her employees started rebelling by asking her for bribes. She assumed they discovered she was an alien, and wanted to be paid for silence. Upon questioning, they explained they had built an ultra-luxurious island outpost, but other than the secretive plane even she herself never used, she lacked a planned revenue-stream from this venture. Thus, they concluded, she had to be planning on flying drugs out of Tahiti into the US. When she laughed this off, they reported these suspicions to French Polynesian authorities, who soon swarmed the island with a raid. When Wocega-23 spotted their ship on her radar, she was forced to take flight to avoid an inspection of her Blackbird unraveling her identity.

Wocega-23 had to figure out her next destination before she could land again in her inconspicuous plane. The Computer had to map a course to avoid the 500,000 pieces of junk flying in orbit around Earth, not because a collision would have damaged the resistant shuttle but because each of these items was being tracked by different Earth governments, so exploding, digesting or knocking one of these down to a lower or higher orbit would have been recorded as an unexplained phenomena, attracting research-attention from every astronomer on Earth. On the bright side, this abundance of debris allowed the Blackbird shuttle to avoid spiking alarm when it was spotted in space, as astronomers assumed it was a piece of debris changing orbit. With the Computer worrying about garbage, Wocega-23 searched online listings and found a new island for sale just off the coast from South Carolina, within a boat-ride's reach of Hilton Head and its luxuries, but far enough away from the nearest town to avoid observation. This location was ideal because we had determined pre-landing the United States of America was a country leading Earth into an environmental catastrophe. The US was a leading emitter of greenhouse gases, and by steps such as failing to ratify the Kyoto Protocol back in 2005, it was preventing efforts by other countries against global warming. The "South" had been blamed for its anti-science and pro-coal or pro-business platform, which gave their Republican party a majority in the country's government. South Carolina was at the heart of this southern region, so Wocega-23 hoped to understand their motivations from within this geographical anomaly. Cadata Potentates' attempt to profit from mining an asteroid had a planet-shattering impact. If Cadata could have been habitable for another 2 billion years without this artificial asteroid diversion, this decision meant 2 billion years of lost revenue, and the nearing extinction of all cadatas. Wocega-23's research project was understanding if there was a way to convince a group like the Potentates of the benefits of long-term planetary survival.

Wocega-23 made a couple of orbits around Earth as she was finalizing the online purchase. She made a rapid-stop landing on the short runway starting at a river dock and ending in front of the main house. From the cockpit she spotted a swarm of mosquitoes enjoying the dense marshlands dominating the island's 15 acres. The main house, guest house, and dock were in working shape, so she slept in a bed, rather than on the shuttle on the first night. The accompanying golf cart and three-skiff boat allowed her to commute from this house to Hilton Head, where she could take a reticent flight from its international Savannah airport to regions she needed to explore. Wocega-23 took a pocket-sized robot with her when she traveled, so it could help her operate a boat, or to perform strange tasks outside of her training. She was careful to avoid massive redevelopments to this new island, having learned of human greed and suspicion of flaunted money.

As Wocege-23 settled into life on her quiet southern island, she began analyzing the data the nanobots had been gathering on Earth's environment. To understand the numbers, Wocega-23 also ventured into the field to confirm the Computer's identifications of rock and species categories. She wrote up a booklet for future cadata explorers on Earth's geology, paleontology, biology, chemistry, geophysics, atmospheric science, and climatology. Wocega-23 did a good deal of this work on her own island, which had a mix of terrains with a diminutive sandy beach, marshes, and an abundance of edible and beatific vegetation. To minimize the odds of discovery, Wocega-23 did not hire a human assistant, and instead had her robots fix household and gardening problems.

After a year of isolation, in December of 2017, Wocega-23 grew weary of island life and began taking frequent boat trips to the mainland. At first, she would just sit on a bench on a boardwalk and watch people, but she received a lot of dirty looks for staring, so she started running along this boardwalk, so she could stare at those she passed without arising suspicion. At first, they all seemed to be doing a set of the same activities (running, skating, walking, strolling, gazing), in the same outfits (same brand names, same types of shirts and pants, same color patterns), and while talking about the same topics (sex, work, death). But as Wocega-23 began starting conversations with these people, she learned they were living the same repeating plot in near-identical houses and wardrobes. Wocega-23 was experiencing a phenomenon her ancestors reported suffering on other planets: an alien species blends into an indistinguishable mass of identical units. Wocega-23 even had difficulty distinguishing humans from their monkey and tree genetic-predecessors: trees' roots mimicked human legs and branches matched the appearance of arms. When walking outside at night, Wocega-23 jumped a few times when she saw a tree, but thought it was an unexpected human visitor. Humans ate, drank, and reproduced in identical patterns. Thus, Wocega-23's case was a bit more severe than her predecessors'. Part of her challenge in distinguishing between humans was the power corporate monopolies exerted, which prevented people from purchasing hand-crafted or otherwise unique goods. The publishing monopolies limited humans' original thought because published books repeated the same plots. Cadata researchers believed Wocega-23 was failing to connect with humans because of her flawed mimicry of the local South Carolina accent, as well as the speech patterns in the other regions she visited. Every time she greeted a new person, and tried to start a conversation, she was asked where she was from. In response, she tried to ask them where they thought she was from, as she did not know what specific mistakes in her drawl they were spotting. In contrast with other alien species, even when Wocega-23 visited a village with a mere ten residents, none of them acknowledged or greeted her first; they pretended they were bored by a visitor and hoped for her prompt departure. Wocega-23 always had to be first to greet these anti-socials. And when she said "hello", she had to be smiling, or the approached human began hollering she was harassing them. Some escalated until they threatened to call the police. Wocega-23 was greeted first on three occasions, and in these cases, the "friendly" humans vandalizing her car, stole her bag, and made fun of her accent. Another difficult lesson was: whenever a woman said "hello" to a man of any age or type, the man assumed she had a sexual interest in him. Thus, the male suggested they engage in a pre-sexual dining, or drinking ritual. Even as she did her best to avoid talking to or direct eye-contact with male humans, they kept staring at her milk-production glands and excretion outlet. She also received requests from predatory males for her to wear paste to color her cheeks in a lighter shade, while brightening her lips in a color such as purple, a natural indicator of a serious human health condition. An outfit seller in a mall also questioned her decision to purchase a suit and tie because women were supposed to purchase a short cone to expose their bare legs to the male gaze. Conversations with other women were also cultural minefields because they asked if she was married. Learning she was single, they looked at her sideways, before recommending she should solve this abnormality by marrying an elderly, obese, and semi-literate male. When cornered, she found refuge in explaining she was wealthy, and living on a private island, without having performed any labor to deserve these riches. Having money via an inheritance, or divorce, instead of a profession was socially-acceptable for human women.

While Wocega-23 was traveling across the country, she encountered a difficult with purchasing whole foods containing the nutrients human biology required. She had been on a vegan diet, which matched medical advice from human researchers. But without fresh choices in an airport where she had a layover, Wocega-23 decided to attempt eating fast food at a fast burger restaurant. The idea of "cooking" foods was strange for cadata researchers; they had encountered few planets with a species who took natural nutrients it was designed by evolution to eat and burned these into cinders, boiled them in explosive oil pools, minced them into microscopic dust, or otherwise processed food until it turned into cardboard. She was glaring at the pictures above the servers for an uncomfortably long time, trying to grasp what each item used to be before it was processed. The signs said "beef" and "chicken", but the first had the appearance of post-digestion roadkill and the other of a cardboard fluffed and compressed into an oval. She decided to order each of the items on the menu to dissect these strange dishes with a scientific eye. She had tasted a whole-cooked chicken from a local store, and a beef steak from a fancy Hilton Head restaurant before, so she was able to guess which was which by taste without peaking at the boxes. While this mystery was solved, many new puzzles surface upon closer scrutiny. She could taste bits of the actual cardboard boxes these sandwiches came in. The cardboard was leaking through because the hot sandwiches were melting it. Wocega-23 contemplated: *Why would humans put them inside trees, if they can't digest tree trunks?* The strongest taste coming across was of salt crystals, as she tried to crunch on yellow, rubbery strings with deflating middles, labeled with the strange term: "fries". Wocega-23 had to look up what this was referring to because she guessed they were fried rubber; it turned out these began as potatoes. The interior of potato shells seeped out during the oil-boiling process, leaving air and oil. This combination was disguised with a spoon of *salt*: a white rock humans are addicted to. Because salt is dehydrating, Wocega-23 needed a drink afterwards. She sipped on a brown soda through a straw. The rush of glucose made her nauseous, so she stopped. The taste and sensation were like drinking fluids out of a gassy animal. As she waited to examine the sensation, she felt a momentary spike in energy prompting her to read the nutrition label. As she was searching for definitions of cryptic words on the label, she felt a violent crash, so she drank a few more gulps of this bubbling sugar bomb to recover from the withdrawal a single hit generated. Dizzy and breaking into a cold sweat, Wocega-23 escaped the restaurant, leaving the samples untested. She explained her failure by arguing she was a bite away from escaping Earth to bypass a lifetime of addiction to indigestible poisons.

There were other bizarre human habits Wocega-23 imitated to grasp their significance before returning to more cadata-science-proven methods. One of these small practices was teeth-brushing. Cadatas did not have teeth as they consumed nutritional gases, but they had encountered species with teeth before, and among these none used the tooth-brushing method humans propagated. Based on Wocega-23's research, toothpaste was a combination of chemicals humans misunderstood, which facilitated the growth of resistant bacteria, blocked taste buds' sensitivity, infused the body with indigestible sugars, blocked digestion with enzyme disrupters, and contained carcinogens. Using this paste was parallel to cadatas choosing to have the same cleaning procedure on their mouths as on a spaceship after a lengthy flight through space. Wocega-23 was also stunned she had to use a device with little hairs on it to stroke her teeth with this paste. The bulk of the bacteria was untouched by these bristles even with a mechanic toothbrush, and yet humans failed to visualize a benefit in inventing a device capable of complete-surrounding of teeth to polish them from all directions in closer and less abrasive contact. This procedure ended with a mouthwash rinse lethal to a fraction of the mouth's bacteria, while 100 trillion bacterial cells remain alive and well elsewhere in the human body. The hairs of a toothbrush had been invented 5,000 years earlier, and nobody designed an update to this technology since. Digital knowledge-sharing across the past century failed to expedite invention because monopolies profited from cheapening old production methods rather than investing in new research.

One day, Wocega-23 waded through her marshes to swim in the ocean surrounding her island, she saw two giant watery eyes pop out from the water. She had been swimming for over an hour as part of her daily exercise routine, so she was too tired to interpret these as a danger. A moment later these eyes dived under and she felt a pain in her leg. She wrestled with what she later determined with an American alligator and freed her foot. Despite the blood in the water near her, she reached the shore. She prepped to utilize an emergency medical device she always carried to reattach the loosened foot. When she switched this device on, the Computer alerted cadatas back on the Mothership, and they tuned in to this unraveling event. Instead of a brisk fix, Nally and the team insisted she dulled the pain and then attempted to fix it with human methods. Nally argued this was a chance for practical research into the humans' medical system. They were curious how far human medical science advanced. Nally asked: "What is a *hospital?*" It was immoral to propose an intentional injury to test this "hospital" system, but since Wocega-23 was already "naturally" injured, it would have been a scientific failure to surrender to discomfort. Wocega-23 agreed. She took precautions to switch off her pain receptors. She also backed up her physical data in case the doctors damaged her body in an unregrowable manner. Rather than driving herself or taking a cab, Wocega-23 called 911 and asked for an ambulance as she would if she was human and in mortal danger. She grew frantic with boredom waiting for over 2 hours for the helicopter. They called her back halfway to confirm her address was on an island without an access-road. Wocega-23 later learned this flight cost $25,000. The medics in the copter bandaged her limp foot with a rag. They connected her to compressing and stinging equipment to measure her body's functions. Once they landed on the roof of the Savannah hospital, she was taken into an emergency operation room, where the doctors applied live leeches to the affected tissue to keep the veins alive while they primed to stitch them together. Then, they cut off some of the bone with a bone chisel and the tissue from the foot around the bite with curettes (silver metal scraping devices with a loop at the end) to shorten the leg. They cauterized the wound to stop it from bleeding with electric shocks from a cautery. While stitching the tissue together they held it in place with a type of scissors with non-piercing ends called forceps. As a finale, they used a piece of metal (called a needle), a metallic string, and a clasping needle-holder to craft the stitches. For some reason, they switched to just stapling the skin back together with an object matching the appearance of an ordinary white stapler. It was distinguished from this office tool by a strange ruler extending from its lower end. If you are imagining Wocega-23 traveled back in time for this procedure, you are too optimistic about your humanoid medical system. This is modern American medicine. While microsurgical steps for connecting arteries are less than 100 years old, the same type of needlework was carried out thousands of years prior, and no modern human inventor has rethought the use of needles in this death-defying procedure. Wocega-23 redid this entire procedure and regrew a new limb as soon as she was allowed to go back home two weeks later (after a painful stay in a foul-smelling room with horrid green-gelatin and mini-beef-ovals for food). Cadatas had not witnessed a scarier medical procedure in terms of risk of infection across their galactic journey, and we all agreed never to let a cadata into a human hospital again.

Just as the hospital-stay was ending, Wocega-23 lost the audio-visual connection to the Mothership. The connection did not reestablish itself as it was supposed to if there was a temporary blackout. Wocega-23 realized she had lost touch with our shuttles across the Solar System. It was a weird problem. The Computer on her Blackbird shuttle proposed a fix for the connection. She had to install a mini satellite in space above Earth's atmosphere, so future interference would not drown-out the signal. The Computer advised her not to take the Blackbird for this mission because if humans caught her in it, and she did not have a way to ask other cadatas to back her up, she might have been in serious danger of being kept captive by humans. To be more inconspicuous, Wocega-23 decided to travel into space in the disguise of a human. Wocega-23 first created two false identities with military piloting experience, and applied to join the American and the Russian space programs. She was rejected from both despite a glowing artificial resume; she learned past astronauts applied for up to fifteen years before being accepted. The next idea they considered was paying a few millions to the Russians for a tourist seat on a flight, but they learned seats were reserved a decade in advance. And the tourist trips never approached the International Space Station. Thus, Wocega-23 decided to mutate her appearance down to the genetic level to become identical to an astronaut or a cosmonaut already in the program, and who was scheduled to fly on an upcoming flight. She encountered a moral dilemma as she thought through how a replacement would work. What could she do with the person being replaced in the interim? She considered putting him or her into a comma, jailing him or her in an island mansion basement, and various other barbaric notions before she settled on a simpler solution. She figured every astronaut and cosmonaut had an innate curiosity about aliens and their gadgets. So, she decided to trade the use of her Blackbird shuttle for a flight on the human shuttle. She approached the sole American woman[6] scheduled to fly to repair the Hubble Space Telescope in the coming months. Wocega-23 tempted Reth with a flight on a Blackbird, and this was sufficient for her to come along. Then, instead of staying at a Blackbirds top speed and height, she rocketed it up and past the atmosphere and took a quick loop in it around the moon. Her human guest was stunned, but calmed as Wocega-23 explained she was an alien visitor offering a trade. Reth had been assigned minor tasks on the mission. Given humanity's stalled space-exploration progress, she anticipated never crossing the Solar System in comfort, without alien intervention. Thus, as cadatas anticipated, she agreed. With Reth' participation, the genetic edit was effortless. Reth received light

---

6   This astronaut's name was changed to *Reth* in this narrative to avoid a libel lawsuit.

training in cadata technologies, opting to rely on the Computer to execute her vocal commands. The single rule was Reth had to avoid Earth until Wocega-23 completed her space program experiment. Reth promised she would be un-tempted to return prior to their demand for the restoration of this marvelous space shuttle. Cadatas were duly concerned Reth might find trouble on this voyage, but with the Computer on-board to supervise her progress, they hoped she would survive it.

In her new body, Wocega-23 had to take a 21-hour flight from Savannah to New York City, then to Frankfurt, Germany, and then to the Almaty Airport in Kazakhstan. And from there, she took a 17-hour drive from the airport to the Baikonur Cosmodrome spaceport on the other side of the country. This was the site of Russia's first Sputnik launch, and most other Russian pioneering space ventures. Americans had not launched a manned space flight since 2011. The astronauts arrived weeks earlier to participate in Russian language classes, necessary because most manuals and signs in the shuttle, and vocal commands sent to the station were in Russian. Reth had attended physical and linguistic training. Now Wocega-23 took her place by slipping into the dormitory and joining the last two days of training. The scheduled launch was for an all-American crew of seven, with two rookies. Reth was considered a rookie despite having spent several years in the program, assisting missions in the background. They were to travel to the Space Station on a servicing mission for the Hubble Space Telescope.

At 3am, on the morning of the flight, the astronauts departed from the complex, where they had been training, in an astro-van headed to the launchpad. Wocega-23 heard two attack helicopters circling in the sky around the base. A dozen Russian special police forces, the OMON, were assigned to our van, and were guarding the launchpad as we arrived. Seeing a couple of these heavily-armed OMONs marching down the van's aisle made Wocega-23 paranoid. She questioned if they discovered an alien invaded their space program. Cadatas failed to prep for an exit-plan, if her alien identity had been detected. She was too nerves now to devise a suitable response to this threatening situation. Her first thought at seeing the spaceship during the drive was it was too small in the circumference of the main living pod in comparison to its three-story-high metal jugs of fuel tanks holding liquid oxygen and hydrogen. Contemplating these characteristics gave Wocega-23 goosebumps as she fathomed, she had volunteered to sit on a giant bomb about to be exploded under them. If the simple metals in the compartment around Wocega-23 had an imperfection, her body might have burned up in moments without even cadatas' repair capabilities. Wocega-23 had tried to leave her cadata spacesuit on under her garments, but the cosmonauts had confiscated her possessions, including the suit and her cadata-issued medical unit back in the changing rooms. The sole alien item she smuggled by hiding it in her ear canal was the mini-satellite she had to install.

They went up an elevator, down a walkway, and then entered through a side door into the main cabin of the shuttle. As they stepped through the thick doorway, they heard steam hissing and metal groaning. This was the first noise she ever heard from one of her cadata ships. Wocega-23 jumped, and asked, in terror, if the ship was being crushed together by the force of some unforeseen explosion. A senior astronaut glared at her askance. He reassured her the noises were standard, indicating the ship's systems had been turned on. As she looked inside this space, it seemed to shrink and the walls seemed to tighten around Wocega-23's head. Her agoraphobia was turning into claustrophobia.

She missed their cadata spaceship and her Blackbird shuttle, but none of these flight machines were this restricting in terms of space per pilot. The shuttle's crew compartment was only 2,325 cubic feet and this space was intended for the seven of them to live and work for over two weeks. Two crewmen were placed in the back on the mid-deck without windows. Four astronauts, including Wocega-23 (who was an assistant pilot) and the head pilot sat on the front flight deck. Wocega-23 sense of entrapment intensified as the cosmonaut crew strapped them into their seats, as if they were invalids. This treatment was required because their spacesuits limited their motion, making self-service buckling awkward. These cosmonauts, grounded for this mission, secured a helmet on Wocega-23's head. This Russian crew was making $6,000 per year on the ground, but could make $60,000 for a six-months on the ISS. They were rough with the American crew during this service-session because they were jealous the Americans misplaced them out of these enriching spots on the prolonged ISS mission. In contrast, American astronauts were given an extra $136 for 120 days in space or $1.13 per diem. This lack of fiscal reward contributed to Reth's refrain from insisting on space-flight prior to this trip. She had a comfortable, middleclass life from her stimulating development research in rocket science.

Unaware of these economics, Wocega-23 was fidgeting because the tight glass orb of the helmet made her imagine how easy it would be for it to shatter on impact during turbulence, so her claustrophobia intensified with it constricting her head. The orange spacesuit was also heavy and uncomfortable, and would have made running challenging in case of a disaster. Once the crew left them alone, Wocega-23 tried to focus on the liftoff set to end this terrifying anticipation. But to her surprise, they all just lay there without the ability to move much more than a finger for the next four hours.

Wocega-23 hoped to avoid speaking with the other astronauts without compulsion, but a couple of hours into this silent tedium she asked a neighbor: "How do we get out of here in an emergency?"

"You don't remember? The parachute? But it's all totally useless. Nobody has ever managed to open the hatch and successfully escape from one of these shuttles. If something catches fire, or blows; we're all dead," the astronaut replied to Wocege-23's horror.

At last, the liftoff procedure was initiated. Wocega-23 had re-read the steps in the plan she was responsible for, but did not feel confident she understood how to flick a switch or press a button. Specifics were even more puzzling. She tried asking the head pilot what definition of "flicking" the handbook was referring to, but the sideways, frightened look she received in response made her add she was joking. Now, she hoped the few bits she understood were sufficient to avoid blowing up the ship by turning on a booster on the runway. Wocega-23 slipped in her attempts to display composure when she asked the pilot what the control stick between her legs was for. The pilot was flabbergasted by her ignorance. He explained it was a useless design feature imbibing the shuttle with a spacy-ambiance: a spaceship has six degrees of freedom, as it can move up and down as well as right, left, back and forward; in contrast, the stick only had two degrees of freedom or back and forward. Neither of these directions could have helped in an emergency as no break could stop the triggered bomb below them, and an accelerator could not have released more than the pre-scheduled volume of fuel. The ship was destined to go up until it burned a stage in its fuel cycle. Wocega-23 was un-reassured. The flight had to be automated because humans failed to invent a stick with six degrees of freedom. They were eons away from a spaceship capable of changing speed or direction to match the commands of a manual operator with a stick.

These reflections happened in flashes. She would be helpless once the pilot initiated the massive engines and the spacecraft started rattling like a metal can with a can opener trapped inside of it. Our Command Module was creaking as if its metal walls were stretching to the limits. Then, Wocega-23 heard the announcement the tie-downs had been released, and this reduced the noise and tribulations. The explosion Wocega-23 was fearing happened next. A ball of orange flame burst from the rocket's base. As it burned, it emitted a cloud of black smoke, which spread so wide it coated and fogged their front windows, canceling their visibility. The shuttle was shaking so much while it was still, the liftoff felt almost less turbulent in the first 20 second. But with every passing second, the g-force on their bodies increased. Their thoughts were fixated on the weight pushing them down. While the liftoff was automated, Wocega-23 was required to watch the monitors for problems. She did not remember what the readings indicated. She hoped the remote, Earth-bound team would alert them if they comprehended a problem in these hieroglyphics. A minute after launch, they broke the speed of sound, reaching Mach 1, as humans call it. The speed of re-entry into Earth's orbit by an average manned shuttle maxes at Mach 25, though it has reached Mach 26 on drops from the Hubble because of its greater elevation. The g-force scrambled from 1 to 4.5 g's. At this peak, the first stage burned out and separated with a bang before falling back down to Earth. The momentary drop in acceleration threw them painfully forward into the thick straps of their restraint harnesses. After the lurch, they became weightless for a stalled moment before the second stage ignited and added forward-thrust. The ship seemed to squirm like a worm as it made a slight adjustment in its direction. Wocega-23 was comfier with the higher g's than the other crewmembers having been exposed to higher pressures as a cadata. Her cadata brain functioned optimally in high ranges, while human brains were oxygen-deprived during these temporary high-g's. When they reached their intended orbit height, the rocket-engine auto-switched off, and the rate of climb dropped to zero. The rocket tilted from a vertical to a horizontal position, and began circling Earth, keeping a steady orbit thanks to GPS-tracking. They were all upside down now and weightless. Wocega-23 studied the other pilots and realized they were scanning the readings for velocity and altitude on the dashboards. So, she mimicked these glances despite their nonsensical appearance to her. Wocega-23 skimmed the manual because she assumed, she had been flying a cadata spaceship for hundreds of years, so human space flight would be easier; she discovered rudimentary machines created more work for a manual operator.

Wocega-23 was hanging there when, to her relief, she spotted the pilot detaching his helmet and gloves. She had been itching to rip these off since the crew put them on her, so she mimicked this removal with a sigh. He let them float in the air afore him to show-off a weightlessness trick. After a lifetime in outer space, Wocega-23 was more curious about weightfulness, or matter being attracted to the ground on Earth. The astronauts floated items their hands encountered. This misguided playfulness contributed to the module designers' preference to tie most objects down. After the floating gloves were motionless for a minute or so, the pilot lost interest, and released his lap belt. He then floated out of his seat. Wocega-23 copied this maneuver. Being weightless in a human body was a bit unusual as her muscles tensed up even when they were not needed without gravity to fight with. Wocega-23 was surprised when she felt a pillow of snout flooding her head to the eardrums. The mucous continued building in her face across the trip instead of flushing downwards as it does on a gravity-laden planet. Wocega-23 repeated blowing her nose on an increasingly sticky handkerchief. When Wocega-23 had to spin to grab this tissue, she discovered another inaptness in the human body to weightlessness. The human inner ear interprets movement with help from gravity, so without its force, it cannot tell if the body is moving, thus creating a disagreement between the eyes seeing the world spinning around and the brain being told by the ear it is still. The outcome of these conflicting messages is space sickness, a feeling Wocega-23 never felt as a cadata.

After the stressful liftoff and adjusting to the discomforts of human space travel, Wocega-23 tried to stretch her muscles, floating between the front and back of the ship. These movements caused her to unconsciously scrape a finger on a handrail. Wocega-23 was used to cadatas' anti-impact spacesuits allowing rapid movement in zero-gravity without fear of unwanted collisions with walls or furniture. She recognized the danger of the impact when she saw a little red liquid ball floating behind her in a window reflection. She realized what it was when she smelled it, it was a drop of her blood. She found a tissue and soaked this drop up to keep it from splattering and leaving a permanent stain on a seat or the ceiling. Given the space-scarcity, she feared injury from the lightest movement. She tried to reason humans had to keep it this tiny because of the extreme cost of fuel to lift every added pound against Earth's gravity. Wocega-23 crept, watching her surroundings, and placing her feet into struts at the bases of couches, for a firm grip, before shifting her point of leverage towards the next restraint.

She was restless again after an hour of trying to sit in place after the scratch-incident, so she ventured back out to serve herself a cup of coffee, this time moving step-by-step. She had been used to zero-gravity on the Mothership mid-flight, but she had never seen a bag of hot water she had to mix with coffee by hand without robotic assistance, or at least a human-made coffeemaker. As she scrounged the cup, some of the darkening hot water spilled out, leaving a red mark on her somewhat burned fingers.

As a couple of days passed, Wocega-23 grew comfortable with the station, so she began leaving her hairbrush and log book floating wherever she had been using them. She had always had a cadata robot at her service to clean loose items moments after she abandoned them, so she lacked the discipline to put items away for herself. The other astronauts started to notice and told her to stop cluttering up the walkways with clouds of floating stuff.

While drinking coffee and eating some of the food was difficult due to weightlessness and a lack of cooking implements on the Space Station, the pilots liked to joke about how much worse space food was back in the Apollo days. As Wocega-23 listened to them, she understood the reason humans were growing fat on hyper-processed, mystery-chemicals food was because of the "innovations" developed to keep food from spoiling for months on a long space mission. Space food service was designed to simplify preparation until it could be achieved in spacesuit gloves. The stereotype against men cooking, and a lack of women in the space program might have contributed to encouraging these extreme cooking-avoidance strategies. Food scientists developed freeze-drying techniques to keep food semi-food-like until water is sprayed in by squeezing the trigger of a wall-hung pistol into the plastic bag, before the concoction is mixed with the fingers, and eaten. Food was stamped into bite-size powder cubes. It could have been candy, or a cookie, or bacon: they were distinguishable by their colors. These shades were achieved with food-coloring because the preservation-process killed most nutrients giving food natural colors. The food scientists were tasked with an economic calculation: what are the cheapest, edible substances per a portion-sized cube? Their answers included salt, sugar, water, wheat, corn, and soy. They mixed these with sprinkles of more expensive ingredients such as beef or fruit flavoring to create cubes for maximum preservation-capacity until they were bought by an advertising-hoodwinked consumer. Freeze-dried nuts could not be reduced to a powder without losing the essence of nuttiness. But spaghetti could be cardboard with a bit of wheat-flavoring. And meatballs could be salt with beef innards. Cadatas manipulated their food when they engineered their nutritional gases, but these included the precise combination of substances cadatas needed, without toxins. Since the Apollo times, the food transitioned from goo to approach Earth's fast food. Despite the improvements in texture, its artificial chemical composition increased. Wocega-23 ate these foods as she could have tried poisons from different venomous animals to check a human body's reactions.

Back on the first day of the flight, soon after the team was free to float around, they began snapping pictures of each other to post on social media. Wocega-23 would have preferred to avoid these since her genetic-edit might have retained a flaw; a small, new freckle or birthmark could have caused suspicion upon later scrutiny from a detail-oriented historian. She said she did not have her makeup on when first asked, and then tried making up some other excuses before the chief pilot ordered her to participate, as social engagement or marketing of NASA's space program was part of the astronauts' assigned duties.

A couple of days into the flight, Wocega-23 encountered a bigger audio-visual challenge when a friend of Reth' working for Mission Control (why not call him "Dave") used a video-call to reach her. Dave was thrilled "Reth" was weightless and at the station. Wocega-23 managed to answer simple questions about how it felt up there. Then, Dave started alluding to NASA parties and conferences, where they shared panels. Wocega-23 nodded, smiled, and made general observations to hide her ignorance of these "shared" events. At one point, she deflected a direct question about where she had left her dog while she was away by asking Dave what weather they were seeing back in DC. He retorted with a joke, asking if she could tell him what his weather was like by looking down from space at the clouds around the DC area. Wocega-23 would have called Reth after this discussion to ask if her dog was homeless or starving during her outing, but cadata communications were still down.

Wocega-23 later reported in her cadata video diaries, sleeping on the station was the most unpleasant part of the trip. Her tiny rectangular cabin was across from the restroom. In quiet moments there, floating in a sleeping bag tied to posts, she heard water splashing against the grating metals of tanks and pipes. Other unsettling noises included the grunting, fluttering fan-vibrations, and the radio static. Broadcast failures were unfixable despite the billions space flight was costing humans. Sounds were not alone in being magnified in this confined space. She woke up every time a shipmate went into the vestibule and generated an eye-watering odor of re-processed-goo. She tried to remedy this problem by finding another place to sleep. Some of the astronauts slept in the back by the suit-closet, or right in the middle of a walkway. But in these spaces, she had to tie-down to the floor, which had a constant foot-odor. They also lacked enclosing walls, so Wocega-23 would hear (with startling magnification) the shipmates working at the command post or doing an inventory of supplies. After a few unsuccessful ideas, Woce-ga-23 solved a couple of her problems by returning to the restroom-neighboring box and putting on an emergency breathing mask. It had the double benefit of providing fresh air and breathing into it generated enough noise for Wocega-23 not to hear astronauts entering the lavatory.

To dispel the gloom that she felt after these failures, Wocega-23 spent much of her free time glaring out of a mini-window, holding onto its rim for a steady gaze, as she floated through the air. She recalled the dot on an Earth constellation map where the Computer indicated Cadata's location. She found this end of the constellation and the spec likely to be her ancestors' star. She glared at it as if it was home. She reflected, Earth could be their last stop before they surrender their quest and return to Cadata. What had transpired in Cadata's environment since their last message? Even if cadatanauts on her spaceship spent generations to return to Cadata, the system might be so lethal, they would have to continue their interstellar migrations without touching down on the molten or charred skeleton of Cadata.

She was also looking out of a window, when they approached the primary mission target, the Hubble telescope. On first sight, it seemed to be just one of the stars in the clear-black sky. Across the following hour, it expanded until it gained the shape of a sparkling metallic cylinder. The chief astronaut in charge of the approach[7] switched to manual at this half-of-a-mile juncture. Ray fired the engines a few times to slow down to a ½mph speed for a smooth "landing", and to course-correct to steer the ship into the pre-planned final position. The jerks these constant changes generated annoyed Woce-ga-23 to the point of insanity. She had never traveled this slowly. Even walking humans achieved 3mph. Her agoraphobia and fear of these types of human shortfalls returned until she came close to pushing the operator aside to grab the controls. Exerting all available self-control Wocega-23 waited through this maneuver until they were thirty feet away. At this juncture, another astronaut (why not call him Hip) took up the robotic arm to bring the telescope into the shuttle.

Wocega-23 noticed small signs of decomposition or stripping of bits of metal from the telescope and asked Hip about it. Hip explained the limited degree of protection the telescope and their shuttle had from radiation was also unimpeded here by Earth's atmosphere. The materials composing the windows and the light metals in the walls were inadequate; astronauts had developed cancers and other signs of radiation-damage from exposure. Wocega-23 panicked. She asked for sunscreen or for an alternate measure of protection on board. Hip laughed, assuming she was joking. Faced with this carelessness, Wocega-23 considered harnessing components from the ship to engineer an anti-radia-tion shield for herself. Then she grasped, she could not execute such a procedure without information and manual assistance from a cadata robot.

---

7   His name has been changed to Ray to protect his identity.

While thinking through the station's design, Wocega-23 asked Hip questions gaining significance at this breathtaking height. Wocega-23 had been used to the smooth millimeter-precision landings in cadata shuttles. She learned: a human shuttle landed in a manner closer to a car being dropped from space, with a petite parachute to soften the violent speed of this descent. Protective measures included toy-like contraptions like a slight angle during atmospheric incursion, rolling like a comet and a piece of metal strapped to the front to keep the ship from disintegrating from friction. offered protection. Despite the control the computer was given to calculate the trajectory, and execute most maneuvers, there was a mile-wide variation in its estimates for the final landing location. To Wocega-23, this was equivalent to a self-driving car a lane off from its intended location in traffic, veering from one side of the road to the other to activate a slow-down. The actual deployment of a parachute in the final minutes prior to impact was as absurd. Imagine a non-reusable car with a mere eject button instead of a break; to stop, the driver must be tossed out in a parachute, while the car lacks stopping-capacity; each of these cars is destined for a single drive always ending in a lethal crash totaling the vehicle. Remaining in a car deploying a parachute as a stopping mechanism is even more absurd, and yet this is how human shuttles land. If it ends up on a mountain slope, it can tumble down the mountain side with the full force of gravity tossing the humans inside, as if it is mixing dressing into a salad. And if it lands on a tree and a parachute is tangled in the branches, the human must climb down, or crash with it, without the failed parachute to assist in the last part of the descent. Another horror landing location is in a swamp, especially if the escape-hatch is on the bottom and the shuttle sinks, blocking the egress. Hip reported: once a Soyuz capsule landed on a mountainside tree-top. They then spent the night skiing down the mountain to the sound of howling wolves. This misadventure prompted the inclusion of a pistol in all Russian landing pods.

At minimum, a pod must have sufficient thrust-power for a gentle landing. If humans invented a helicopter; why have they lagged on adding a helicopter's features to a space shuttle for landings? Hip's advice for surviving this looming disaster was for Wocega-23 to keep her mouth firmly closed and her head flat against the headrest. The landing was equivalent to a car-crash in the rapid deacceleration the body experiences and the back-and-force whiplash movement. All astronauts experienced a minor whiplash injury. Instead of a pillow or a head wrap to safeguard against this, human astronaut administrators instructed astronauts to fluid-load on 64 ounces two hours prior to landing. This water was taken with salt tablets cancelling a good portion of all this work. This was akin to eating a whole turkey and then exercising for two hours prior to a landing. This was recommended under the guise of maintaining the energy level, but just allowed supervisors to watch with glee as astronauts tried but failed to keep from soiling themselves. If astronauts suffered dehydration in microgravity; then, the time to combat this was across the flight, not in the minutes before landing; once on the ground, they could drink without limits to replenish a spike in water-loss from the high-g stress.

The software on the shuttle remained unchanged since the 1960s. The firm responsible for designing the original ancient version secured a monopoly on it by arguing it managed to survive a trip to the moon, whereas other firms' alternatives had not yet achieved this. Competitors could never test their superior designs because the monopoly squeezed them out of the one-horse space race. The special programming language this dude developed is as slow as… well, do you know of, or have you ever used a program created in the 1960s? Any modern coding language would outpace it by thousands, or millions of times. Imagine this programmer convinced businesses to utilize his 1960-model computer, and programs forever to protect their ventures from the explosions guaranteed to accompany innovation. With every horrifying word escaping from Hip's mouth, Wocega-23 was approaching a decision to expedite the communication-re-establishment, and to depart from the microwave-deathtrap-of-a-shuttle. An escapade was preferable to waiting for a crash landing at the end of the four-months space mission Reth signed up for.

Wocega-23 had chosen Reth' post for the switch because she was scheduled to do a reparatory spacewalk, an important inspection point for humanity's capabilities. When the time for this spacewalk arrived, Wocega-23's agoraphobia manifested in the jerkiness of her hands, complicating the glove and spacesuit donning maneuver. This fear was spiked when she was ordered to slip into a maximum-absorbency diaper, with a pair of long underwear over it, thick white socks, and a strange liquid-cooling garment. These items were as bulky as human engineering could muster. The white jetpack alone weighed 350 pounds back on Earth, but was close to weightless in space. Even without the need to deadlift over 700 pounds, these items' bulkiness still made it difficult to move with all this stuff attached.

Her anxiety intensified, as she again questioned her teammates on the specifics of their outfits. They told her the suit was composed of nylon and spandex: common materials used in ordinary clothing. Thus, any sharp object could rip a hole in this suit. And it was as flammable as a tree. At the precipice of the void of space, she was petrified the gloves would not form a full seal with the main body of the suit, leaving the possibility they might slip off, letting air escape, and exposing Wocega-23's body to below-freezing space temperatures. Wocega-23 was also told Reth had chosen the suit-style nobody else wanted: the huge one-size-fits-all penguin suit, to enter which one had to climb inside through its back with help from two assistants. It was bulkier than the others because it was conceived as a muscle-building tool with cables and belts to deliberately and isometrically stiffen muscles in microgravity. Now Wocega-23 had to use this torture-device on her spacewalk. She inspected the horrors of this suit during this conversation, clanking the boots on the floor, turning the pockets inside out and deciphering the survival gear inside. Other than useless survival trinkets, she had 20 minutes of oxygen and a parachute attached to the suit's exterior for emergencies. She tried to imagine how many years it would take for her body to be pulled down by Earth's gravity to an altitude low enough for a parachute to become relevant.

Talking through these lethal ideas, Wocega-23 broke out in torrents of sweat. Without gravity to drag the water down her face, salty and stinging tears and sweat collected where they appeared, in and around her eyes. She took off her gloves a couple of times to wipe these drippings off while she was still in the cabin, but once she began depressurization in an airlock cylinder with one other astronaut before opening the hatch for the walk, she had to tolerate this blinding water-buildup. She missed her cadata skin-tight suit, which allowed for easy wiping. The sole relief from a need to scratch was a mini foam pad within reach of her nose at the bottom of the helmet. As she put on her gloves for the last time after a final good scratch, she spotted her face in a window and realized her hair was sticking out like a porcupine, as her ponytail seemed to have exploded from static electricity, spilling out of its scroungy. She also had red marks across her chest and arms in the places where she was rubbing her suit from anxiety. She was unconcerned with appearing beautiful, but her façade and actions were being recorded with her helmet camera, so if she looked inhuman, it would be on the record. It was too late to grab a hairbrush to tame some of her appearance, so she just rubbed her wet face against the little scratch-pillow.

Wocega-23 had done a few practice spacewalks outside the Mothership during the deacceleration years on their approach towards Earth. She was used to slipping into a body-tight suit, popping through the wall, and floating around without a harness because the ship auto-pulled cadatanauts back towards itself. Astronauts' space experience presented mountainous challenges at each of these steps. A human would die if he or she attempted to hop out of a station without being depressurized first. The sweaty delay this hold created allowed too much time to imagine dangers, other than low or high pressure, being in space presented unknown to a species new to this transportation method. The first maneuver Wocega-23 had to execute once she took the little step outside was connect an 85-foot braided cable hooked wire to a smaller hook in her spacesuit and to an exit point on the airlock hatch. As she was securing these rings into basic hooks, it was clear the ring could slip out accidentally as easily as it fit in. The panel was just within this 85-feet range. If it was a bit farther; Wocega-23 would have had to swing around to reattach an end to a closer tether-point.

Step one was completed. Now, Wocega-23 realized she forgot most of the Hubble-fix procedure instilled in Reth. Wocega-23 was too worried about hiding her mini-satellite in the collection of knickknacks in her pockets, and about executing its attachment to the Hubble without her helmet camera spotting it. She had stepped out with Hip, who had trained for the second half of the procedure. He was to keep contact with the station while she worked at the end of the arm holding the Hubble. She heard the astronauts on the communication system discussing her every hand motion as they watched them in real-time. When she reached the panel to be removed for the mechanical fix on the human side to be employed, they repeated the first step of the power-supply replacement procedure she was about to carry out. The first order was to use a pre-designed screw-removing-device to extract 111 millimeter-sized screws. Since Wocega-23 had never held a screwdriver before, the power-tool was super-intimidating. She tried to slap the device against the screws and flip on a switch to activate the tool, but she may have secured it upside-down or off-level, as the device vibrated with such violence it might have shattered the screws. Still, none of them budged in the desired circular motion. How much force was necessary to achieve detachment was not specified in the instructions, so Wocega-23 decided to exercise her maximum pressure-potential by digging her feet behind a set of metal bars and pulling in the opposite direction, while once again trying to drill the screws out. One pull just shook the panel, so she kept tugging it. When she stopped to scratch her nose on the pad, she spotted a miniscule slash in his spacesuit's outer fabric. If the hole had penetrated the Kevlar right below it, Wocega-23 would have been leaking air and pressure for a couple of minutes by that point. Wocega-23 was cheered with the assumption her fears were mere paranoia. After human technology proved it could fail with lethal consequences, she acknowledged her graver fears could also be realized. She had tested her cadata spacesuit in simulations against nuclear explosions and a burial in lava. In contrast, a human spacesuit failed to handle the first step of a procedure humanity spent years designing and practicing. As cold sweat added to the existing flooding, she feared the cool sensation indicated space-exposure. Petrified, she asked the astronauts to read out the full instructions as a reminder. They read them methodically, a line-at-a-time. Wocega-23 had not anticipated each line would involve a feat as strange to an alien as learning to spear-hunt while a shark is stalking you. The entire panel matched the size of her padded gloves at 14-inches wide, so dimensions of the components of the panel neared the microscopic in comparison. The pilot was reading to Wocega-23: the black electric wire had to be cut. With electric-shock-odds reduced, Wocega-23 was to remove two torque set screws, a handrail, and four hex-head screws with a few of the other special devices, which had seemed to be irrelevant clutter in her pockets. With blocking parts removed, the old power-cell had to come out, a new one had to go in, and the previously-removed tiny parts had to be returned. The removals could not damage the fragile components. Wocega-23 was silent for a moment, imaging reasons to continue tugging at a unit promising to relieve her of her oxygen. Flabbergasted, she switched off her video and audio connections to the crew. With the feed out, Wocega-23 pressed the cadata mini-satellite into a spot to the side of the panel hidden behind a rod. She then pressed it in a pattern to turn it on. It had to be activated in open space rather than inside the Space Station to work. As soon as it began sending signals, Wocega-23's location came up on the cadatas main screens, as they had been searching for her whereabouts since the outage began. She said a few words to explain where she was and what she had to do to reconnect. Since she started complaining, she also insisted she needed an out, or help from a robot to complete the brainteaser humans were waiting on her for. There were no votes supporting bringing in a robot to a Space Station with several security cameras and hundreds of human overseers. Finally, Wocega-23 pitched them picking her up if she managed to escape the station. Given the lack of roadways nearby, the team was perplexed, but then Wocega-23 just disconnected her harness and pushed away from the ship and into space. Wocega-23 waved sheepishly to Hip, who tried to grab onto her, but failed to catch her due to the speed and force of her drift. Wocega-23 saw the palpitations in Hip's reddening face. The humans inside the station and back on Earth were in a raging panic as they contemplated with bamboozlement how they might rescue an unhooked astronaut. They did not have a shuttle ready with enough gas to retrieve her and land back on Earth, and even if they did, they would have needed years to program it to execute this maneuver. Wocega-23 was nervous cadatas might encounter challenges with the pickup, so the station's cameras recorded her genuine worry. The loss of a crewman was worldwide news minutes later because they had been doing a live-broadcast of the fix. Conspiratorial reasons for the disconnection sprung up in forums and then from news anchors. Reth showed up for a new bout of training two days later, reporting she lacked recollection of the drift, or how she fell safely to Earth with a mere parachute. The most popular emerging theory was she had been kidnapped by aliens. The second was she was a spy for the Russians. Apparently, they had asked her to disconnect to show the weakness of Americans' space program. Then, they picked her up, and told her to feign memory-loss.

During this maneuver, Nally-23 was waiting in a cloaked shuttle on the darkside of

Earth's Moon. When Wocega-23 showed signs of distress, Nally-23 began a disguised approach towards the Space Station. And when Wocega-23 left the Station's field of vision, Nally-23 maneuvered her shuttle to retrieve her. Wocega-23 glided through the shuttle's wall, and ripped her bulky spacesuit off. Nally-23 had been studying asteroids and other smaller near-Earth objects from the Moon base. She put on a cadata spacesuit for the rendezvous with Wocege-23 because the leadership required this safety percussion to prevent contamination between Nally-23 and Wocega-23. Wocega-23's exposure to human germs had not yet reached the five-year test-phase minimum. The Computer on Nally-23's shuttle ran a molecular-level cleaning right after she dropped Wocega-23 back on her South Carolina island. Wocega-23 blew her nose while also emptying her bowls in Nally-23's restroom across the half-an-hour it took to plummet through the atmosphere, and to park on her island's runway. The pressure, air, and temperature measures in Nally-23's shuttle were set at ideal levels for a human. This allowed Wocega-23 to recover, regaining her walking-balance, after the trauma of that drift through the void of space. Wocega-23 transformed back into her original human form once she caught-up on sleep that night. The expedience of this re-transformation was designed to avoid one of her neighbors, or employees reporting a spotting of an abducted astronaut in South Carolina.

With the communications restored, Wocega-23 explored ways for cadatas to build lives on Earth, with or without informing humans of our alien background. Her spy-role meant she had to talk with her wealthy neighbors in the resorts, beach-front houses, and plantation-style suburban mansions. She also had to question the waiters, construction workers, and maintenance staff, who she encountered as she built her alias façade as a socialite with a well-manicured island. As she questioned them, she had to avoid questions about her own background. If cornered; she had to repeat the same vague description of her origins-story. She even had to create social media profiles because her neighbors questioned a lack of an online-presence as a mystery with a dark undertone. Wocega-23 learned to exaggerate, and bend the truth to satisfy doubts.

She described herself as a business owner in the real estate and development sector. She further developed her corporate identity within the Cadata Real-estate Holdings LLC, which she started to sell her first unsuccessful island. Incorporation meant she would not be personally liable if other cadatas engaged in illegality. The lost island had been somewhat improved. It retained infrastructural and access problems that were likely to anger the next owner.

Wocega-23 led the cadata research team in developing schemes to build the necessary elements for all cadatas to move to Earth. Cadata Holdings was at the center of most plans. The team had been inspired by the robotic miotail hackers to create their own electronic currency, or to divert a money supply via loopholes in humans' financial systems. They experimented with printing billions in paper money by having nanobots duplicate machines used at the Bureau of Engraving and Printing in Washington DC. The electronic credit cards they created with fake identities were repaid through a complex network of new cards, and an infusion from a fictional, cadata-invented blockchain currency. However, humans invested in this cadata currency, spiking its value. Thus, the resulting profits were more than cadatas anticipated generating. These tactics are morally acceptable to cadatas.

Another experiment exploited the fact that a pound of graphite costs $5 on Earth. We can turn this graphite into a pound of pure diamonds, worth $1.4-million, with our robotic technology in minutes. To make this conversion convincing, Wocega-23 guiding robot replicators to insert imperfections of tiny cracks, clouds, chips, black crystal deposits, and grains of other stones (including the originating graphite) into the crystals. However, she learned diamond buyers were watchful regarding the source of diamonds. There was a planet-wide warning against purchasing diamonds produced by regions that had rebelled against the established government. The term war-diamond seems to have been designed to bias international purchasers away from diamonds produced in Africa because 99% of modern diamonds are from conflict-free zones, and yet the terminology persists as a major "humanitarian" policy-point. British-American films about "blood diamonds" began appearing in 2001, just as the size of this illicit trade dropped from 21% in 1980s to 1% by 2004. Creating a fake stone production trail for diamonds proved more difficult than printing paper money, or generating blockchain imaginary electronic money.

Learning from Wocega-23 being duped into buying an uninhabitable island, cadatas commenced buying land in isolated spots in liquidation sales, foreclosures, and from near-bankrupt local jurisdictions. They resold these after investing in small marketing campaigns to make these undeveloped, swampy, or deforested, semi-habitable places appear to be idyllic paradises. These were lucrative deals. Cadatas' real estate holdings, and financial assets swelled. Wocega-23 even bought penthouses in New York and London, and a mansion in Los Angeles. City domiciles were inconvenient because they required flying human-speed helicopters to private landing strips to access the space-capable Blackbird.

She began a socializing campaign by inviting celebrities and business owners to parties at her mansions. These connections to power and wealth were necessary for cadatas' campaign to eventually find acceptance among humans. She learned people were searching for investors in their schemes, businesses, and media projects. If she drove an expensive car with cadata-enhanced features, and lived next-door to the wealthy; they welcomed her money without investigating the origin.

Our mission's cadatas agreed on hoping to avoid engaging in the non-creative hard-labors most humans performed "for a living". They did not want to pilot rickety planes, teach rowdy kids, or take non-automated messages in physical offices. If we could employ trickery to meet our survival needs; then, we would have free time to focus on creatively developing cadata science, technology, and art. Thus, Wocega-23's central objective became securing entire neighborhoods, or towns for development into isolationist, humanless cadata outposts.

As Wocega-23 purchased several enormous plots of land, and began developing infrastructure for essential cadata-based services, a split emerged among cadatas. We had failed to identify a major obstacle to the eventual establishment of a permanent cadata settlement on Earth. We agreed that miotails were right: Earth is the most habitable planet in the Milky Way galaxy. The split was between: 1. those who hoped to settle on Earth without disclosure, and thus without conflicts with humans, and 2. those who insisted on disclosing our arrival, and then defending our bases against subsequent attacks. When our entire planet's population eventually migrated to Earth, it would be impossible to keep this enormous flow of migrants secret. By that juncture, the preceding issue of an advanced warning would serve as a contract against Earth's potentially violent defense. Thousands, or millions of spaceships dropping to a planet from the sky, and billions of cadatas emerging from them (including some without gene-editing into humans) would invite perpetual warfare. This outcome can be logically avoided with advance diplomacy. As a compromise, we decided to send the type of pre-arrival message that miotails issued to cadatas upon observing our entry into their solar system. Placing this plan in practice surfaced a problem with this geopolitical maneuver. We had received a single interplanetary massage from a non-cadata species amidst millennia of silence. In contrast, human bureaucrats receive thousands of messages daily. They would not have understood the meaning, if we had sent it in our native Cadat. Sending it in the native languages of humans meant it blended with junk-email. Here is the message template we copied, and sent to countries across the planet:

Dear President X:

We write to inform you we are an alien species called cadatas. We are aware that most of Earth's governments do not accept refugees of natural disasters. Our home planet, Cadata, is suffering from a potentially life-ending environmental catastrophe. However, its initial cause was an unnatural redirection of the KoraFePub asteroid by a profiteering business-government entity, which we believe makes this an alien-rights issue. We further hope to convince you that our current extreme distance from this home-world deserves for us to be classified as a "special humanitarian concern". We have spent the last 100,000 years traveling across the Milky Way in search for a new habitable planet. We have conducted preliminary experiments that indicate Earth is uniquely suitable for cadatas' settlement. A single cadata, who is called Wocega-23, has landed on Earth to perform an inspection of your facilities. Other members of the cadata crew might soon join her. This forthcoming arrival prompted us to send this informative letter to gauge if we might be violating your laws, in case we decide in favor of this larger settlement. If a few hundred of us can live in harmony within your world; billions of cadatas could migrate to Earth to be saved from the worsening disaster on Cadata. It would take them 25,000 years to reach Earth at light-speed. This interval would be sufficient for cadatas to improve Earth's decrepit infrastructure. We would like to schedule an appointment, at your earliest convenience, to establish a line of communication, and to negotiate the terms of our resettlement. Aerial photos of our main Cadata island before, and after the disaster are attached, as proof of our claims for asylum. We are also attaching photos of how Wocega-23 looked as a cadata prior to her gene-edit into a human, and then after the edit. We cannot mail a sample of our advanced technologies to you as proof because of previous cataclysmic outcomes of such exchanges. In one incident, most of the surface of a planet was engulfed in flames. We would be happy to attempt to answer questions you might have to confirm our alien—though also very friendly—nature.

Sincerely,

Wocega Lapovy-23, Chief Pilot
Earth Examen Team Director
Cadata Spaceship Mission
+882 (869) 568-6894
wocega@cadatas.com
https://cadatas.com

We only received a dozen automated out-on-vacation, and message-received responses to this inquiry sent to officials in refugee-processing centers, or those responsible for country-wide refugee policy. The highest officials we reached on follow-up phone calls were minor secretaries, who hung up a sentence into our explanation. There are apparently some African princes who occasionally file with USCIS, who have prompted an automatic disposal system. The photos of Wocega-23 and Cadata must have been presumed to be a science-fiction digital drawing. Alternatively, countries with weak anti-virus scanners might have refrained from opening these attached photos. Additionally, the photo of Cadata was just a shot of an Earth-like ocean covered with islands. Thus, again, humans must have assumed it was a digital manipulation of a picture taken on Earth, and enhanced with slight alien features. We had hoped at least the smallest countries would be receptive to a meeting to discuss humanity's demands. However, even Monaco and Vatican City, while tiny in acreage, control such prime real estate, they refuse to recognize, or negotiate with aliens. The countries listed as statistically accepting the most refugees tend to specialize in helping neighboring countries (most refugees in Turkey are from Syria), or those suffering from publicized human wars (i.e. Germany accepted many refuges during the Balkan wars), as opposed to alien invaders.

A couple of newspapers have reported our alien refugee applications are a hoax to protest international anti-immigration movements. While this was flattering press, it did not help us convince authorities we are genuine.

Faced with this wall of silence, the cadata leadership decided to pursue a legally standard asylum application with 195 countries. The Computer filled out hundreds of these applications for refugee status in each locale. This monumental task of filling-in 60,000 applications for all members of our cadata mission was assisted by our Computer's processing power. Each country had a unique application process the Computer had to incorporate into its program.

As it began computing the process in the United States, it encountered mountains of problems barring entry. US's online system is only for those physically residing in the US. Other countries have a similar residence requirement; and we cannot reside in all simultaneously, without these cross-references contradicting. Another American problem is that to be considered as a refugee, a group had to have been previously referred by the UN Refugee Agency, or be included on America's list of priority groups. To be added to this list, cadatas needed to run a media campaign that explained our plight (i.e. the disasters on our home planet), and why we needed a new home. However, our press releases attempting to garner free publicity were dismissed as hoaxes. When we attempted to purchase infomercial time on cable TV networks, we were asked to give pitches of our idea; and after seeing our disaster-narrative, the administrators complained that our science-fiction movie was fit for a commercial, and not for an infomercial that had to have an informative programming component. Without public acceptance of cadatas as victims, this venue was a non-starter. And even if we could win this publicity battle; we were not seeking asylum because we were a member of a persecuted group based on race, religion, or nationality, but rather mostly because of our environmental problems.

These forms tend to require documented identity evidence. This turned out to be a vulnerability for Wocega-23 that nearly ended tragically. While most agencies discarded our applications as a prank, US's UNHCR office in DC invited Wocega-23 for an interview to determine her eligibility. They seem to have skimmed over the application, and invited her automatically because they showed surprise after she summarized her asylum claim during this interview. Meanwhile, she could not report her success as a business-woman because this income was generated under false identities. Upon hearing Woce-ga-23 was an "alien", the interviewer recorded in her laptop that Wocega-23 was suffer-ing from delusions of alien-origin. She kept asking if Wocega-23 was a danger to herself, or others. Our hackers reported in Wocega-23's listening-device that the interviewer was attempting to solicit a motive to confine Wocega-23 on a mandatory psychiatric hold. Wocega-23 immediately changed her story, explaining that she was a reporter who was researching the treatment of the mentally ill by asylum-agencies. She gave a copy of one of her alias passports that proved she was already a US "citizen"; this was enough for the interviewer to let her go to attend to her next appointment.

We attempted several other approaches. For some applications, we forged a series of documents confirming membership to a recognized human political, or social con-flict. These applicants were all invited for interviews. To test what would happen if we proceeded with this strategy, Wocega-23 applied through a new US refugee agency: the Office of Refugee Resettlement. In this application, under a different alias-name, she described herself as a Cuban who was opposed to Cuba's communist policies. She wrote that she was being discriminated against and needed asylum. All Cubans can request asylum because Cuba is recognized as a place in need of resettlement. She developed a complex story about the jobs she had been refused, impoverished housing conditions, and the times she spoke out against the Cuban government. She kept repeating she hated communism, and loved capitalism. This time the interviewer accepted her victim-hood. This merely allowed Wocege-23 to fill out a new set of forms. This case remains under a prolonged review. We are considering bribing authorities to expedite this pro-cess. However, it would be cheaper, and more logical to simply fabricate false aliases for the team. And we cannot commence asylum interviews for the rest of our cadatas in this Solar System who want to migrate to Earth because the five-year pre-settlement-research hold is still in place. And we now must wait until September 21, 2026 before the P/2016 BA14 comet makes its third pass to allow us to disembark from the Mothership to Earth in a covert manner.

We have exhausted our customary stages of integration into Earth's civilization. We now must decide between: staying on Earth in disguise, attempt a blatant disclosure by hovering our Mothership above a busy city square, or commencing a new journey to the next potential planet. The latter option is least likely because we have not yet suffered a catastrophe on Earth equivalent to what it took for us to abandon other planets.

One recent development that might change our decision about Earth has been our detection of cross-galactic fast radio burst signals. These are equivalent to 80 years of

Sun's energy production. These originated from an intelligent species in the neighboring Large Magellanic Cloud satellite galaxy. This ceups species is issuing a warm invitation to species across the universe to come to share their solar system. Ceups complain that they are experiencing depopulation due to a lack of sexual interest, and hope interspecies friendship will ignite this spark. Their sun has the outwards appearance of a semi-black hole. A scientific study determined that matter was gravitating into this system at a higher rate than it would towards merely a star. Though the cause of this attraction is not matter being drawn into a compressed natural black hole. This matter is being consumed by ceups' technological activities. Ceups are using incoming matter to construct planets for potential visitors. This is one of the rare cases where cadatas have received a direct invitation from an alien species. However, this planet was not already in our *Catalog* because it is more distant that the length between our previous stops. This distance introduces several intergalactic problems. The light our technology is detecting from Magellanic Cloud was sent 160,000 years ago. In the interim, this civilization could have suffered an implosion from their planet-building hyper-development. And we would be signing-up for a non-stop trip longer than our entire 23-generation journey through our own Galaxy. Another concern is that the artificial-black-hole properties of Ceup have changed this galaxy from a regular and flattened spiral galaxy, to a loosely-held cloud of stars. If the ceups' activities have already destabilized their entire galaxy, a faction in our leadership doubts the odds of a peaceful and profitable coexistence.

Meanwhile, as we have been doing at each of our stops, we sent a full report on our findings about Earth to Cadata. As usual, this report cautioned the rest of Cadata from joining us on Earth, until at least a few hundred years have passed, and we have enough data to at least judge it to be biologically safe for long-term-habitation for mutated cadatas.

We received the latest communication from Cadata during this Earth exploration. It was dispatched 25,000 years ago. It observes that, back in 22,976 BC, our old home's environment was already venomous. Cadata scientists had devised methods for mutating toxic gases in the atmosphere into safe air in enclosed artificial spaces, such as buildings and other inhabited structures. They were working on trying to fix the air on the entire planet. However, they lacked an imaginative, visionary scientist, who might have reshaped their existing technology to realize this goal. Thus, the task fell to our own mission's trained-by-experience scientists to solve this atmospheric problem, if we hoped to return to Cadata. Even if we design a solution; there is a chance Cadata has experienced new unanticipated catastrophes in the past 25,000 years, which might eliminate Cadata from the star-map, before our potential return. If a variant of the latter has happened; the surviving population of cadatas might already be on their way to Earth, or to another potential planet from the *Catalog*.

Meanwhile, our cadata team within the Solar System has been frozen by indecision. The only logical solution seems to be gently cohabiting on Earth, while running the full set of durability experiments. This book is an attempt to disclose our presence to establish an open-dialogue with Earth's governments. Though this goal will surely be missed because it will be dismissed as fiction. Some of our past missions failed solely because our alien identities were discovered. Thus, the lesson from those might be that secrecy is necessary for survival.

Chronology

c. 7,000,000 BC	First cadataids evolved the body-mutation gene, and sail-wings in response to violent environmental stressors on Cadata.
c. 4,000,000 BC	Early cadataid societies first form complex governing structures with help from a sculptural language.
c. 1,006,000 BC	Beginning of the Mineral-Manipulation Age: learn how to use various minerals (stones, metals, and liquids) to improve their lives.
1,004,876 BC	The Scientific Innovation Age: gene editing, computing, and gas-processing scientific fields discovered.
1,003,650 BC	The Millenia Age: lifespan grows to 5,000 years; a cap is placed on lifespan potential to allow for gene improvements to be introduced into the next generation.
1,002,876 BC	Age of Progress: the self-reassembly gene allows for resistance to most accidents and incidents. Murder becomes obsolete because it is impossible to destroy every cell in a cadata's body to cause unnatural death. Brain size is maximized, and its location is decentralized. Full-spectrum vision offered to the new generation.
584,350 BC	Last sleeping-pill miscalculation incident.
537,283 BC	Sleep edited out of cadatas' genetic code.
512,023 BC	Last cadata infant born "naturally" through asexual reproduction; afterwards, all infants develop in incubators.
100,163 BC	KoraFePub asteroid strikes Cadata, forcing the need for resettlement.
99,781 BC	Examen wins the CSEA competition for the Ganeir mission.
99,531 BC	First exploration of a habitable planet outside of the cadatas' solar system.
99,028 BC	Arrival in orbit around Ganeir.
98,317 BC	De-acceleration begins towards Nerore.

98,023 BC	Beginning of Mothership's exploration of the Milky Way galaxy.
95,620 BC	De-acceleration begins towards Likos.
85,951 BC	De-acceleration begins towards Nomuat.
77,497 BC	Arrival in Ragile's solar system.
66,372 BC	Arrival to Nefo.
60,947 BC	De-acceleration begins towards Bhasab.
59,904 BC	Mothership departs from Bhasab.
50,068 BC	Arrival in the Pallos system.
39,365 BC	Arrival in the Eekalah solar system.
29,398 BC	Mothership arrives and establishes an orbit around Ahutababha.
29,355 BC	Departure from Miotail.
18,675 BC	Arrival in Netos' solar system.
8,352 BC	Arrival in the Bulshiyket system.
8,347 BC	Beginning of journey towards Earth.
721 AD	Birth of Ortack-23.
1124 AD	Birth of Wocega-23.
2008 AD	Beginning of deacceleration towards Earth.
2024 AD	*The History of Cadatas* released in English, and in Cadat in the Cadata Share archive.

Bibliography

Hominid Works

Clayton C. Anderson. *It's a Question of Space: An Ordinary Astronaut's Answers to Sometimes Extraordinary Questions.* Lincoln: University of Nebraska Press, 2015.

Stephen Blackmore. *How Plants Work: Form, Diversity, Survival.* Princeton: Princeton University Press, 2018.

Alan Boss. *Universal Life: An Inside Look Behind the Race to Discover Life Beyond Earth.* Oxford: Oxford University Press, January 2019.

Bradley W. Carroll and Dale A. Ostlie. *An Introduction to Modern Astrophysics, Second Edition.* Cambridge: Cambridge University Press, 2018.

Eric H. Cline. *Three Stones Make a Wall: The Story of Archeology.* Princeton: Princeton University Press, October 23, 2018.

Donn Eisele. *Apollo Pilot: The Memoir of Astronaut Donn Eisele.* Lincoln: University of Nebraska Press, 2017.

Epictetus; A. A. Long, translator. *How to Be Free: An Ancient Guide to the Stoic Life: Epictetus: Encheiridion and Selections from Discourses.* Princeton: Princeton University Press, 2018.

Donald Goldsmith. *Exoplanets: Hidden Worlds and the Quest for Extraterrestrial Life.* Cambridge: Harvard University Press, 2018.

Gregory Hakim and Jerome Patoux. *Weather: A Concise Introduction.* Cambridge: Cambridge University Press, 2018.

Andreas Hofmann and Samuel Clokie, editors. *Wilson and Walker's Principles and Techniques of Biochemistry and Molecular Biology, Eighth Edition.* Cambridge: Cambridge University Press, 2018.

William B. Irvine. *You: A Natural History.* Oxford: Oxford University Press, 2018.

Robert L. Jaffe and Washington Taylor. *The Physics of Energy.* Cambridge: Cambridge University Press, 2018.

Michio Kaku. *The Future of Humanity: Terraforming Mars, Interstellar Travel, Immortality, and Our Destiny Beyond Earth.* New York: Penguin Random House LLC, 2018.

Andrew Kahn and Kelsey Rubin-Detlev, Translators. *Catherine the Great: Selected Letters.* Oxford: Oxford University Press, 2018.

Mike Massimino. *Spaceman: An Astronaut's Unlikely Journey to Unlock the Secrets of the Universe.* New York: Crown Publishers, 2016.

Jean-Jacques Rousseau; Victor Gourevitch, editor/ translator. *Rousseau: The Discourses and Other Early Political Writings, Second Edition: Cambridge Texts in the History of Political Thought.* Cambridge: Cambridge University Press, 2018.

Jean-Jacques Rousseau; Victor Gourevitch, editor/ translator. *Rousseau: The Social Contract and Other Later Political Writings, Second Edition: Cambridge Texts in the History of Political Thought.* Cambridge: Cambridge University Press, 2018.

Alan Stern and David Grinspoon. *Chasing New Horizons: Inside the Epic First Mission to Pluto*. New York: Picador, 2018.

Cadata Works

Chapter 1

Magy Gundy. *The History of the Occoros*. Bonne: Holograph Studios, 1,003,965 BC.

Pirth Lopar. *Speech of Commander Pirth Lopar at the Post-Flight Press Conference*. Alleges: Propulsion Space Agency, 1,003,650 BC.

Thpbsohphe. *Answers Are Death*. Bonne: Shellmar Bar Press, 1,002,773 BC.

Onefound Mageplume. *My Lowly Croaking Devotee*. Bonne: Shellmar Bar Press, 1,002,775 BC.

Rood Nennote. *No, No, NO*. Bonne: Shellmar Bar Press, 1,002,771 BC.

Onefound Mageplume, Thpbsohphe and Rood Nennote. *There Existed Our Matter!* Bonne: Shellmar Bar Press, 1,002,753 BC.

Chapter 4

Wern Lexause, Doctor. *Diseases in a Horrifying Universe*. Alleges: Science and Other Stuff Publishing, 100,133 BC.

Wern Lexause, Doctor. *Diseases in a Horrifying Planet*. Alleges: Science and Other Stuff Publishing, 100,145 BC.

Chapter 7

Gesoth Torfar. *Wocega Lapovy: A Spacefaring Story*. Earth: Mothership Press, 2014.

Index

BRRAM

20-Volume Series

Proves with computational linguistics, handwriting and biographical analysis:

6 GHOSTWRITERS

Created All British Renaissance Texts

First translations of inaccessible books, with annotations, introductions

https://AnaphoraLiterary .com/Attribution

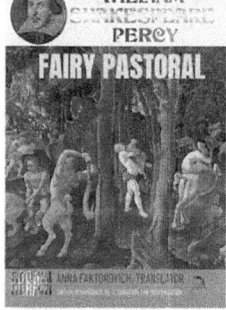

www.ingramcontent.com/pod-product-compliance
Lightning Source LLC
Chambersburg PA
CBHW081141020726
47504CB00009B/1947